SEMICONDUCTORS AND SEMIMETALS

VOLUME 21

Hydrogenated Amorphous Silicon

Part C

Electronic and Transport Properties

Semiconductors and Semimetals

A Treatise

Edited by *R. K. WILLARDSON*
WILLARDSON CONSULTING
SPOKANE, WASHINGTON

ALBERT C. BEER
BATTELLE COLUMBUS LABORATORIES
COLUMBUS, OHIO

Willardson Robert K

SEMICONDUCTORS AND SEMIMETALS

VOLUME 21

Hydrogenated Amorphous Silicon

Part C

Electronic and Transport Properties

Volume Editor

JACQUES I. PANKOVE

RCA/DAVID SARNOFF RESEARCH CENTER
PRINCETON, NEW JERSEY

1984

ACADEMIC PRESS, INC.
(Harcourt Brace Jovanovich, Publishers)

Orlando San Diego New York London
Toronto Montreal Sydney Tokyo

ACADEMIC PRESS, INC.
Orlando, Florida 32887

United Kingdom Edition published by
ACADEMIC PRESS, INC. (LONDON) LTD.
24/28 Oval Road, London NW1 7DX

Library of Congress Cataloging in Publication Data
(Revised for volume 21B-21D)
Main entry under title:

Semiconductors and semimetals

Includes bibliographical references and indexes.
Contents: v. 1-2. Physics of III-V compounds --v. 3. Optical properties of III-V compounds-- --v. 21, pt. B, C, D. Hydrogenated amorphous silicon.
1. Semiconductors--Collected works. 2. Semimetals--Collected works. I. Willardson, Robert K. II. Beer, Albert C.
QC610.9.S47 537.6'22 65-26048
ISBN 0-12-752149-6 (21C)

PRINTED IN THE UNITED STATES OF AMERICA

84 85 86 87 9 8 7 6 5 4 3 2 1

Contents

Chapter 8 **Doping Effects in a-Si:H**

W. Beyer and H. Overhof

Chapter 9 **Electronic Properties of Surfaces in a-Si:H**

H. Fritzsche

Chapter 10 **The Staebler–Wronski Effect**

C. R. Wronski

Chapter 11 **Schottky Barriers on a-Si:H**

R. J. Nemanich

List of Contributors

Numbers in parentheses indicate the pages on which the authors' contributions begin.

B. ABELES, *Corporate Research Science Laboratories, Exxon Research and Engineering Company, Annandale, New Jersey 08801* (407)

W. BEYER, *Institut für Grenzflächenforschung und Vakuumphysik, Kernforschungsanlage Jülich, D-5170 Jülich, Federal Republic of Germany* (257)

J. DAVID COHEN, *Department of Physics, University of Oregon, Eugene, Oregon 97403* (9)

J. DRESNER, *RCA/David Sarnoff Research Center, Princeton, New Jersey 08540* (193)

H. FRITZSCHE, *Department of Physics and James Franck Institute, University of Chicago, Chicago, Illinois 60637* (309)

ARNOLD R. MOORE,* *RCA/David Sarnoff Research Center, Princeton, New Jersey 08540* (239)

K. MORIGAKI, *Institute for Solid State Physics, University of Tokyo, Roppongi, Tokyo 106, Japan* (155)

R. J. NEMANICH, *Xerox Corporation, Palo Alto Research Center, Palo Alto, California 94304* (375)

H. OVERHOF, *Fachbereich Physik, Universität Paderborn, D-4790 Paderborn, Federal Republic of Germany* (257)

JACQUES I. PANKOVE, *RCA/David Sarnoff Research Center, Princeton, New Jersey 08540* (1)

P. C. TAYLOR, *Department of Physics, University of Utah, Salt Lake City, Utah 84112* (99)

T. TIEDJE, *Corporate Research Science Laboratories, Exxon Research and Engineering Company, Annandale, New Jersey 08801* (207, 407)

C. R. WRONSKI, *Corporate Research Science Laboratories, Exxon Research and Engineering Company, Annandale, New Jersey 08801* (347)

* Present address: Institute of Energy Conversion, University of Delaware, Newark, Delaware 19711.

Foreword

This book represents a departure from the usual format of "Semiconductors and Semimetals" because it is a part of a four-volume miniseries devoted entirely to hydrogenated amorphous silicon (a-Si : H). In addition, this group of books — Parts A – D of Volume 21 — has been organized by a guest editor, Dr. J. I. Pankove, an internationally recognized authority on this subject. He has assembled most of the who's who in this field as authors of the many chapters. It is especially fortunate that Dr. Pankove, who has made important original contributions to our understanding of a-Si : H, has been able to devote the time and effort necessary to produce this valuable addition to our series. In the past decade, a-Si : H has developed into an important family of semiconductors. In hydrogenated amorphous silicon alloys with germanium, the energy gap decreases with increasing germanium content, while in alloys with increasing carbon content the energy gap increases. Although many applications are still under development, efficient solar cells for calculators have been commercial for some time.

In Volume 21, Part A, the preparation of a-Si : H by rf and dc glow discharges, sputtering, ion-cluster beam, CVD, and homo-CVD techniques is discussed along with the characteristics of the silane plasma and the resultant atomic and electronic structure and characteristics.

The optical properties of this new family of semiconductors are the subject of Volume 21, Part B. Phenomena discussed include the absorption edge, defect states, vibrational spectra, electroreflectance and electroabsorption, Raman scattering, luminescence, photoconductivity, photoemission, relaxation processes, and metastable effects.

Volume 21, Part C, is concerned with electronic and transport properties, including investigative techniques employing field effect, capacitance and deep level transient spectroscopy, nuclear and optically detected magnetic resonance, and electron spin resonance. Parameters and phenomena considered include electron densities, carrier mobilities and diffusion lengths, densities of states, surface effects, and the Staebler – Wronski effect.

The last volume of this miniseries, 21, Part D, covers device applications, including solar cells, electrophotography, image pickup tubes, field effect transistors (FETs) and FET-addressed liquid crystal display panels, solid state image sensors, charge-coupled devices, optical recording, visible light emitting diodes, fast modulators and detectors, hybrid structures, and memory switching.

R. K. WILLARDSON
ALBERT C. BEER

Preface

Hydrogenated amorphous silicon, a new form of a common element, is a semiconductor that has come of age. Its scientific attractions include a continuously adjustable band gap, a usable carrier lifetime and diffusion length, efficient optical transitions, and the capability of employing either *n*- or *p*-type dopants.

Furthermore, it can be fabricated very easily as a thin film by a technology that not only inherently escapes the expense of crystal perfection but also requires significantly smaller amounts of raw materials.

The discovery of a new material endowed with wondrous possibilities for very economical practical applications naturally attracts many researchers who invariably provide new insights and further vision. Their meditation and experimentation build up rapidly and lead to a prolific information flow in journals and conference proceedings.

The initial cross-fertilization generates an overload of data; books are written that attempt to digest specialized aspects of the field with state-of-the-art knowledge that often becomes obsolete by the time the books are published a year or two later.

We have attempted to provide this book with a lasting quality by emphasizing tutorial aspects. The newcomer to this field will not only learn about the properties of hydrogenated amorphous silicon but also how and why they are measured.

In most chapters, a brief historical review depicts the evolution of relevant concepts. The state of the art emerges, and a bridge to future developments guides the reader toward what still needs to be done. The abundant references should be a valuable resource for the future specialist.

We hope that this tutorial approach by seasoned experts satisfies the needs of at least one generation of new researchers.

CHAPTER 1

Introduction

Jacques I. Pankove

RCA/DAVID SARNOFF RESEARCH CENTER
PRINCETON, NEW JERSEY

This introductory chapter reviews the contents of this volume and introduces some new information that is relevant to the topic of electronic and transport properties of a-Si : H. The authors of this volume have been chosen for their eminent contributions in the various approaches that deepen our insights into this material. Each author was asked to provide a tutorial presentation of his technique, to review the current knowledge, and to suggest what more needs to be done.

In Chapter 2, Cohen discusses the determination of the density of states by several techniques: field effect, capacitance, and deep-level transient spectroscopy (DLTS). He defines the limitations of each method and describes the sophistication resorted to in order to circumvent the limitations. The field-effect method is very sensitive to surface or interfacial conditions and therefore is not suitable to probe bulk properties. Capacitance–voltage coupled to capacitance–temperature and capacitance–frequency measurements is a powerful technique that is suitable to study undoped a-Si : H. Deep traps can be studied by thermally stimulated processes after the traps have been filled at low temperature by means of either a pulse of voltage or a pulse of light. As the temperature is ramped up, successively deeper levels are emptied, signaling this event as a change in either the capacitance or the current, or both. In DLTS, repetitive pulses are applied to a diode from forward or zero bias to reverse bias. During forward bias the traps are filled. They empty thermally during reverse bias, thus changing the capacitance. The transient response of capacitance reflects the emptying of deep levels. The DLTS signal, the difference in capacitance at two different times Δt apart, is maximum when Δt is equal to the temperature-dependent trap-emptying time. An algorithm that is particularly suitable in the case of a quasi-continuous distribution was developed to convert DLTS data into a density of states. Depth profiling is possible by varying the depletion width via the reverse bias. Cohen presents several variations of the DLTS method.

If the distribution of localized states is assumed to be exponential, its characteristic shape, defined by an effective characteristic temperature T_c

ISBN 0-12-752149-6

can be determined from the $I(V)$ characteristics (Furukawa *et al.*, 1982). This technique consists in measuring the space-charge-limited current (SCLC) in a planar structure on one side of the film with the voltage applied across a gap between two electrodes. In this way, a high voltage can be applied across the gap so that the drop at the contacts is negligible compared to the voltage drop in the bulk across the gap. The $I(V)$ data contain an ohmic region and a region of SCLC. The transition voltage between the two regions is the critical voltage V_c. A power law for $I(V)$ is obtained: $I \propto V^m$ $(m > 2)$ when $V > V_c$. The nature of the SCLC is confirmed by the fact that V_c is proportional to the square of the gap length. The characteristic T_c is derived from the short circuit photocurrent I_{sc}, where $I_{sc} \propto V^{T_c/(T+1)}$ (Furukawa and Matsumoto, 1983). The characteristic temperature is greatly affected by doping. However, there is no problem with impurity diffusion from the contact to the a-Si:H layer.

A magnetic field splits each state into $2I + 1$ and $2S + 1$ energy levels (I and S are spin quantum numbers). The energy separation of the split levels is small. Electrons can be excited to the split levels by pumping with rf or microwaves having a quantum energy equal to the splitting. In practice, it is often simpler to use an oscillator generating waves of constant frequency and to achieve resonance by adjusting the magnetic field. Once the various resonances have been identified, this technique becomes a powerful analytical tool.

In Chapter 3, Taylor reviews the information available from magnetic resonance measurements. These fall into two categories: nuclear magnetic resonance (NMR) and electron spin resonance (ESR). Nuclear magnetic resonance gives insights on the environment near a given atom (H or D) because it senses how the nuclear field is distorted by that of neighboring atoms. Two types of configuration are found: isolated SiH that gives a narrow resonance line riding on top of a broad resonance due to clusters of hydrogen atoms, probably in microvoids. Nuclear magnetic resonance data of boron suggest that B is mostly threefold coordinated (e.g., not surrounded by four Si atoms), thus explaining the low doping efficiency of B in a-Si:H. There is also evidence that B is bonded to at least one H and that 1% of the hydrogen is in the molecular H_2 form. Electron spin resonance probes the resonance of unpaired electrons. Silicon dangling bonds have a characteristic resonance corresponding to a gyromagnetic ratio $g = 2.0055$. There are about 10^{19} cm^{-3} dangling bonds in a-Si; their concentration drops by several orders of magnitude in a-Si:H, becoming as low as 10^{15} cm^{-3}. The evolution of H and the restructuring of bonds can be followed by ESR. Phosphorus doping yields a resonance at $g = 2.004$, whereas that of B occurs at $g = 2.013$. Optical excitation produces a light-induced ESR that decays in time. High-intensity optical excitation creates metastable states owing to dangling bonds that can be annealed thermally.

A variation of this method is the optically detected magnetic resonance (ODMR) that is discussed by Morigaki in Chapter 4. In ODMR the ESR is tuned to a known resonance, thus homogenizing the population of the split levels; then one compares the luminescence spectrum with and without the microwaves. Generally, if the luminescence signal increases in the presence of the microwaves, the resonance involves a radiative center. If the luminescence decreases, the center is a nonradiative recombination center. Since the various states (dangling bonds, donors, acceptors, etc.) have been identified, one can assess their role in the luminescence process. Thus, dangling bonds are nonradiative recombination centers. Their concentration increases when light-induced metastable states are formed. A center identified with trapped holes is radiative at ~ 0.8 eV; it also can be a light-induced metastable state. A refinement of the ODMR technique consists in adding to the microwave a tunable rf pump that accesses the nuclear fine splitting. This technique is called electron nuclear double resonance (ENDOR).

In Chapter 5, Dresner distinguishes between three kinds of mobilities: microscopic mobility, drift mobility, and Hall mobility. He considers the effects of disorder and of long dielectric relaxation times that allow a gross distortion of the electric field. There is a profound problem when the low mobility corresponds to a mean free path shorter than the distance between adjacent atoms. Dresner explains the conditions under which the wrong sign of the Hall effect can be obtained. He describes the enormous difficulties in measuring low mobilities. Charge transport can occur by tunneling between localized states (impurity-band conduction) and, at higher temperature, by hopping or by motion via extended states above the mobility edge.

In Chapter 6, Tiedje uses the time-of-flight experiment to determine the density of states in the energy gap. The time-of-flight experiment follows the motion of electrons or holes produced by a short flash of light in a strong electric field. The current decays in time according to two different power laws, first with a power smaller than one and then with a power larger than one. This behavior indicates dispersive transport. The holes have a drift mobility two orders of magnitude lower than electrons, and they suffer a great dispersion. The model that accounts for the observed time dependence following two power laws involves an assumed exponential density of states decreasing from the mobility edges toward the center of the energy gap. Within 10^{-12} sec after the optical excitation, all the carriers thermalize to traps in the tails. Only carriers above the mobility edge (and those thermally excited to the mobility edge) are available for transport and can drift across the sample in the classical way. But along their travel down the field, their number is depleted by the still empty traps, where the optical excitation did not penetrate. Thus, the occupation probability in the tail decreases, making this trapping process dominate the current after one transit time. From the formalism of this model, Tiedje derives values of 42 and 27 meV for the characteristic shapes of the exponential valence-band and conduction-band tails, respectively, and he finds that these values compare reasonably with those determined by other types of measurements.

In Chapter 7, Moore discusses a method he has perfected for determining the diffusion length of carriers in a-Si:H. In this surface photovoltage method, carriers are photogenerated at various depths by a chopped monochromatized light. The photovoltage results from carriers diffusing to the surface, where an electric field separates the electron–hole pairs. The analysis of the functional dependence of the surface photovoltage on the absorption coefficient α reveals a simple relationship if one keeps the photovoltage constant by adjusting the light intensity I. Then, a plot of I versus $1/\alpha$ forms a straight line that extrapolates to an intercept with the abscissa at the diffusion length L_D. Moore discusses the limitations of this simple technique and circumvents the major limitation—the assumption of a depletion width narrower than L_D—by adding a penetrating dc biasing light that flattens the bands in the bulk without eradicating some potential drop at the surface Schottky barrier. The effect of impurities such as oxygen and of light-induced defects on L_D is demonstrated in a manner that provides further insights (for example, that oxygen must enter the structure in pairs).

The band-flattening effect of strong illumination was used by Goldstein *et al.* (1981) and Goldstein and Szostak (1984) to profile the potential distribution in p–i–n solar cells. To determine the potential, a Kelvin probe measures the photoinduced change in the surface potential with respect to the back contact. This measurement is repeated after known amounts of a-Si:H have been removed by sputter etching. Thus, the band bending can be followed as a function of position from the original surface to the back contact. Auger analysis helps in identifying the various layers. The photovoltage measurements are done in high vacuum and the Kelvin probe does not touch the surface (a clean open-circuit condition). The barrier height is determined at high light intensity and low temperature. The depletion width is derived from the photovoltage profile. From these measurements the profile of the built-in electric field is readily obtained. The initial surface photovoltage has the same magnitude and intensity dependence as the open-circuit voltage of the solar cell obtained with the same illumination. These measurements indicate that in p–i–n solar cells most of the open-circuit voltage is generated by the p–i transition. This profiling method was used to identify factors that distinguish good solar cells from poor cells and from degraded ones. The main deficiencies were in low photovoltage at the $i-n$ transition and the presence of low field regions in the i layer due to narrow depletion widths caused by high space-charge densities.

Beyer and Overhof in Chapter 8 discuss how one can extract information about the transport mechanism in a doped disordered semiconductor, how one can tell whether the carriers migrate inside the bands or tunnel between impurities, or whether it is the intermediate case of variable hopping. By combining classical conductivity and thermoelectric power relations, they derive a factor Q that should be temperature independent if temperature affects only the Boltzmann distribution. Therefore, any temperature-induced change in Q signals the onset of a new transport process. Beyer and Overhof show the influence of a great number of impurities both substitutional and interstitial and discuss their compensation. They point out not only that the impurities act as donors or acceptors in affecting conductivity, but also that they influence the binding of hydrogen. The formation of compensating defects at high doping levels appears as a universal observation.

In Chapter 9, Fritzsche considers the importance of surfaces in determining the properties of a thin film. Surface states and states inside the unavoidable surface oxide, and also those at the interface to the substrate, contribute to band bending, to charge transport and recombination, and to the kinetics of absorption and desorption of various species from the ambient. However, the surface of a-Si:H is not as readily oxidized as that of c-Si because its bonds are saturated with hydrogen. Transport can be dominated by charges in the two-dimensional potential wells at the boundaries of the film. A number of techniques can be used to study the surface: The Kelvin probe measures the surface potential with respect to a standard; the surface photovoltage would measure band bending if one could flatten the bands by intense light—care must be taken because penetrating light would generate an opposing photovoltage at the other surface. In some films, such as oxidized B-doped a-Si:H, illumination causes metastable (Staebler–Wronski) effects to occur at the surface. The effect is accounted for by trapped charges that can be removed either thermally or by exposure to H_2O vapor at room temperature. Photoconductivity is strongly dependent on surface recombination that increases in importance as the film is made thinner. Fritzsche presents several other techniques for studying band bending and charged states at the surface, e.g., photoconductivity under a gate electrode in a field-effect transistor (FET) configuration. Transport near the surface can be studied independently from bulk properties as follows: An acoustoelectric surface wave is propagated in a piezoelectric crystal near the surface of the a-Si:H. From the dc voltage induced along the film by the fringing electric field, one can determine the sign of the majority carrier at the surface and obtain the time dependence of its drift mobility at a time $t = 1/\nu$ (ν is the frequency of the driving wave).

In Chapter 10, Wronski reviews his joint discovery of the Staebler–Wronski effect wherein the dark conductivity of a-Si:H can drop by orders of magnitude after prolonged exposure to visible light. The lowered value of dark conductivity persists until the material is heated above 150°C. Evidence is presented that the Fermi level moves close to midgap, which leads to the conclusion that light creates $10^{16}-10^{17}$ new states near the middle of the energy gap. It is demonstrated that the effect is a bulk phenomenon. Extensive measurements lead to the conclusion that, in the temperature dependence of conductivity, the preexponential factor and the activation energy are related by the Meyer–Neldel rule. The Meyer–Neldel rule is also obeyed if the conductivity changes are induced by doping rather than by light. Wronski examines the consequences of light-induced centers on the performance of solar cells. For a more extensive discussion of those light-induced metastable effects that appear in nonelectrical measurements, see the chapter by Schade in Volume 21B.

In Chapter 11 Nemanich explores the many uses of Schottky barriers to study properties of a-Si:H and also to make good contacts to devices. Schottky barriers allow the probing of band bending near the surface and the determination of the density-of-states distribution inside the energy gap. The various techniques of Chapter 2 use Schottky barriers. The height of a Schottky barrier can be obtained from the forward $J(V)$ characteristics in the case of thermionic emission in nearly ideal diodes. The $C(V)$ measurements yield the value of the built-in potential and the charge density in the depletion region, if it is uniform (and the deviation from uniformity if it is not). Additional information can be obtained by internal photoemission experiments. Time-of-flight measurements with Schottky barriers can give the potential profile in the depletion layer. The built-in potential can also be obtained from that forward bias at which the collection efficiency of photogenerated carriers tends to zero (the flat-band condition). Another area of study discussed by Nemanich is the structural change that occurs under the electrode due to atomic migration leading to compound formation (silicides) and to phase segregation. Structural changes affect the $I(V)$ characteristics. In some cases, interdiffusion appears to eliminate interfacial defects. Heavy doping of a-Si:H results in thin tunnelable barriers that can be used as ohmic contacts to devices. Schottky barriers deserve special attention in view of their frequent use in large arrays of devices, as will be seen in Volume 21D.

In Chapter 12 Abeles and Tiedje describe a novel layered structure alternating between a-Si:H and a-SiN$_x$:H with nearly atomically abrupt interfaces. Although the material is amorphous, a periodic superlattice is formed in one dimension. The superlattice leads to changes in optical and transport properties. As the a-Si:H layers are made thinner, the resistivity and the optical band gap increase, but little change is observed in the emission spectrum; however, the luminescence efficiency at room temperature increases. The larger increase in optical band gap and resistivity when the a-Si:H layer thickness is reduced from 40 to 8 Å results from the confinement of electrons and holes in quantum wells. When the a-Si:H layer is thicker than 40 Å, electron transfer doping from SiN_x:H results in a low resistivity material (10^3 Ω cm) with a density of defects at least one order of magnitude lower than in phosphorous-doped material of comparable resistivity.

REFERENCES

Furukawa, S., Kagawa, T., and Matsumoto, N. (1982). *Solid State Commun.* **44,** 927.
Furukawa, S., and Matsumoto, N. (1983). *Phys. Rev. B* **27,** 4955.
Goldstein, B., and Szostack, D. J. (1984). *J. Appl. Phys.* To be published.
Goldstein, B., Redfield, D., Szostak, D. J., and Carr, L. A. (1981). *Appl. Phys. Lett.* **39,** 258.

CHAPTER 2

Density of States from Junction Measurements in Hydrogenated Amorphous Silicon

J. David Cohen

DEPARTMENT OF PHYSICS
UNIVERSITY OF OREGON
EUGENE, OREGON

I. Introduction

The starting point for a detailed understanding of the basic properties of hydrogenated amorphous silicon (a-Si:H) is the knowledge of the energy

ISBN 0-12-752149-6

distribution of states in the mobility gap. In the last several years significant advances have been achieved by relating most measured properties of a-Si:H to specific properties of these states. Included are most kinds of transport measurements, photoluminescence, spin properties, and optical absorption, as well as detailed device performance (most notably in solar cell, thin film transistor, and xerographic applications). Because so many measurements are directly influenced by the density of gap states $g(E)$ it is possible to obtain direct information about them through a great variety of experimental techniques. However, by far the most fruitful of the methods employed to date have been those that investigate the properties of the space-charge region that forms at a-Si:H interfaces.

The most widely used of these methods in the study of a-Si:H have been field-effect, capacitance, and deep level transient spectroscopy (DLTS) measurements. Capacitance measurements actually include quite a number of variations such as capacitance versus applied voltage ($C-V$), frequency ($C-\omega$), or temperature ($C-T$), and also several kinds of distinct capacitance profiling techniques. The technique referred to as DLTS normally includes both capacitance-transient as well as current-transient measurements and will also be used as a generic term for such recent variations as isothermal capacitance transient spectroscopy (ICTS), constant capacitance methods, and the like.

The manner in which each of these kinds of methods lead one to the distribution of mobility gap states with energy $g(E)$ is the subject of the bulk of the discussion in the sections that follow. It may be helpful if we categorize these techniques in the following way. Field-effect and $C-V$ measurements are each sensitive to a particular single property of the space-charge region. Both are steady-state measurements of an essentially static space-charge layer. The information about $g(E)$ is obtained in both cases by repeating the particular measurement for a range of applied voltages. A distinct second category is the measurement of capacitance versus applied frequency or temperature. Here one typically uses one value of applied voltage (often zero in the case of Schottky barriers) but varies the temperature or frequency. These measurements are dynamic in the sense that the particular spatial region where the space charge is being probed is determined by the details of carrier capture and emission (and, in some cases, also transport) on the time and energy scale determined by ω and T. However, the information about $g(E)$ comes from mapping out the band-bending function from a primarily static distribution of space charge. The third category includes all of the DLTS kinds of measurements listed above, as well as such methods as thermally stimulated capacitance (TSCAP) or current (TSC) measurements. Here the characteristics of the barrier region are changed by a substantial amount by introducing a nonequilibrium occupation of gap states and then by monitoring the capacitance or current during the recovery process. The time and temperature dependence of that recovery is used to obtain $g(E)$. Such techniques differ from the other categories mentioned in that they specifically involve observation of a dynamic space-charge configuration.

At this point one additional technique, recently introduced, that should be mentioned briefly is the carrier drift method of R. A. Street (1983). Although it is not appropriate to include the details of this method in this chapter, it is worth pointing out that this technique has been used to map out the band-bending function in the depletion layer of Schottky barriers formed on a-Si : H films. Such a technique therefore belongs to the second category given above and one can obtain information about $g(E)$ in a manner that is entirely similar to, for example, a capacitance versus temperature measurement. This carrier drift method may prove to be quite useful since it is often more readily performed in highly insulating material. However, it is important to stress that this technique is subject to many of the same advantages and limitations that will be discussed below and that the kind of information achieved will be qualitatively similar.

In the sections that follow we shall first turn to the physics that is common to all of the techniques: namely, to a discussion of the structure of the space-charge region in amorphous semiconductors and to a brief description of the mathematical methods used to solve for the band-bending function from a knowledge of $g(E)$. We shall then discuss the field-effect, capacitance, and DLTS methods in turn. We shall emphasize the manner in which the density of states is obtained from the experimental data and give a number of representative experimental results. We shall also discuss some of the limitations and problems associated with each of these kinds of techniques. We hope that this will give the reader the basis for assessing the relative advantages and problems of the various methods.

II. General Considerations

1. Solution of the Barrier Potential Problem

The interpretation of any experimentally measured quantity that depends on the structure of the space-charge region near an amorphous semiconductor interface begins with a solution of the static band-bending problem. If $\psi(x)$ denotes the dc band potential for electrons relative to the neutral bulk as shown in Fig. 1, Poisson's equation becomes

$$d^2\psi/dx^2 = \rho(x)/\epsilon, \tag{1}$$

where in thermal equilibrium $\rho(x)$ is given by an integral over the density of

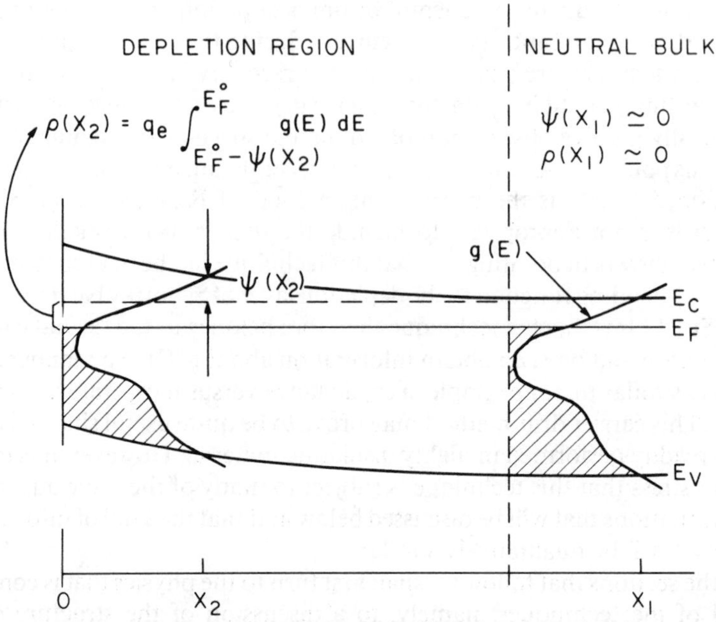

FIG. 1. Depletion region under equilibrium conditions for a material with a continuous density of gap states. The charge density is related to an integral involving the amount of band bending at each point x as indicated. The definition of various symbols is given in the text.

states as

$$\rho(x) = q \int [f(E', E_F^0, T) - f(E', E_F^0 - \psi(x), T)]g(E', x)\, dE'. \tag{2}$$

Here E_F^0 denotes the position of Fermi energy in the bulk relative to the band (or mobility) edges, f is the Fermi function, and q the electronic charge. We include in Eq. (2) the possibility that the density-of-states function $g(E)$ may vary spatially within the sample.

If we neglect any x dependence in $g(E)$, it is then possible to simplify the above equation for $\psi(x)$ by a transformation of the independent variable from x to ψ. Thus for fixed temperature we have $\rho = \rho(\psi)$ and by using the identity

$$\frac{d^2\psi}{dx^2} = \frac{1}{2}\left(\frac{d\psi}{dx}\right)^{-1} \frac{d}{dx}\left(\frac{d\psi}{dx}\right)^2,$$

we integrate Eq. (1) to obtain

$$\frac{2}{\epsilon}\int_0^{\psi} \rho(\psi')\, d\psi' = \left(\frac{d\psi}{dx}\right)^2 = F^2. \tag{3}$$

Thus the electric field at the interface $F_s = F(\psi_s)$ is directly related to the total barrier charge from $x = 0$ ($\psi = \psi_s$) to far into the bulk ($\psi = 0$). Equation (3) is sometimes used to aid in numerical analysis of the band-bending problem for field effect and low frequency capacitance measurements (Powell, 1981; Hirose *et al.*, 1979) and will be examined further in later sections. Note that $F(\psi)$ must in general be solved for each temperature of interest.

In certain cases a simpler low temperature form of $\rho(x)$ may be employed by replacing the Fermi functions by step functions so that

$$\rho(\psi) = q \int_{E_F^0 - \psi}^{E_F^0} g(E')\, dE' \tag{4}$$

may be considered independent of temperature. It has been shown, however, that this approximation is generally *not* valid for an accurate interpretation of field-effect measurements (Powell, 1981; Goodman and Fritzsche, 1980).

For capacitance measurements it has been shown (Cohen and Lang, 1982) that Eq. (4) may be used *provided* the density of states near E_F does not vary appreciably over an energy range on the order of kT. This situation would not, therefore, be expected to hold at moderate temperatures for heavily doped a-Si:H where E_F generally lies in one of the band tails. However, detailed calculations indicate that the discrepancy caused by using the low temperature approximation instead of Eq. (2) is fairly small even for doped material if the statistical shift of the Fermi energy is also included in the calculation (Cohen *et al.*, 1983a). This follows, in part, from the fact that charge neutrality in the bulk requires that the integral of $g(E)f(E, E_F^0, T)$ be a constant. Thus by using the zero-temperature value of E_F^0 together with the zero-temperature form of the Fermi function, the value of ρ given by the integral in Eq. (2) will not be appreciably changed.

To properly treat spatially nonuniform material, $\psi(x)$ must be obtained explicitly from Eqs. (1) and (2). One successful approach has been through an adaption of the Noumerov technique, which provides a rapidly converging numerical solution for the equilibrium diode problem (Cohen and Lang, 1982). Furthermore, this technique has been found extremely useful in the numerical analysis of the time-dependent spectra discussed in Part V.

2. Deep Depletion; Thermal Emission and Capture at Gap States

For small values of applied bias or for metal–oxide–semiconductor (MOS) structures, the Fermi energy may be expected to extend through the barrier region as shown in Fig. 1. However, for larger reverse bias applied to a Schottky barrier this picture is modified and leads to the formation of a quasi-Fermi level near midgap in the deep-depletion region as shown in Fig. 2.

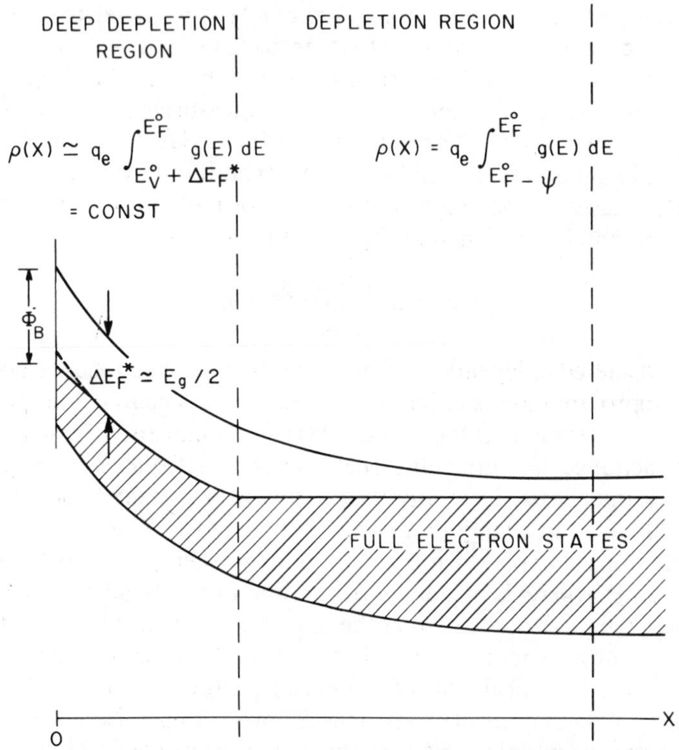

FIG. 2. Space-charge region for deep depletion. In the deep-depletion region the charge density is approximately constant as determined by the position of the quasi-Fermi energy. Parameters are defined in the text.

In general, the occupation of states in the gap is dominated by majority- and minority-carrier emission and majority-carrier capture. The emission rate for electrons at temperature T is usually given as

$$e_n = 1/\tau = \nu_n e^{-E/kT}, \tag{5}$$

where k is the Boltzmann factor and E is the energy *difference* between the particular localized gap state and conduction-band edge E_c at which carriers will move out of the depletion region in response to the electric field. A similar expression applies to hole emission:

$$e_p = 1/\tau_p = \nu_p e^{-E'/kT}, \tag{6}$$

where E' is the energy difference to the valence-band edge.

For most deep levels in crystalline semiconductors the exponential pre-factors in Eqs. (5) and (6) lie in the range $10^{11} < \nu < 10^{13}$ sec^{-1}. The largest values in this range are roughly of the order of acoustic phonon frequencies. However, in a more realistic treatment of the thermal emission process, the ν_n or ν_p prefactor is related to the product of the matrix element for the bound-to-free transition and the effective density of final states within approximately kT of the bottom of conduction band. From detailed balance one can relate ν_n to these more fundamental parameters as

$$\nu_n = \sigma_n \langle v_n \rangle N_c, \tag{7}$$

where σ_n is the electron capture cross section of the defect, $\langle v_n \rangle \simeq (3kT/m)^{1/2}$ is the average thermal velocity of conduction electrons with effective mass m, and N_c is the weighted integral of the conduction-band density of states times a Boltzmann factor. An analogous expression applies for the emission of holes to the valence band.

Since the terms in ν_n or ν_p may themselves be temperature dependent, one must exercise caution in interpreting experimental values of E derived from an Arrhenius plot. It is customary (Lang, 1979) to take account of the T^2 temperature dependence of $\langle v_n \rangle N_c$ by subtracting $2kT_m$ from the measured value of E, where T_m is the measured temperature on the Arrhenius plot for each feature in the spectrum. The temperature dependence of σ_n is a more serious problem and, without an independent measurement of the carrier capture process, cannot be unambiguously determined. Unless this temperature dependence is explicitly taken into account it will automatically be included in the measured activation energy E. Simply put, such cases require that a distinction be made between a "thermal equilibrium" energy scale and a "thermal emission" energy scale. To avoid undue complication, since for a-Si:H these differences seem to be small (Lang *et al.*, 1982a), we shall generally ignore this distinction in the discussion that follows. We shall, however, return to discuss this issue briefly in Section 16.

Thermal equilibrium requires that the capture rate *equal* the emission rate at E_F. Therefore, majority-carrier capture follows the same dependence on energy and temperature as Eq. (5). However, for capture the characteristic energy E denotes the height of the conduction-band edge above the bulk Fermi level, $E = \Delta E_F + \psi(x)$, where $\Delta E_F = E_c - E_F$ in the bulk material. This energy is the average energy of the potential barrier that electrons must travel up to be captured at the point x under consideration.

When sufficient reverse bias is applied to the Schottky diode, the potential barrier $\psi(x)$ depletes the region near the interface of free carriers to the point that the carrier concentration arises predominantly either from the leakage over the barrier from the metal contact or from the intrinsic thermalization of electrons and holes from the valence and conduction bands. This leads to the formation of a quasi-Fermi level E_F^* in the deep-depletion region. For the case that barrier leakage dominates, E_F^* will lie below the conduction band (for n-type material) at a distance roughly equal to the Schottky-barrier height Φ_B (Cohen and Lang, 1982) given by

$$E_c - E_F^* = \Delta E_F^* = \Phi_B - kT \ln(\beta F_s/F(x)). \tag{8}$$

This formula follows directly from isothermal diffusion theory (Schottky, 1938) in which the parameter β denotes the transmission coefficient for majority carriers across the barrier interface ($\beta = 1$ implies a transparent barrier).

For a large barrier height or small transmission coefficient, E_F^* is determined by the electron and hole thermalization from the bands and is given by

$$\Delta E_F^* = \tfrac{1}{2}E_g + \tfrac{1}{2}kT \ln[\nu_n/\nu_p]. \tag{9}$$

This second equation applies if

$$\left[\frac{\nu_p}{\nu_n}\right]^{1/2} \exp\left(-\frac{E_g}{2kT}\right) \gg \frac{\beta}{2}\frac{F_s}{F(x)} \exp\left(-\frac{\Phi_B}{kT}\right).$$

Experimentally, the use of Schottky barriers with large barrier heights is advantageous for a number of reasons and we usually take $\Delta E_F^* \simeq E_g/2$.

Figure 2 illustrates the situation for a large barrier height Φ_B. The Fermi level follows the bulk Fermi energy into the bulk until it reaches an energy ΔE_F^* below E_c as given by Eq. (9). Thereafter it remains at a relatively constant level below E_c [giving rise to a nearly constant charge density $\rho(x)$] until very near the metal interface where, due to a nonnegligible free-hole concentration and hole capture rate, it moves away from $E_g/2$ to join E_F in the metal.

III. Field Effect

3. Historical and Mathematical Background

a. Historical Background

The earliest pictures of the density of states in a-Si : H were obtained from field-effect (FE) measurements (Spear and LeComber, 1972; Madan *et al.*, 1976; Madan and LeComber, 1977). These results indicated a surprisingly low $g(E)$, particularly near midgap, and disclosed two distinct defect bands within about 0.4 and 0.6 eV of the conduction- and valence-band mobility edges, respectively, as well as a fairly sharp band tail near E_c. A major factor in these results, of course, was that *hydrogenated* amorphous silicon was a new and distinct material compared to other forms of amorphous silicon and that indeed $g(E)$ was very different and much lower than for the other kinds of material people had studied. The ability to perform field-effect measurements at all on this material gave perhaps the clearest indication of this difference.

Most of the earliest analyses (Spear and LeComber, 1972) required certain approximations to deduce $g(E)$ from the FE data. These included a definite form of the potential band-bending function (parabolic or exponential), zero-temperature statistics, and restrictive assumptions about the charge density. Later work (Madan and LeComber, 1977; Hirose *et al.*, 1979) removed the latter two assumptions and some recent work (Goodman and Fritzsche, 1981; Weisfield *et al.*, 1981; Goodman, 1982) removes all of them. In addition, several groups have presented theoretical modeling of the detailed correspondence between FE data and $g(E)$ (Goodman and Fritzsche, 1980; Powell, 1981; Weisfield and Anderson, 1981).

Although there is now general agreement that $g(E)$ can be quite low at midgap and that the band tails are, indeed, quite sharp, other detailed aspects of $g(E)$ deduced by FE measurements have been at the center of considerable controversy. For example, the existence of the two sharp defect bands is often called into question, although they have been confirmed in part by other workers (Balberg, 1980; Jan *et al.*, 1980; Nakashita *et al.*, 1981; Weisfield *et al.*, 1981). On the other hand, determinations of $g(E)$ using other techniques (see Parts IV and V) and even other FE measurements (Hirose *et al.*, 1979; Goodman, 1982) indicate the lack of such features. Another issue that has been raised concerns the dependence of FE measurements on the preparation of the insulator–semiconductor interface (Solomon *et al.*, 1978; Tanielian *et al.*, 1980) and therefore questions whether bulk properties of a-Si : H are actually being resolved.

In the following sections we shall describe the field-effect technique in some detail and present a sample of the experimentally determined densities of states. Some of the issues related to the overall sensitivity and reliability of the FE method to determine a bulk density of states in a-Si : H will be examined. We shall also mention some of the recent applications of the FE technique to the study of a great variety of phenomena in a-Si : H and related materials. Indeed, in spite of their shortcomings, FE measurements continue to be widely applied in the study of amorphous semiconductors and hence still qualify as one of the primary techniques to determine $g(E)$ in these materials.

b. General Equations

The field-effect method measures the surface sheet conductance as a function of applied voltage at the semiconductor–insulator interface in MOS device structures. This surface conductance depends on the degree of carrier accumulation or depletion at the interface, which in turn depends on the band-bending function for a given applied gate voltage. This band bending is determined by $g(E)$ within the sample. In Fig. 3 the situation that exists for a positive gate bias V_G for an n-type a-Si : H sample is indicated schematically. The surface potential for electrons ψ_s is determined by V_G and the flat-band gate voltage V_{FB} by the relation

$$\psi_s = (V_G - V_{FB}) - (d_i/\epsilon_i)(\epsilon F_s + \Sigma_s), \tag{10}$$

where ϵ_i and d_i are the insulating layer dielectric constant and thickness, respectively, F_s is the field at the semiconductor interface, and Σ_s is the surface (interface) charge density. In most analyses of FE measurements Σ_s is taken to be zero or at least is assumed constant so that it may be included in the definition of V_{FB}. The surface sheet conductance G is given by an integral of the free carrier concentration in the vicinity of the interface; namely,

$$G = q\int_0^t dx \int_{-\infty}^{+\infty} g(E)\mu(E)f(E, E_F^0 - \psi(x), T)\, dE, \tag{11}$$

where t is the film thickness and $\mu(E)$ is the microscopic mobility. The flat-band conductance G_0 is given by Eq. (11) by setting $\psi(x) = 0$ and may be written, provided conduction is taking place remote from the Fermi level, in

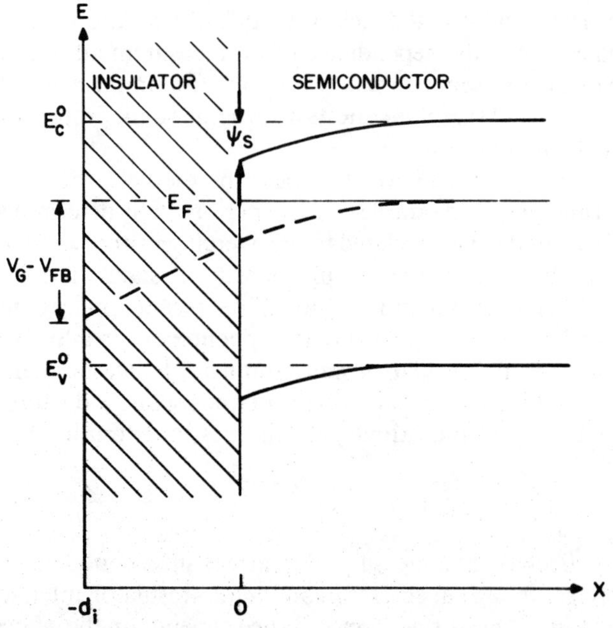

FIG. 3. Barrier region for MOS device structure shown in accumulation. The semiconductor–insulator interface is at $x = 0$ and the metal gate contact is at $x = -d_i$. Other parameters are defined in the text.

the familiar form

$$G_0/t = \mu^{e}_{\text{eff}} N_c(T) \exp(-(E_c - E^0_F)/kT) + \mu^{h}_{\text{eff}} N_v(T) \exp(-(E^0_F - E_v)/kT), \tag{12}$$

where $N_c(T)$ and $N_v(T)$ are expected to be only weakly temperature dependent and μ^{e}_{eff} and μ^{h}_{eff} are *effective* mobilities. In the vicinity of a given point x we may similarly express the contribution to the sheet conductance δG from Eq. (11) as

$$\delta G = \mu^{e}_{\text{eff}} N_c(T) \exp[-(E_c + \psi(x) - E^0_F)/kT] + \mu^{h}_{\text{eff}} N_v(T) \exp[-(E^0_F - \psi(x) - E_v)/kT]$$

or

$$t\frac{\delta G}{G_0} = \frac{1}{1+\alpha}\exp\left(-\frac{\psi(x)}{kT}\right) + \frac{\alpha}{1+\alpha}\exp\left(\frac{\psi(x)}{kT}\right), \tag{13}$$

where α is the ratio of the hole to electron terms in Eq. (12). Thus the change in conductance from the flat-band conditions due to the band potential $\psi(x)$ is given by

$$t\left(\frac{G - G_0}{G_0}\right) = \frac{1}{1+\alpha}\int_0^t \left[\exp\left(-\frac{\psi(x)}{kT}\right) - 1\right] dx + \frac{\alpha}{1+\alpha}\int_0^t \left[\exp\left(+\frac{\psi(x)}{kT}\right) - 1\right] dx, \tag{14}$$

which is identical to the usual expression except that we have used the electronic potential energy ψ instead of $-V$. Following Powell (1981) we may carry the development slightly farther and substitute

$$\int_0^t \left[\exp\left(\pm\frac{\psi(x)}{kT}\right) - 1\right] dx = \int_0^{\psi_s} \frac{\exp(\pm\psi/kT) - 1}{F(\psi)}\, d\psi,$$

so that we obtain

$$\frac{d}{d\psi_s}\left[\left(\frac{G - G_0}{G_0}\right) t(1+\alpha)\right] = \frac{\exp(-\psi_s/kT) - 1}{F_s} + \alpha\,\frac{\exp(\psi_s/kT) - 1}{F_s}, \tag{15}$$

which relates quantities that may be determined experimentally on the left hand side to the quantities F_s and ψ_s on the right hand side, which are directly determined by the solution of the junction band-bending problem in the form of Eq. (3) in combination with Eq. (10). The only significant approximation made in this derivation has been to use nondegenerate statistics in place of the full Fermi function.

4. Density of States Obtained from Experiments

Obtaining the density of states from experimental field-effect data requires using an iterative procedure in which a trial $g(E)$, with a given V_{FB} and α, is input to produce simulated field-effect data using Eq. (14) or (15). Then $g(E)$ is adjusted until the calculated FE spectrum matches the experimental data. Various algorithms for obtaining $g(E)$ in this manner have been discussed in detail (Madan *et al.*, 1976; Goodman and Fritzsche, 1980). Figure 4 shows the now classic field-effect density-of-states obtained by the Dundee group (Spear and LeComber, 1976). The defect bands near $E_c - 0.4$ eV and $E_v +$ 0.6 eV are clearly present. This $g(E)$ was actually obtained from a series of samples with different phosphorus and boron doping; data from such a series of samples is necessary to deduce the density of states over such a large energy

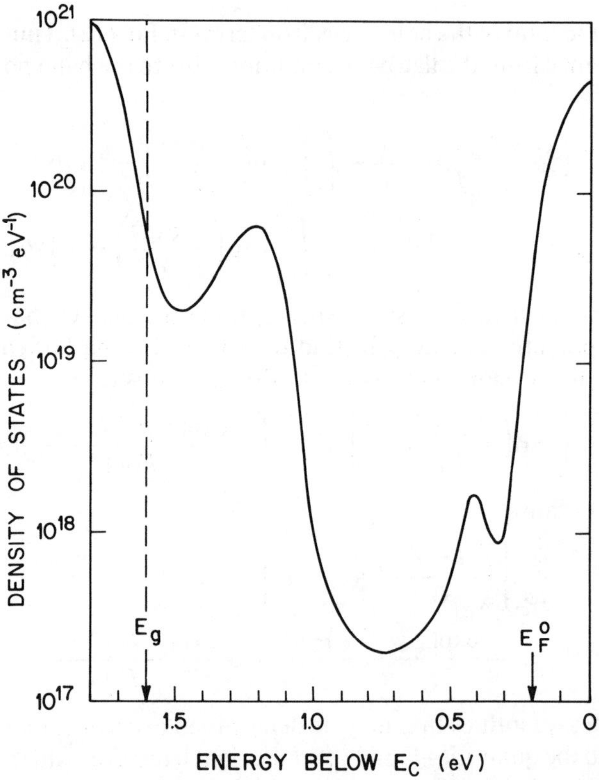

FIG. 4. Density of states for a-Si : H determined from field-effect measurements of the Dundee group. [After Spear and LeComber (1976).]

range because the sheet conductance is generally the most sensitive to the surface potential and hence to $g(E)$ for the regime of moderate majority-carrier accumulation.

Some raw field-effect data taken by the Harvard group (Weisfield *et al.*, 1981) is shown in Fig. 5 indicating the relatively large gate voltages that must be applied to produce the more modest changes in ψ_s at the semiconductor–insulator interface. The flat-band voltage and the value of the source-drain voltage V_{sd} are indicated in the figure. The flat-band voltage in this case is determined by comparing the behavior of the transconductance versus gate voltage at different temperatures (see Weisfield and Anderson, 1981). Alternatively, V_{FB} may be obtained by determining the gate voltage condition at which the photoconductivity through the sample changes sign. The value of

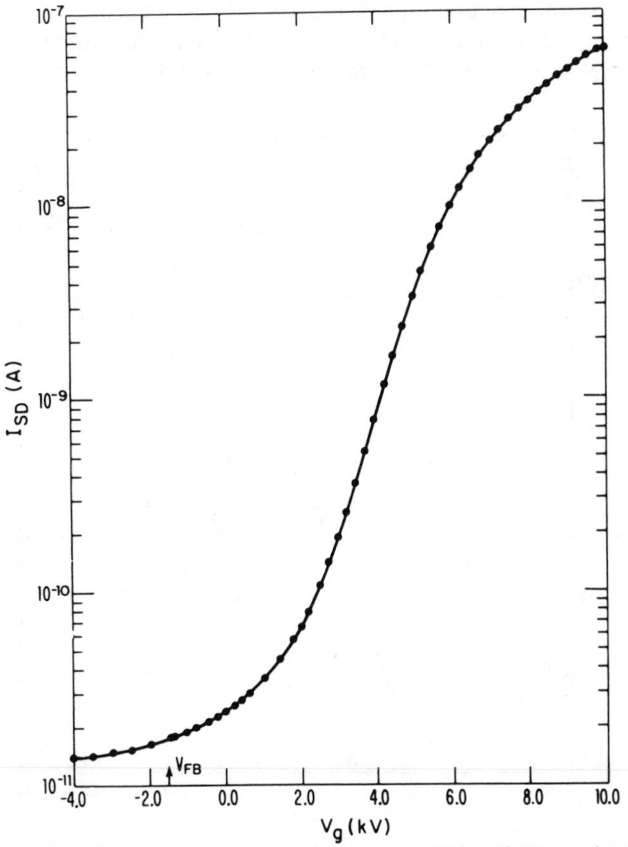

FIG. 5. Field-effect measurements obtained for a 1.1-μm-thick a-Si:H sample at room temperature with a source-drain voltage of 15.0 V taken by the Harvard group. Plotted is the source-drain current I_{sd} versus gate voltage V_g. [From Weisfield *et al.* (1981).]

V_{sd} is kept as small as possible consistent with a good signal-to-noise ratio. Using too large a value of V_{sd} gives rise to distortion of the sheet conductance at large gate voltages (see Goodman, 1982). The density-of-states curve obtained from Fig. 5 is displayed in Fig. 6 and, like Fig. 4, shows a pronounced peak in $g(E)$ at roughly 0.6 eV below E_c.

In Fig. 7 some FE data taken by the Chicago group (Goodman, 1982) are shown for samples grown simultaneously on three different types of quartz insulator. Here the abscissa has been converted into a total barrier charge (per unit area). These data were obtained to investigate the sensitivity of the FE densities of states to the semiconductor–insulator interface properties and hence to the short-range spatial variation within the a-Si : H layer itself. The densities of states derived from the FE data, shown in Fig. 8, indicate that considerable variation does in fact occur and particularly affects the prominence of the defect bands displayed in Figs. 4 and 6.

These latter experimental results indicate some of the potential problems that exist in interpreting the FE data to give a bulk density of states in a-Si : H.

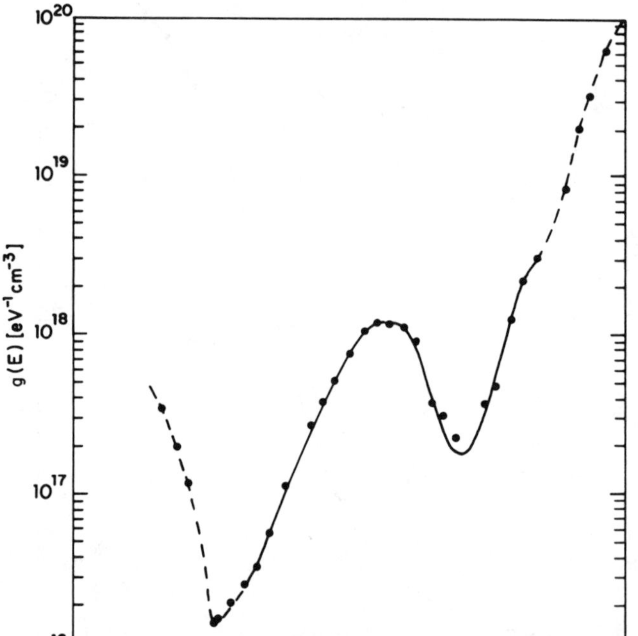

FIG. 6. Density of states $g(E)$ derived for the field-effect data in Fig. 5 for $E_c - E_F = 0.75$ eV. The dashed portions of the curve are not considered reliable estimates of the bulk density of states. [From Weisfield *et al.* (1981).]

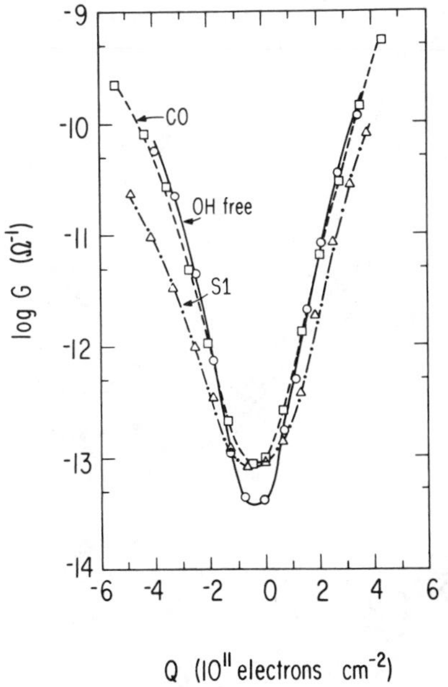

FIG. 7. Field-effect data for three a-Si:H samples deposited simultaneously on different types of quartz: Commercial optical (CO), Suprasil 2 (S1), and Suprasil W (OH free). These substrates had been cleaned, etched, and annealed prior to deposition. Plotted is the conductance G versus the interface surface charge density. [From Goodman (1982).]

In the following section we shall discuss these problems somewhat further. We shall also review briefly some other recent interesting results that have been obtained using the field-effect technique in amorphous silicon.

5. LIMITATIONS TO THE FIELD-EFFECT TECHNIQUE

Representative results of the densities of states obtained by field-effect measurements discussed above indicate a fair amount of variation and disagreement between the results of different groups and also indicate possible problems associated with substrate preparation conditions. Important questions therefore arise concerning the sensitivity of this method and also its ability to determine the true bulk properties of a-Si:H.

The issue of the sensitivity of the method and its ability to resolve fine scale features in $g(E)$ is investigated in Figs. 9 and 10. In this study (Goodman and Fritzsche, 1980) the three hypothetical densities of states in Fig. 9, which differ in the degree of fine-scale structure, were used to calculate the field-effect "data" in Fig. 10. Very little difference is exhibited in the expected spectra. This indicates that the ability of FE measurements to distinguish such features is severely limited. In some sense this result should be apparent

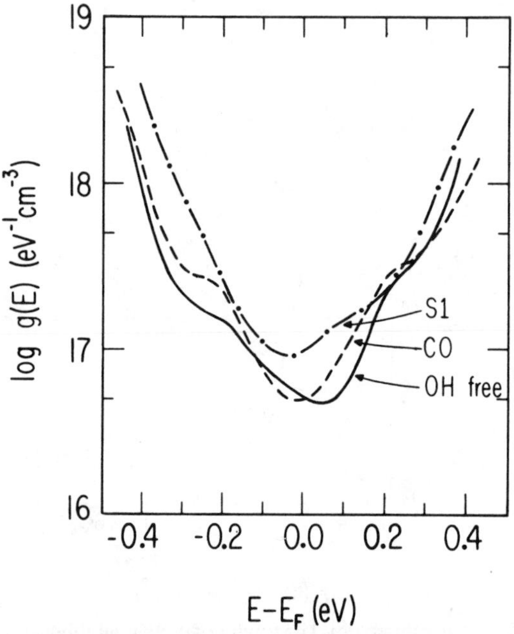

FIG. 8. Densities of states deduced from the field-effect data of Fig. 7. These curves illustrate the effects of the insulator interface on the field-effect results. [From Goodman (1982).]

from Eq. (15). Each of ψ_s and F_s are related to an integral over $\rho(x)$, which in turn is given by an integral over $g(E)$ (see Part II). Thus the experimentally measured quantity G is related to $g(E)$ through *three* levels of integration.

Results of a similar study (Powell, 1981) are shown in Fig. 11. This study compares simulated data not only for the two densities of states shown in Fig. 12 with and without a peak $E_c - 0.4$ eV, but also for various approximations used in the data analysis by different workers. The results of using two of the most common approximations—the assumption of a parabolic profile and the use of zero-temperature statistics—are displayed in Fig. 11. This figure specifically demonstrates that (1) by using the parabolic band approximation the density of states deduced by the FE data will be too large (curves e and d versus curves a and b) and (2) by employing zero temperature statistics the deduced $g(E)$ will also be too large (curve c versus curve a). This figure also indicates the ability of FE measurements to distinguish between the two densities of states shown in Fig. 12 to be quite poor (curve a versus b or curve d versus e).

Even though recent analyses of FE data have been carried out using none of the approximations mentioned above (Goodman and Fritzsche, 1981; Tanielian *et al.*, 1981; Weisfield *et al.*, 1981; Goodman, 1982) the general

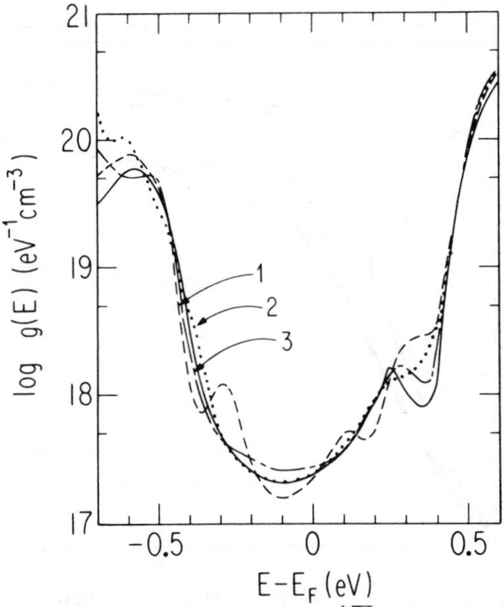

FIG. 9. Three density-of-states curves that fit the hypothetical field-effect data of Fig. 10. The density of states of Spear and LeComber (1976) (Fig. 4) is shown by the solid curve. [From Goodman and Fritzsche (1980).]

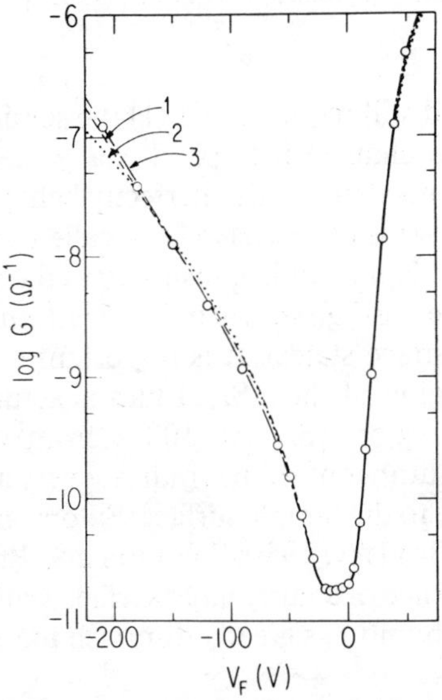

FIG. 10. Field-effect "data" generated from the density-of-states curves of Fig. 9. [From Goodman and Fritzsche (1980).]

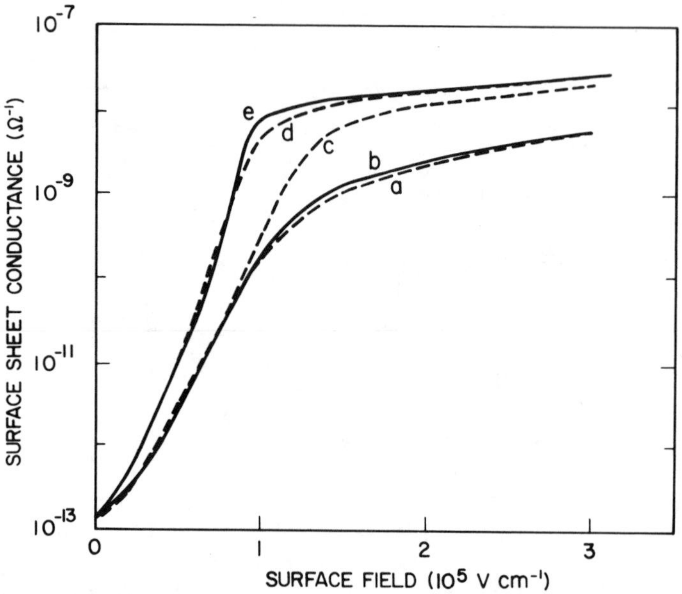

FIG. 11. Calculated field-effect conductance for the structureless density of states and a density of states with a peak in the vicinity of E_c, shown in Fig. 12, as indicated by the solid and dashed lines, respectively. Different methods of analysis were used: curves a and b, with no approximations; curve c, using zero temperature statistics; curves d and e, employing an assumed parabolic band-bending profile. [After Powell (1981).]

validity of the method still requires that (1) the semiconductor be perfectly homogeneous, (2) no change in trapped charge density Σ_s occurs at the interface, and (3) the insulator is also perfectly behaved and has no trapped charges. All of these assumptions have been called into question. The nonuniformity of the a-Si : H film and/or insulator in the vicinity of the interface is strongly suggested in the data shown in Fig. 7 (alternatively, these data may suggest changes in surface states). It is important to recall how extremely spatially limited a region of the a-Si : H film is actually probed by the FE method, typically a region only 20 – 100 Å from the insulator interface (Fritzsche, 1980). A number of studies indicate very anomalous behavior of a-Si : H films this close to the sample surface (Solomon *et al.*, 1978; Tanielian *et al.*, 1978; Ast and Brodsky, 1980; Solomon and Brodsky, 1980; Rehm *et al.*, 1980) or the existence of a fairly large surface state density (Knights and Biegelsen, 1977). Such findings greatly diminish the overall accuracy of the field-effect results.

In spite of these problems, the field effect method continues to be applied to the investigation of a wide range of phenomena in amorphous semicon-

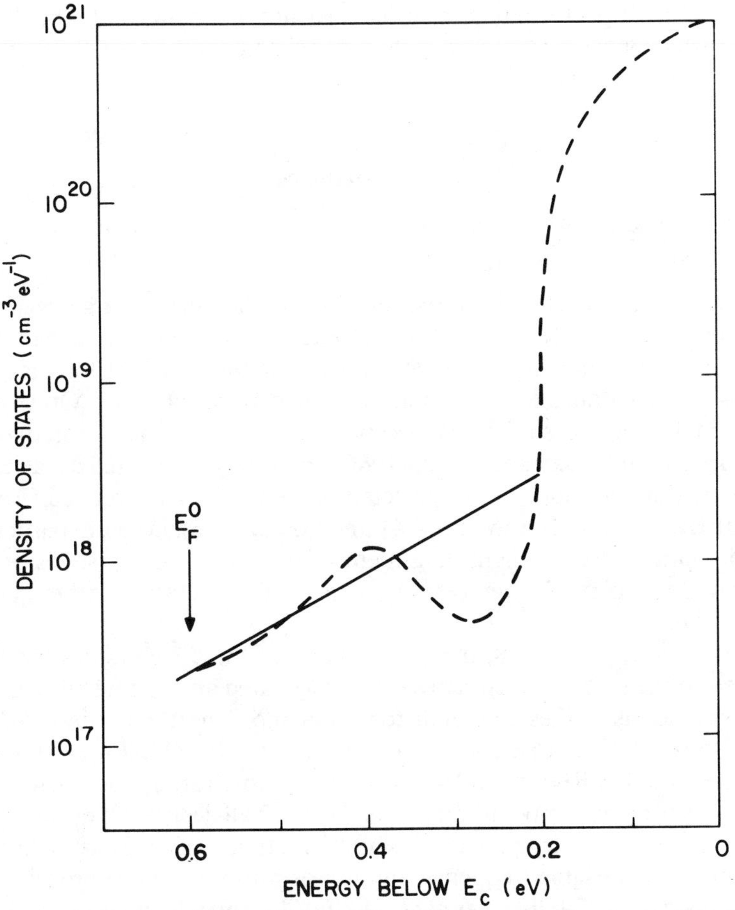

FIG. 12. Two densities of states with and without a well-developed peak near $E_c - 0.4$ eV used to calculate the field-effect "data" in Fig. 11. [After Powell (1981).]

ductors. These include studies of the effects of hydrogen content on $g(E)$ in a-Si:H (Goodman and Fritzsche, 1981), studies of metastable light-induced effects (Tanielian *et al.,* 1981), investigations of changes in $g(E)$ associated with doping (LeComber *et al.,* 1980; Jan *et al.,* 1980; Nakashita *et al.,* 1980), and studies of other amorphous silicon alloys such as a-Si:F:H films (Madan *et al.,* 1979). Indeed, because of its inherent sensitivity to surface properties, the field effect method is particularly suited to the study of interface properties (Goodman, 1982) and hence to the development of thin film field-effect transistors (LeComber *et al.,* 1981). However, based on the discussion above, it is important to recognize that of the various methods discussed in this chapter, field-effect measurements are undoubtedly the *most* surface sensitive and hence the least reliable method for obtaining the bulk density of states.

IV. Capacitance

6. Development of Capacitance Techniques to Study Gap States

The use of capacitance measurements of the space-charge region in Schottky barriers or MOS structures to obtain information about gap states in semiconductors was first discussed by Goodman (1963) for measurements of capacitance as a function of applied voltage ($C-V$). Other early work, including measurements of capacitance as a function of frequency, was carried out by Sah and Reddi (1964) and these and related capacitance methods were developed in great detail in studies by Roberts and Crowell (1970), Beguwala and Crowell (1974), and Losee (1975). Losee, in particular, has presented a very general treatment of the effects of gap states on the measured complex admittance as a function of temperature and frequency ($C-T-\omega$).

In the last several years, the use of many of these techniques has been applied to the study of deep states in hydrogenated amorphous silicon. The earliest analysis of these methods for MOS and Schottky barriers (Döhler and Hirose, 1977; Hirose *et al.*, 1979; Spear *et al.*, 1978) followed along the lines presented by Roberts and Crowell (1970), which are applicable for a low frequency (or high temperature) regime in which all deep levels may respond to the ac voltage. More recently, several methods of analysis have been developed for variable frequency and temperature. In one approach (Viktorovitch *et al.*, 1979; Beichler *et al.*, 1980; Viktorovitch and Moddel, 1980; Tiedje *et al.*, 1980; Viktorovitch, 1981) an equivalent network analysis is used. This is perhaps the more appropriate analysis for fairly insulating (undoped) material in which the time scales of measurement are comparable to the dielectric relaxation time. A somewhat different approach (Snell *et al.*, 1979; Cohen and Lang, 1982; Abram and Doherty, 1982) incorporates a more direct role for the majority-carrier thermal emission and capture processes. For capacitance measurements these processes are manifested primarily by imposing a frequency and temperature dependent cutoff to which the gap state charge can respond. In the treatment of Cohen and Lang (1982) the complex admittance is obtained through the detailed solution of Losee's equation (Losee, 1975) under certain conditions that generally apply for the case of doped amorphous semiconductors or undoped material at sufficiently high temperatures.

In the following discussion we shall examine some of the various ways in which junction capacitance measurements for an amorphous semiconductor may be interpreted. Experimental results for a-Si:H are presented for several representative studies mentioned above. We shall then discuss the limitations of capacitance techniques to deduce densities of states and also mention some of the capacitance profiling techniques that have been used to aid in the interpretation of capacitance measurements.

7. Understanding the Junction Capacitance

Junction capacitance measurements are employed routinely to characterize crystalline semiconductors. In the case of crystals, or in fact any semiconductor in which the space charge is dominated by the shallow donor or acceptor concentrations, capacitance is largely independent of applied frequency or temperature, provided that the dielectric relaxation time is short enough to allow charge to move into and out of the edge of the depletion region on the time scale of the measurement frequency. At the same time, the contribution to the capacitance due to deep levels varies markedly with temperature and frequency depending on the degree of participation of these deep levels in the total alternating charge distribution. Although such contributions are small in most crystals, they can be very large in radiation damaged or amorphous semiconductors in which the concentration of deep levels dominates the material properties.

To understand how this occurs we consider how the total barrier charge is changed by a small change in the applied voltage. By using Eq. (1), we have

$$\frac{d}{dx}\left(x\frac{d\psi}{dx}\right)=\frac{d\psi}{dx}+x\frac{\rho(x)}{\epsilon}.$$

By integrating both sides of this equation from $x=0$ (the semiconductor interface where $\psi=\psi_s$) to infinity (or far into the bulk where ψ and $d\psi/dx=0$), we obtain

$$\psi_s=\frac{1}{\epsilon}\int_0^\infty x\rho(x)\,dx. \tag{16}$$

By considering the simpler case of a Schottky diode first, with $\delta\psi_s=\delta V_A$, and putting in the total barrier charge for a diode of area A

$$Q=A\int_0^\infty \rho(x)\,dx,$$

we then obtain

$$\frac{\delta Q}{\delta V_A}=C=\frac{\epsilon A\int_0^\infty \delta\rho(x)\,dx}{\int_0^\infty x\,\delta\rho(x)\,dx}=\frac{\epsilon A}{\langle x\rangle}, \tag{17}$$

where $\langle x \rangle$ is the *first moment* of the ac charge distribution $\delta\rho$

$$\langle x \rangle = \int_0^\infty x\, \delta\rho(x)\, dx \Big/ \int_0^\infty \delta\rho(x)\, dx. \tag{18}$$

Equation (17) is quite generally valid. The difference between a low frequency or dc capacitance and a higher frequency ac measurement at frequency ω lies entirely in the fraction of the space-charge region over which the charge density is able to change on the time scale $1/\omega$. This, in turn, is determined by the limitations imposed by emission and capture times to states at the Fermi energy, which lie at increasingly deeper energies in the gap as one moves into the depletion region from the bulk semiconductor.

Figure 13 illustrates schematically a dc low frequency case as well as a high frequency situation both for a crystalline semiconductor with a single discrete deep level and an amorphous semiconductor with a continuous large distribution of deep states. For the crystal, the charge density is essentially a constant equal to $q(N_D - N_A)$, so that $\delta\rho \neq 0$ only at the edge of the depletion region located at $x = W$. Thus $\langle x \rangle \simeq W$ and $C = \epsilon A/W$. A small difference will exist for low frequency (or high temperature) measurements due to the deep level that crosses the Fermi energy closer to the junction interface, because it will then contribute to $\delta\rho$ and shift $\langle x \rangle$ slightly toward the interface (see Fig. 13a), giving rise to a slightly larger capacitance.

For the amorphous semiconductor at low frequency (Fig. 13b) $\delta\rho$ is nonzero throughout the depletion region and gives rise to a smaller $\langle x \rangle$ and a larger capacitance for the same spatial extent of the space-charge region as in Fig. 13a. Actually, the concept of *depletion width* is somewhat ill-defined in the amorphous case. For the ac measurement the emission time limitation at the Fermi energy produces a cutoff to $\delta\rho$, which increases $\langle x \rangle$ and decreases the capacitance.

The proper interpretation of capacitance is thus intimately connected to the relationship between the measurement time $1/\omega$ and the thermal emission process from deep gap states. For example, a measurement of dc capacitance may only be so considered if the measurement time is greater than the equilibration time for the entire depletion region. This is determined by the thermal emission time from the Fermi energy to the mobility edges near the junction interface. Since for the best Schottky barriers a nearly 1 eV barrier must be overcome for thermal emission and capture, such equilibration times can be long indeed even at slightly elevated temperatures. At the other extreme, at large ω, we eventually reach a situation in which the thermal emission time in the tail region of the depletion layer becomes comparable to dielectric relaxation times in the bulk. At this point, details of carrier transport through the bulk must be included for a proper understanding of the complex admittance of the device. This is a very difficult situation to ana-

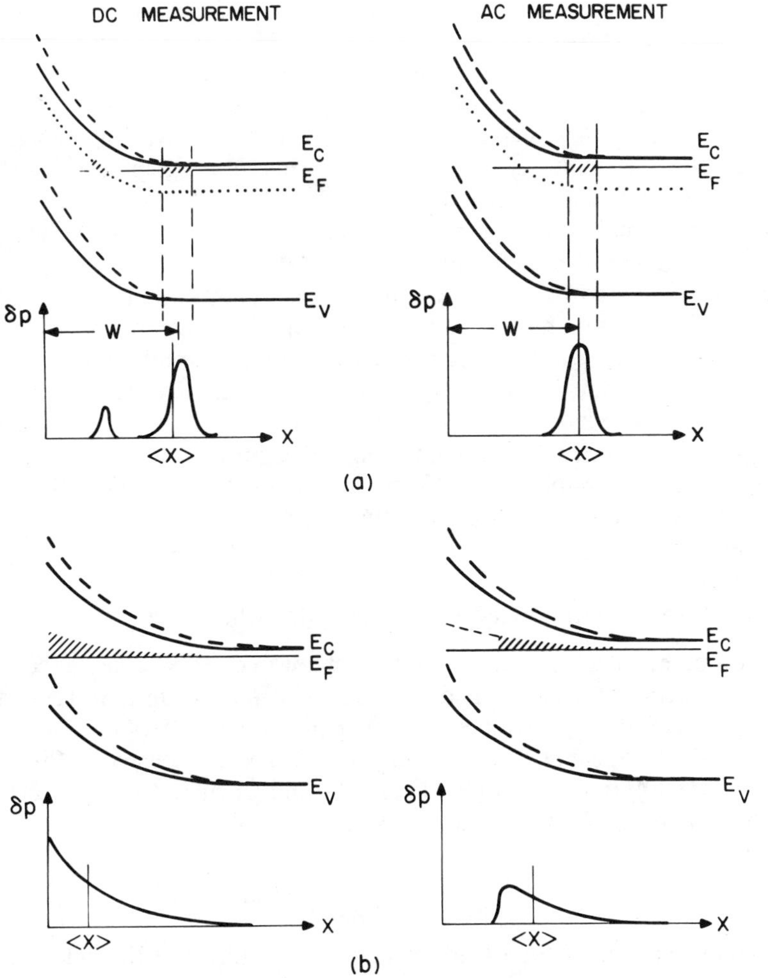

FIG. 13. Schematic diagram showing the qualitative differences between a dc and an ac capacitance measurement for (a) a crystalline semiconductor with a single discrete deep level and (b) an amorphous semiconductor with a continuous distribution of gap states. The measured capacitance in each case is inversely proportional to the value of $\langle x \rangle$. Other parameters are explained in the text.

lyze, although some attempts have been made for a-Si : H (Tiedje *et al.*, 1980; Viktorovitch, 1981). For intermediate frequency regimes we must also include the contributions of thermal emission and capture processes to the *real part* of the admittance for a complete description of the capacitance. The method of assessing this contribution is not apparent from our discussion thus far; fortunately, it is relatively straightforward and will be discussed in Section 9.

For capacitance measurements on sandwich MOS structures the preceding picture applies as well with two small modifications. Equation (17) is still applicable, although we must include the contribution to $\delta\rho$ on the semiconductor–insulator interface. For measurements with negligible change in the surface charge density this reduces to having the insulator capacitance $\epsilon_i A/d_i$, in series with the junction capacitance as treated above using the appropriate semiconductor surface potential ψ_s as given by Eq. (10). The second difference between capacitance for Schottky barriers and MOS structures occurs in the behavior at large applied bias due to the formation of a quasi-Fermi energy in the former case versus an accumulation or inversion layer for the latter. In high frequency measurements at high bias in MOS devices one then has to consider whether this charge accumulation near the semiconductor–insulator interface has sufficient time to contribute to $\delta\rho$. However, since almost all capacitance measurements on MOS devices have been performed at low frequency, this issue has not usually had to be addressed.

8. DC Capacitance versus Voltage Measurements

Compared to field effect, the analysis of low frequency capacitance versus voltage measurements to yield a density of gap states in a-Si:H is rather straightforward. Such studies were first presented by Döhler and Hirose (1977) and by Spear *et al.*, (1978) and have also been discussed by Singh and Cohen (1980), Powell and Döhler (1981), and Chen and Lee (1982). If we differentiate Eq. (3) twice with respect to ψ, we obtain

$$\frac{d}{d\psi}\left(F\frac{dF}{d\psi}\right) = -\frac{1}{\epsilon}\frac{\partial\rho}{\partial\psi} \simeq \frac{q}{\epsilon}\, g(E_F^0 - \psi), \tag{19}$$

where we have used Eq. (4) to relate $\partial\rho/\partial\psi$ to $g(E)$. But the field F at the semiconductor interface is related to the total barrier charge Q by integrating Eq. (1); namely, $F_s = Q/\epsilon A$. Thus we obtain

$$g(E_F^0 - \psi_s) = -\frac{\partial}{\partial\psi_s}\left[\frac{Q}{\epsilon A^2}\left(\frac{\partial Q}{\partial\psi}\right)_{\psi_s}\right]. \tag{20}$$

Since Q may be obtained experimentally as

$$Q = \int_0^{\psi_s} \frac{dQ}{d\psi_s}\, d\psi_s = \int_0^{\psi_s} C_B\, d\psi_s, \tag{21}$$

where C_B is the barrier capacitance (as distinct from the measured capacitance for MOS structures, which includes the oxide), Eq. (20) relates the density of states directly to experimentally derived quantities. The semiconductor interface potential ψ_s is readily determined as $-(V_A - V_{BI})$, where $V_{BI} = \Phi_B - \Delta E_F$ is the "built-in" potential for a Schottky barrier or, by Eq. (10), for MOS devices. The determination of ψ_s and Q depend, however, on knowing V_{BI} or the flat-band potential fairly accurately.

Figure 14 shows typical results from this kind of measurement on two MOS devices for both *n*- and *p*-type a-Si : H (Hirose *et al.*, 1979). Plotted as well are results of field-effect measurements on the same samples. Generally speaking these curves indicate a smooth variation of $g(E)$ with E. There is a minimum indicated in the density of states of between 10^{16} and 10^{17} cm^{-3} eV^{-1} near midgap. The density of states derived from the FE analysis is somewhat higher than for $C-V$ and indicates a hint of the $E_c - 0.4$ eV peak reported in some of the other field-effect measurements. Since these FE results are based on an approximate analysis (parabolic profile) the differences between the densities of states displayed in Fig. 14 are probably not significant. Indeed, the parabolic profile analysis for field-effect data is expected to yield a $g(E)$ that is somewhat too large (see Part III).

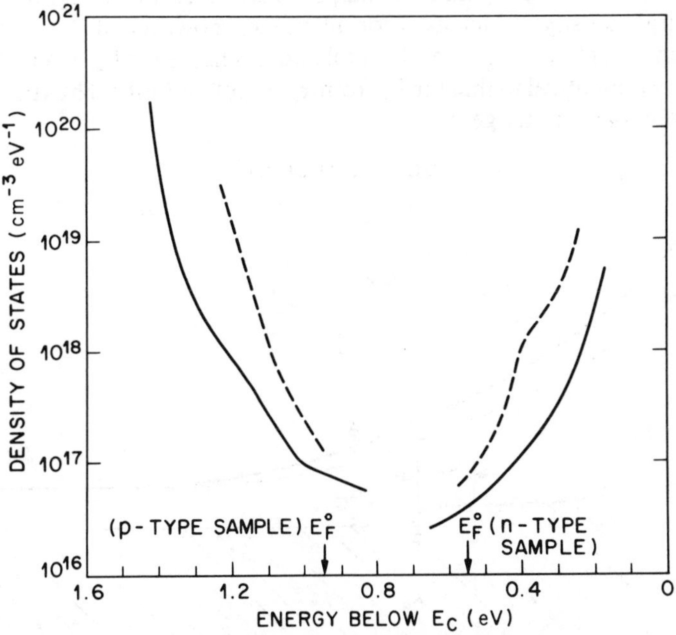

FIG. 14. Density of states deduced from the low frequency (dc) capacitance versus voltage characteristics of glow-discharge a-Si : H MOS devices (solid curves). Dashed curves are the result of an approximate analysis for field-effect measurements taken on the same MOS devices. [After Hirose *et al.* (1979).]

9. Capacitance at Arbitrary Frequency and Temperature

a. *Analysis*

As discussed previously, the junction capacitance at high frequency depends on the thermal emission and capture times of trapped gap-state charge. In effect, this imposes an energy cutoff E_e such that states at E_F deeper than E_e cannot respond to the applied ac voltage. As discussed in Part II, this energy cutoff is given by

$$E_e = kT \ln(\nu/\omega), \tag{22}$$

where ν is the exponential prefactor for majority carrier emission and capture and ω is the ac measurement (angular) frequency.

A simple understanding of the resulting measured capacitance comes from Fig. 15. This figure shows barrier profiles for two values of applied bias that differ by ΔV_A. Occupied states for V_A lie below the Fermi level at E_F. For $V_A + \Delta V_A$ this holds only for $x > x_e$ because states deeper than E_e have not had time to be emitted. Hence the occupied states are as indicated by the shaded region, indicating that the charge density does not change for $x < x_e$. The charge density *is* increased for all $x > x_e$; however, if the material is homogeneous for $x > x_e - \delta x$, the total barrier charge for $V_A + \Delta V_A$ from x_e to infinity is identical to that for V_A from $x_e - \delta x$ to infinity. Thus the change in the total barrier charge is

$$\Delta Q = A\, \delta x\, \rho(\psi_e), \tag{23}$$

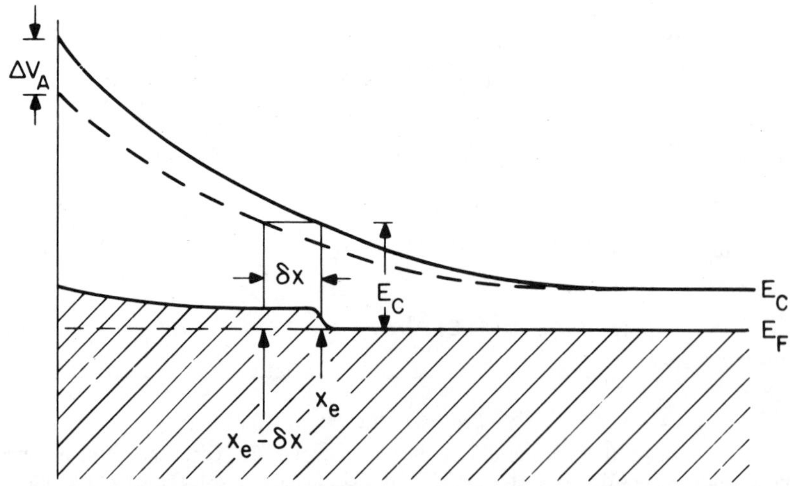

FIG. 15. Schematic diagram used to derive an approximate expression for the ac capacitance of an amorphous semiconductor diode junction as described in the text.

where $\psi_e = E_e - \Delta E_F = \psi(x_e)$. One can readily compute ΔQ in terms of ΔV_A, x_e, and the electric field $F(x_e)$ by matching the appropriate boundary conditions at x_e. Taking the limit for small V_A we obtain

$$dQ/dV_A = C \simeq \epsilon\rho_e/(\epsilon F_e + x_e\rho_e), \tag{24}$$

where $\rho_e = \rho(\psi_e)$ and $F_e = F(x_e)$. Equation (24) indicates that there are two experimental means to vary C: (1) by changing the applied dc bias, which changes x_e *but not* ρ_e or F_e, and (2) by changing either the temperature or frequency, which, for a fixed dc bias, will change all 3 quantities by changing ψ_e. This suggests an important advantage of including temperature and frequency variation in capacitance measurements, namely, the ability to investigate spatial variation within the sample and energy variation within $g(E)$ to some degree independently.

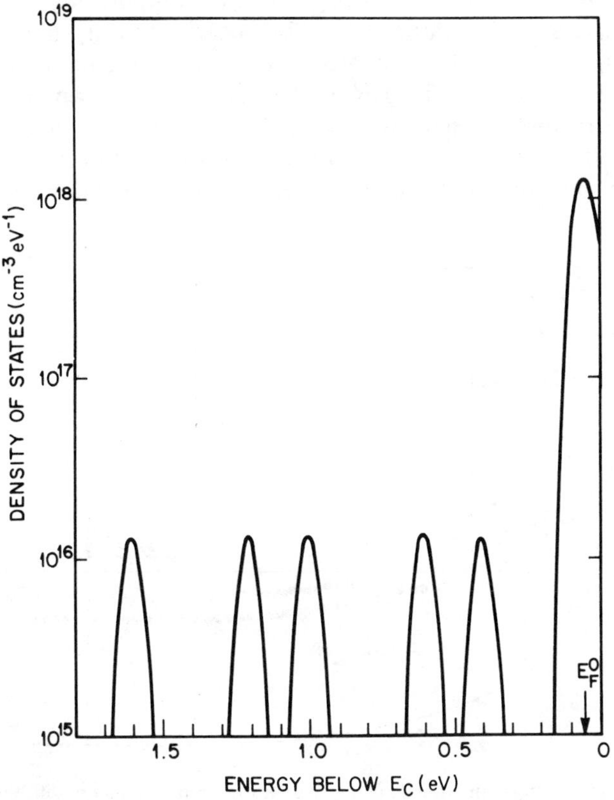

FIG. 16. Density of states showing five Gaussian "trap" levels, each with total integrated density 1×10^{15} cm^{-3}, plus shallow "donor" level corresponding to a net donor concentration of 5×10^{16} cm^{-3}. The position of the bulk Fermi energy is taken to be 50 meV below E_c and the gap energy is assumed to be 1.8 eV.

Equation (24) was derived by Snell *et al.* (1979) in a slightly different form by using arguments similar to those given above. This derivation is clearly of a somewhat approximate nature and does not, unfortunately, allow us to assess the discrepancies caused by neglecting finite temperature statistics; nor does it allow us to evaluate the real part of the ac admittance G/ω. A more complete derivation was later demonstrated by Cohen and Lang (1982) who solved the very general equations for ac admittance given by Losee (1975) for a semiconductor with a continuous density of states. This solution was demonstrated to be accurate to within a few percent for a wide class of densities of states of interest. It also gave the real part of the admittance as

$$G/\omega = kT(\tfrac{1}{2}\pi)C^2(F_e/\rho_e^2)\, qg(E_F^0 - \psi_e). \tag{25}$$

By using Eqs. (24) and (25), it is possible to compute, say, the complex admittance versus temperature curves for a nearly arbitrary given density of states $g(E)$, which may be compared to experimental data on a-Si:H. An illustrative example is provided by the density of states shown in Fig. 16 for a series of nearly discrete (narrow Gaussian) levels of concentration N_T with a dominant 50-meV-deep shallow (dopant) level of concentration $N_D \gg N_T$. The barrier potential and deep-level occupation is shown in Fig. 17 for a

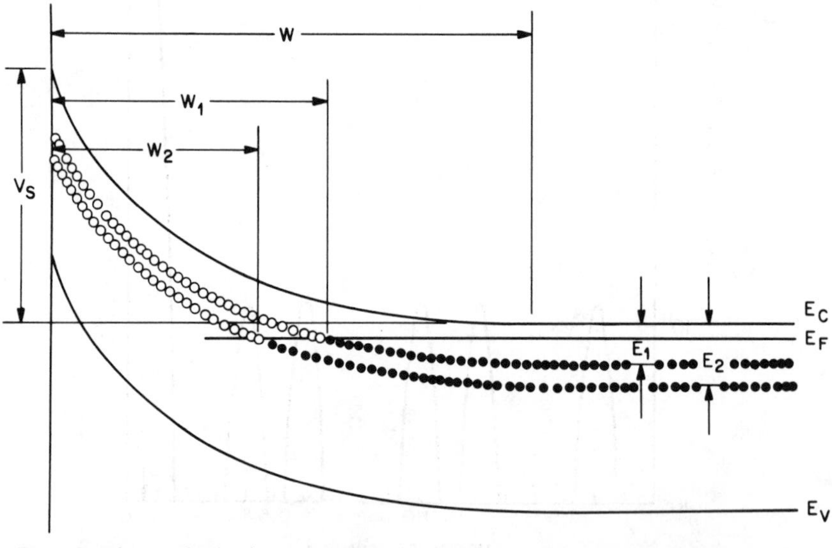

FIG. 17. Diagram indicating the depletion region for a semiconductor with the density of states shown in Fig. 16. The open and closed circles indicate the uppermost two trap levels as empty or full, respectively. The total depletion width is W and the position of the ac charge density due to the deep levels is at W_1 and W_2, respectively, if the temperature is sufficiently high to allow the thermal emission and capture of trapped charge from these levels.

Schottky-barrier diode under high reverse bias. In this case, because the levels below midgap lie below E_F^* and are always occupied, only the two deep levels above midgap at $E_c - 0.4$ eV and $E_c - 0.6$ eV will give rise to an ac charge distribution within the depletion region. The capacitance versus temperature curve for an assumed constant emission rate prefactor of $\nu \sim 10^{13}$ sec^{-1} and $\omega = 2\pi \times 10^4$ sec^{-1} is shown in Fig. 18. Note that the vertical capacitance scale is greatly expanded so that in fact the capacitance is very nearly constant above the "turn-on" temperature at roughly 50°K. The two small capacitance steps at 225°K and 340°K correspond to the temperatures at which E_e [see Eq. (22)] reaches the values 0.4 and 0.6 eV, respectively. In this case, the band-bending curve is nearly parabolic, so that the depletion width is given by the usual expression

$$W = (2\epsilon V_s / qN_D)^{1/2}, \tag{26}$$

with the corresponding capacitance $C = \epsilon A/W$. This is consistent with Eq.

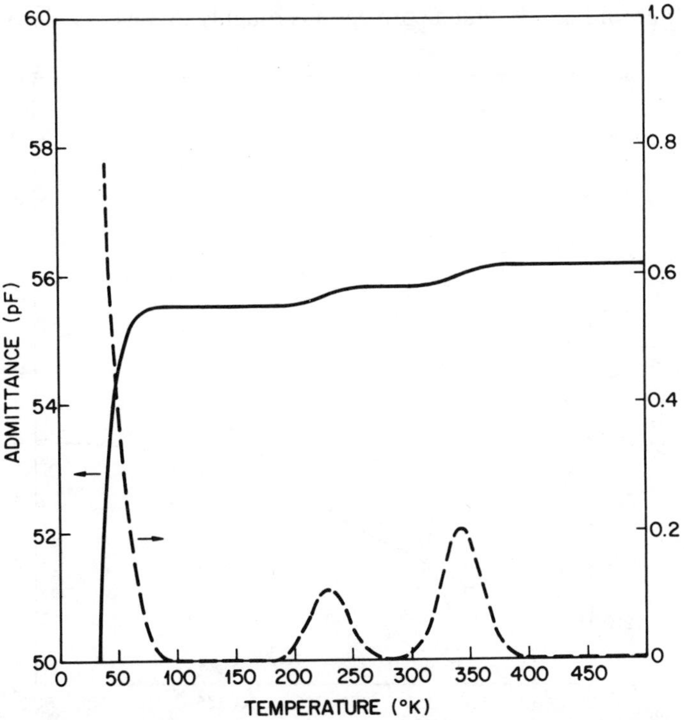

FIG. 18. Imaginary (solid) and real (dashed) part of 10 kHz admittance obtained for density of states in Fig. 16 under 5 V reverse bias ($\Phi_B = 0.55$ eV). Note that the real part G/ω is also expressed in picofarads (right-hand scale) and is multiplied by a factor of 10 compared to the capacitive part (which is offset). The junction area A is taken to be 2×10^{-3} cm^2.

(24) since $\rho_e = qN_D$ = const and $F(x_e)$ is equal to the total integrated charge from x_e to W divided by ϵ; i.e., $F_e = qN_D(W - x_e)/\epsilon$. When each of the deep traps is able to respond at the ac frequency the value of ρ_e increases slightly without greatly affecting F_e; hence the capacitance increases.

The density of states proposed from FE measurements by Spear and LeComber (1976), as shown in Fig. 4, provides a second example. The capacitance versus temperature curve for n-type material ($E_c - E_F^0 =$ 0.2 eV) is shown in Fig. 19. In this case the capacitance is not so nearly constant but exhibits a large step increase similar to those in Fig. 18 due to the fairly narrow level at $E_c = 0.4$ eV. The capacitance at high temperatures becomes fairly constant due to the deficit of states near midgap.

Two final examples are for the densities of states shown in Fig. 20, one of which is very similar to those derived from low frequency $C-V$ methods (see Fig. 14). The other is similar to those obtained from DLTS measurements on n-type doped samples (see Part V). The capacitance versus temperature curves in Figs. 21 and 22 are both generally smooth but clearly distinct from each other. In Fig. 21, which corresponds roughly to the $C-V$ derived den-

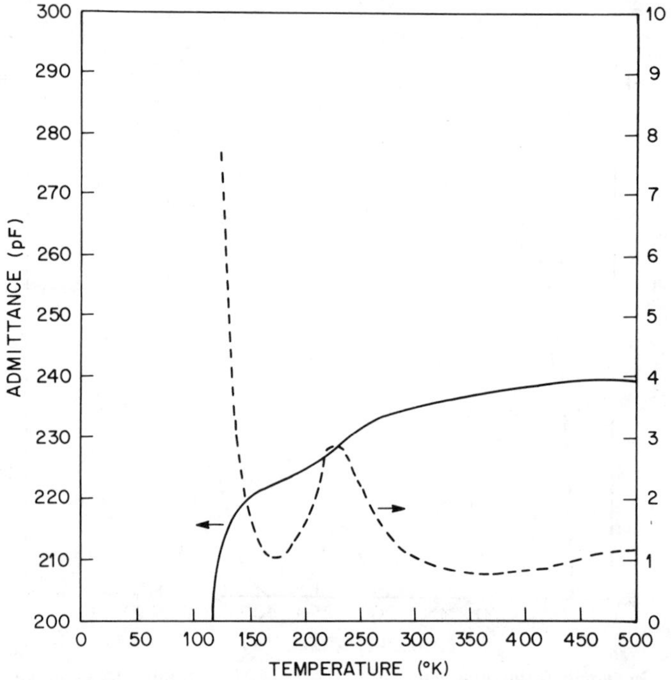

FIG. 19. Imaginary (solid) and real (dashed) parts of 10 kHz admittance obtained for the density of states in Fig. 4 and 5 V reverse bias with $\Phi_B = 0.7$ eV. Other parameters used in the calculation are as in Fig. 18.

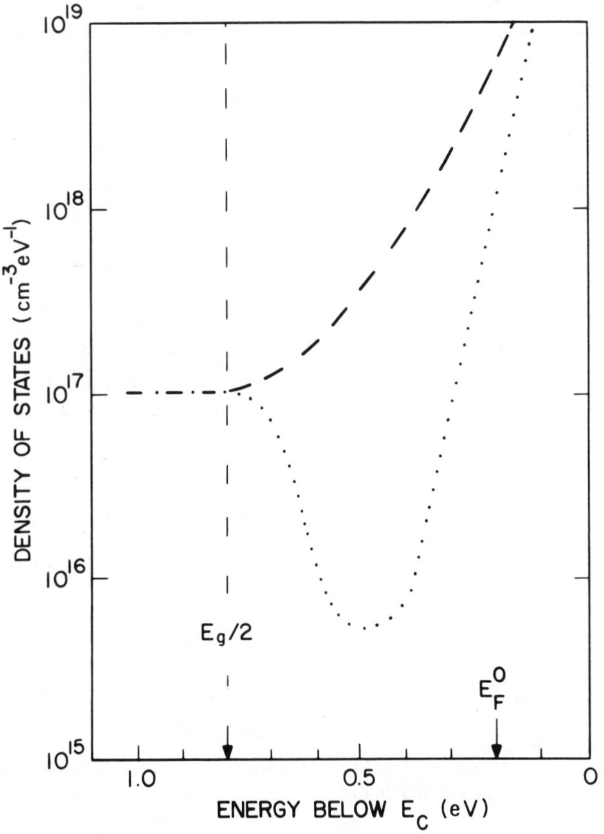

FIG. 20. Two densities of states used to calculate the capacitance versus temperature curves shown in Figs. 21 and 22. The dashed curve is similar to $g(E)$ results obtained from dc $C-V$ analysis as in Fig. 14 and the dotted curve is similar to $g(E)$ as obtained from DLTS (see Fig. 52).

sity of states, the capacitance rises very rapidly with T at first, then continues to rise more slowly. In Fig. 22, on the other hand, the capacitance above turn-on is fairly constant and then begins to rise at high temperatures due to the participation of the large number of states near midgap.

b. *Experimental Results*

With the calculated examples of the previous discussion in mind we turn to the experimental data in Fig. 23a, which shows capacitance and G/ω versus temperature curves at a series of measurement frequencies (Cohen *et al.*, 1983b). These curves are for an *n*-type a-Si:H sample (100 vppm PH_3) that has been specially prepared to have a physically well-delineated area. The curves are very similar to those shown in Fig. 22 and indicate that $g(E)$

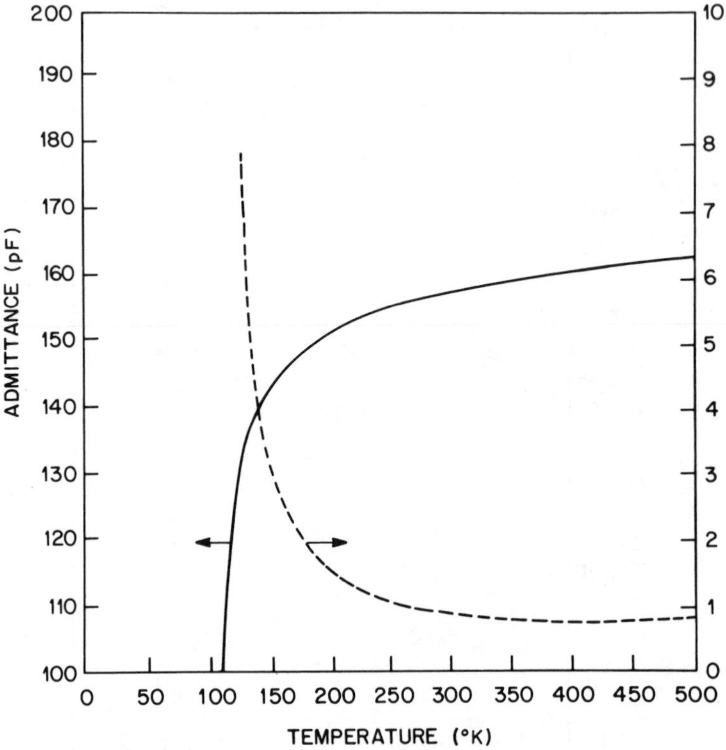

FIG. 21. Imaginary (solid) and real (dashed) parts of 10 kHz admittance obtained for the dashed curve density of states shown in Fig. 20. Parameters used in the calculation are as in Fig. 19.

has a significant shallow-state concentration but very few deep states so that there is a reasonably large constant capacitance (as in Fig. 18). At high temperatures the curves start to increase, indicating a possible contribution from very deep states. These data indicate that the total number of occupied shallow states (in neutral bulk material) is roughly $1 \times 10^{17}\ \mathrm{cm}^{-3}$.

An additional piece of information may be obtained from such $C-T-\omega$ measurements. The temperature T_0 at which the capacitance abruptly increases from its low temperature value defines the temperature at which the RC time constant of the film is comparable to ω^{-1}. Relating this resistance to the bulk resistivity we have

$$\rho(T_0) \simeq (\omega C d)^{-1}, \tag{27}$$

where d is on the order of the film thickness. Note that if d is given in centimeters and the depletion width capacitance C is given in farads, then

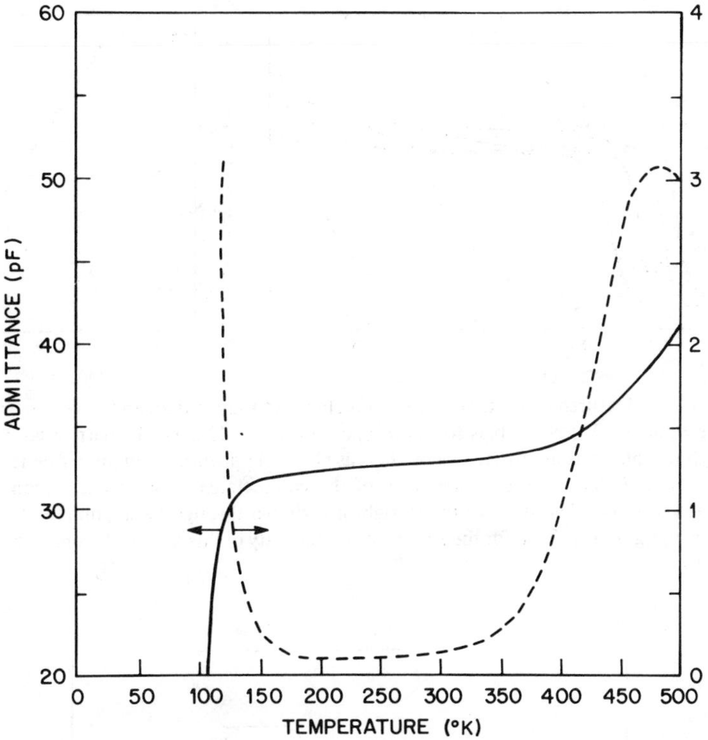

FIG. 22. Imaginary (solid) and real (dashed) parts of 10 kHz admittance obtained for the dotted curve density of states shown in Fig. 20. Parameters used in the calculation are as in Fig. 19.

this formula gives the resistivity of the film directly in ohm-centimeters. Plotting the logarithm of the measurement frequency versus $1/T_0$ at each frequency gives the activation energy of the bulk conductivity as shown in Fig. 23b. This method for determining the conductivity of a-Si : H is noteworthy in that it is greatly devoid of possible problems encountered in dc measurements associated with the quality of ohmic contacts. This method of determining conductivities has been used to good advantage in several recent studies (Lang *et al.*, 1982a; Cohen *et al.*, 1983a).

A somewhat different kind of behavior is exhibited in Fig. 24, which shows the results of $C-T$ measurements for Schottky barriers on undoped a-Si : H grown at various substrate temperatures (Jousse and Deleonibus, 1983). This behavior is quite consistent with the 0.6 eV activation energy for the onset of the capacitance, since the Fermi energy at $E_c - 0.6$ eV will lie in a region where $g(E)$ is more nearly a constant for most of the a-Si : H densities of states discussed so far. The lower conductivity of undoped films necessi-

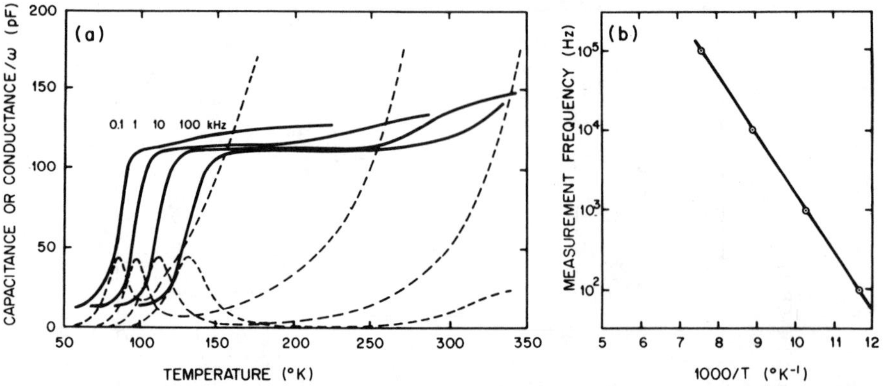

FIG. 23. (a) Measured capacitance and conductance at four ac frequencies versus temperature taken under 4 V reverse bias for an n-type doped a-Si:H Schottky-barrier sample in a cleaved chip configuration. (b) Semilog Arrhenius plot of the ac measurement frequency versus $1000/T$, where T denotes the temperature of the conductance peak at each frequency as determined from (a). The slope of the straight line drawn through these points indicates an activation energy of 0.147 eV for the electrical conductivity of this sample. [From Cohen *et al.* (1983b).]

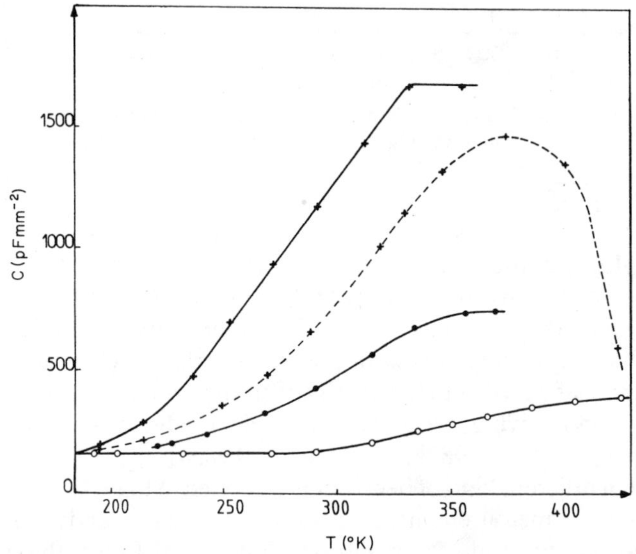

FIG. 24. Capacitance–temperature characteristics of Pt Schottky diodes of undoped a-Si:H samples deposited at substrate temperatures $T = 450°C$ (+), 400°C (O), and 350°C (●). Solid line curves were taken at 1 Hz; the dashed curve was taken at 30 Hz. The high temperature decrease in the 30 Hz data is due to leakage over the barrier. [From Jousse and Deleonibus (1983).]

tates using considerably lower measurement frequencies, as indicated, than for the doped a-Si : H results in Fig. 23. Distinguishing a particular detailed model is thus quite difficult for nearly intrinsic films because only a small region of $g(E)$ lying very close to midgap is revealed by these measurements.

As a final example we consider the capacitance versus temperature and capacitance versus frequency curves shown in Figs. 25 and 26. These measurements were made on metal–insulator–semiconductor (MIS) structures with undoped a-Si : H grown by rf sputtering (Weisfield *et al.*, 1981). These two figures give a very nice illustration of the reciprocal relationship, as expected from Eq. (22), between scanning temperature at fixed frequency versus scanning frequency at fixed temperature. The capacitance rise at 60°C in Fig. 25 corresponds to the high frequency drop in capacitance near 0.1 and 10 kHz shown for the two curves in Fig. 26. The three sets of data indicate a Fermi-level position (conductivity activation energy) lying between 0.7 and 0.8 eV below E_c. The capacitance plateau in the $C-T$ curve between 100 and 150°C is consistent with the value of capacitance corresponding to the low frequency, high temperature (dc) limit in which all gap states in the space-charge region above E_F may respond to the ac voltage. The value of the plateau capacitance indicates a midgap density of approximately 10^{17} cm^{-3} eV^{-1}. The continued rise in capacitance at higher temperatures (or lower frequencies) is attributed to states in the SiO_x insulating layer (Weisfield *et al.*, 1981).

For both of the studies on undoped a-Si : H that we have mentioned, the primary information obtained concerning the density of states consists of an

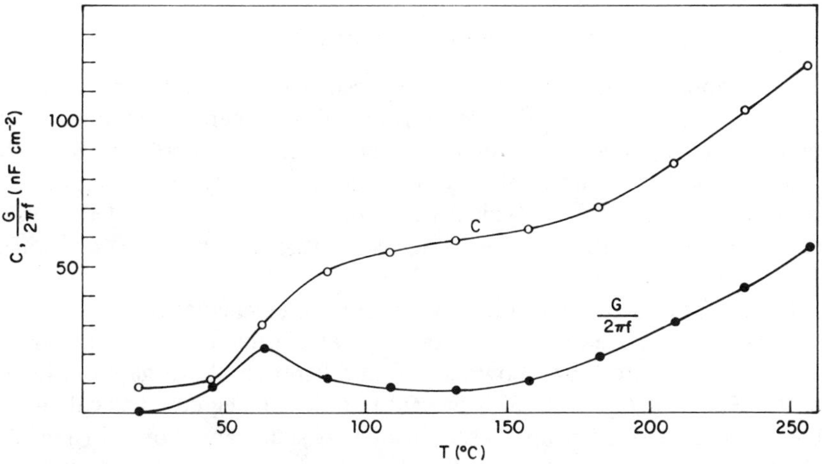

FIG. 25. 2-Hz capacitance and conductance versus temperature measurements for undoped a-Si : H MIS devices (nichrome–SiO_x–sputtered SiH). [From Weisfield *et al.* (1981).]

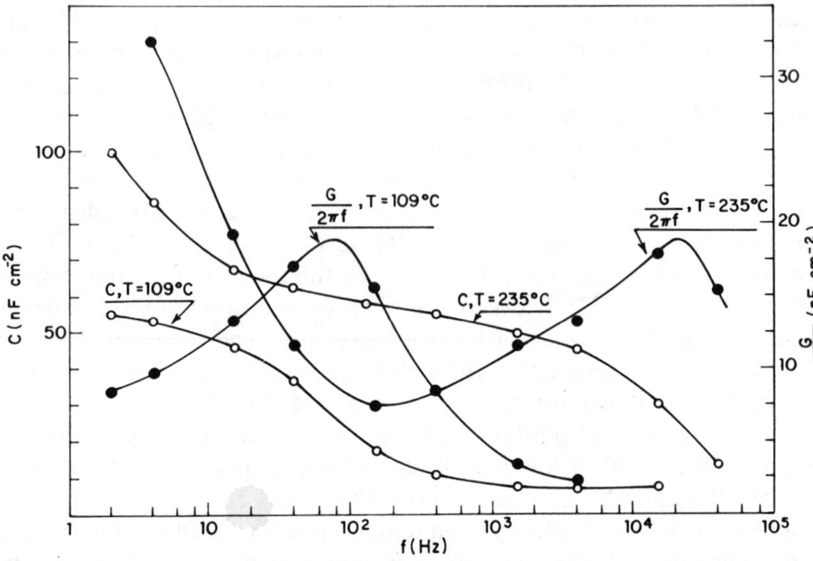

FIG. 26. Capacitance versus frequency measurements at two temperatures for the sample described in Fig. 25. [From Weisfield *et al.* (1981).]

activation energy indicating the Fermi-level position and the value of $g(E)$ near midgap. In the case of the doped films a somewhat better assessment can be made between the various proposed models for $g(E)$. However, even in these cases, a detailed picture of $g(E)$ based on $C-T-\omega$ measurements is fairly difficult to achieve.

10. LIMITATIONS OF CAPACITANCE TECHNIQUES

One significant difference between the results of the $C-V$ technique and of the $C-T-\omega$ methods is that the former measurements have been interpreted to give $g(E)$ over a fairly large energy range, whereas most of the proponents of the latter method seem to deduce a fairly modest amount of information about $g(E)$—typically the value of $g(E)$ over a limited range of energy. Because of this, we turn primarily to an assessment of the precision of $C-V$ method and analysis.

In Figs. 27 and 28 the results of some computer calculations based on an a-Si:H structure are presented (Goodman and Fritzsche, 1980). In Fig. 27 four densities of states are shown with four different mobility gaps in which a possible peak at $E_c - 0.4$ eV is present to differing degrees. The calculated $C-V$ curves derived from these densities of states are shown in Fig. 28. Clearly there is no significant difference among the different curves. Since experimental data is somewhat more subject to uncertainty than the com-

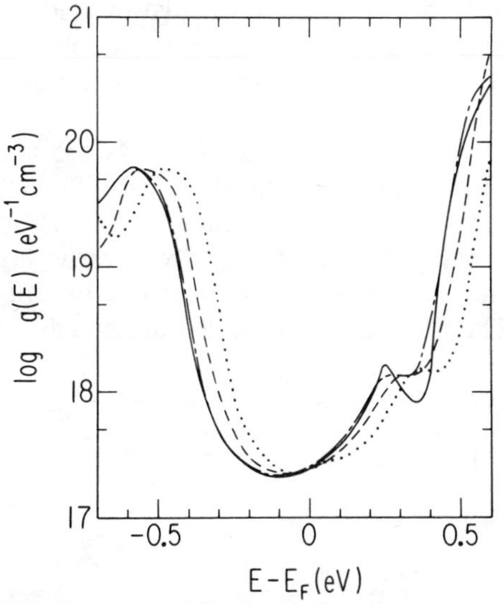

FIG. 27. Three density-of-states curves with different mobility gaps of 1.44, 1.54, and 1.64 eV corresponding to the values of α [see Eq. (13)] of 1.2×10^2 (dotted), 2.0×10^4 (dotted-dashed), and 1.8×10^6 (dashed), as indicated. The original density of states of Spear and LeComber (1976) is shown by the solid curve. Each of these densities of states fits the hypothetical "data" of Fig. 28. [From Goodman and Fritzsche (1980).]

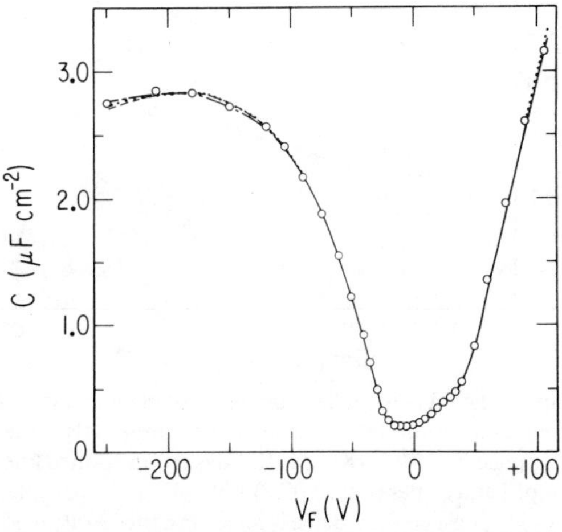

FIG. 28. Hypothetical dc capacitance–voltage "data" for an a-Si:H MOS device derived from the densities of states (and respective mobility gaps) displayed in Fig. 27. (Key as in Fig. 27.) [From Goodman and Fritzsche (1980).]

puter generated curves displayed, these calculations indicate that it would be very difficult to use $C-V$ measurements to, say, resolve the apparent differences between the FE results of Fig. 4 and the more featureless $C-V$ results of Fig. 14.

A second example of some calculated $C-V$ data is shown in Fig. 29 for high frequency measurements for hypothetical a-Si:H Schottky barriers (Cohen and Lang, 1982). Here the four curves displayed are based on (1) the FE density of states of Fig. 4, (2) the two $g(E)$ curves displayed in Fig. 20, and (3) a density of states that is completely flat across the mobility gap with $g = 10^{17}$ cm^{-3} eV^{-1}. The four curves are scaled to have the same value at 0 V

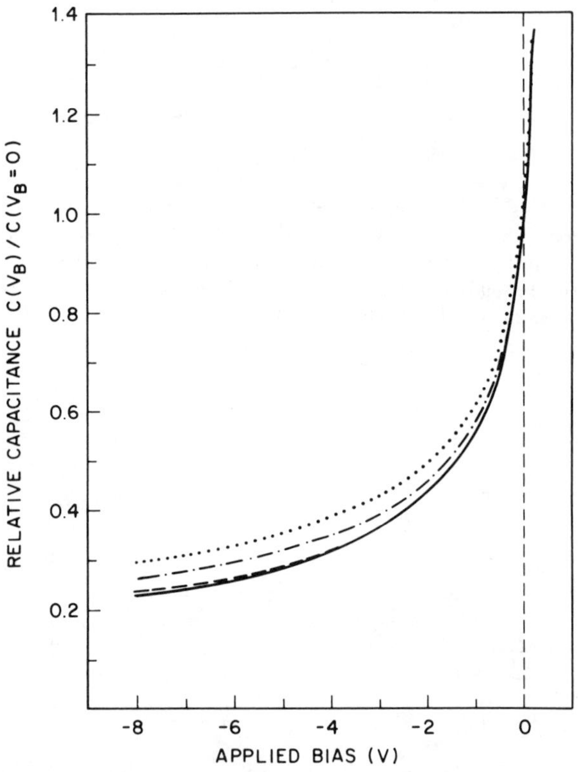

FIG. 29. 10-kHz capacitance versus applied voltage at room temperature for a Schottky-barrier diode with 4 different densities of states. The solid line corresponds to the $g(E)$ shown in Fig. 4 [capacitance at zero applied bias $C(0)$ is 811 pF]; the dashed and dotted lines correspond to the U-shaped [$C(0) = 544$ pF] and center peaked [$C(0) = 93$ pF] to the densities of states shown in Fig. 20, respectively, and the dashed-dotted line is for a constant density of states: $g = 10^{17}$ cm^{-3} eV^{-1} [$C(0) = 187$ pF]. In each case the Fermi energy E_F^0 is assumed to lie 0.2 eV below E_c. These $C-V$ curves have been rescaled to have the same value at zero applied volts (the flat-band voltage is +0.5 volts) with the actual value of zero-voltage capacitance as given above.

applied bias. This scaling is not significant since it is related only to the overall magnitude of $g(E)$. The differences among the four curves representing vastly different densities of states are only modest at best. The situation is worse if the flat-band potential is not accurately known or varies across the sample. This could shift each curve horizontally by an unknown amount of perhaps up to 0.2 V, which would, if one also adjusted the scale factor, make these curves practically indistinguishable over a wide range of applied voltage. The importance of and difficulties in determining the flat-band voltage in interpreting $C-V$ and FE data has been discussed at length in the literature (Solomon *et al.,* 1978; Tanielian *et al.,* 1978). One additional problem, as discussed in connection with the field effect, concerns the assumption of sample uniformity. This assumption is crucial to the proper interpretation of these $C-V$ data because there is no way to sort out effects due to the spatial variation within the sample from the energy variation of $g(E)$.

Although such remarks cast doubt on the detailed picture of $g(E)$ obtained by $C-V$ analysis, the overall magnitude of $g(E)$ obtained by these methods is undoubtedly reasonably accurate and is consistent with those obtained by alternative techniques. A significant improvement in capacitance measurements comes from combining $C-V$ with the $C-T-\omega$ techniques. This allows the spatial and energy variation of $g(E)$ in a-Si:H samples to be properly distinguished. Such methods have not been fully exploited; however, one suspects that the kinds of detailed pictures presented by the early FE measurements or the recent DLTS results (see Part V) may still be difficult to achieve from capacitance measurements alone. In the following section one such method ($C-T$ slope method) is presented as an example of how $C-T$ measurements can be combined with measurements at different applied bias to yield direct information about $g(E)$ in a-Si:H with a minimum of numerical analysis.

11. Other Capacitance Methods

For the case of crystalline semiconductors in which the dopant introduces a well-defined, shallow discrete level in the gap, the $C-V$ plot is routinely used to evaluate the net dopant concentration using the relationship (Miller, 1972)

$$N_{CV} = -\frac{C^3}{\epsilon q A^2}\left(\frac{dC}{dV_A}\right)^{-1}. \tag{28}$$

For Schottky barriers this equals $N_D - N_A$ or $N_A - N_D$ for *n*- or *p*-type material, respectively, where N_D (N_A) is the donor (acceptor) concentration. Dopant concentration profiles are obtained by plotting N_{CV} versus $W = \epsilon A/C$. For uniform doping N_{CV} as given above will be independent of the applied voltage.

For a-Si:H, or any other semiconductor with a large concentration of deep levels, N_{CV} will depend on both the applied frequency and the rate of the bias voltage sweep because of the long time response of the space charge. Under ideal conditions, for which the bias is changed sufficiently slowly, the quantity N_{CV} will still vary with the applied bias even for material that is spatially uniform because of the nonuniform charge density within the depletion region as given by Eq. (2). However, at sufficiently large applied reverse bias the value of N_{CV} obtained from $C-V$ measurements gives a good estimate of the integrated density of states from midgap to E_F^0. This is illustrated in Table I for a hypothetical semiconductor with $g = \text{const} = 10^{17}$ $\text{cm}^{-3}\,\text{eV}^{-1}$, and with $E_g = 1.8$ eV and $E_c - E_F^0 = 0.2$ eV. Although N_{CV} varies markedly with V_A at low bias, at higher bias it approaches the value 7×10^{16} cm^{-3} expected. This follows from the fact that in deep depletion, for which $E_F^* \simeq E_g/2$, the charge density is constant and has this value (multiplied by q).

Another technique is based on the $C-T$ technique through the temperature dependence of the quantity $C^2(dC/dT)^{-1}$ (Cohen and Lang, 1982). Using Eq. (24) one finds that

$$C^2(dC/dT)^{-1} = A\rho_e^2/qg(E_F^0 - \psi_e)F_e, \tag{29}$$

which is independent of the parameter x_e provided the material is spatially uniform for $x > x_e$. Since for a fixed frequency and temperature the varia-

TABLE I

CALCULATED VALUE OF BARRIER CHARGE DENSITY N_{CV} FOR A MATERIAL WITH A CONSTANT DENSITY OF STATES AS DESCRIBED IN THE TEXT[a]

Applied bias (V)	N_{CV} (cm^{-3})		
	100 Hz	10 kHz	1 MHz
0	3.7×10^{16}	2.9×10^{16}	2.2×10^{16}
−1	5.2×10^{16}	4.5×10^{16}	3.8×10^{16}
−2	5.6×10^{16}	5.0×10^{16}	4.4×10^{16}
−5	6.1×10^{16}	5.6×10^{16}	5.0×10^{16}
−10	6.3×10^{16}	5.9×10^{16}	5.5×10^{16}
−15	6.4×10^{16}	6.1×10^{16}	5.7×10^{16}

[a] The value N_{CV} is calculated for a variety of applied bias and measurement frequencies. The correct value of the number density in deep depletion for this hypothetical Schottky-barrier sample is 7×10^{16} cm^{-3} (assumed barrier height of 0.7 eV; exponential prefactor is 10^{13} sec^{-1}).

tion of C on applied bias depends totally on the variation of x_e (see Section 9), this implies that $C^2(dC/dT)^{-1}$ will be independent of V_A.

By plotting $C^2(dC/dT)^{-1}$ versus temperature we may furthermore extract the value of the density of states at the neutral bulk Fermi energy $g(E_F^0)$ as well as the value of E_F^0 itself in a particularly straightforward manner. If we expand $g(E)$ in a Taylor series around E_F^0

$$g(E) = g_0 + g_1(E_F^0 - E) + g_2(E^0{}_F - E)^2/2 + \cdots, \tag{30}$$

then it follows from Eq. (27) and the definitions of ρ_e and F_e that

$$C^2\left(\frac{dC}{dT}\right)^{-1} = A(\epsilon q g_0)^{1/2}(T - T_0) + \frac{1}{3}\frac{g_1}{g_0}(T - T_0)^2\, k \ln(\nu/\omega) + \cdots, \tag{31}$$

where T_0 is the turn-on temperature of the ac response defined previously. Thus the first and second derivatives of $C^2(dC/dT)^{-1}$ are directly related to the magnitude and slope of $g(E)$ at E_F^0.

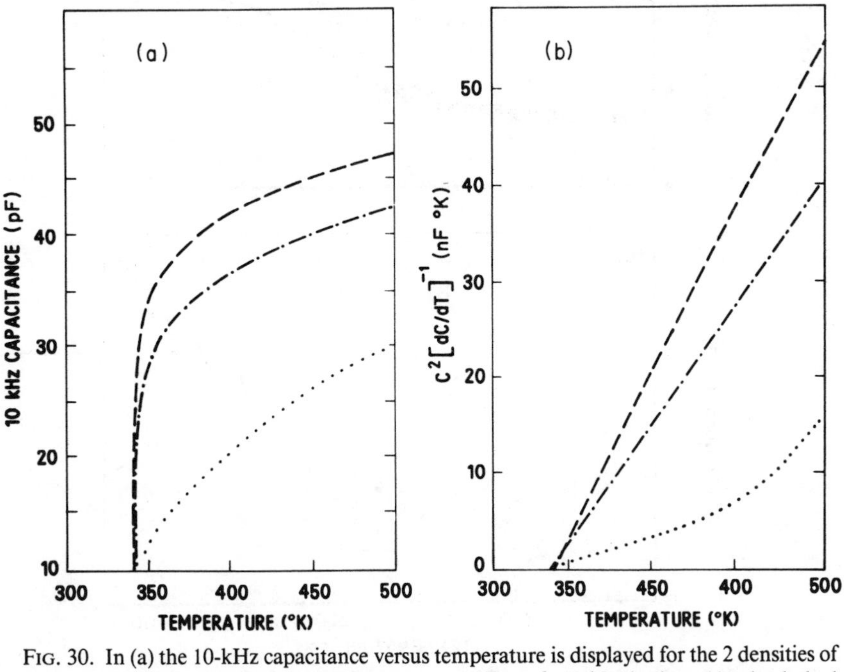

FIG. 30. In (a) the 10-kHz capacitance versus temperature is displayed for the 2 densities of states shown in Fig. 20 as well as a constant $g(E) = 10^{17}$ cm^{-3} eV^{-1} [as indicated by the dashed and dotted (as in Fig. 20) and dash-dotted lines, respectively] at which now the bulk Fermi level E_F^0 is assumed to lie 0.6 eV below E_c. In (b) the ratio $C^2[dC/dT]^{-1}$ is plotted versus temperature. As explained in the text, the slope and curvature of these curves are directly related to the value and slope, respectively, of $g(E)$ at E_F^0.

Figure 30 illustrates this method for the densities of states shown in Fig. 20 plus the constant density of states ($g = 10^{17}$ cm^{-3} eV^{-1}), assuming that in each case E_F^0 lies 0.6 eV below E_c. The capacitance versus temperature curves in Fig. 30a resemble data taken on undoped a-Si:H such as those displayed in Fig. 24. The quantity $C^2(dC/dT)$ versus temperature is plotted in Fig. 30b. Here by comparing $g(E)$ with these curves one can immediately identify the correspondence between the magnitude of the slope and the value of $g_0 = g(E_F^0)$ as well as the sign and degree of curvature with g_1. Because the curvature displayed in Fig. 30b is so small, however, it is unlikely that one could actually extract g_1 very reliably from experimental data. It is

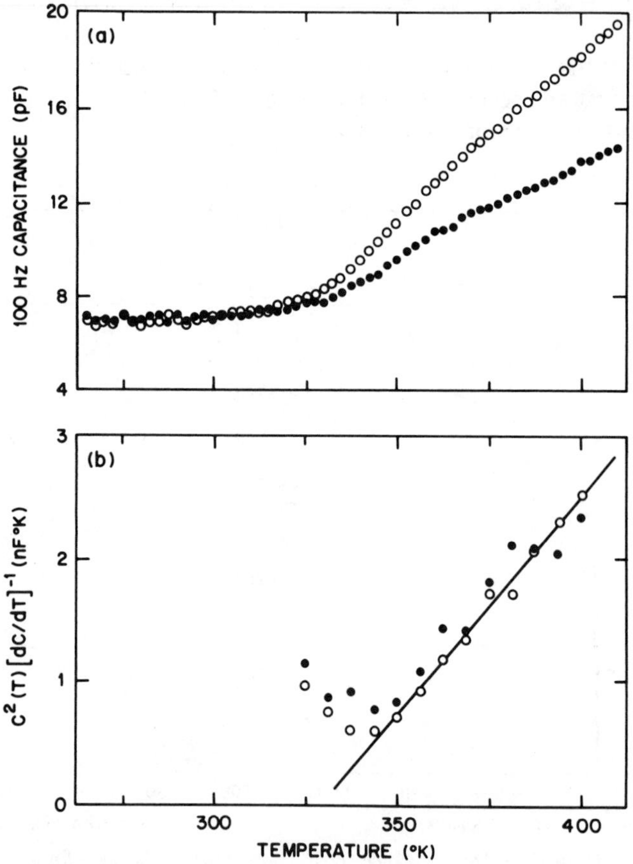

FIG. 31. Illustration of the $C-T$ profiling technique for obtaining $g(E_F)$ in undoped films. (a) 100 Hz capacitance versus temperature data for reverse bias voltages 2 V (○) and 4 V (●) in a typical high-resistivity undoped a-Si:H film. (b) Replot of the above data in the form $C^2[dC/dT]^{-1}$ versus T. The slope of a line drawn through the data (35 pF) gives a density of states at the Fermi level of 1.9×10^{15} cm^{-3} eV^{-1} as explained in the text.

also very important when dealing with experimental measurements to apply this analysis only to data sufficiently above the turn-on temperature T_0.

Figure 31 shows this method applied to undoped a-Si:H for $C-T$ data taken at two values of applied bias (Lang *et al.*, 1982a). Although the value of capacitance in Fig. 31a depends strongly on the bias used, the quantity $C^2(dC/dT)^{-1}$ is independent of bias as expected for reasonably homogeneous material. The slope of the line in Fig. 31b gives the value of $g(E_F^0)$, which indicates a very low midgap density of states of 1.9×10^{15} cm^{-3} eV^{-1} for this sample. Such low values of $g(E)$ have also been reported for undoped a-Si:H grown in other laboratories (Tiedje *et al.*, 1981; Crandall, 1981). The temperature intercept of the line in Fig. 28b gives the best estimate of the value of E_F^0. From an Arrhenius plot of measurements at a few frequencies one finds that $E_c - E_F^0 = 0.87$ eV for this sample.

The capacitance–temperature slope method described above is one manner in which $C-T$ curves at different applied bias may be used to independently study the spatial and energy variations of the density of states in a-Si:H. This and similar methods, we believe, currently offer the most reliable means of learning about the bulk density of states in *undoped* a-Si:H.

V. Deep Level Transient Spectroscopy

12. Development of Thermal Transient Spectroscopic Techniques

The possibilities of using capacitance-transient measurements to study traps in crystalline semiconductors were first discussed by Williams (1966) and by Furukawa and Ishibashi (1967). During the same period, methods were being proposed in which a single voltage or light pulse excitation would be used to fill levels at low temperature, and then this trapped charge would be allowed to return to equilibrium by thermal emission from deeper and deeper levels as the temperature was scanned upward. Such temperature scanning methods using *current detection* of the release of trapped charge were proposed first (Driver and Wright, 1963), and later the method of *capacitance detection* was proposed (Carballas and Lebailly, 1968). Both of these methods, now designated as thermally stimulated current (TSC) and thermally stimulated capacitance (TSCAP), respectively, have been employed with great success to study defects levels in crystalline silicon by Sah and co-workers (1970) and by Buehler and Phillips (1976).

The development in 1974 of deep level transient spectroscopy (DLTS) by Lang and co-workers (Lang, 1974; Miller *et al.*, 1977; Lang, 1979) coupled the idea of the transient measurement method with the temperature scanning methods. In principle, DLTS simply obtains the same information as the previous methods but uses a different measurement algorithm. In practice, however, the development of DLTS brought the principles of thermal evolution spectroscopy to its full power and, as a result, this technique has become the standard method for learning about deep levels in crystalline semiconductors. Although the applications of these techniques to crystals are now too numerous to mention, some of the pioneering work using these methods in the latter 1970s include observation and characterization of radiation damage in silicon and III–V semiconductors by Miller, Lang, and Kimmerling (for a review see Miller *et al.*, 1977); the study of complex trapping and thermal emission involving multiphonon processes by Lang and Henry (1975); the study of silicon–SiO_2 interface states by the constant capacitance DLTS technique by Johnson *et al.* (1978); and numerous other studies identifying the varied nature of defects and defect complexes that are responsible for energy gap levels in silicon and III–V compounds.

In many ways, the study of mobility gap states in a-Si:H has followed a similar historical development as for crystals, with quasi-static capacitance–voltage and admittance versus frequency being tried first, followed by current DLTS by Crandall (1980) and capacitance DLTS by Cohen *et al.* (1980a). Related techniques soon followed including TSCAP (Lang *et al.*, 1982a), TSC (Vieux-Rochaz and Chenevas-Paule, 1980; Fuhs and Milleville, 1980), isothermal capacitance-transient techniques (Okushi *et al.*, 1981, 1982, 1983), and constant capacitance DLTS (Johnson, 1983). Due to the low conductivity inherent in many amorphous semiconductors and the predominant interest in intrinsic a-Si:H samples in connection with solar cell development, methods such as DLTS have been slow to catch on and still compete with many other techniques in the determination of $g(E)$ in the mobility gap of these materials. However, for samples that are sufficiently conducting (such as doped a-Si:H), DLTS enjoys the same advantages over many of the other techniques as it does in crystalline semiconductors and gives spectra that most directly reflect the density of states in the mobility gap over the widest range of electron energies.

In the sections that follow, we shall first discuss the physics common to all thermal emission spectroscopic techniques, then examine several of the more popular techniques in detail with the major emphasis on DLTS. Since the application of DLTS to the study of deep levels in crystals has been reviewed extensively (Miller *et al.*, 1977; Lang, 1979), this discussion will focus on the special problems involved in applying DLTS to materials that contain a large continuous distribution of gap states. We shall then outline the procedure used to obtain a quantitative density of states from the measurement, and finally discuss the experimental results. As in the previous sections, we shall end with a discussion of the limitations and unresolved issues associated with thermal transient measurements.

13. Basic Concepts; Analysis of the Dynamic Junction Problem

a. *Voltage Pulse Filling of Gap States*

The basic physics of thermal emission measurements performed on semiconductor junctions is illustrated in Fig. 32. This figure displays the evolu-

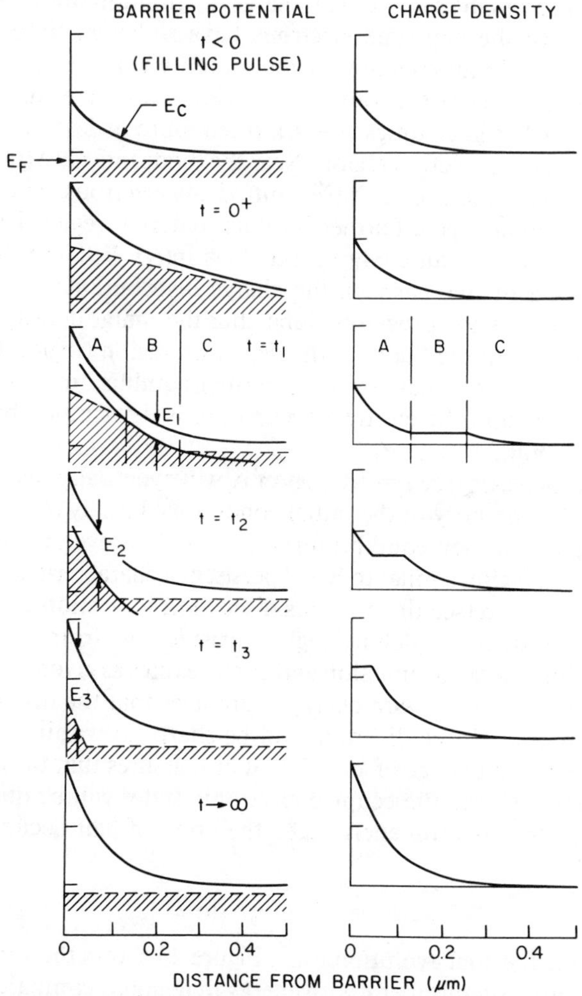

FIG. 32. Sequence of diagrams showing time evolution of the space-charge region following an abrupt change of applied voltage. The curves for the barrier potential and charge density were obtained from a numerical calculation for a hypothetical material with a constant density of states $g(E) = 10^{17}$ cm^{-3} eV^{-1}. A detailed description and the definition of all parameters is found in the text.

tion back to equilibrium of a Schottky-barrier diode following a change of the applied diode voltage. This sequence of diagrams therefore describes the situation following application of a voltage pulse of sufficient duration in a DLTS, TSCAP, or TSC measurement.

In the top diagram the diode is in a steady-state condition during the latter part of the applied voltage pulse. Immediately after the ambient voltage bias is restored ($t = 0^+$) the change in voltage simply produces an additional uniform field across the sample. The charge distribution ramains essentially unchanged until the gap state electrons have sufficient time to undergo thermal emission to an energy near the mobility edge at which they have sufficient mobility to be swept out of the depletion region by the large electric field. At somewhat later times ($t = t_1$), three more or less distinct regions develop within the depletion region. Near the diode interface (region A), the occupied states are too deep to have emitted any electrons. Thus the charge distribution is unchanged. Farther into the material (region B) charge has been emitted down to an energy depth $E_1 \propto \ln(t_1)$. Because this region is greatly depleted of free carriers, the charge distribution in this region is determined by emission processes alone; thus the charge density is relatively constant. Finally, in the "tail" of the depletion region (region C) emission and capture rates are balanced; i.e., thermal equilibrium has been established. In this region the electron occupation is determined by the Fermi energy in the bulk.

As time progresses (see $t = t_3$), region A will eventually disappear as the system loses its memory of the initial conditions. Finally ($t \to \infty$), region B will disappear as the new equilibrium is attained. However, for larger reverse ambient bias, a region similar to B will persist at a characteristic quasi-Fermi energy near midgap (see the previous discussion in Section 2).

The fact that such well-defined regions exist follows from the exponential dependence on energy of emission and capture rates as given by Eqs. (5) and (6). This rapid variation with energy guarantees that the division between regions A, B, and C (see diagram) will be sharp (to within a distance Δx corresponding to a change of kT in ψ). It also assures that the demarcation between occupied and unoccupied electronic states will be quite sharp. In region B the demarcation energy, i.e., the point of half occupancy will be given by

$$E_e = kT \ln(\nu_n t_e / \ln 2), \tag{32}$$

where t_e is the thermal evolution time. Figure 33 shows the variation of the electron occupation number n with energy in region B compared to a Fermi function with $E_F = E_e$ at $T = 300\,°K$. We see that the demarcation between "empty" and "full" states is actually slightly sharper than a Fermi function at the same temperature. We denote the occupation function appropriate to

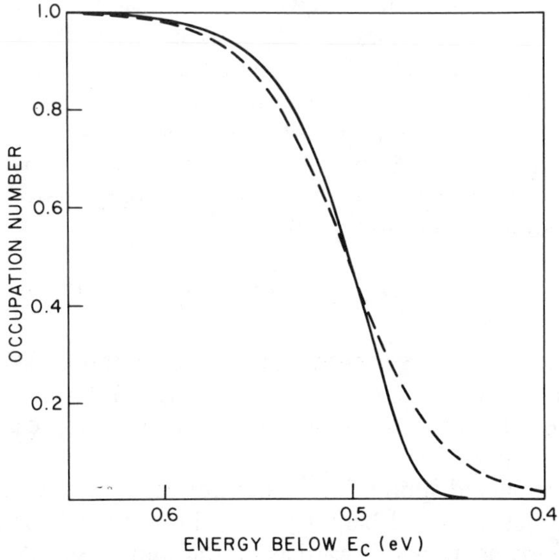

FIG. 33. Comparison of a $\Delta E_F = 0.5$ eV Fermi function (dotted line) at room temperature with the function describing the electron state occupancy in region B of Fig. 32 (solid line) for $E_e = 0.5$ eV also at room temperature.

region B by f^* as

$$f^*(E, E_e, T) = \exp(-e_n t_e), \tag{33}$$

where E_e is given by Eq. (32) and $e_n(E)$ by Eq. (5).

Figure 32 shows that the redistribution of gap state occupancy during the time evolution of the space charge can be rather complicated. For example, at roughly 0.25 μm from the interface the charge density first *increases* with time, then *decreases*. This occurs because electrons are first emitted far from the interface, greatly increasing the charge density in the tail of the depletion region. At later times a positive charge builds up near the interface, which necessarily shrinks the tail region in order to maintain a constant dipole moment of the Schottky-barrier space charge [as mandated by Eq. (16)].

b. The Barrier Potential Problem under the Nonequilibrium Conditions

The numerical analysis of the dynamic space-charge region begins, as with the static barrier, with a solution of Poisson's equation for the band-bending function. However, now the charge density will have a very different form in each of the regions discussed above. This poses no particular problem; the space charge $\rho_{eq}(x)$ in region C is given by Eq. (2) since in this region the capture and emission processes are fully in equilibrium. Region A contains the equilibrium charge distribution ρ_0 appropriate to the junction under the bias present during the voltage pulse. In region B, the space-charge density is nearly a constant (see Fig. 32) because the occupation of states is dominated by emission only, which will have emptied the occupied levels down to a constant energy E_e below E_c. We may write the charge distribution in this region in a manner similar to Eq. (2) as

$$\rho^* = q \int [f(E', E_F^0, T) - f^*(E', E_e, T)]g(E')\, dE'. \tag{34}$$

Computing the above charge density can often be simplified as before by the use of step functions for $f(E, E_F^0, T)$ and $f^*(E, E_e, T)$. This is appropriate as long as the densities of states of interest do not vary appreciably over an energy scale of size kT.

Since the region and appropriate charge density is readily determined by the magnitude of the band bending $\psi(x)$, the numerical solution of the problem is nearly as straightforward as for the static case. Namely, if $\psi_0(x)$ and $\rho_0(x)$ denote the static band bending and charge density, respectively, during the filling pulse, then we can solve for $\psi(x)$ by choosing $\rho(x)$ such that

$$\rho = \begin{cases} \rho_{eq}(x) & \text{computed via Eq. (2) for} \quad \psi(x) < E_e, \\ \rho^*(E_e) & \text{computed via Eq. (34) if} \quad \psi(x) \geq E_e \quad \text{and} \quad E_e > \psi_0(x), \\ \rho_0(x) & \text{if} \quad E_e \leq \psi_0(x). \end{cases}$$

With this procedure we obtain $\psi(x)$ and $\rho(x)$ as functions of temperature and evolution time [for more details see Cohen and Lang (1982)]. From the above prescription we see that this time and temperature dependence is primarily contained in one parameter E_e. Although the integrals for the charge density also depend on temperature through the occupation functions f and f^*, these integrals are largely independent of temperature for most actual cases of interest if the statistical shift of the Fermi energy E_F^0 is also taken into account (see the discussion in Section 1).

The determination of thermal emission currents or capacitance transients to compare with experiment is straightforward. For example, the total change in barrier charge that occurs in a time δt at temperature T is simply

$$\begin{aligned} \delta Q(E_e) &= Q(E_e + \delta E_e) - Q(E_e) \\ &= \int_0^\infty \rho(x, E_e)\, dx - \int_0^\infty \rho(x, E_e + \delta E_e)\, dx \end{aligned} \tag{35}$$

with E_e and δE_e obtained from Eq. (32). Since in this case the change in barrier charge occurs by the emission of electrons that must flow through the external circuit to restore total charge neutrality to the junction, the experimentally observed current is given by

$$\frac{dQ}{dt} = \left[\frac{\partial}{\partial E_e}\int_0^\infty \rho(x, E_e)\, dx\right]\left(\frac{\partial E_e}{dt}\right). \tag{36}$$

Similarly one can compute the barrier capacitance evolution from a knowledge of $\psi(x)$ and $\rho(x)$ using Eq. (24), provided that the capacitance measurement frequency ω is large compared to $[t_e]^{-1}$.

c. *Laser Pulse Excitation*

The discussion so far has been restricted to the application of voltage pulses to change the gap-state's population. This technique in Schottky-barrier diodes is limited to producing changes in the electronic occupation only for levels above the deep depletion quasi-Fermi level. Because of this, voltage pulse methods, like the steady-state capacitance techniques described previously, may only be used to study $g(E)$ in the upper half of the mobility gap.

In studies of n-type crystalline semiconductors thermal emission measurements have been extended to study hole trapping levels below midgap by using voltage pulses of sufficient forward bias to provide hole injection. In a-Si : H such methods have not been successful primarily because of the low effective hole mobilities and the lack of good ac ohmic contacts. Instead, an alternative method has been developed that uses optical excitation to produce electrons and holes in the barrier region. This allows the observation of hole emission from gap states below midgap.

Figure 34 gives a schematic picture of the probable evolution for a Schottky-barrier space-charge region following light pulse excitation. Initially, as the optically produced electrons and holes are captured, the gap state population reflects the relevant capture cross section of electrons and holes into each type of center. This gives rise, in general, to a partial occupation of levels throughout the mobility gap. A short time later ($t = t_1$) the capture of free electrons restores equilibrium except within the depletion region itself. Here the return to equilibrium must again wait for the relatively slow processes of electron and hole emission to the appropriate mobility edges or for the slow capture due to the few majority carriers that exist within the depletion region.

This situation is indicated for a series of "snapshots" at three successively greater times following the light pulse shown in Fig. 34. As in the voltage pulse case described above, region C denotes the part of the depletion region in which equilibrium has been reestablished, whereas region B is determined by thermal emission only—in this case down to an energy E_e^n from the

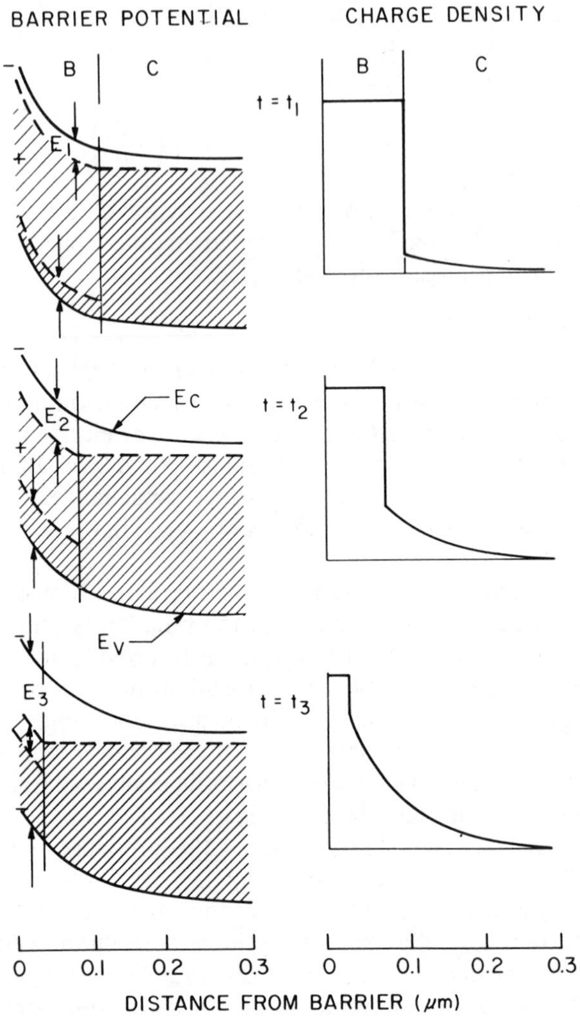

FIG. 34. Sequence of diagrams showing time evolution of the space-charge region following a laser pulse excitation. The gap states for this material, again with $g(E) = \text{const} = 10^{17}\ \text{cm}^{-3}\ \text{eV}^{-1}$, are assumed to be half occupied at $t = 0$. A detailed description and definitions of parameters are given in the text.

conduction band as given by Eq. (32) and up from the valence-band edge, due to hole emission, to an energy

$$E_e^p = kT \ln(\nu_p(E_e^p)t/\ln 2). \tag{37}$$

For $\nu_p = \nu_n$ these two energies will be equal; otherwise the difference is given by

$$E_e^n - E_e^p = kT \ln[v_n(E_e^n)/v_p(E_e^p)]. \tag{38}$$

This energy difference will still be small as long as v_n and v_p are within an order of magnitude.

Obtaining the charge density in region B from the density of states thus requires some knowledge about the relative size of the capture rates v_n and v_p, which is somewhat difficult to obtain experimentally. Whereas capture rates for majority carriers are easily measurable (even to states involved in *minority-carrier* emission), minority capture rates are not so easily obtained. A related difficulty lies in determining the post-light-pulse occupation function $\eta(E)$. On the one hand, it is possible to saturate the experimental signals at sufficiently high laser intensity, which indicates that a uniform spatial distribution of trapped charge exists initially within the space-charge region. However, the energy distribution is primarily determined by the relative capture cross section for electrons and holes and is, therefore, probably not uniform. Since the total space charge is given by the integral of the *product* of the occupation function and the density of states, a nonuniform and unknown $\eta(E)$ is extremely difficult to deconvolve from the unknown $g(E)$.

For the purposes of our analysis we usually assume $v_n = v_p$ and $\eta(E) = \text{const} = \frac{1}{2}$. Utilizing such assumptions in the analysis of experimental spectra (see Section 15) probably limits the accuracy in obtaining $g(E)$ below midgap to within a factor of two of its calculated value. It is important to stress that this kind of uncertainty does *not* exist in the interpretation of $g(E)$ above midgap obtained from voltage pulse measurements since in that case all the relevant parameters may be experimentally verified.

Given the electron and hole capture rates and $\eta(E)$, ρ^* in region B is readily computed and we can solve the dynamic barrier problem as in the voltage pulse case. The only additional complication for the calculation of experimental quantities arises for the case of current measurements. This is due to the fact that with minority-carrier emission present the change in the barrier charge need not involve the external circuit (see Cohen and Lang, 1982). Thus to calculate the experimentally measured current for the laser pulse case, we must keep track of electron and hole emission processes separately. The calculation of other experimental quantities such as capacitance (or spin density) is unaffected by such considerations.

14. Description of Experimental Techniques

a. Thermally Stimulated Capacitance and Current Measurements

The techniques of thermally stimulated capacitance (TSCAP) and thermally stimulated current (TSC) are the simplest methods for surveying the distribution of gap states in semiconductors and have been widely used over the past 15 years in crystals. In both of these methods the barrier region is prepared in a metastable condition usually by cooling the sample under zero bias and then restoring the reverse bias at low temperature. The system then returns to equilibrium as it is warmed to moderate temperatures. To appreciate the manner in which such methods allow the experimenter to control and record the recovery process we need only glance at Table II to see that the nonequilibrium conditions will persist for arbitrarily long times depending on the temperature of the sample. As the temperature is raised, states of increasing energy depth will be able to emit on the time scale of experimental observation. In the notation of the preceding section, these methods cause E_e to increase roughly linearly in time by linearly sweeping the temperature.

Figure 35 shows the behavior for our example of a discrete level system. The TSCAP experiment takes at least two temperature scans, one at the ambient bias in the usual manner to give a capacitance baseline, then another following a cool-down in zero bias. In the latter situation both deep levels at E_1 and E_2 are completely filled above E_F at low temperatures (see Fig. 17) and therefore do not contribute positive charge to the depletion region. Thus the capacitance is due exclusively to the shallow donors alone and will differ from its usual equilibrium value C_0 by ($N_1, N_2 \ll N_D$)

$$\Delta C \simeq \tfrac{1}{2} C_0 (N_1 + N_2)/N_D. \qquad (39)$$

This equation follows directly from Eq. (26). As the temperature is raised, the electrons at level E_1 are emptied above the Fermi energy, which gives rise to a capacitance change at temperature $T_1 = E_1[k \ln(\nu_n \tau_{\text{eff}}^{(1)})]^{-1}$. The effective measurement time τ_{eff} is determined by the temperature scan rate and may be defined in terms of the temperature at which the fractional occupation of

TABLE II

THERMAL EMISSION TIMES AS A FUNCTION OF ENERGY DEPTH AT A VARIETY OF TEMPERATURES[a]

Temperature (°K)	Energy depth (eV) 0.3	0.5	0.7	0.9
200	3.6 μsec	0.4 sec	12 hr	150 yr
250	0.1 μsec	1.2 msec	13 sec	1.6 d
300	11 nsec	25 μsec	58 msec	2.2 min
350	2 nsec	1.6 μsec	1.2 msec	0.9 sec
400	0.6 nsec	0.2 μsec	66 μsec	22 msec

[a] The exponential prefactor ν is taken to be 10^{13} sec^{-1} independent of temperature.

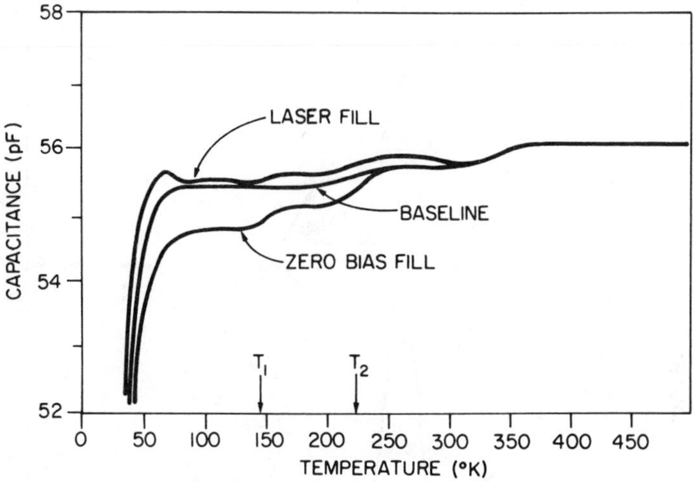

FIG. 35. Calculated set of thermally stimulated capacitance (TSCAP) scans for the quasi-discrete-level $g(E)$ of Fig. 16 under voltage pulse and laser pulse filling conditions as described in the text.

that level n_1 decreases to one-half. That is, the rate of change of n_1 due to the thermal emission from an energy depth E_1 is

$$dn_1/dt = n_1 \nu_n \exp(-E_1/kT), \tag{40}$$

so that the elapsed time t_s for the level to be half emptied is given by

$$\int_{t_0}^{t_s} \nu_n \exp(-E_1/k\zeta t)\, dt = \ln 2, \tag{41}$$

where the level is assumed full ($n = 1$) at $t = t_0$ and ζ is the temperature scan rate. Therefore the equivalent measurement time for the fixed temperature ζt_s, where the step change in capacitance is recorded, is defined by

$$\tau_{\text{eff}}^{(1)} \nu_n \exp(-E_1/k\zeta t_s) = \ln 2. \tag{42}$$

For a scanning rate of $\zeta = 20\,°\text{K min}^{-1}$ and $\nu_n = 10^{12}\ \text{sec}^{-1}$ one finds $\tau_{\text{eff}}^{(1)} \simeq 13$ sec. Further details of the temperature scan analysis for both TSCAP and TSC are given by Buehler and Phillips (1976).

Similarly, we shall observe a capacitance step at temperature T_2 due to the emission of electrons from the level at E_2. At this point the curve joins the equilibrium baseline. Note that these step changes *relative* to the baseline curve occur at positions different from the step changes in the baseline curve itself, which are determined by the shorter time scale $1/\omega$, as discussed in Part IV.

The TSCAP signal is often plotted as the difference with respect to the

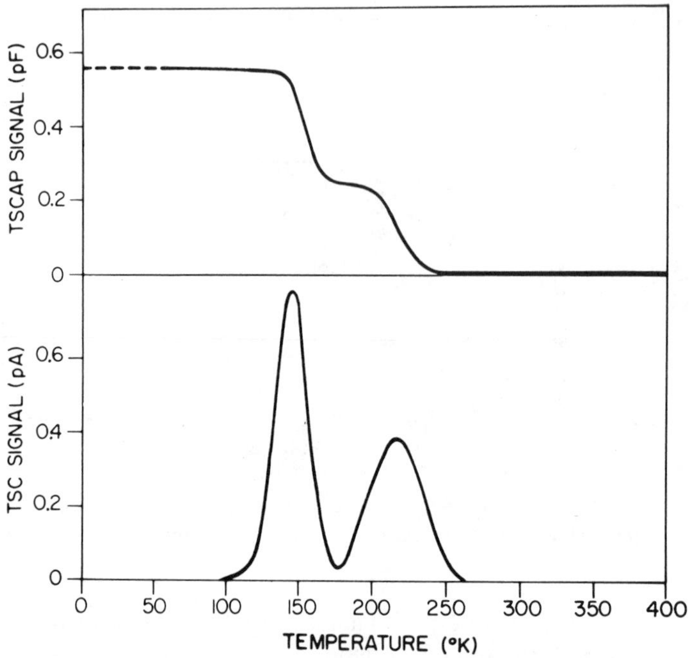

FIG. 36. Top diagram: Voltage pulse *difference* TSCAP signal obtained from Fig. 35. Bottom diagram: Calculated TSC signal for same conditions.

baseline curve; this is shown in Fig. 36 (top) along with the TSC signals in Fig. 36 (bottom). In contrast to TSCAP, the TSC signal magnitude depends directly on the rate of trapped charge emission so that large signals require using the fastest possible temperature scans. One advantage of TSC measurements is that the data is obtained directly in a single scan (the baseline is zero).

The TSC or TSCAP measurements may also be made by scanning in temperature following light pulse excitation at low temperatures (see Fig. 35). For our discrete level example, assuming that the laser pulse causes each deep level to be half full initially, the TSC and TSCAP signals will appear as shown in Fig. 37. Note that in this case the TSCAP is initially positive due to the net greater number of hole traps. The signal then increases or decreases depending on whether electron or hole emission predominates at that temperature. The TSC signal, on the other hand, always has the same sign. This illustrates one important difference between junction capacitance and current measurements, namely, that the capacitance distinguishes between majority- and minority-carrier emission, whereas current measurements do not. Thus the magnitude and sign of the light pulse TSCAP at low tempera-

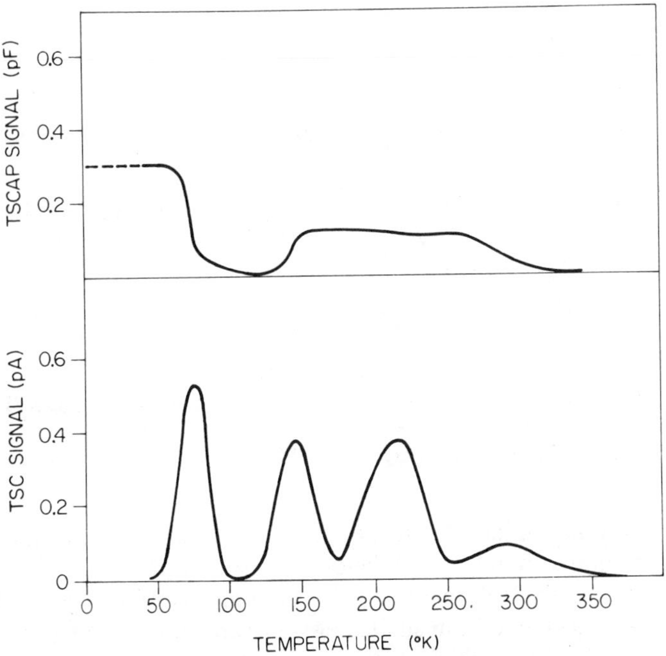

FIG. 37. Top diagram: Laser pulse *difference* TSCAP signal obtained from Fig. 35. Bottom diagram: Calculated TSC signal for same conditions.

tures gives a measure of the net hole minus electron emitting states; the integral over the TSC spectrum gives a measure of the total number of deep gap states. However, there also exist differences in the spatial sensitivity between current and capacitance thermal-transient methods that should be kept in mind and that we discuss in Section 14d.

Both TSC and TSCAP methods have been applied to a-Si : H. Results for TSCAP for both the voltage and laser pulse filling techniques (Lang *et al.*, 1982a) are shown in Fig. 38. This figure displays 3 curves in the manner of Fig. 35: the "baseline" curve, which is just the standard $C-T$ scan under -2 V bias, the scan labeled "zero bias fill," which corresponds to cooling the sample in zero bias and warming in -2 V bias, and the "laser fill" curve as described above. From these data one immediately obtains a reasonably good idea of the number and rough distribution of gap states. The quantity ΔC_e is a measure of the total number of states above midgap and below a cutoff energy determined by the dielectric relaxation time (in this case about 0.5 eV). The quantity ΔC_{e+h} is a measure of the net hole minus electron states, as discussed above, and indicates the predominance of states below midgap in this a-Si : H sample. In this case, the region of the mobility gap that

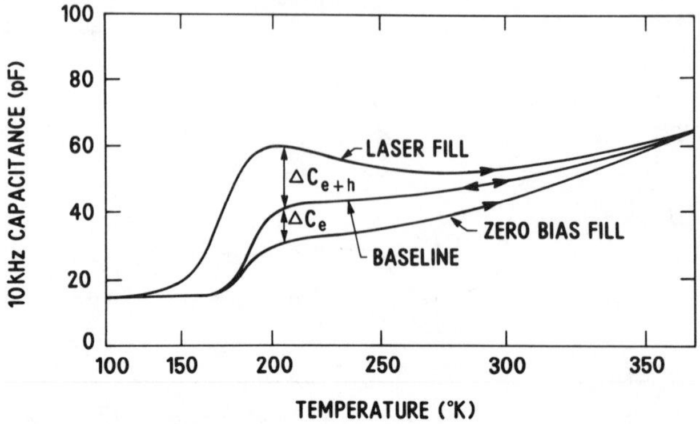

FIG. 38. Typical set of measured TSCAP spectra for one *n*-type doped (60 ppm PH_3) a-Si:H Schottky-barrier sample under 2 V reverse bias. The data were obtained in a manner analogous to that described for the calculated discrete-level TSCAP data in Fig. 35. Other parameters are described in the text.

contributes to the difference signal ΔC_{e+h} observed just above turn-on extends from approximately 0.5 eV below E_c to 0.5 eV above E_v. The effects of gap states on the TSCAP signal outside of this energy range are not observed because trapped charge from shallower states are emitted too quickly at the lowest temperatures at which the junction capacitance may be measured.

Other workers (Vieux-Rochaz and Chenevas-Paule, 1980; Fuhs and Milleville, 1980; Thompson *et al.*, 1981) have also carried out TSC studies on a-Si:H. The results of such studies to date are generally inconsistent with the results of the TSCAP measurements discussed above. A major concern in such measurements in these materials are sources of extraneous currents that obscure the thermal emission phenomena. These include thermal emission over the barrier itself and other kinds of leakage, high sensitivity to surface rather than bulk states, thermoelectric currents, and secondary photocurrent in the case of laser pulse studies. Capacitance measurements are largely insensitive to such effects and are thus more easily interpreted. For these reasons the results of TSC measurements must be carefully checked (for example, by verifying that the features scale properly with applied bias) before they are interpreted in terms of thermal emission from bulk states.

One could in principle obtain $g(E)$ in a-Si:H from the TSCAP measurements we have discussed. The DLTS spectra discussed in the next section are very closely related to the temperature derivative of the TSCAP spectra. For a number of practical reasons, including signal-to-noise considerations, the DLTS spectra are believed to be generally more reliable. However, because we are not time-constrained in TSCAP measurements to the length of the voltage pulse or degree of light exposure that we may employ to prepare the prescan initial conditions, it is much easier to verify that the trapping of electrons and/or holes into gap states has reached its steady-state value so that the subsequent transient signal reflects the true magnitude of $g(E)$. Thus in most cases, TSCAP is more reliable in setting the overall quantitative scale of $g(E)$.

A possible shortcoming of the TSCAP method, in addition to some of those mentioned, is that since TSCAP is a single shot method, any competing mechanism that leads to the equilibration of the metastable state other than thermal emission will present a serious problem. Thus weak effects such as capture of barrier leakage current or deep-state hopping conduction can accumulate over the typically long duration (~ 10 min) of the TSCAP scan. Such concerns should not, however, affect the reliability of the initial phase of the temperature scan, so that we may still use the values of ΔC_e and ΔC_{e+h} to deduce overall integrals of $g(E)$ as described above.

In DLTS, as we shall see, the time scales of measurement are typically much shorter (10–300 msec) and fixed over the entire spectrum. Thus weak effects due to hopping conduction and leakage may be discriminated against more readily. Indeed, it will probably be possible to study these alternative kinds of equilibrating mechanisms by comparing TSCAP with DLTS methods.

b. The DLTS Algorithm

Deep level transient spectroscopy involves the synchronous detection of repetitive, thermally activated transients of a given time constant while the temperature is scanned. Under ideal conditions in which the barrier-charge density is dominated by shallow level concentration, the DLTS spectrum is directly related to the Laplace transform of the transient time dependence. Such spectra are readily transformed into density-of-states functions of a thermal emission energy variable (see Lang, 1979). Even for more complex systems such as a-Si:H in which the number of deep states is large and continuously variable, or for systems with energy dependent capture cross sections, the DLTS spectrum approximates such a density of states to a surprisingly good degree.

The DLTS algorithm can be illustrated and explained most simply for the case of a discrete trap level in which a double boxcar (dual-gated integrator) is used as the DLTS signal processor. This is shown schematically in Fig. 39 with the two gate times denoted as t_1 and t_2. The DLTS signal is defined in this case as the difference between the transient amplitude at time t_1 minus the amplitude at time t_2. For the case of a capacitance transient this difference is denoted by $C(t_1) - C(t_2)$. As a function of temperature this difference signal for a discrete level with a well-defined activation energy goes through a maximum at the temperature for which the transient time constant is on the order of the gate spacing. For this very simple case, the time constant τ_{max}

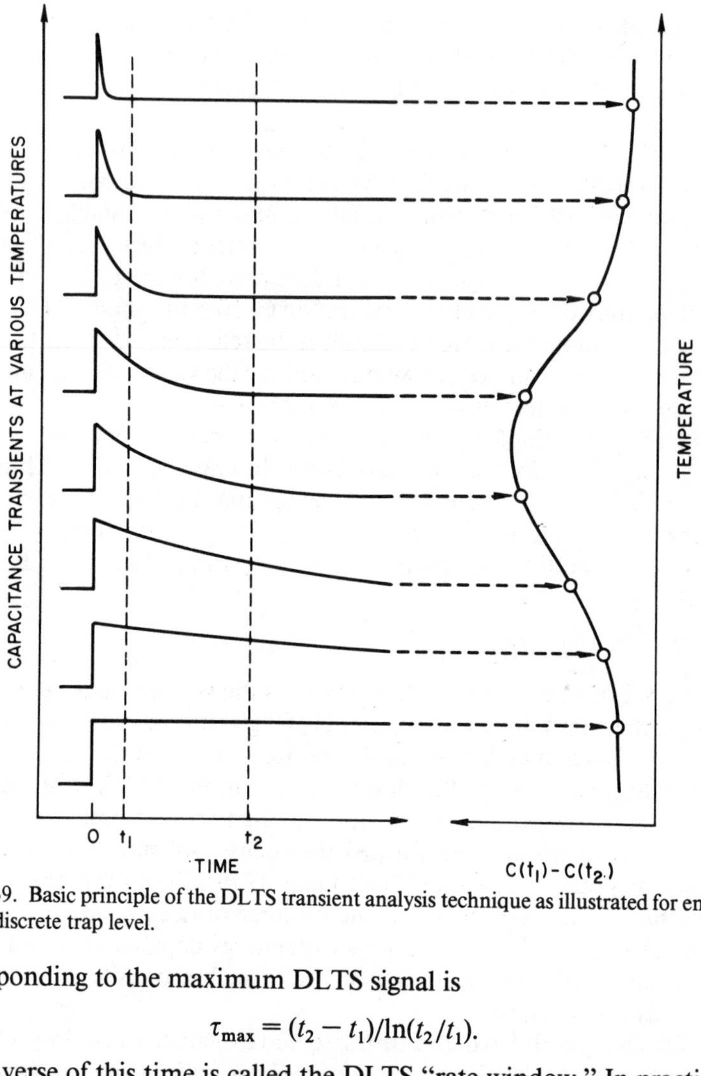

FIG. 39. Basic principle of the DLTS transient analysis technique as illustrated for emission from a discrete trap level.

corresponding to the maximum DLTS signal is

$$\tau_{\max} = (t_2 - t_1)/\ln(t_2/t_1). \tag{43}$$

The inverse of this time is called the DLTS "rate window." In practice, of course, the boxcar gates at t_1 and t_2 must have finite widths. In fact, for optimum signal-to-noise the gate widths should be as large as possible (Lang, 1979). In this case, the simple derivation of Eq. (43) is no longer valid. It has been shown, however, that even in the wide gate limit, Eq. (43) is still reasonably close to the true solution provided that t_1 and t_2 are taken as the *midpoints* of the gates (Day *et al.*, 1979).

A DLTS *spectrum* is defined as the $C(t_1) - C(t_2)$ difference signal plotted as a function of temperature. A typical experimental timing sequence for the voltage pulse and laser pulse excitation used with a-Si : H is shown in Fig. 40. The excitation pulses are applied repetitively and the boxcar enhances the overall signal-to-noise ratio by exponential averaging over many such transients. Because of the repetitive excitation, DLTS spectra may be taken by scanning either up or down in temperature and a signal may be recorded even at fixed temperature. This is to be contrasted with the situation for TSCAP or TSC measurements discussed previously.

For our case of several sharp levels of Fig. 16 the capacitance DLTS spectra for voltage pulse and laser pulse excitation are shown in Fig. 41. In this case the spectra consist of several well defined peaks. The energy depth of each level with respect to the closer band edge is proportional to temperature according to Eq. (32). The sign of the signal indicates, for the capacitance transient case, whether one is observing hole emission to the valence band (positive signal) or electron emission to the conduction band (negative sign). The size of each signal is related to the number of states in each level. For the case that the number of deep levels N_T is much smaller than the number of

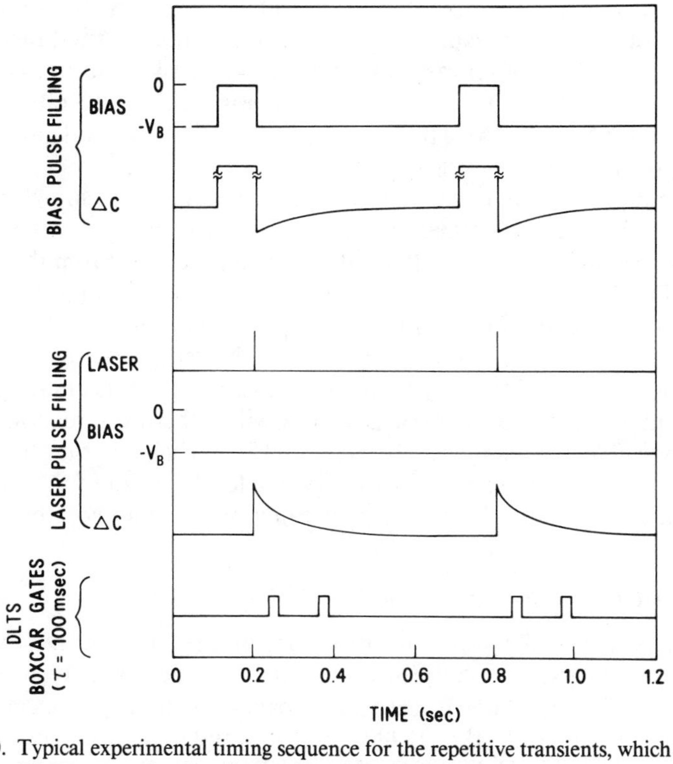

FIG. 40. Typical experimental timing sequence for the repetitive transients, which are converted into DLTS spectra by the principle illustrated in Fig. 39.

shallow states N_D, one deduces from Eq. (39) that

$$N_T/N_D \simeq 2\,\delta C/C_0, \tag{44}$$

where δC is the size of the capacitance DLTS signal and C_0 is the steady-state capacitance value. This formula is strictly valid for $N_D \gg N_T$ in the limit of large reverse bias and has been discussed thoroughly in the literature (see Lang, 1979). The laser pulse induced signals are roughly half as large as the voltage pulse signals due to the (assumed) initial half occupancy of each level as compared with full occupancy of electron states for voltage pulse filling. Note the lack of a feature at 250°K in the laser pulse spectrum due to cancellation of the electron emission from the level 0.6 eV below the conduction band and the hole emission from the level 0.6 eV above the valence band. The temperature width of each peak is primarily due to the *intrinsic* energy resolution of the DLTS method (of roughly 10% of the temperature at which each peak occurs). For defect levels such as those found in a-Si : H, the features in the DLTS spectra are generally much broader than this and therefore more likely to reflect true energy widths.

From this simple example it is immediately clear how the energy level structure of Fig. 16 can readily be reconstructed by the spectra shown in Fig. 41. The conversion of temperature to an energy scale is accomplished by making an Arrhenius plot for a series of experimental DLTS rate windows (see Fig. 48 in Section 15). This will give not only the energy depth but also the prefactor, ν_n or ν_p, for each level.

The interpretation of more complex spectra is not quite so straightforward due to the absence of such sharp features and the greater variation of the size of the space-charge region as thermal emission proceeds. From the TSCAP data of Fig. 38 it is clear that we are not in the $\delta C/C_0 \ll 1$ limit for a-Si : H; simple formulas like Eq. (44) do not apply. Capacitance itself is a more complicated function of the barrier region. Nevertheless, the correspondence between the DLTS spectra and the density of states is unique and can be determined provided that one is dealing with a thermal emission process with known matrix elements. In Section 15 we shall see that the simple approach we have illustrated for the discrete level case still forms a convenient starting point for the interpretation of more complex spectra to yield a density of states.

c. *Related Techniques: ICTS and CC-DLTS*

There exist several variations of the capacitance-transient DLTS procedure described above. A couple of these that have been applied recently to the study of a-Si : H are isothermal capacitance transient spectroscopy (ICTS) by Okushi *et al.* (1981, 1982, and 1983) and the constant capacitance DLTS method (CC-DLTS) utilized by Johnson (1983).

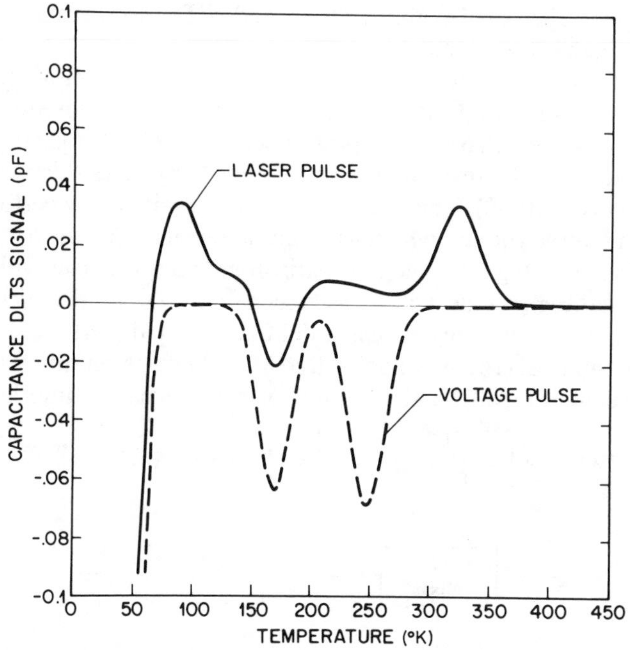

FIG. 41. Capacitance DLTS spectra for the quasi-discrete-level density of states of Fig. 16 for a 10 sec^{-1} rate window. A 5 V reverse bias and an area of 2×10^{-3} cm^2 were assumed in the calculation.

As discussed in Section 13, all thermal transient methods monitor the change in barrier charge as the energy depth from which trapped charge can be emitted is increased. In DLTS this is accomplished using a fixed evolution time and variable temperature. In ICTS this is accomplished using a fixed temperature and variable time. Since one typically records DLTS spectra for a series of measurement times and ICTS spectra at a series of temperatures, the two methods contain completely equivalent information. Thus it is even slightly misleading to call this method a distinct experimental technique.

Proponents of this method claim that by using a fixed temperature the variation of conductivity within the sample as well as the occupation statistics are also fixed so that these spectra are more easily interpreted. Proponents of the standard DLTS method point out the enhanced ability to vary the thermal emission energy by sweeping temperature (a DLTS scan with 10 sec^{-1} rate window taken over the temperature range 100–450°K in a-Si:H is equivalent to a time scale range at 300°K of from 3 nsec to 65 hr. To the extent that the DLTS spectra for a variety of rate windows can be fit consistently to a single density of states, there should be no difference between these two approaches. We shall examine DLTS and ICTS spectra in Section 15 to ascertain whether this is the case.

The constant capacitance method (Goto *et al.*, 1974) is a more interesting variation. In this method the applied voltage is adjusted subsequent to the excitation pulse such that the junction capacitance is kept constant. A schematic diagram for the instrumentation that accomplishes this is shown in Fig. 42 (Johnson, 1982). The transient signal analysis is now performed on the applied sample voltage, which will show a transient similar to the capacitance transient in Fig. 40. The comparison between the fixed voltage and fixed capacitance methods is most easily seen for a crystalline semiconductor for which the depletion region has a definite width W and for which $C = \epsilon A/W$. By using Eq. (16), we imagine that $n(x)$ electrons per unit volume are emitted over an interval Δx at a distance x from the barrier interface. Then

$$\delta V = \delta\left[\frac{1}{\epsilon}\int_0^W x\rho(x)\,dx\right] = \frac{q}{\epsilon}\,[n(0)W\,\Delta W - n(x)x\,\Delta x], \quad (45)$$

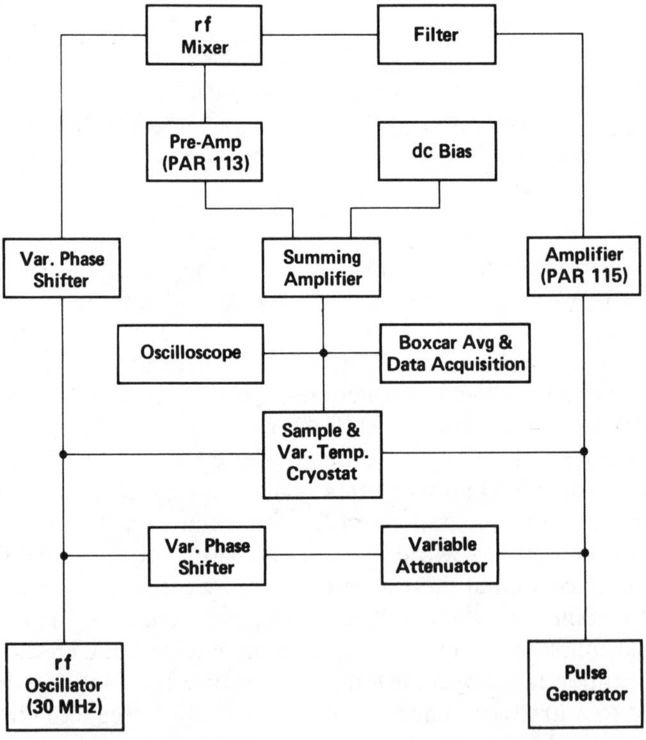

FIG. 42. Block diagram of apparatus for constant-capacitance, deep level transient spectroscopy. [From Johnson (1982).]

where $qn(0)$ is the charge density in the semiconductor at the barrier interface. Standard DLTS forces $\delta V = 0$ and lets W and therefore C respond to changes in the depletion region charge. Constant capacitance DLTS maintains $\Delta W = 0$ and sees the same physical effect from changes in the applied voltage.

For many studies the constant capacitance method offers a distinct advantage because a constant spatial region (of fixed extent W) is involved during the entire charge evolution process. This method is particularly useful in the study of interface charge in MOS type devices since the total device capacitance consists of an oxide capacitance (which is constant) in series with the depletion width capacitance. Holding the total capacitance constant allows one to equate the voltage change across the device directly to the change in charge at the oxide–semiconductor interface.

The constant-capacitance method was applied to a-Si : H in order to obtain a similar expected advantage in the interpretation of the experimental spectra to give $g(E)$. Since the band-bending function and the capacitance itself are much more complicated than for crystals with a shallow dominant dopant level, some care must be taken. The approach taken by Johnson has been to work at constant temperature and variable rate window (like ICTS) so that the variation in capacitance is entirely due to changes in the space charge within the depletion region. He then simplifies his analysis by ignoring changes in the band-bending curvature as thermal emission takes place. This reduces the analysis to a solution of the static barrier problem. Such an assumption is not fully justified; however, we should recall that the density of states obtained for n-type doped a-Si : H in the upper half of the mobility gap is obtained fairly accurately with the very simplified analysis described by Eq. (44). This accuracy is due to the shape of $g(E)$ in this energy regime, which consists of a region of very shallow states plus a distinct region of very deep states. For CC-DLTS the amount of distortion using such a simple analysis should be even smaller, particularly because it includes the correct numerical solution to the dc barrier problem. Indeed, the density of states from this analysis of CC-DLTS measurements agrees very well with those obtained from DLTS using a more complete numerical analysis (see Section 15).

d. Capacitance versus Current Transients; Spatial Sensitivity of DLTS

Both current and capacitance transient measurements are related to the same basic physical phenomenon. The current DLTS measurement corresponding to the discrete-level density of states of Fig. 16 is shown in Fig. 43. By comparing these spectra with the capacitance DLTS spectra in Fig. 41, we see an immediate difference, which was pointed out previously with regard

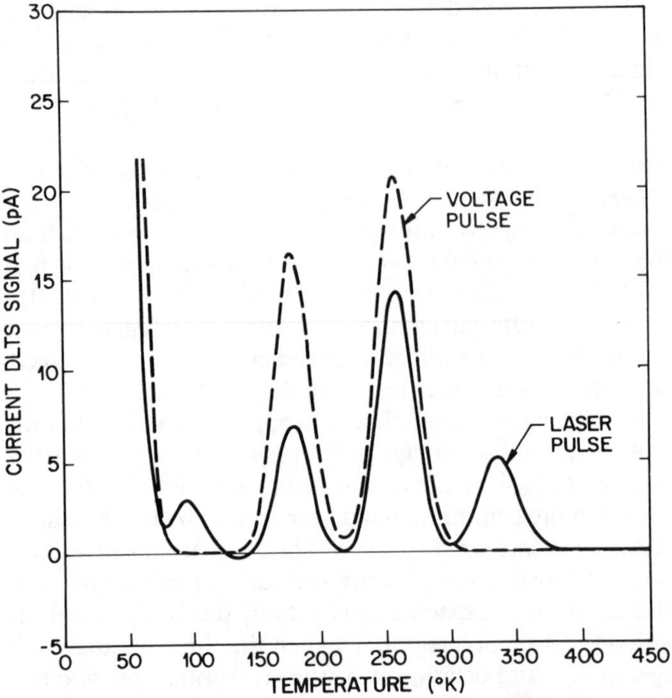

FIG. 43. Current DLTS for the quasi-discrete-level density of states of Fig. 16. Other parameters are the same as those given for Fig. 41.

to TSCAP and TSC, namely, the sign of the signals due to electron or hole emission is the same. Thus the signals due to electron and hole emission at 0.6 eV, which cancelled each other in the laser pulse capacitance-transient spectra of Fig. 41, now add constructively. This ability of capacitance measurements to distinguish between the sign of the trapped charge emitted is a distinct advantage compared to current measurements. However, by comparing the laser and voltage–current DLTS spectra we could again deduce the underlying level structure without too much difficulty.

As also pointed out previously in connection with TSC, care must be taken with current transient measurements to distinguish between true thermal emission processes and effects such as leakage currents, thermoelectric currents, and secondary photocurrent (in the case of laser excitation) all of which may also produce current transients. Such effects may often be identified by comparing a series of DLTS spectra obtained using a wide range of measurement parameters.

Perhaps one of the most important differences between capacitance and current transients is their relative sensitivity to emission from states near the barrier interface. Suppose that there exists an anomalous region in our fictitious discrete-level sample that lies within 500 Å of the barrier interface that, in addition to the two bulk levels at 0.4 and 0.6 eV below E_c contains an additional level 0.5 eV below E_c with a density of ten times that of the bulk levels (see Fig. 44). There are two questions we wish to address: (1) How does this level affect the capacitance and current transient measurements, and (2) is it possible to distinguish between such a state localized near the barrier interface and the uniformly distributed bulk levels? Such issues have considerable relevance to the study of a-Si:H particularly when comparing the various measurement techniques.

The answer to the second question is made obvious by studying Fig. 44. By varying the voltage during the filling pulse we may access different spatial regions in the sample. The variation of signal with filling pulse amplitude distinguishes the spatial variations of the energy levels within the sample. In this case the bulk-level signals rise monotonically with V_P, whereas the interface-localized level produces a signal that appears abruptly at the value of V_P necessary to bring it below E_F at the surface. In a similar manner, by a judicious choice of the ambient bias and pulse bias, we may choose the spatial region in which carrier capture occurs. Thus the density of states can be studied as a function of position *as well as* energy.

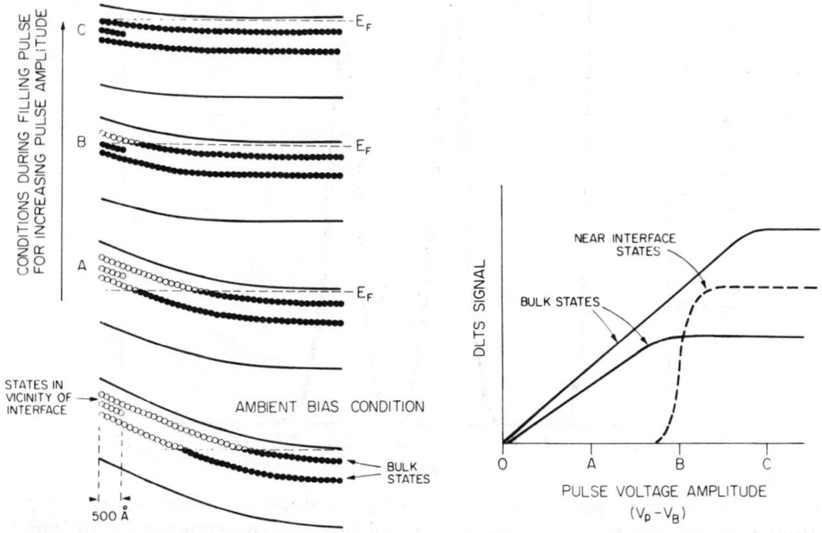

FIG. 44. Occupation of discrete level gap states under ambient conditions and during filling pulse voltage for two bulk levels (solid curves) as well as an interface-localized level (dashed curve) that exists within the first 500 Å of the material. The DLTS signal is proportional to the difference in state occupation between the voltage pulse and ambient bias condition and varies with pulse amplitude as shown at the right for each of the three levels indicated.

Figure 45 gives the answer to the first question. Here the capacitance and current signal response is plotted for a series of voltage pulse values as described above. The horizontal temperature scale has been converted to an energy scale. These diagrams indicate again how the bulk levels rise much more monotonically in *both* kinds of measurements. However, for current DLTS the surface localized level produces a signal that is ten times larger than for capacitance DLTS in comparison with the bulk-level signals.

In fact, the sensitivity of the capacitance-transient signal increases linearly with x for a fixed quantity of thermally emitted charge and actually vanishes at the barrier interface ($x = 0$). This follows from Eq. (16) since it is the change in the total dipole moment that causes the depletion width and hence the capacitance to change. The current-transient signal, on the other hand, is maximum at the surface ($x = 0$) and falls linearly to zero at $x = W$. This is due to the cancellation of the charge motion by the displacement current,

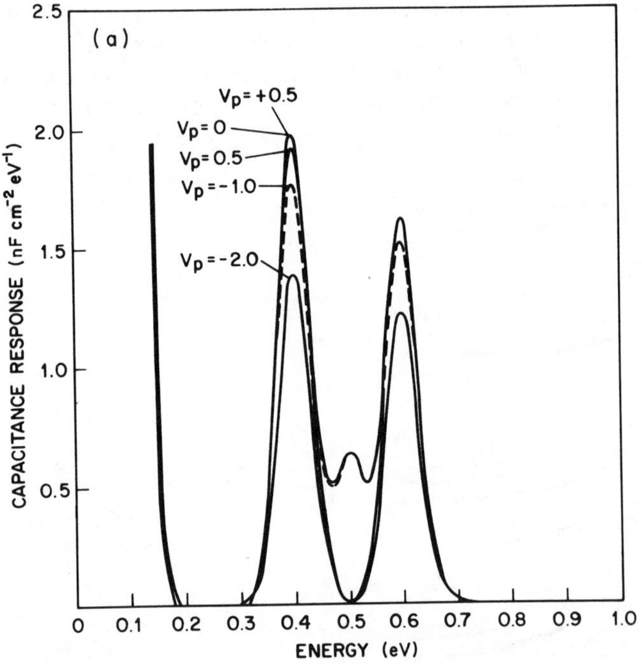

FIG. 45. (a) Capacitive signal response for a sample that has a "surface" region of 500 Å containing a single trap level 0.5 eV below E_c of total integrated concentration 1×10^{16} cm^{-3}. Beyond this 500 Å region $g(E)$ is given by Fig. 16. Note the variation of the surface trap signal with filling pulse amplitude similar to that shown in Fig. 44 and the relatively low sensitivity to this level compared with the 2 bulk levels. (b) Current signal response for the same hypothetical sample as described in (a). Note the relatively large sensitivity to the surface trap level in this case.

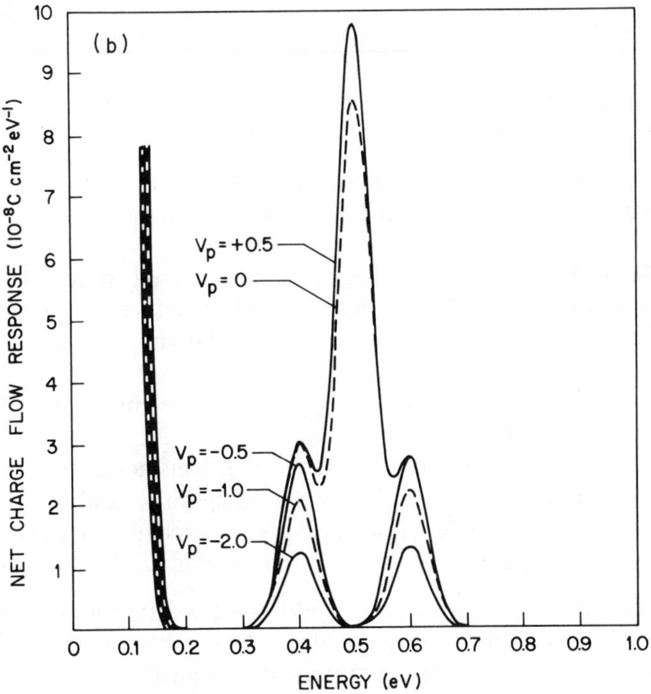

FIG. 45 *(Continued)*

which is largest for those emitted carriers that contribute most to the total dipole moment.

This fundamental surface sensitivity for current transient measurements also applies to TSC and low frequency capacitance measurements. The latter case follows from the fact that the low frequency capacitance is proportional to the total barrier *current* that flows through the external circuit as a result of small voltage changes. The capacitance DLTS and TSCAP methods, on the other hand, are largely *insensitive* to surface effects. Thus features in current DLTS that do *not* appear in the corresponding capacitance DLTS spectra are likely candidates for states localized near the barrier interface.

15. Experimental Results

a. Analysis of Experimental DLTS Spectra

In this subsection, we describe one manner in which a density of states is obtained by the thermal emission spectroscopic data. These data usually consist of both voltage pulse and laser pulse DLTS spectra plus TSCAP. A full discussion of the analysis has been presented previously (Lang *et al.*,

1982a; Cohen and Lang, 1982) so that we shall restrict the current discussion to a quick review of the key assumptions and a brief illustration of the analysis for one set of experimental spectra.

The primary assumptions used in the analysis of a-Si:H data are the following:

(1) the dynamic response above the dielectric turn-on temperature is considered to be limited by thermal emission from gap states and not by carrier transport,

(2) the electric properties (E_g, ρ, E_F) of the material are considered to be laterally uniform on a scale greater than the Debye screening length,

(3) the exponential prefactor ν in Eqs. (5) and (6) is taken to be a constant times T^2,

(4) the final-state quasi-Fermi energy in the space-charge deep-depletion region is taken as $E_g/2$,

(5) the initial state for voltage pulse filling is considered to be in equilibrium with the bulk Fermi level during the filling pulse, and

(6) the initial state for laser pulse filling is assumed to be uniformly at half occupancy for gap states in the range of observable DLTS energies.

We conservatively class these six conditions as assumptions since we feel that they have not been rigorously verified. However, most of them seem to be nearly true on the basis of measurements designed to test their validity. We shall go into details of some of these auxiliary measurements in Section 16. In particular, we believe that conditions (3), (4), and (5) are close enough to being true that any slight deviations should not lead to discrepancies beyond our stated limits of error. Condition (6) is one of the most difficult to assess. We know that the laser pulse induced signals are *saturated* but we do not *a priori* know the occupation fraction. Thus if states in the lower half of the gap were to capture electrons more readily than holes ($\sigma_n \gg \sigma_p$) the concentrations of these states would be grossly underestimated by our procedure. Fortunately this situation is not likely since for most deep levels in crystals the larger capture cross section for a given level is associated with carriers in the nearest band. Thus we might more reasonably expect $\sigma_p \gg \sigma_n$ for states below midgap. In this case assumption (6) would cause us to overestimate the trap concentration by a factor of 2. We stress that such kinds of ambiguities do *not* arise for the density of states in the upper half of the gap deduced from the voltage pulse DLTS spectrum [condition (5)].

Condition (1) has already been discussed in Part IV in connection with our description of capacitance methods. For DLTS this condition is valid if the time for thermal emission is much longer than the time required for the emitted charge to be swept out of the depletion region. Because of the high electric fields in the depletion region ($\sim 10^5$ V cm^{-1}) this carrier sweep-out time is on the order 1 nsec for a mobility of 1 $cm^2\ V^{-1}\ sec^{-1}$. This is to be compared with a typical thermal emission time of 10–300 msec for DLTS experiments. Thus the drift mobility would have to be less than $10^{-7}\ cm^2\ V^{-1}\ sec^{-1}$ for condition (1) to be violated. Retrapping of the emitted carrier is also not expected to cause any serious problems because the empty states available to retrap are generally *shallower* than the emitting states. This happens because the deeper states have not yet emitted their trapped charge and are still full (see Fig. 32). If retrapping *were* a serious problem, it would manifest itself as a severe distortion of the DLTS spectra taken under conditions where initially the various levels were partially filled. Such distortions are not seen (see Section 16).

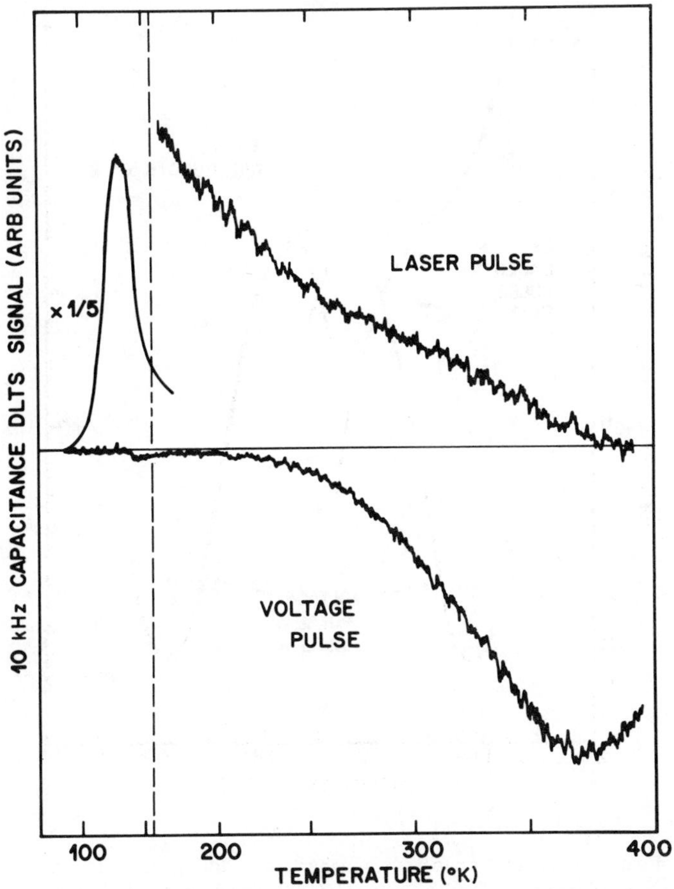

FIG. 46. Typical DLTS hole and electron emission spectra for an a-Si : H sample doped with 300 ppm PH_3. (10 sec^{-1} rate window, 5 V reverse bias on rear junction.)

Condition (2) is necessary to keep our analysis from becoming too cumbersome. However, there is a large body of evidence from structural studies (Knights and Lujan, 1979; Knights, 1980; Nemanich *et al.*, 1980) that a-Si:H films are not laterally uniform but exhibit a columnar growth morphology. How such morphology will affect the transport and barrier formation properties of a-Si:H is an interesting question that has not been answered at the present time.

Utilizing our six assumptions we demonstrate the method for extracting $g(E)$ from the experimental DLTS spectra displayed in Fig. 46. From our previous illustration of DLTS based on the sharp level density of states (Fig.

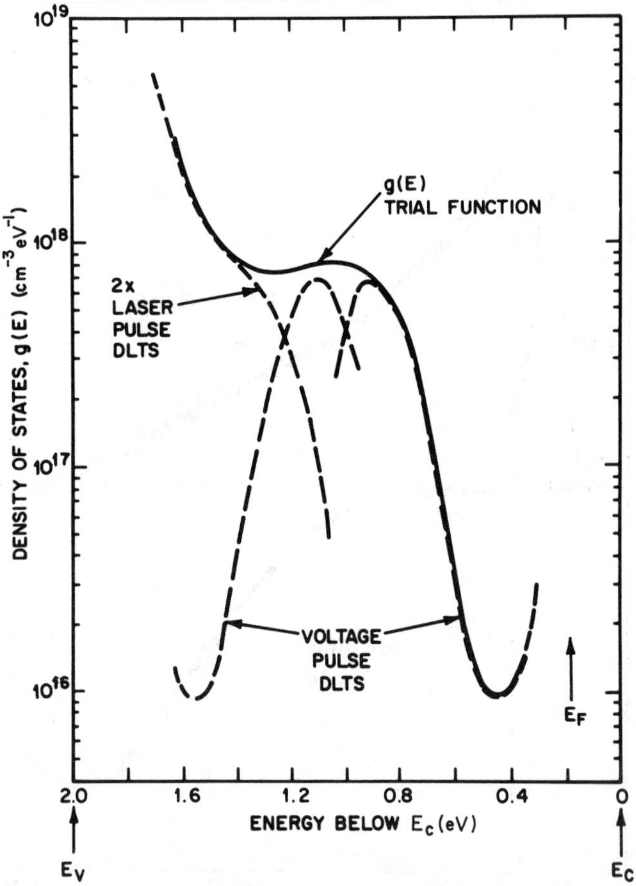

FIG. 47. First $g(E)$ trial function to be used in the iterative fitting procedure which ultimately gives the bulk density of states for this sample (JH152). The dashed lines indicate how the raw DLTS data in Fig. 46 is used to construct the first trial function (solid line) as described in the text.

41) we may immediately interpret the voltage pulse spectrum as indicating a low density of states in the upper half of the gap (small signal at low temperatures) increasing as we move toward midgap. The spectrum falls off at the highest temperatures due to the midgap quasi-Fermi energy cutoff. The laser pulse spectrum indicates that $g(E)$ is larger in the lower half of the gap (net positive signal) decreasing as we move deeper in energy.

We construct our initial trial $g(E)$ following these remarks as shown in Fig. 47 by superimposing the voltage pulse spectrum in the upper half gap and

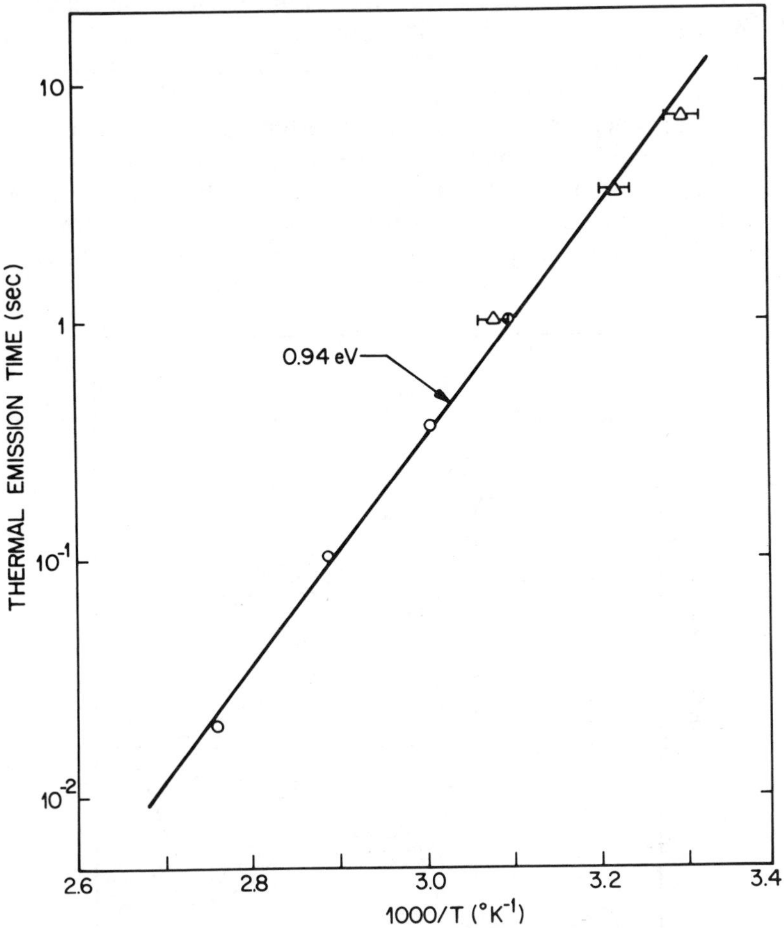

FIG. 48. DLTS measurement time constant τ (inverse of rate window) versus the inverse temperature of the peak in the electron emission spectrum at that τ for a typical sample. The activation energy of 0.94 eV is the slope of the raw data without the $2kT$ correction (see text). The circles are from DLTS spectra; the triangles are calculated from directly recorded capacitance transients.

adding twice the laser pulse spectrum to the voltage pulse signal in the lower half of the gap. This combination below midgap follows from the assumed half occupancy of gap states following laser excitation and the fact that the laser pulse spectrum arises from the net hole *minus* electron emission. The overall magnitude of the trial $g(E)$ is usually taken from the magnitude of the TSCAP spectrum for the reasons discussed previously (Section 14a) and the value of E_g from the high temperature cutoff in the DLTS spectrum at E_F^* assumed to lie at $E_g/2$. The shape of the conduction-band tail in the vicinity of E_F is not given directly by the DLTS spectrum. Rather, we assume it to be exponential with an intercept at E_c in the vicinity of 10^{21} cm^{-3} eV^{-1} and adjust its characteristic energy E_0 to give the correct value for capacitance TSCAP data just above dielectric turn-on.

The overall energy scale is found from several spectra taken at different DLTS rate windows by making an Arrhenius plot of the high temperature peak in the voltage pulse spectra as shown in Fig. 48. The 0.94 eV slope of the raw data is corrected for the probable T^2 dependence of the exponential prefactor by subtracting $2kT = 0.06$ eV to give $E = 0.88$ eV $\pm$ 10%. Thus we

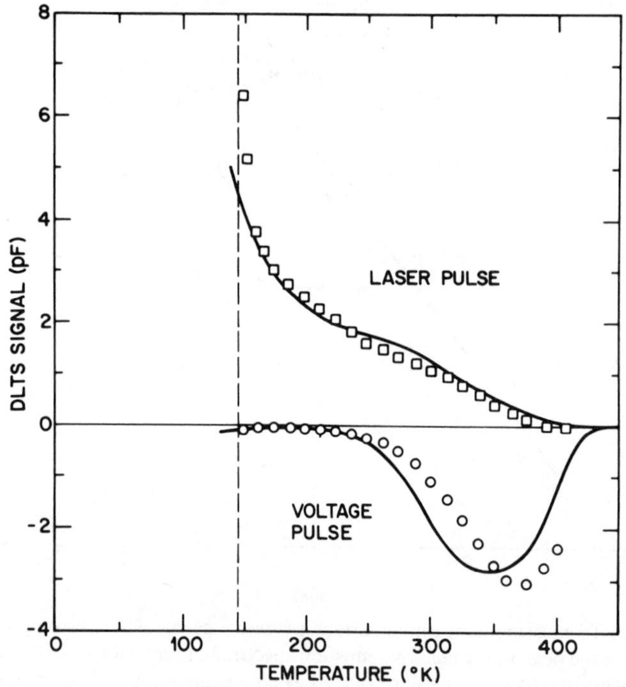

FIG. 49. DLTS spectra for sample JH152 calculated from the first $g(E)$ trial function in Fig. 47 (solid curve). The DLTS data points (□, O) are taken from the spectra in Fig. 46.

associate an energy of roughly 0.9 eV with a temperature of 360°K for the data taken with a 100 msec rate window as shown in Fig. 46. This implies an exponential prefactor of 10^{13} sec^{-1}, which is assumed to vary as T^2 across the entire measured spectrum. A one order-of-magnitude deviation of ν for different parts of the spectrum will actually affect the thermal emission energy scale by only ~0.07 eV as determined by Eqs. (5) and (6). The issue of whether the exponential prefactor varies more markedly with energy is discussed further in Section 16.

The approximate $g(E)$ in Fig. 47 is used to numerically calculate an associated DLTS spectrum, which is compared to the experimental data in Fig. 49. Although the derived spectrum of the trial $g(E)$ is not perfect, it is clearly close enough to the general shape of the experimental spectrum to be a useful starting point. This trial $g(E)$ is then varied point by point until good

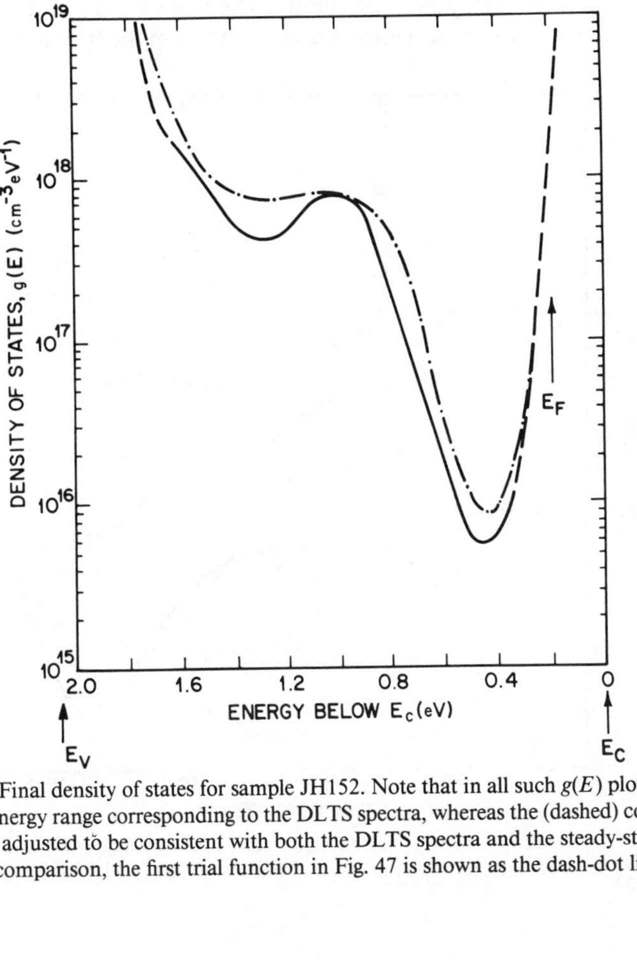

FIG. 50. Final density of states for sample JH152. Note that in all such $g(E)$ plots the solid line is the energy range corresponding to the DLTS spectra, whereas the (dashed) conduction-band tail is adjusted to be consistent with both the DLTS spectra and the steady-state capacitance. For comparison, the first trial function in Fig. 47 is shown as the dash-dot line.

agreement is achieved with experiment. This is a somewhat complex procedure because increasing $g(E)$ at, say, $E_c - 0.7$ eV has an effect not only at the corresponding temperature of the DLTS spectrum but to a lesser extent also at most other temperatures. However, with some experience, a good fit can typically be achieved in four iterations. In Fig. 50 the final $g(E)$ is displayed along with the first trial $g(E)$; the agreement of the derived DLTS spectra with the experimental spectra is shown in Fig. 51.

It is important to stress that the density of states shown in Fig. 50 is not a fit to theory; it is merely a different representation of the DLTS spectra shown in Fig. 46 that removes the nonlinearities inherent in the response of the space-charge capacitance due to thermal emission from states at different energy depths. These nonlinearities are negligible for crystals with $N_D \gg N_T$ but large for a-Si : H because of the large number of deep states. The derived density of states follows directly from Poisson's equation based on our assumptions and auxiliary measurements. The energy scale is the superposition of two thermal emission energy scales: one from the thermal emission of

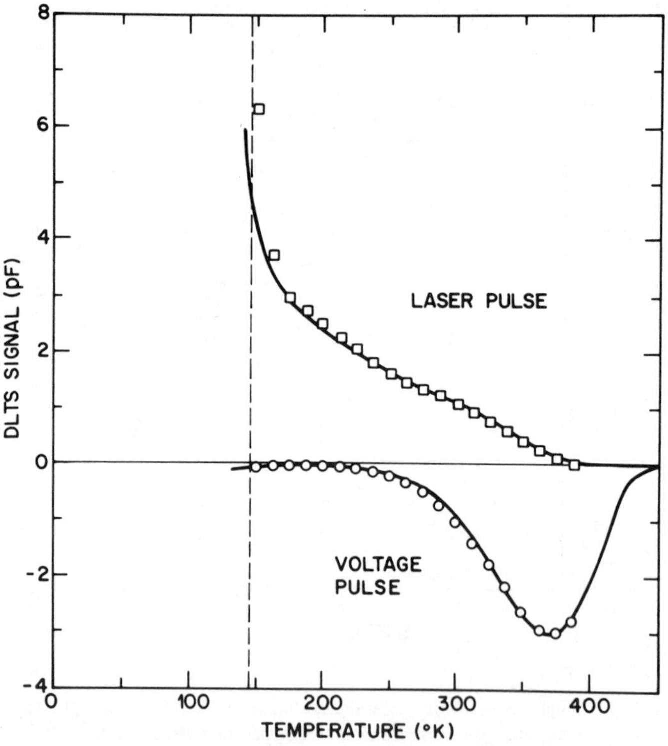

FIG. 51. DLTS spectra for sample JH152 calculated from the final $g(E)$ in Fig. 50 (solid curve). The DLTS data points (□, ○) are taken from the spectra in Fig. 46.

electrons to the conduction band and the other from the emission of holes to the valence band. Displaying the DLTS spectra as a density of states is extremely useful because it provides a direct basis for comparison with other measurements. It also points out small scale structure in $g(E)$ not immediately apparent in the raw DLTS data.

b. Densities of States Derived from Capacitance DLTS

A composite of DLTS-derived densities of states is shown in Fig. 52 for five samples grown with varying amounts of PH_3 doping (Lang *et al.*, 1982a). Sample 1 was grown without PH_3, samples 3 and 4 were grown with 300

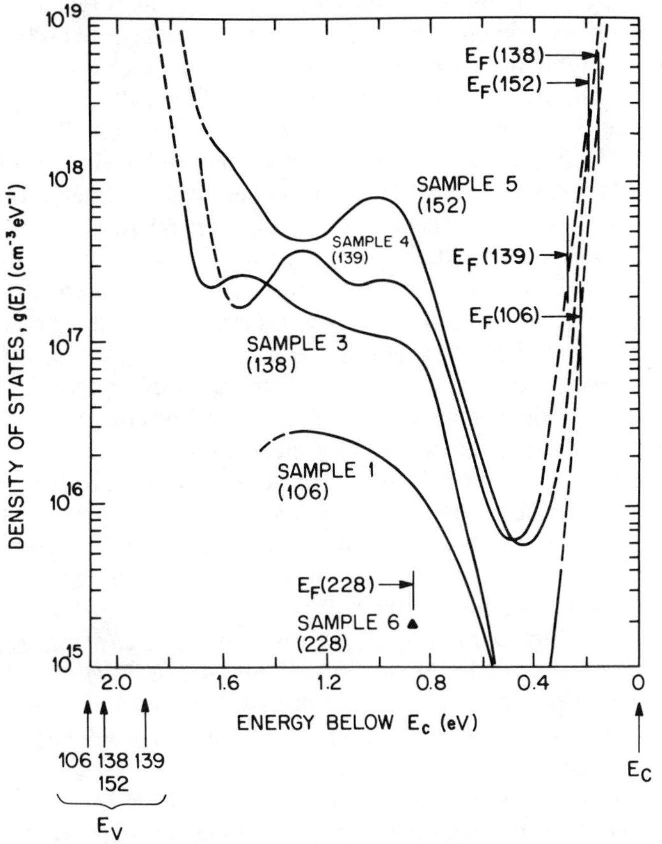

FIG. 52. Summary of the density-of-states results for five different a-Si : H films with differing amounts of phosphorus doping (see description in text). The films have slightly different deduced band gaps, and for purposes of comparison the curves are all normalized to the conduction-band (mobility) edge E_c. The position of the bulk Fermi level E_F^0 in each film is indicated. [From Lang *et al.* (1982a).]

vppm PH_3, and sample 4 with 60 vppm PH_3. Sample 1 behaved like a doped sample (had a shallow Fermi energy) for unknown reasons, which enabled a reasonably complete DLTS spectrum to be obtained. Sample 6 was also grown without PH_3 but exhibited the more typical fairly insulating behavior with $E_c - E_F^0 = 0.87$ eV. The single point at $g(E_F^0)$ for sample 6 was obtained by the $C-T$ profiling method described in Section 11. Generally speaking, these data indicate an increase in $g(E)$ with increased phosphorus doping. Such an increase in $g(E)$ with n-type doping has been confirmed by light-induced electron spin resonance (LESR) measurements (Biegelsen *et al.*, 1981).

Samples 3, 4, and 5 in Fig. 52 each exhibit a couple of fairly distinct defect bands at midgap and also at roughly $E_v + 0.5$ eV. The midgap feature has now been positively identified as arising from thermal emission of one electron from a doubly occupied dangling-bond state (Cohen *et al.*, 1982). The lower energy feature has not been positively identified, although corroborating evidence for its existence in undoped samples comes from infrared quenching (Persans and Fritzsche, 1981) and photoinduced absorption measurements (Vardeny *et al.*, 1982). Field-effect measurements on boron-doped samples also show a peak in $g(E)$ in this energy regime (see Part III).

Another more recent DLTS study that investigates the dependence of $g(E)$ on added dopant impurities is shown in Fig. 53 (Cullen *et al.*, 1983). These samples, all grown with roughly a 300 vppm PH_3/SiH_4 gas ratio, vary in the amount of B_2H_6 counterdoping as indicated. The results show a general increase in $g(E)$ in the vicinity of the valence-band tail and a small decrease in the midgap feature. Such conclusions from these data are somewhat tentative because of unavoidable differences in $g(E)$ just due to the fine scale variations in run-to-run growth conditions, which produce considerable scatter in the observed spectra. In addition, these samples exhibited a particularly strong tendency to suffer light-induced metastable changes (Staebler and Wronski, 1977) during the course of recording the laser pulse DLTS spectrum. Nonetheless, the decrease in the concentration of the midgap defect level with counterdoping is statistically significant and agrees with previous LESR counterdoping measurements (Biegelsen *et al.*, 1981). The degree of variation in $g(E)$ near $E_v + 0.5$ eV between the samples grown without boron and the heaviest counterdoping is also much larger than the sources of uncertainty named above.

A rather good demonstration of the sensitivity of the DLTS method for discerning small changes in $g(E)$ is shown in Fig. 54 in which a couple of measurements were made on the same sample but with differing amounts of prolonged light exposure and subsequent dark annealing (Lang *et al.*, 1982b; Cohen *et al.*, 1983b). A detailed discussion of such metastable effects in a-Si:H is given in Chapter 10 by Wronski. Treatment of samples in this

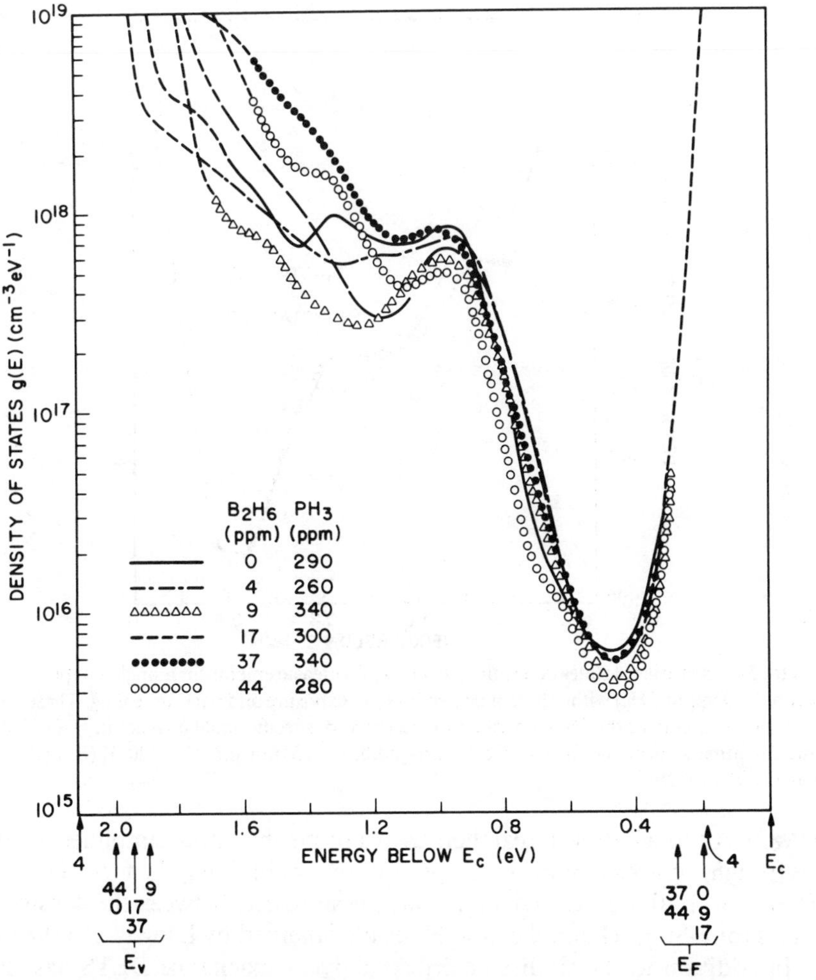

FIG. 53. Densities of states derived from DLTS measurements for six phosphorus doped a-Si : H samples containing varying amounts of boron counterdoping. The gas-phase PH_3 and B_2H_6 concentrations are indicated. Note the variation in $g(E)$ in the lower half of the gap and the general decrease with boron counterdoping of the midgap feature. [From Cullen *et al.* (1983). Copyright North-Holland Publ. Co., Amsterdam, 1983.]

manner are now widely believed to alter the shape of the density of states. Indeed, the marked increase in only the lower defect band and close reproducibility of the rest of the $g(E)$ curve indicates the degree to which DLTS can be used to isolate and study individual features in the density of states.

As a final example we display in Fig. 55 the density of states derived from a fluorinated amorphous silicon sample obtained by Hyun, Shur, and Madan

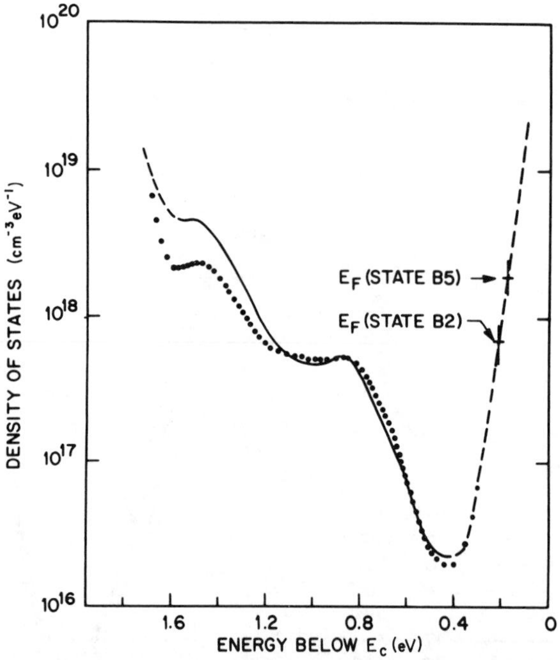

FIG. 54. Densities of states determined from DLTS measurements for a single *n*-type a-Si : H sample (30 vppm PH_3) with different degrees of light exposure and dark annealing. These data indicate the ability of the DLTS method to discern very specific small changes in $g(E)$. [Solid line, 60 min anneal at 380°K (B2). Dotted line, additional 5 min anneal at 420°K (B5).] [From Lang *et al.* (1982b).]

(1982). Although these researchers used a somewhat more simplified analysis [roughly corresponding to the procedure for obtaining the first trial $g(E)$ in Section 15a] these data demonstrate the similarity between the density of states for a-Si : F : H and the a-Si : H results obtained by Lang *et al.* (1982a).

In addition to the studies described above, capacitance DLTS has also recently been employed by Crandall and Staebler (1983) to study the dynamic properties of the Staebler–Wronski effect (Staebler and Wronski, 1977) in a-Si : H. The junction-capacitance related thermal emission methods of ICTS and constant capacitance DLTS have recently also been used to determine $g(E)$ in a-Si : H. These will be discussed in Section 15d.

c. Current DLTS

One of the earliest applications of thermal transient spectroscopic methods to amorphous silicon was through the current DLTS measurements of R. Crandall (1980). Current DLTS measurements were carried out during the same period by Cohen *et al.* (1980b), and more recently by

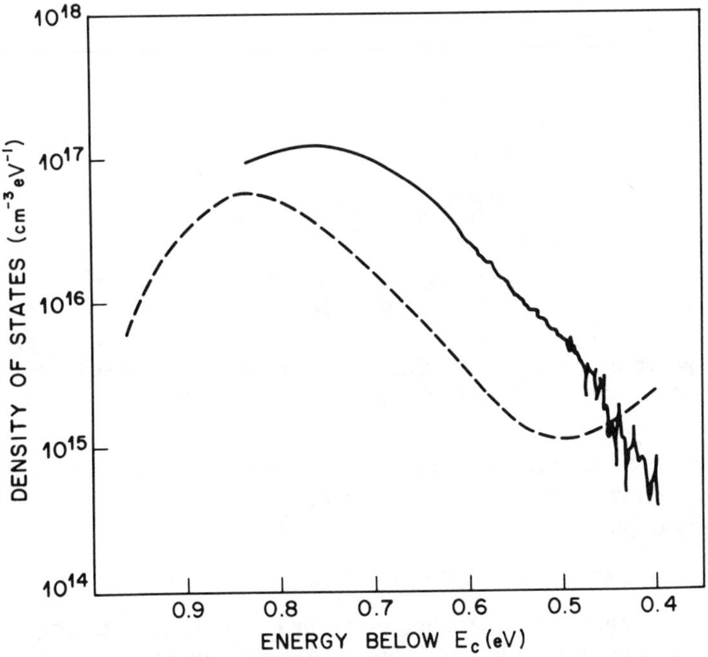

FIG. 55. Capacitance DLTS-deduced density of states for a *fluorinated* amorphous silicon sample (solid line). For comparison a typical density of states for a-Si : H from Fig. 52 is included (dashed line). Note the general similarity between the densities of states for these two kinds of materials deduced by capacitance DLTS. [From Hyun *et al.* (1982).]

Thompson *et al.* (1981). Crandall's measurements were actually isothermal current transient measurements made at one fixed temerature and thus extend over a very narrow energy range. In this energy range they agree very well with the results of the capacitance DLTS method described above.

The current DLTS measurements shown in Fig. 56 (Cohen *et al.*, 1980b), were made by scanning temperature with a fixed rate window for an unintentionally doped sample very similar to sample 1 in Fig. 52. Although these spectra indicate general agreement with $g(E)$ obtained by capacitance methods, there is disagreement with regard to the extra structure seen at low temperatures in Fig. 56. On the basis of our earlier discussion (Section 14d) we now attribute this structure to surface or near-interface states. Also, because we sometimes observed extraneous current transients found to be most likely associated with barrier tunneling, we eventually phased out our studies of current DLTS in favor of the more reliable capacitance DLTS.

Thompson *et al.* (1981) used current DLTS methods to study undoped a-Si : H films. They concluded that the sources of current-transient signals in

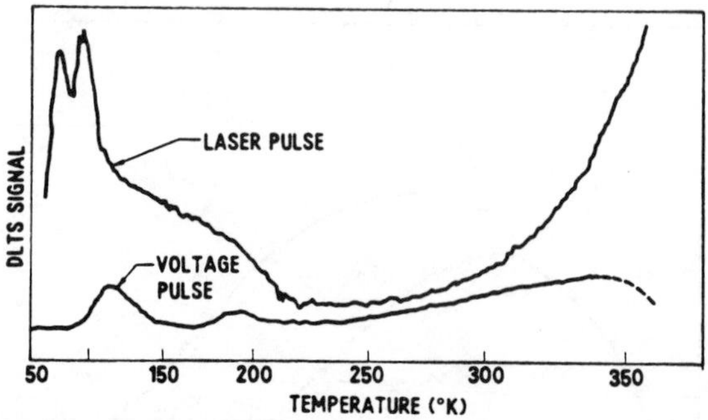

FIG. 56. Current DLTS for an a-Si:H sample similar to sample 1 in Fig. 52. Although in general agreement, these data indicate additional structure that is attributed to near-interface states. [From Cohen *et al.* (1980b). Copyright North-Holland Publ. Co., Amsterdam, 1980.]

their samples were primarily due to dielectric relaxation effects and that, therefore, current DLTS was not a viable method for obtaining $g(E)$ in intrinsic samples

d. Constant Capacitance DLTS and ICTS

Recent results of measurements using the constant capacitance (voltage-transient) method described in Section 14 are shown in Fig. 57 (Johnson,

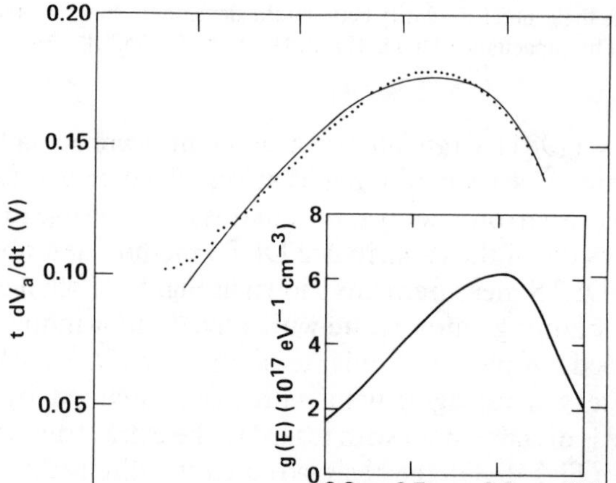

FIG. 57. Constant-capacitance DLTS spectrum for a *n*-type doped a-Si:H Schottky-diode sample. The insert shows the density of states derived from this spectrum. Dotted, from experiment; solid, from analysis. [From Johnson (1983).]

1983). These measurements are also made isothermally since otherwise a constant capacitance would not necessarily reflect a constant depletion region. As described earlier, the analysis of these data was made using a somewhat simpler algorithm than the full solution of the time evolution of the band-bending potential that is applied for capacitance DLTS. Nonetheless, the density of states derived for the data shown in the figure inset is probably reasonably accurate. The results agree in detail with the results obtained by the more traditional method.

A different conclusion was reached by workers using the isothermal capacitance (ICTS) method (Okushi *et al.*, 1982, 1983). Although they found a similar peak in the electron emission at larger energies (longest emission times) they concluded that this represented a peak in $g(E)$ at $E_c - 0.6$ eV rather than a midgap cutoff at $E_c - 0.9$ eV. They also concluded, based on these measurements and/or the dependence of the filling of gap states with voltage pulses of different time duration, that the exponential prefactor for thermal emission and capture was much lower at this peak, only 10^9 sec^{-1} and increased exponentially for states closer and closer to E_c.

It should be pointed out that a value of ν_n of 10^9 sec^{-1} is at least two orders of magnitude smaller than any ν associated with the trapping of electrons in gap states in crystals (Land, 1979) and must be regarded as very unusual. We will examine the evidence for the energy scale and prefactor assignments in the following section.

16. Advantages and Limitations of DLTS in the Study of Amorphous Semiconductors

There are two principal advantages of the DLTS method. First of all, it is very sensitive to the underlying density of states. Indeed, to a considerable degree there is a direct correspondence between $g(E)$ and the DLTS spectra. This is best illustrated by comparing two very similar densities of states shown in Fig. 58 with the corresponding calculated DLTS spectra shown in Fig. 59. The differences between the fits to the experimental data generated by these two very similar $g(E)$ curves can clearly be distinguished within the signal-to-noise of the experimental data. Thus fine features, such as those indicated in Figs. 52–54, are well resolved.

The second major advantage is the ability to resolve the sample's spatial uniformity. One manner in which this may be explored is indicated in Fig. 60 (Lang *et al.*, 1982a). In this case, for a fixed ambient bias V_B the size of the voltage filling pulse is varied from 0.5 V above V_B to a 1 V forward bias condition. These measurements correspond to filling gap states over an increasingly larger fraction of the total depletion width as indicated schematically in Fig. 44 of Section 14d. Figure 61 shows the variation of the peak signal with the amplitude of the filling pulse. A density of states in the upper

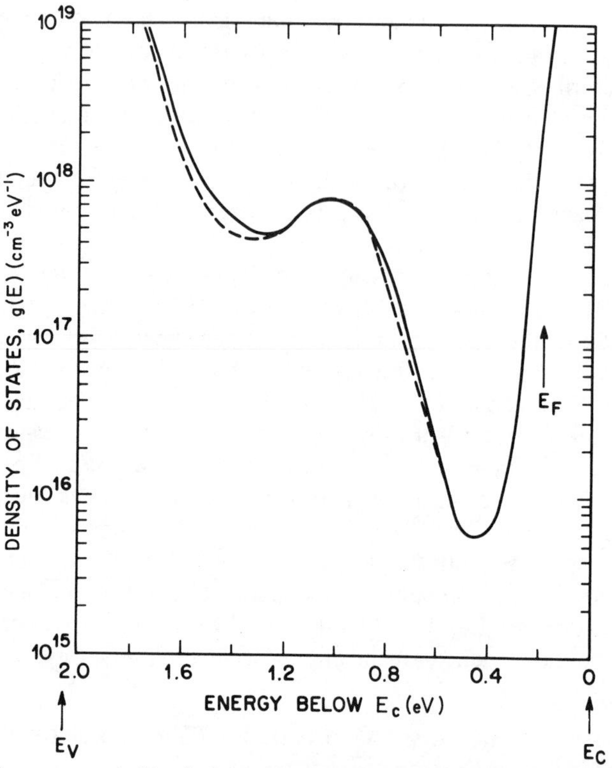

FIG. 58. Two very nearly identical densities of states employed as trial functions to fit the DLTS spectra of Fig. 46 for sample JH152. (Solid, trial 3; dashed, trial 4.)

half of the gap is calculated by the 5 V filling pulse DLTS signal alone. This $g(E)$ is then used to calculate the expected dependence on the filling pulse amplitude assuming a spatially uniform sample. This is the solid line shown in Fig. 61. The agreement is excellent. Furthermore, our calculation also predicts a slight shift in the peak of the electron emission signal for the different size pulses. The *calculated* positions are indicated by the small arrows in Fig. 60. Again there is almost perfect agreement. In addition to attesting to the homogeneity of this sample, such close agreement between the numerical model and the detailed behavior of DLTS measurements provides very strong evidence as to the validity of our analysis and assumptions.

It should also be pointed out that capacitance DLTS is, in any case, very *insensitive* to states near the semiconductor interface. This is an additional advantage when attempting to measure bulk properties. Other methods, on

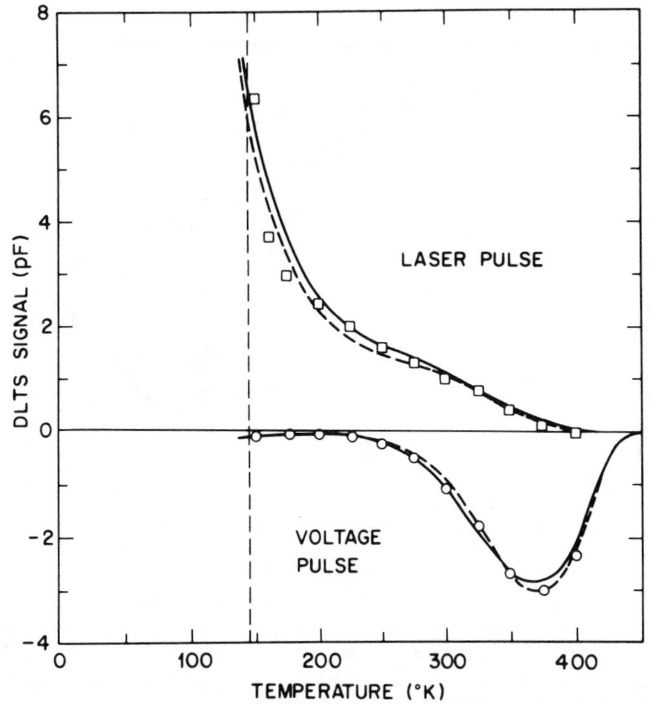

FIG. 59. Calculated DLTS spectra for sample JH152 from the densities of states in Fig. 58. Solid, trial 3; dashed, trial 4. Note that the differences between the two calculated spectra are readily discernible. The symbols □, ○ denote the experimental data.

the other hand, such as current DLTS, are sensitive to such interface states and could be used to study material properties in this region.

There are also some drawbacks to DLTS as it has been applied to a-Si : H. First of all, the instrumentation and analysis is rather complex compared with, for example, capacitance methods. In addition, DLTS is restricted for the most part to the study of *doped* samples. Although the picture of $g(E)$ obtained for reasonably n-type samples by using voltage and laser pulse measurements together extends over a wide energy range, it is restricted approximately to the range $E_v + \Delta E_F^0$ to $E_c - \Delta E_F^0$. This is due to the limitations put upon the transient time response due to dielectric relaxation. Since many very interesting issues concern intrinsic amorphous semiconductors, being limited to doped samples is unfortunate.

Another problem is that several of the key assumptions, used to extract $g(E)$ from the DLTS spectra cannot be verified or have been called into question by other workers. We have already discussed these issues in some

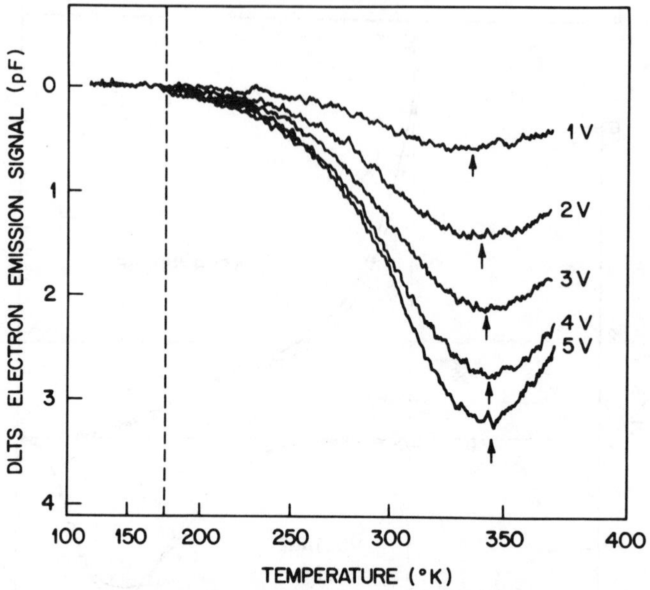

FIG. 60. Set of five DLTS electron emission spectra corresponding to different amplitudes of the 10 msec bias voltage pulse used to fill the gap states with electrons. The arrows indicate the DLTS peak positions calculated for pulse amplitudes less than 5 V using a spatially uniform $g(E)$ fit to the 5 V pulse data. (10 sec^{-1} rate window, 5 V reverse bias on rear junction, sample JH139.) [From Lang *et al.* (1982a).]

detail in Section 15a in which we concluded that the most difficult issue to resolve concerned the post laser pulse gap-occupation profile. This uncertainty affects our derived $g(E)$ only in the lower half of the gap and probably only by a factor of two. We hope that some of this uncertainty may eventually be resolved by further study.

The issue that has caused the greatest dispute among other workers is the assumption that the energy dependence of the exponential prefactor is weak and that it has a value of roughly 10^{13} sec^{-1}. This assumption is based on direct experimental evidence but has recently been disputed by ICTS measurements (see the preceding section). Because the thermal emission energy scale is based on these quantities, the issue is a very important one to resolve.

The activation energy derived for the data displayed in Fig. 48 needs no further comment. However, let us consider some additional evidence. In Fig. 62 the DLTS spectra for one sample using two different rate windows are shown (Lang *et al.*, 1982a). The exponential prefactor is precisely the quantity that determines how the DLTS should change with rate window. The solid lines through the solid data points are spectra calculated from the density of states derived from the 100 msec rate window data. The solid lines

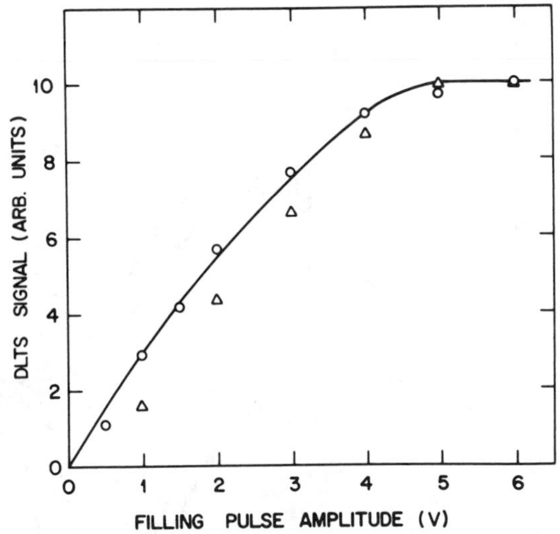

FIG. 61. Magnitude of DLTS electron emission peak versus bias pulse amplitude for sample JH139 from data similar to that in Fig. 60. The ambient reverse bias was 5 V. The data represented by ○ and △ correspond to voltage pulse durations of $\frac{1}{2}$ and $\frac{1}{10}$ of the DLTS emission time constant, respectively. The solid line was calculated assuming a spatially uniform density of states. Note the good agreement between theory and experiment for the longer pulse data. [From Lang *et al.* (1982a).]

drawn through the open data points are calculated from the same density of states using a 20 msec rate window. There are no adjustable parameters used in the calculation for the 20 msec data. The agreement in this case is totally inconsistent with a variation of ν more than one order of magnitude from its assumed 10^{13} $\sec^{-1}$ value.

We also measured the variations of the voltage pulse DLTS signal as a function of filling pulse time (Lang *et al.*, 1982a). This is shown in Fig. 63. Although one can observe that different regions capture carriers slightly more readily, most of this variation is accounted for by the difference in carrier densities at the different temperatures. Thus the variation of capture times from roughly 180 μsec at 322°K to 46 μsec at 368°K indicates a 0.3 eV activation energy, which is consistent with the 0.29 eV conductivity activation energy of this sample. Again there is little indication of more than one order-of-magnitude variation in ν. Furthermore, some variation is expected even for a truly constant ν because the filling voltage pulse does not bring the sample quite to a flat-band condition.

The capture cross section derived from Fig. 63 is, as also found by the ICTS measurements, several orders of magnitude too low compared with what is expected from detailed balance based on the emission data. As discussed

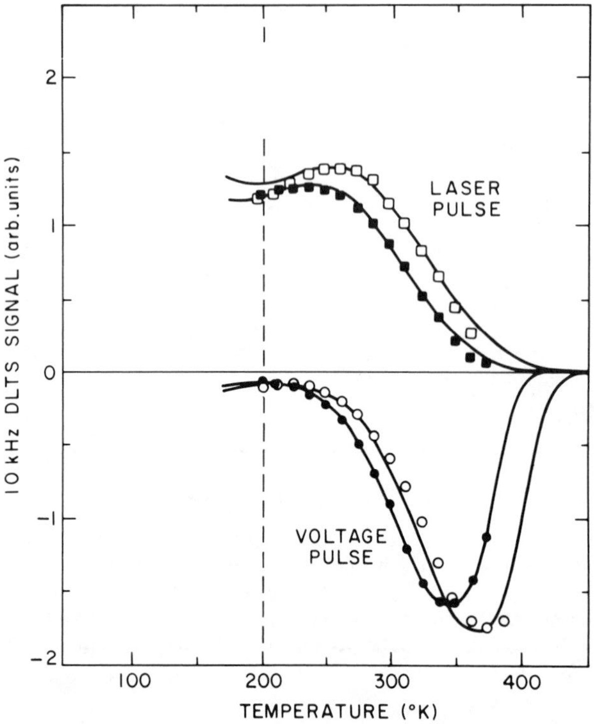

FIG. 62. DLTS hole and electron emission spectra for the two rate windows 10 sec^{-1} (closed figures) and 50 sec^{-1} (open figures) for sample JH139. The solid lines are calculated from the density of states indicated for this sample shown in Fig. 52, which is derived solely from the 10 sec^{-1} DLTS data. (5 V reverse bias on rear junction.) [From Lang *et al.* (1982a).]

previously (Lang *et al.*, 1982a) we can give several possible reasons for this. First of all, a value of ν of 10^{13} sec^{-1} determined by the emission measurements should correspond, by detailed balance, to a filling time of only 0.1 nsec at 350°K. The dielectric relaxation time at this same temperature is at least two orders of magnitude longer, which limits the time scale in which the band potential can respond to the filling pulse. This does not, however, explain the observed 100 μsec capture time. Perhaps the relatively large current surge (10 A cm^{-2} for 1 nsec) needed to fill all the electron states in the time predicted is impeded by contacts that are not perfectly ohmic. In any case, it is by no means obvious that for a-Si:H detailed balance between capture and emission applies when emission occurs in an electric field of $\sim 10^5$ V cm^{-1} and capture occurs close to zero field.

The ICTS measurements also indicate a different Arrhenius behavior for thermal *emission* than that shown in Fig. 48. The reasons for this disagree-

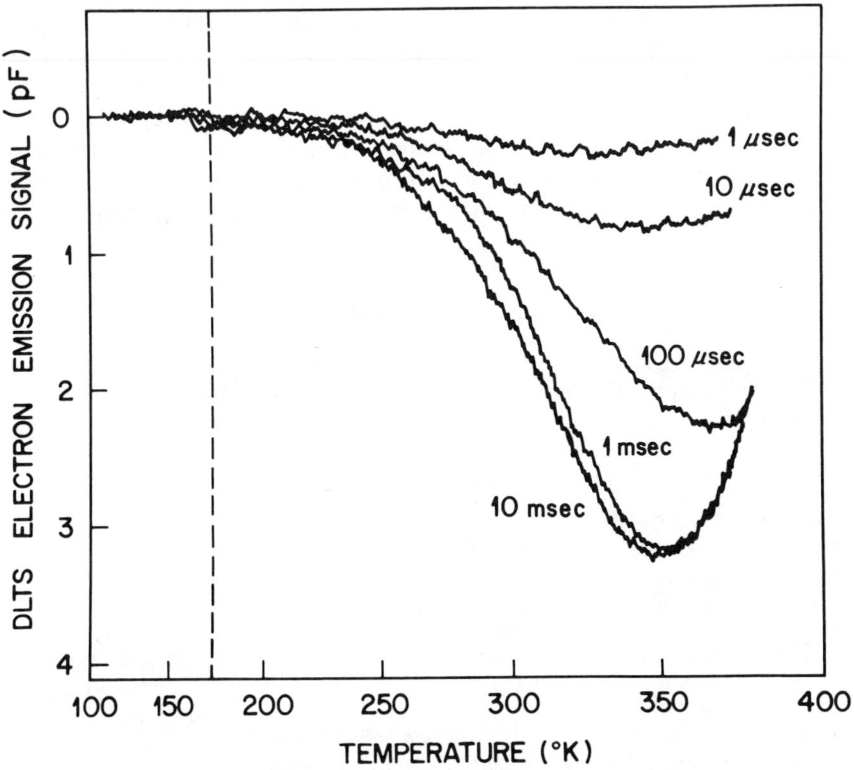

FIG. 63. Set of five DLTS electron emission spectra for sample JH139 corresponding to different durations of the bias voltage pulse used to fill the gap states with electrons. (10 sec^{-1} rate window, 5 V reverse bias on rear junction.) [From Lang *et al.* (1982a).]

ment are not entirely clear. However, as discussed in Part II, a low Schottky-barrier height will cause E_F^* to lie higher in the gap and thus cut off the emission spectrum prematurely. Also, if care is not taken to anneal away all metastable effects due to light exposure or field-induced doping (Lang *et al.*, 1982c) then the Fermi-level position might not be uniform throughout the sample and an anomalously low apparent exponential prefactor can be the result. A detailed investigation to help explain such kinds of apparent inconsistencies is being undertaken and will be reported by the author and collaborators in the near future.

Whereas issues such as these attest to some of the complexity in understanding the results of thermal-transient measurements, they also indicate the great wealth of information that such measurements can provide. The detailed gap-state properties obtained using methods like DLTS are ultimately important for the interpretation even of many of the steady-state measurements (such as ac capacitance). Since the aim of all of the junction measurements discussed in this chapter has been to obtain a basic detailed description of mobility gap states, the dynamic thermal-transient measurements undoubtedly offer some of the greatest possibilities because of the large number of measurement parameters. From the results presented in this discussion, one can see that a much more detailed and broader understanding of the density of states in a-Si : H has already emerged as a result of such techniques.

Acknowledgments

I would like to thank David V. Lang for many valuable discussions and comments. Research at the University of Oregon was supported by National Science Foundation grant DMR-8207437.

References

Abram, R. A., and Doherty, P. J. (1982). *Philos. Mag. B* **45,** 167.
Ast, D. G., and Brodsky, M. H. (1980). *J. Non-Cryst. Solids* **35–36,** 611.
Balberg, I. (1980). *Phys. Rev. B* **22,** 3853.
Beguwala, M., and Crowell, C. R. (1974). *Solid-State Electron.* **17,** 203.
Beichler, J., Fuhs, W., Mell, H., and Welsch, H. M. (1980). *J. Non-Cryst. Solids* **35–36,** 587.
Biegelsen, D. K., Street, R. A., and Knights, J. C. (1981). *In* "Tetrahedrally Bonded Amorphous Semiconductors" (R. A. Street, D. K. Biegelsen, and J. C. Knights, eds.), p. 166. Amer. Inst. Phys., New York.
Buehler, M. G., and Phillips, W. E. (1976). *Solid-State Electron.* **19,** 777.
Carballas, J. C., and Lebailly, J. (1968). *Solid State Commun.* **6,** 167.
Chen, I., and Lee, S. (1982). *Appl. Phys. Let.* **40,** 487.
Cohen, J. D., and Lang, D. V. (1982). *Phys. Rev. B* **25,** 5321.
Cohen, J. D., Lang, D. V., and Harbison, J. P. (1980a). *Phys. Rev. Lett.* **45,** 197.
Cohen, J. D., Lang, D. V., Harbison, J. P., and Bean, J. C. (1980b). *J. Non-Cryst. Solids* **35–36,** 581.
Cohen, J. D., Harbison, J. P., and Wecht, K. W. (1982). *Phys. Rev. Lett.* **48,** 109.
Cohen, J. D., Lang, D. V., and Harbison, J. P. (1983a). *Bull. Am. Phys. Soc.* **28,** 259.
Cohen, J. D., Lang, D. V., Harbison, J. P., and Sergent, A. M. (1983b). *Solar Cells* **9,** 119.
Crandall, R. S. (1980). *J. Electron. Mater.* **9,** 713.
Crandall, R. S. (1981). *J. Phys. Colloq. Orsay, Fr.* **42,** C4-413.
Crandall, R. S., and Staebler, D. L. (1983). *Solar Cells* **9,** 63.
Cullen, P., Harbison, J. P., Lang, D. V., and Adler, D. (1983). *J. Non-Cryst. Solids* **59/60,** 261.
Day, D. S., Tsai, M. Y., Streetman, B. G., and Lang, D. V. (1979). *J. Appl. Phys.* **50,** 5093.
Döhler, G. H., and Hirose, M. (1977). *In* "Amorphous and Liquid Semiconductors" (W. E. Spear, ed.), p. 372. CICL, Univ. of Edinburgh.
Driver, M. C., and Wright, G. T. (1963). *Proc. Phys. Soc. London* **81,** 141.
Fritzsche, H. (1980). *Solar Energy Mater.* **3,** 747.
Fuhs, W., and Milleville, M. (1980). *Phys. Status Solidi B* **98,** K29.
Furukawa, Y., and Ishibashi, Y. (1967). *Jpn. J. Appl. Phys.* **6,** 503.
Goodman, A. M. (1963). *J. Appl. Phys.* **34,** 329.
Goodman, N. B. (1982). *Philos. Mag. B* **45,** 407.

Goodman, N. B., and Fritzsche, H. (1980). *Philos. Mag. B* **42,** 149.
Goodman, N. B., and Fritzsche, H. (1981). *In* "Tetrahedrally Bonded Amorphous Semiconductors" (R. A. Street, D. K. Biegelsen, and J. C. Knights, eds.), p. 176. Amer. Inst. Phys., New York.
Goto, G., Yanagisawa, S., Wada, O., and Takanashi, H. (1974). *Jpn. J. Appl. Phys.* **13,** 1127.
Hirose, M., Suzuki, T., and Döhler, G. H. (1979). *Appl. Phys. Lett.* **34,** 234.
Hyun, C. H., Shur, M. S., and Madan, A. (1982). *Appl. Phys. Lett.* **41,** 178.
Jan, Z. S., Bube, R. H., and Knights, J. C. (1980). *J. Appl. Phys.* **51,** 3278.
Johnson, N. M. (1982). *J. Vac. Sci. Technol.* **21,** 303.
Johnson, N. M. (1983). *Appl. Phys. Lett.* **42,** 981.
Johnson, N. M., Bartelink, D. J., and Schultz, M. (1978). *In* "The Physics of SiO_2 and Its Interfaces" (S. T. Pantelides, ed.), p. 421. Pergamon, New York.
Jousse, D., and Deleonibus, S. (1983). *J. Appl. Phys.* **54,** 4001.
Knights, J. C. (1980). *J. Non-Cryst. Solids* **35/36,** 159.
Knights, J. C., and Biegelsen, D. K. (1977). *Solid State Commun.* **22,** 421.
Knights, J. C., and Lujan, R. A. (1979). *Appl. Phys. Lett.* **35,** 244.
Lang, D. V. (1974). *J. Appl. Phys.* **45,** 3014.
Lang, D. V. (1979). *In* "Topics in Applied Physics" (P. Bräunlich, ed.), Vol. 37, Chap. 3. Springer-Verlag, Berlin, and New York.
Lang, D. V., and Henry, C. H. (1975). *Phys. Rev. Lett.* **35,** 1525.
Lang, D. V., Cohen, J. D., and Harbison, J. P. (1982a). *Phys. Rev. B* **25,** 5285.
Lang, D. V., Cohen, J. D., Harbison, J. P., and Sergent, A. M. (1982b). *Appl. Phys. Lett.* **40,** 474.
Lang, D. V., Cohen, J. D., and Harbison, J. P. (1982c). *Phys. Rev. Lett.* **48,** 421.
LeComber, P. G., Spear, W. E., Muller, G., and Kalbitzer, S. (1980). *J. Non-Cryst. Solids* **35–36,** 327.
LeComber, P. G., Snell, A. J., MacKenzie, K. D., and Spear, W. E. (1981). *J. Phys. Colloq. Orsay, Fr.* **42,** 423.
Losee, D. L. (1975). *J. Appl. Phys.* **46,** 2204.
Madan, A., and LeComber, P. G. (1977). *In* "Amorphous and Liquid Semiconductors" (W. E. Spear, ed.), p. 377. CICL, Univ. of Edinburgh.
Madan, A., LeComber, P. G., and Spear, W. E. (1976). *J. Non-Cryst. Solids* **20,** 239.
Madan, A., Ovshinsky, S. R., and Benn, E. (1979). *Philos. Mag. B* **40,** 259.
Miller, G. L. (1972). *IEEE Trans. Electron Devices* **ED-19,** 1103.
Miller, G. L., Lang, D. V., and Kimmerling, L. C. (1977). *Annu. Rev. Mater. Sci.* **7,** 377.
Nakashita, T., Taniguchi, M., Hirose, M., and Osaka, Y. (1980). *Bull. Fac. Eng. Hiroshima Univ.* **28,** 41.
Nakashita, T., Hirose, M., and Osaka, Y. (1981). *Jpn. J. Appl. Phys.* **20,** 471.
Nemanich, R. J., Biegelsen, D. K., and Rosenblum, M. P. (1980). *J. Phys. Soc. Jpn.* **49** Suppl. A, 1189.
Okushi, H., Tokumaru, Y., Yamasaki, S., Oheda, H., and Tanaka, K. (1981). *Jpn. J. Appl. Phys.* **20,** L549.
Okushi, H., Tokumara, Y., Yamasaki, S., Oheda, H., and Tanaka, K. (1982). *Phys. Rev. B* **25,** 4313.
Okushi, H., Takahama, T., Yokumaru, Y., Yamasaki, S., Oheda, H., and Tanaka, K. (1983). *Phys. Rev. B* **27,** 5184.
Persans, P. D., and Fritzsche, H. (1981). *J. Phys. Colloq. Orsay, Fr.* **42,** C4-597.
Powell, M. J. (1981). *Philos. Mag. B* **43,** 93.
Powell, M. J., and Döhler, G. H. (1981). *J. Appl. Phys.* **52,** 517.
Rehm, W., Fischer, R., and Beichler, J. (1980). *Appl. Phys. Lett.* **37,** 445.
Roberts, G. I., and Crowell, C. R. (1970). *J. Appl. Phys.* **41,** 1767.

Sah, C. T., and Reddi, G. K. (1964). *IEEE Trans. Electron Devices* **ED-11,** 345.
Sah, C. T., Forbus, L., Rosier, L. L., and Tasch, Jr., A. F. (1970). *Solid-State Electron.* **13,** 759.
Schottky, W. (1938). *Naturwissenschaften* **26,** 843.
Singh, J., and Cohen, M. H. (1980). *J. Appl. Phys.* **51,** 413.
Snell, A. J., MacKenzie, K. D., LeComber, P. G., and Spear, W. E. (1979). *Philos. Mag. B* **40,** 1.
Solomon, I., and Brodsky, M. H. (1980). *J. Appl. Phys.* **51,** 4548.
Solomon, I., Dietl, T., and Kaplan, D. (1978). *J. Phys. Paris* **39,** 1241.
Spear, W. E., and LeComber, P. G. (1972). *J. Non-Cryst. Solids* **8/10,** 727.
Spear, W. E., and LeComber, P. G. (1976). *Philos. Mag.* **33,** 935.
Spear, W. E., LeComber, P. G., and Snell, A. J. (1978). *Philos. Mag. B* **38,** 303.
Staebler, D. L., and Wronski, C. R. (1977). *Appl. Phys. Lett.* **31,** 292.
Street, R. A. (1983). *Bull. Am. Phys. Soc.* **28,** 258.
Tanielian, M. H., Fritzsche, H., Tsai, C. C., and Symbalisty, E. (1978). *Appl. Phys. Lett.* **33,** 353.
Tanielian, M. H., Chatani, M. Fritzsche, H., Smid, V., and Persans, P. D. (1980). *J. Appl. Phys.* **51,** 3262.
Tanielian, M. H., Goodman, N. B., and Fritzsche, H. (1981). *J. Phys. Colloq. Orsay, Fr.* **42,** C4-375.
Thompson, M. J., Johnson, N. M., and Street, R. A. (1981). *J. Phys. Colloq. Orsay, Fr.* **42,** C4-617.
Tiedje, T., Wronski, C. R., and Cebulka, J. M. (1980). *J. Non-Cryst. Solids* **35–36,** 743.
Tiedje, T., Moustakis, T. D., and Cebulka, J. M. (1981). *Phys. Rev. B* **23,** 5634.
Vardeny, Z., Strait, J., Pfost, D., Tauc, J., and Abeles, B. (1982). *Phys. Rev. Lett.* **48,** 1132.
Vieux-Rochaz, L., and Chenevas-Paule, A. (1980). *J. Non-Cryst. Solids* **35–36,** 737.
Viktorovitch, P. (1981). *J. Appl. Phys.* **52,** 1392.
Viktorovitch, P., and Moddel, G. (1980). *J. Appl. Phys.* **51,** 4847.
Viktorovitch, P. Jousse, D., Chenevas-Paule, and Vieux-Rochas, L. (1979). *Rev. Phys. Appl.* **14,** 201.
Weisfield, R. L., and Anderson, D. A. (1981). *Philos. Mag. B* **44,** 83.
Weisfeld, R., Viktorovitch, P., Anderson, D. A., and Paul, W. (1981). *Appl. Phys. Lett.* **39,** 263.
Williams, R. (1966). *J. Appl. Phys.* **37,** 3411.

CHAPTER 3

Magnetic Resonance Measurements in a-Si:H

P. C. Taylor

DEPARTMENT OF PHYSICS
UNIVERSITY OF UTAH
SALT LAKE CITY, UTAH

I. Introduction

The application of magnetic resonance techniques to the study of amorphous silicon (a-Si) and hydrogenated amorphous silicon (a-Si:H) has greatly increased our understanding of the microscopic properties of these films. In this chapter we shall review the types of magnetic resonance measurements that have been performed on a-Si and a-Si:H and discuss the significance of the results.

The term magnetic resonance includes both nuclear magnetic resonance (NMR) and electron spin resonance (ESR). One of the important uses of NMR is as a detailed local probe of the bonding configurations in amorphous silicon and amorphous-silicon-based alloys. This technique is particularly useful for studying hydrogen atoms in these films. In addition to local bonding information, NMR measurements are useful for investigating hydrogen diffusion and evolution. In some cases NMR experiments provide

ISBN 0-12-752149-6

information about defects and impurities through analyses of spin–lattice relaxation mechanisms.

The ESR experiments primarily examine the microscopic nature of defects or other localized electronic states that lie below the energy gap. Coupled with optical excitation, ESR techniques also provide useful information concerning electronic metastabilities such as an optical rearrangement of the electrons in existing localized states or the optically induced creation of new states. These experiments often provide data for useful comparisons with transport or optical properties.

In Part II we discuss NMR measurements in intrinsic and doped a-Si:H and include results on hydrogen, deuterium, silicon, boron, phosphorus, and fluorine. The topics discussed include local structural order, hydrogen evolution and diffusion, and spin–lattice relaxation. In Part III we present ESR results in a-Si, intrinsic and doped a-Si:H, and some complicated alloy systems. The topics discussed include the characterization of the observed ESR responses, hydrogen diffusion and evolution, optically induced changes in the ESR, the influence of impurities, and spin–lattice relaxation.

II. Nuclear Magnetic Resonance

In a-Si:H NMR techniques have been useful primarily for the determination of local structural arrangements. Studies of ^{1}H NMR at elevated temperatures, or at lower temperatures after annealing at elevated temperatures, have probed the details of hydrogen evolution in a-Si:H. In undoped films both hydrogen (in the form of ^{1}H and ^{2}D) and silicon (^{29}Si) have been investigated. In doped films studies have been reported for boron (^{11}B), phosphorus (^{31}P), and fluorine (^{19}F). A second useful result of the ^{1}H NMR measurements in a-Si:H has been the inferred existence of trapped H_2 molecules from spin–lattice relaxation measurements. The presence of these molecules has been very difficult to observe by other experimental techniques.

1. Local Structural Order and Bonding

Measurements that employ NMR often provide a very detailed, albeit very local, probe of structural arrangements and bonding. The important terms in the Hamiltonian for most situations of interest in NMR in a-Si:H are given by

$$H = H_{nz} + H_d + H_c + H_Q. \qquad (1)$$

In Eq. (1) the first term is the nuclear Zeeman interaction, which is of the form

$$H_{nz} = -\gamma\hbar\mathbf{I} \cdot \mathbf{H}, \qquad (2)$$

where γ is the nuclear gyromagnetic ratio, $\hbar$ is Planck's constant divided by 2π, $\mathbf{I}$ is the nuclear spin operator, and $\mathbf{H}$ is an externally applied magnetic field.

The second term in Eq. (1) is the dipolar interaction between nuclear spins, which can be written as

$$H_d = \frac{1}{2} \sum_j{}' \sum_k{}' \left[\frac{\boldsymbol{\mu}_j \cdot \boldsymbol{\mu}_k}{r_{jk}^3} - \frac{3(\boldsymbol{\mu}_j \cdot \mathbf{r}_{jk})(\boldsymbol{\mu}_k \cdot \mathbf{r}_{jk})}{r_{jk}^5} \right]. \tag{3}$$

In Eq. (3) the prime on the sum denotes $j \neq k$. The magnetic moments $\boldsymbol{\mu}_j$ are given by $\boldsymbol{\mu}_j = \gamma_j \hbar \mathbf{I}_j$, and $\mathbf{r}_{jk}$ is the radius vector from $\boldsymbol{\mu}_j$ to $\boldsymbol{\mu}_k$.

The third term in Eq. (1) is the so-called chemical shift interaction, which results from a complicated interplay between the electronic Zeeman contribution to H and the interaction between electronic and nuclear spins. In an applied field, the electronic Zeeman contribution polarizes the electronic spins, which can then interact via both their orbital and spin components with the nuclear spins. These interactions in general produce both diamagnetic and paramagnetic contributions to the local magnetic field experienced by a nucleus in a solid. Because all of the electrons are in principle involved in this interaction, the situation is complicated to the point where one usually writes H_c in parameterized form as

$$H_c = \gamma \hbar \mathbf{I} \cdot \boldsymbol{\sigma} \cdot \mathbf{H}, \tag{4}$$

where $\boldsymbol{\sigma}$ is the chemical shift tensor whose principal components depend, in some complicated way, on the local structural order.

In nonspherical nuclei ($I > \frac{1}{2}$) there exists a nonzero quadrupole moment for the nuclear charge density that contributes to the quadrupolar term (fourth term) in Eq. (1). In the expression for the electrostatic interaction energy for the nucleus in the potential produced by the electrons, the electric quadrupole moment of the nucleus is coupled to the gradient of the electric field at the nuclear site. This field gradient is, of course, due to the electrons. In diagonalized form one may write

$$H_Q = \frac{e^2 Q q}{6I(2I-1)} [(3I_z^2 - I^2) + \eta(I_x^2 - I_y^2)], \tag{5}$$

where the parameters eq and η are related to the principal components of the electric field gradient tensor V_{xx}, V_{yy}, V_{zz} ($V_{xx} < V_{yy} < V_{zz}$) by

$$eq = V_{zz}, \qquad \eta = (V_{xx} - V_{yy})/V_{zz}. \tag{6}$$

Thus, in parameterized form

$$H_Q = \mathbf{I} \cdot \mathbf{Q}' \cdot \mathbf{I}, \tag{7}$$

where the diagonal matrix elements of $\mathbf{Q}'$, which can be obtained from Eq.

(5), depend on the details of the electronic wave functions (predominantly the bonding wave functions) in the neighborhood of the nuclear site.

It is apparent from the functional forms for H_c and H_Q [Eqs. (4) and (7)] that the NMR spectrum observed from a single crystal depends on the orientation of the crystal with respect to the applied magnetic field. The energy levels of the spin system, and hence the NMR absorptions, are determined by diagonalizing Eq. (1). In a polycrystalline or amorphous sample the nuclear (or paramagnetic) sites are randomly oriented with respect to the applied magnetic field. In these cases the magnetic resonance spectrum, which is referred to as a "powder pattern," is an average over all orientations (Taylor *et al.*, 1975). Often it is a lengthy numerical exercise to obtain the eigenvalues for a representative sampling of the members of an ensemble of randomly oriented sites by exact diagonalization of the Hamiltonian [Eq. (1)]. In many cases the Zeeman term dominates to the extent that perturbation theory can be employed to derive approximate solutions.

a. ^{1}H *NMR*

For the hydrogen atom, where $I = \frac{1}{2}$, the nuclear Zeeman term in the Hamiltonian is by far the most important ($H_{nz} \gg H_d \gg H_c$). The position of the resonance in frequency for a given field is determined by H_{nz} and H_c, whereas the shape and width of the resonance are determined predominantly by H_d.

For a system of nuclei of spin I, the second moment due to the dipole–dipole interaction of Eq. (3) is given by

$$M_2 = \frac{3}{5}\gamma\hbar^2 I(I+1)\sum \frac{1}{r_{ij}^6}, \tag{8}$$

where r_{ij} is the distance between nuclei i and j. For a Gaussian line shape, which is expected for concentrated spins (> 5–10 at. %), the full width at half maximum (FWHM) is given by

$$\sigma = (8(\ln 2)M_2)^{1/2}. \tag{9}$$

For more dilute systems, the line shape is more closely approximated by a Lorentzian, in which case a statistical approach yields for the FWHM (Carlos and Taylor, 1982b)

$$\sigma = (4\pi^2/3\sqrt{3})\gamma^2 hn, \tag{10}$$

where n is the density of spins.

Experimentally, the line shape of ^{1}H NMR, where $I = \frac{1}{2}$, is most often determined from a Fourier transform of the free induction decay (FID) following a 90° pulse. Investigations using NMR of a-Si:H were first reported by Reimer *et al.* (1980) who concluded from the line shapes and linewidths that the hydrogen atoms were spatially distributed in an inhomogeneous fashion in the films. This basic conclusion, which is perhaps the most important feature of ^{1}H NMR in a-Si : H, has been confirmed in subsequent studies reported by others (Carlos and Taylor, 1980; Jeffrey and Lowry, 1981) and reviewed by Reimer (1981). The ^{1}H NMR spectrum consists of two distinct lines superimposed on one another as shown in Fig. 1. The rapid decay that dominates the FID before ~ 30 μsec constitutes one line and the slower exponential decay that dominates at longer times constitutes the other.

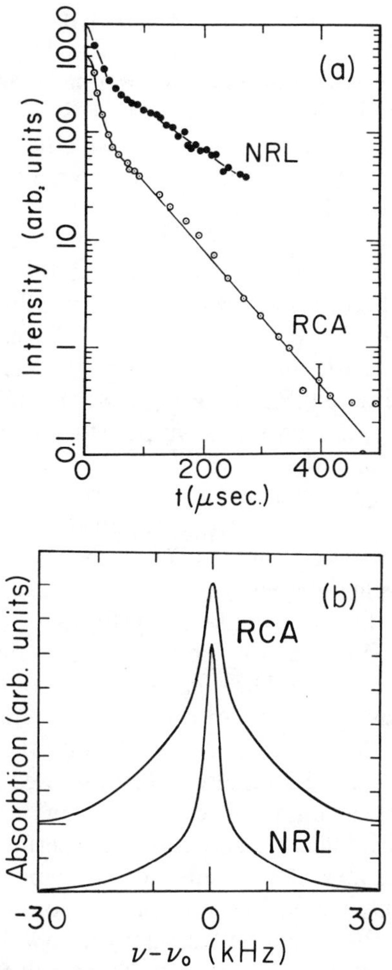

FIG. 1. (a) Free induction decays (FIDs) for two representative samples of a-Si : H made by the glow-discharge technique. (b) The Fourier transform of the free induction decay, which yields the line shapes for the two samples in (a). [From Carlos and Taylor (1982b).]

The line shapes shown in Fig. 1b are determined by Fourier transforming the FID curves shown in Fig. 1a. Although some subtle differences can be seen in the two FIDS, they are quite similar. The FID curves of Fig. 1a are both composed of two components: an exponential decay at long times, which transforms to a narrow Lorentzian line in the frequency domain (Fig. 1b), and a Gaussian shape at short times, which transforms to a Gaussian line in frequency.

Early studies (Reimer *et al.*, 1980) indicated a broad Gaussian component superimposed on a narrower Gaussian, but most recent studies (Reimer *et al.*, 1981b; Carlos and Taylor, 1980, 1982b; Taylor and Carlos, 1980; Jeffrey and Lowry, 1981) suggest that the narrow line has a Lorentzian shape. Microcrystalline films of Si : H also exhibit a broad Gaussian superimposed on a narrower Lorentzian line (Kumeda *et al.*, 1983). This fact demonstrates that the microstructure that gives rise to the two different sites exists in both the crystalline and amorphous forms. All studies confirm that the widths of the two lines are governed by the dipolar term [Eq. (3)] in the Hamiltonian.

The nature of the two ubiquitous NMR lines that are exhibited by hydrogen in nearly all a-Si:H films regardless of the hydrogen content or the growth technique (Reimer *et al.*, 1980, 1981b,d; Jeffrey and Lowry, 1981; Jeffrey *et al.*, 1981a; Carlos and Taylor, 1980, 1981b) has been the subject of some debate, but the basic interpretation is now generally accepted. From "hole-burning" experiments Reimer *et al.* (1981b) found that on a time scale of ~ 1 msec the hydrogen atoms that compose the broad line do not interact with the hydrogen atoms that compose the narrow line. Figure 2 summarizes the hole-burning experiments of Reimer *et al.* (1981b). In these experiments a weak pulse of frequency ω_{SAT} is applied to the sample over a time such that only those spins within ~ 1 kHz of the frequency ω_{SAT} are directly excited. If ω_0 is the center frequency of the two hydrogen lines and σ_N the line width of the narrow line as given by Eq. (10), then $|\omega_{SAT} - \omega_0| \gg \sigma_N$ describes the situation in which only those hydrogen atoms in the wing of the broad line are directly excited. The effect of this pulse is to saturate only those spins that the dipolar interaction of Eq. (3) can couple on a time scale of ~ 1 msec. This coupling is, of course, a strong function of the distance between spins. If this saturation pulse is then followed by a short 90° pulse, only those spins that were unsaturated by the first pulse will contribute to the FID (see Fig. 1a). The difference between this experiment and the ordinary FID experiment (Fig. 2b and 2c) yields the line shape that was saturated by the hole-burning pulse. Figure 2d shows that the Fourier transformed difference spectra are composed only of Gaussians whose widths fit the broad lines observed in the ordinary FID experiments. This result demonstrates that the hydrogen atoms in the broad line are in rapid spin–spin communication with one another but not with those spins in the narrow line.

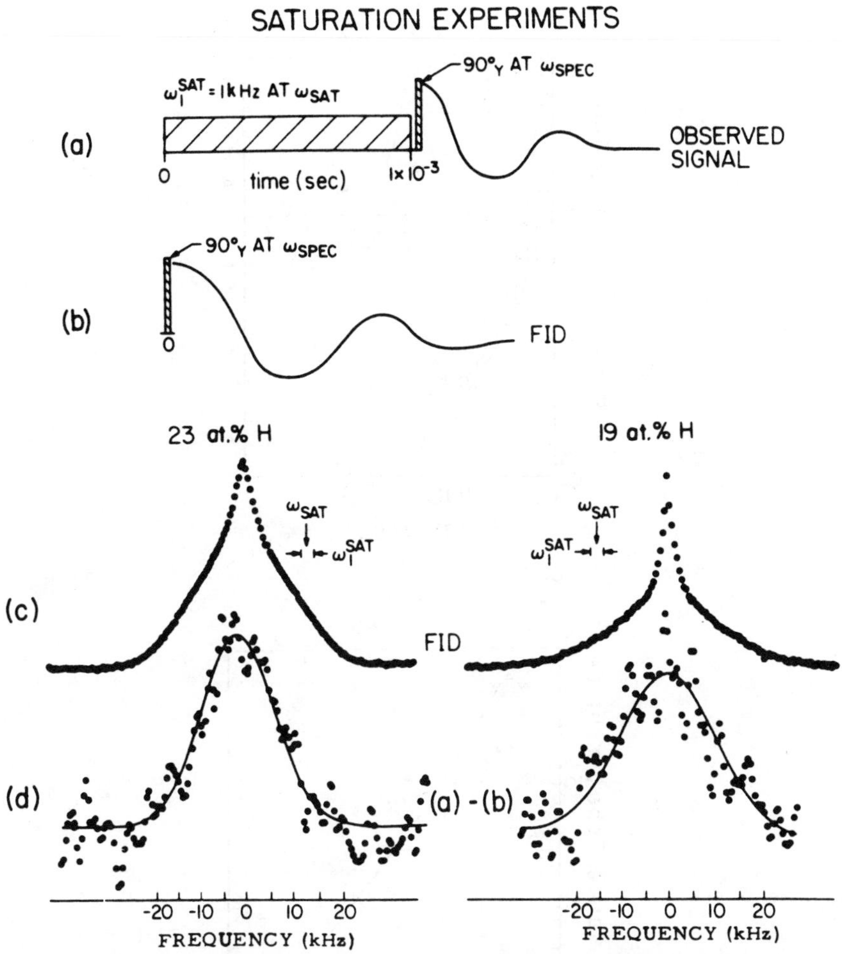

FIG. 2. (a) Schematic diagram of pulse sequence used in hole-burning experiments as described in the text. (b) Usual FID pulse sequence. (c) Unsaturated FID spectra for two representative samples. (d) Difference spectra. The solid lines are Gaussian fits to the data. [From Reimer *et al.* (1981b).]

A similar conclusion was reached by Carlos and Taylor (1982b) who employed "solid echo" experiments to determine that the two hydrogen environments were decoupled on a scale of several hundred microseconds. It can be shown that for an ensemble of nuclei with $I = \frac{1}{2}$, an echo results at time $t = 2\tau$ after the application of two 90° pulses at $t = 0$ and $t = \tau$, whenever the phase of the second pulse is 90° with respect to that of the first. A representative echo is shown in the insert of Fig. 3a. The main portion of Fig.

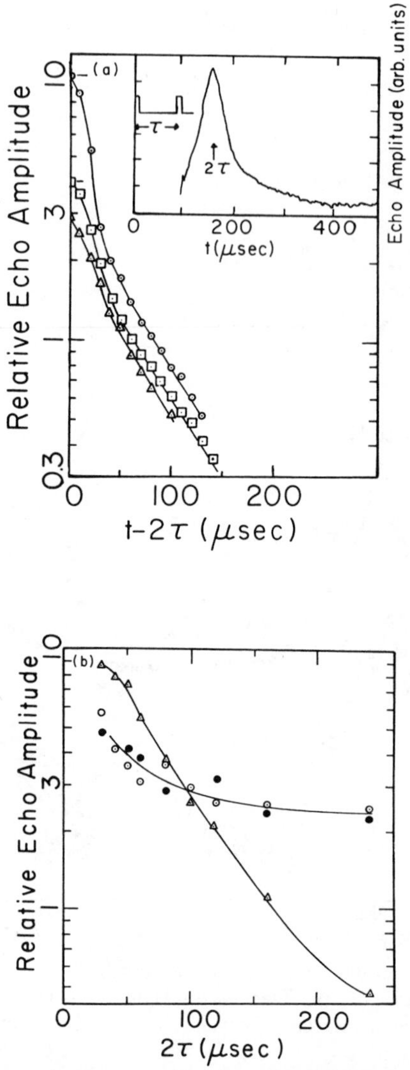

FIG. 3. (a) Solid echo amplitudes as obtained from the pulse sequence described in the text. The inset shows a representative echo and a schematic of the pulse sequence with the pulse widths exaggerated. The main figure displays results for three different values of the pulse separation: $2\tau = 40$ μsec (⊙), 120 μsec (⊡), 240 μsec (△). (b) Decay in the echo amplitude as a function of pulse separation for the two components of the echo. △, broad line; ⊙ and ●, narrow line. [From Carlos and Taylor (1982b).]

3a indicates one half of the symmetric echo as a function of the pulse separation. Figure 3b shows the relative heights of the two ^1H NMR components as a function of the pulse separation. Because the two components decay at markedly different rates, one may conclude that there is negligible spin communication (spin diffusion) between the two lines on the time scales of the experiments. This decoupling implies that the hydrogen atoms that compose one line are physically separated from those of the other line. Reimer *et al.* (1981b) estimate this separation to be > 5 Å, although the exact value is unknown and may vary from film to film.

Perhaps the most important challenge presented by the ^1H NMR experiments concerns an understanding of the origin of the two distinct NMR lines and their ubiquitous appearance in films of very different morphologies (as determined by electron microscopy) and local hydrogen bonding arrangements (as determined by infrared and Raman spectroscopy). It is generally accepted that the narrow Lorentzian line is due to hydrogen atoms either randomly placed (Reimer *et al.,* 1981b; Jeffrey *et al.,* 1981a; Jeffrey and Lowry, 1981) or slightly clustered (Carlos and Taylor, 1982b) in the a-Si network at monohydride sites. The term *monohydride* refers to the bonding of only one hydrogen atom on a given silicon atom, and this situation is sometimes indicated by the shorthand notation SiH. In films with high hydrogen content, there sometimes exist silicon atoms that are bonded to two hydrogens (dihydride or SiH_2 groups) or three hydrogens (trihydride or SiH_3 groups). Infrared absorption measurements are useful in discovering which groups are present in the films.

The narrow line constitutes at most ~12 at. % of the sample in any a-Si:H film regardless of the hydrogen content or method of preparation (Reimer *et al.,* 1980; Shanks *et al.,* 1981; Jeffrey and Lowry, 1981). Typical results are indicated for sputtered films in Fig. 4 (Jeffrey and Lowry, 1981) for films that showed no infrared evidence for anything other than monohydride hydrogen sites. The x's indicate the rising contribution of the broad line as the total hydrogen content is increased.

Although the relative proportions of the broad and narrow lines vary with the hydrogen content and method of preparation (Carlos *et al.,* 1981), the linewidths of each site are constant within factors of approximately two ($\sigma \sim 2-3$ kHz for the narrow line and $\sigma \sim 15-30$ kHz for the broad line). Furthermore, these linewidth variations are greater between different methods of preparation than they are between films with varying hydrogen content that are all made in the same manner. These results suggest that the degree of clustering (both the average H–H separation within a cluster and the average cluster size) is essentially independent of the hydrogen content. With increasing H content one simply increases the number of clusters. This two-phase compositional inhomogeneity appears to be independent of film

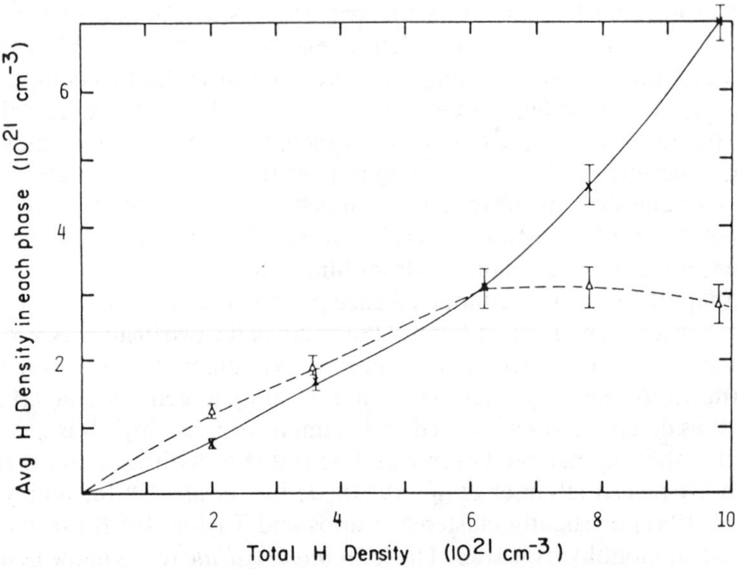

FIG. 4. Variation in the concentrations of the broad (clustered phase) (solid curve with crosses) and narrow (less clustered or distributed phase) (dashed curve with triangles) NMR components as functions of hydrogen concentration for samples that exhibit only monohydride bonding in infrared absorption measurements. [From Jeffrey and Lowry (1981).]

thickness as well (Reimer *et al.,* 1981b) at least down to thicknesses of $\sim 1\ \mu$m.

The interpretation of the broad line is somewhat more ambiguous than the interpretation of the narrow line. Because the broad line occurs in films with dihydride or trihydride sites as well as in films with only monohydride sites, there is no direct correlation between the sites that constitute the broad line and the local bonding configurations indicated by the infrared absorption measurements. Evidently many kinds of clustered environments can contribute to the broad line. The most commonly accepted interpretation of this line is that it is due to the hydrogen atoms bonded on the internal surfaces of small voids (Reimer *et al.,* 1981b; Jeffrey *et al.,* 1981a; Carlos and Taylor, 1982b).

These conclusions are based on calculations of the expected second moment M_2 as given by Eq. (9) for model distributions of H atoms. Calculations of M_2 for mono- and divacancies (Reimer *et al.,* 1981b), even after lattice relaxation (Carlos and Taylor, 1982b), yield linewidths that are too large to be consistent with the observed linewidths of the broad line. Therefore, most of the hydrogen atoms do not reside in sites of these types. Completely hydrogenated cyrstalline surfaces, such as (111) or (110), have hydrogen atoms bonded only to next-nearest-neighbor silicon atoms and yield linewidths that are too small. Carlos and Taylor (1982b) have shown that a broad, random distribution of H–H separations on internal surfaces (including H atoms bonded to nearest-neighbor Si atoms) can account nicely for the width of the broad line. Shanks *et al.* (1981) have shown that such a clustered arrangement is consistent with a Si–H stretch mode at 2090 cm^{-1}. Although there is no compelling reason to invoke such a broad distribution of H–H separations, recent model calculations to be discussed below are capable of producing the appropriate linewidths for such a distribution (Mackinnon, 1980). Such a broad distribution does consistently explain why no new hydrogen NMR line is observed when SiH_2 and SiH_3 groups are present in the films. That is, the addition of polyhydride groups to a broad distribution of nearest-neighbor monohydride separations will not appreciably affect the linewidth.

The two-phase hydrogen clustering persists in films of a-Si:H that are alloyed with significant amounts of C, N, F, or B (Reimer *et al.,* 1981d; Shimizu *et al.,* 1983). Even in pure a-C:H without silicon the two distinct hydrogen NMR lines remain.

At least one important mystery remains concerning the ubiquitous appearance of the narrow and broad ^1H NMR lines in a-Si:H films. What is the extent and the nature of the buffer regions that must exist between these two types of hydrogen environments in order that two inhomogeneous lines be observed by NMR? It has been suggested (Reimer and Knights, 1981) that this separation arises from regions of high strain during film growth that "getter" H from the neighboring a-Si network and subsequently form voids to relive the local strain. Although this hypothesis is plausible, there is currently little specific evidence with which to test it.

Model calculations (Mackinnon, 1980; Mackinnon and Kramer, 1981) that employ a random network (Weaire *et al.,* 1979) have been utilized to calculate the expected ^1H NMR linewidths. In these calculations the atomic positions of the original "hand-built" model were relaxed to the energy minimum given by a Keating Hamiltonian that was modified to include a repulsive term between nonbonded atoms and a coulombic term to simulate the polarization of the Si–H bond. With these two extra terms, linewidths appropriate to the broad ^1H NMR line can be obtained, but unfortunately these calculations provide no indication of the origin of the narrow line. As mentioned above, it is interesting to note that the broad line is produced from an H–H distribution that is broad and essentially structureless.

Carbon is sometimes added to a-Si:H alloys to increase the optical band gap. Studies of ^1H NMR in these films (Reimer *et al.,* 1981d; Nakazawa *et al.,* 1982) have shown that the hydrogen incorporated in the alloys increases with the carbon content. This additional hydrogen is incorporated almost exclusively in the more clustered (broad line) component. As might be expected, the fraction of H bonded to C increases with carbon content as well.

b. ^{2}D *NMR*

The substitution of deuterium for hydrogen in amorphous silicon alloys provides a very useful probe of bonding because of the extreme sensitivity of the nuclear quadrupolar interaction [Eq. (5)] to changes in local structural order. Unfortunately, the results on deuterated samples are rather limited. Leopold *et al.* (1982) have investigated a sample of a-Si:D,H that was made by plasma deposition from 5% silane and 95% molecular deuterium on a room temperature substrate. Both ^{2}D and ^{1}H NMR measurements indicated that this sample contained 7 at. % H and 24 at. % D. This film exhibits clear structural differences from the more stable a-Si:H films that are deposited on high temperature substrates and contain ~ 5 – 10 at. % H; nonetheless, the results are enlightening.

Leopold *et al.* (1982) observe two distinct deuterium lines, one of which is relatively narrow and one of which exhibits a resolved quadrupolar splitting. Figure 5 shows the Fourier transform of the ^{2}D echo observed after a $90° - \tau - 90°$ pulse sequence in which the rf phase of the second pulse is 90° with respect to that of the first. The solid line in the 39°K trace of Fig. 5 represents a calculated powder pattern of the broad (Pake doublet) component of the ^{2}D line. This component comprises 21 at. % of the sample. The NMR signal from the rest of the deuterium (3 at. %) is contained in the narrow portion of the echo, which has a temperature-dependent width. It is tempting to conclude that these two components of the ^{2}D NMR correspond to the narrow and broad components of the ^{1}H NMR.

The broad line is easily explained because the width corresponds to a quadrupole coupling constant $e^2qQ/h = 88$ kHz, which is consistent with the observed infrared vibrational frequency for D in a-Si:H,D of ~ 1460 cm^{-1}. This consistency is obtained through the existence of an empirical scaling relation between the quadrupole coupling constants and the vibrational force constants for hydrogen-bonded systems (Leopold *et al.*, 1982). In other words, the broad line of Fig. 5 is consistent with deuterium bonded to Si in the a-Si network.

The narrow line is less easy to understand. At 4.2°K the quadrupolar contribution to this line shape is the most important, but it is significantly less than that of the broad line. This fact has lead Leopold *et al.* (1982) to term this component "weakly bound," although the most conservative statement one can make concerning this component is that it results from deuterium that is not bonded to silicon in the fashion commonly observed in

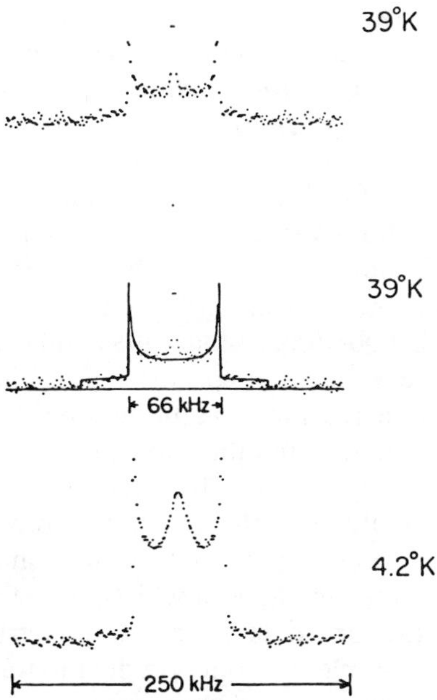

FIG. 5. Fourier transforms of deuterium NMR at 30 MHz for a quadrupolar echo sequence (90-τ-90_{90} pulse sequence). The doublet has a splitting of 66 ± 1 kHz, and the solid lines in the middle trace indicate the fit for the doublet portion only, assuming axial symmetry ($\eta = 0$). The narrow signals at the center of the traces correspond to 3 at. % D, whereas the doublet component corresponds to 21 at. % D. [From Leopold *et al.* (1982).]

diatomic molecules and the usual hydrogen-bonded solids. The exact nature of the narrow deuterium line, remains a mystery.

c. ^{29}Si *NMR*

In principle, the study of those 5% of the silicon nuclei that possess a magnetic moment (^{29}Si) should yield valuable information concerning the silicon bonding through an analysis of the chemical shift interaction [Eq. (4)], but the results are ambiguous for several reasons. First of all, the NMR signals are weak so that thick samples (~50 μm) must be employed and special "cross polarization" techniques with ^{1}H adopted in order to obtain results. In addition, "magic angle spinning" must be utilized to eliminate dipolar [Eq. (3)] contributions to the NMR line shape. Finally, the most important reason for the ambiguities is that the results depend dramatically on the samples that are employed.

Several groups have examined ^{29}Si NMR in both sputtered (Lamotte *et al.*, 1981a,b; Jeffrey *et al.*, 1981b) and glow-discharge (Reimer *et al.*, 1981a) films of a-Si : H. In all cases the chemical shift contributions to the lines are much broader ($\Delta H/H_0 \sim 50$ ppm, where H_0 is the applied magnetic field and ΔH is the FWHM) than in typical crystalline solids, so that different silicon sites cannot be resolved in most cases. More importantly, the chemical shift itself [Eq. (4)] is very dependent on the sample preparation conditions. Lamotte *et al.* (1981a,b) find in sputtered samples that the peak of the resonance is shifted 80 ppm with respect to a standard reference sample (tetramethyl silane) as shown in Fig. 6. Jeffrey *et al.* (1981b) find also in sputtered samples that the peak is shifted 62 or 75 ppm depending on whether there is only SiH bonding or a mixture of SiH and SiH_2 bonding (as determined by infrared absorption spectroscopy). Finally, Reimer *et al.* (1981a) find in glow-discharge films that the peak is shifted 42 or 50 ppm for films deposited on the cathode and the anode, respectively.

Lamotte *et al.* (1981b) report resolvable structure in some spectra such as that exhibited in Fig. 6, although this structure has not been reported by other groups. These authors suggest that the resonance of Fig. 6 can be resolved into five components, three of which can be attributed to Si–Si, SiH, and SiH_2 configurations. Given the major discrepancies between measurements on different samples it is only prudent to wait for further results before any definite conclusions are drawn.

There is general agreement that the frequency (or field) shift increases as one adds di-, tri-, or polysilane [$(SiH_2)_n$] groups to the films, but the absolute shift is apparently not well determined in a-Si : H made under different conditions. This fact contrasts dramatically with the ^1H NMR results, in which the two hydrogen lines are present essentially unaltered in all the films that have been investigated. The reason for these variations in the case of ^{29}Si is undoubtedly that the silicon bonds are distorted in a-Si : H and that these

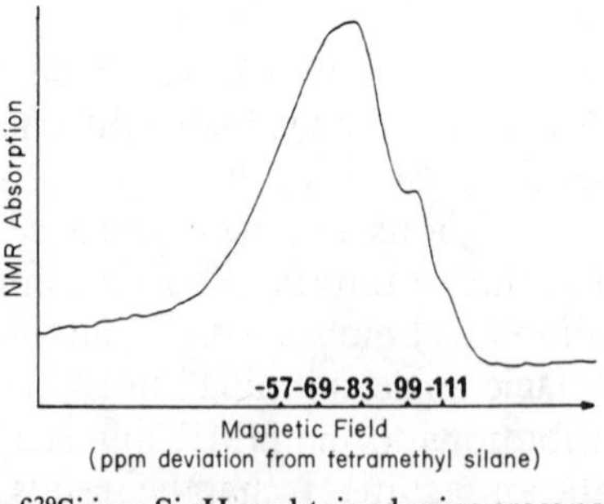

FIG. 6. NMR spectrum of ^{29}Si in a-Si : H as obtained using cross polarization and magic angle spinning techniques. The numbers represent a suggested decomposition into five separate contributions. [From Lamotte *et al.* (1981b).]

distortions vary dramatically with sample preparation. Because all of the films employed in these experiments were rather thick (~50 μm) and were removed from the substrates, it is questionable what conclusions can be drawn concerning silicon bonding in thin (~1 μm) films deposited on high temperature substrates.

d. ^{19}F *NMR*

Fluorine is sometimes employed in addition to hydrogen as a "passivation agent" in amorphous silicon films. The first ^{19}F NMR experiments on fluorinated films were performed by Ueda *et al.* (1981a) who found a single broad line in as-deposited, sputtered samples. After annealing above ~450°C, a narrow ^{19}F line appeared in addition to the broad component. Similar effects have been observed in glow-discharge films (Ueda *et al.*, 1982). The linewidths for these two components are in the ranges 2–5 kHz and 15–34 kHz for the narrow and broad components, respectively. As in the case of ^{1}H NMR, the broad line is evidence for fluorine clustering in the films (Ueda *et al.*, 1981a; Shimizu *et al.*, 1982a, 1983).

Unlike the hydrogen situation, the narrow ^{19}F NMR line is usually observed to have a temperature-dependent linewidth. The "motional narrowing" effect has been attributed to fluorine atoms bonded to SiF_4 or $(SiF_2)_n$ groups, which can undergo local motion in these films (Ueda *et al.*, 1981b, 1982). In some cases, especially when hydrogen is not present (i.e., a-Si:F), a narrow ^{19}F line is observed that is temperature independent at low temperatures (Ueda *et al.*, 1982). This line may be analogous to the narrow line observed in ^{1}H NMR experiments and represent fluorine atoms in less clustered environments.

e. ^{11}B *NMR*

In addition to the major constituents of the amorphous silicon alloys such as silicon, hydrogen or deuterium, carbon, and fluorine, dopants have also been investigated using NMR techniques. Boron is routinely employed as a *p*-type dopant in a-Si:H, but only a small fraction (≤0.1%) of the boron atoms is effective in doping. The ^{11}B NMR experiments (Greenbaum *et al.*, 1982, 1983) provide a reasonable explanation for the low doping efficiency because the line shape indicates that the vast majority of the boron atoms is three coordinated. Elementary bonding considerations would indicate that three-coordinated boron atoms, which have all of their valence requirements satisfied, would not produce electrically active states in the gap, although Robertson (1983) has suggested that three-coordinated boron in a-Si:H can indeed produce deep-gap states.

Two samples of B-doped a-Si:H have so far been investigated using the quadrupolar spin echo technique—one containing ~10 at. % B and one containing ~0.7 at. % B (Greenbaum *et al.*, 1983). Figure 7 shows the echo envelope (a) and the Fourier transformed line shape in frequency space (b) for the sample containing ~ 10 at. % B. There are two components to the line shape of this figure, the broader line appearing as shoulders on the narrow line. The major source of broadening for both of these lines is the nuclear quadrupolar interaction of Eq. (5) ($I = \frac{3}{2}$ for ^{11}B).

For the central transition ($m = +\frac{1}{2} \rightarrow -\frac{1}{2}$) for ^{11}B, second-order perturbation theory yields a linewidth proportional to $[(e^2qQ)]^2/\nu_0$, where ν_0 is the spectrometer operating frequency. Greenbaum *et al.* (1982) assumed an

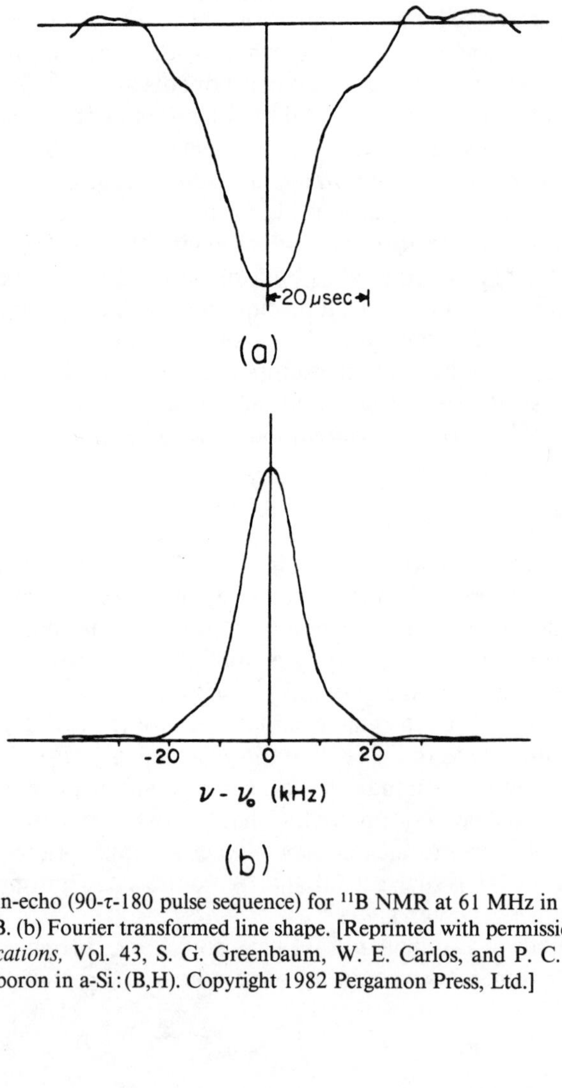

FIG. 7. (a) Spin-echo (90-τ-180 pulse sequence) for ^{11}B NMR at 61 MHz in a sample that contains ~ 10% B. (b) Fourier transformed line shape. [Reprinted with permission from *Solid State Communications*, Vol. 43, S. G. Greenbaum, W. E. Carlos, and P. C. Taylor, The coordination of boron in a-Si:(B,H). Copyright 1982 Pergamon Press, Ltd.]

axially symmetric field gradient and demonstrated the expected dependence of the linewidth on ν_0^{-1} for both the broad and narrow features in Fig. 7b. By comparison with quadrupolar echoes observed in materials with known coupling constants, these authors estimated values of e^2qQ/h of ~5.8 and ~3.5 MHz for the broad and narrow lines, respectively. For the sample with ~0.7 at. % B, Greenbaum *et al.* (1983) observed only the narrow line.

The magnitude of these quadrupolar interactions for ^{11}B in a-Si:H,B films provides convincing evidence that most of the ^{11}B nuclei (≥90%) are not tetrahedrally coordinated. Even in highly distorted tetrahedral environments, boron coupling constants in borate minerals and boron-containing molecules are always less than ~1 MHz.

Greenbaum *et al.* (1982) suggest that the two boron sites observed in the NMR result from planar, three-coordinated boron in BSi_3 (broad line, 5.8 MHz coupling constant) and Si_2BH (narrow line, 3.5 MHz coupling constant) structural groupings. If this interpretation is correct, then the absence of the broad line in the films with 0.7 at. % B indicates that essentially all the boron in this sample is bonded to at least one hydrogen atom. The interpretation also suggests that hydrogen and boron bond preferentially, a fact that is corroborated by infrared absorption measurements (Tsai, 1979; Shen and Cardona, 1981).

The decay of the 90° – 180° spin echo as a function of the pulse separation τ is a measure of the dipolar interaction between mutally resonant boron atoms in the a-Si:H,B films. By comparison with samples in which the boron is known to be randomly distributed, Greenbaum *et al.* (1982, 1983) have shown that, like the hydrogen atoms, the boron atoms in both sites are clustered in these inhomogeneous materials. From estimates of boron–hydrogen heteronuclear dipolar interactions, Reimer *et al.* (1981b) have also inferred that the boron atoms are clustered in heavily B-doped films. Although a detailed description of the clustering is as elusive as in the case of hydrogen, the ^{11}B echo decay experiments do place limits on the degree of boron clustering. In particular, these decays suggest that few B–B bonds, either in the a-Si network or in B_2H_6 molecules, exist in the films.

f. ^{31}P *NMR*

The dominant *n*-type dopant used in a-Si:H films is phosphorus. Reimer and Duncan (1983) have investigated the local bonding of this dopant using ^{31}P NMR techniques. Different bonding environments for phosphorus can be deduced on the basis of the chemical shift interaction [Eq. (4)]. Heteronuclear (P–H) dipolar effects [Eq. (3)] can be employed to suggest the type of atom to which the phosphorus is bonded.

A typical free induction decay spectrum for ^{31}P in a sample of P-doped (~2 mol %) a-Si:H is shown at the top of Fig. 8. It can be shown by magic

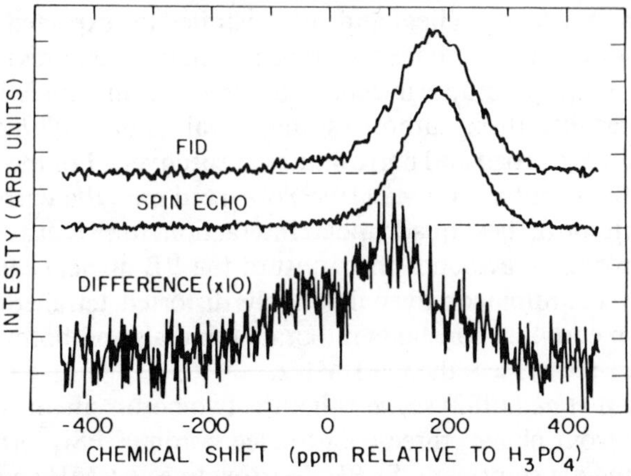

FIG. 8. NMR of ^{31}P in P-doped a-Si : H with 1.8 mol % P. The top trace is from the FID and the middle trace from an echo (90-τ-180 pulse sequence) experiment. The bottom trace represents the difference between the top two traces as described in the text. [From Reimer and Duncan (1983).]

angle spinning and echo experiments that this line, which is broad and slightly asymmetric, is primarily the result of a distribution of chemical shifts at different P sites in the film. This conclusion is similar to that which was drawn in the case of silicon (see Subsection 1c).

The middle trace in Fig. 8 is the line shape observed using the spin echo technique at a pulse separation of $\tau = 0.25$ msec ($180° - \tau - 90°$ pulse sequence). Reimer and Duncan (1983) have determined that the FID and echo line shapes of Fig. 8 are different, and the difference spectrum is plotted at the bottom of Fig. 8. Furthermore the T_2 as measured by the echo technique is very different for these two components. The spectrum with the longer T_2 ($\simeq 1.4$ msec) in the middle of Fig. 8 is best fit with a Lorentzian line shape centered at 175 ppm with respect to a standard reference (^{31}P in 85% H_3PO_3). The broad line at the bottom of Fig. 8 is best fit with a Gaussian line shape centered at 70 ppm.

A comparison with known chemical shifts in liquids, just as was attempted in the case of silicon (Section 1c), suggests that the narrow peak at 175 ppm is probably due to threefold-coordinated phosphorus and the broad peak at 70 ppm to fourfold-coordinated phosporus. The ratio of three- to four-coordinated phosphorus atoms in this sample is approximately 4 to 1 and is typical of samples containing small amounts ≤ 2 mol %) of phosphorus.

The echo decay of the three-coordinated phosphorous is well accounted for by P–P dipolar coupling for a random distribution of P atoms in the film. This fact suggests that the three-coordinated P atoms are bonded exclusively to Si atoms. The faster decay of the four-coordinated site is attributed by

Reimer and Duncan (1983) to P–H heteronuclear dipolar coupling, which requires clustered H atoms in the neighborhood of the four-coordinated site. For this reason these authors suggest that the four-coordinated sites form primarily near the surfaces of microvoids where the clustered hydrogen is thought to occur. In samples compensated with boron, a slight narrowing of the line attributed to the four-coordinated sites is observed. This narrowing is taken as an indication of the existence of B–P complexes in the compensated films (Reimer and Duncan, 1983).

2. Comparison with Other Experiments

The most significant gap in our understanding of the two ^1H NMR sites and their relationship to the morphology of the a-Si:H films concerns the lack of detailed correspondence between these two sites and the local bonding configurations as probed by infrared absorption measurements (Brodsky *et al.,* 1977; Lucovsky *et al.,* 1979; Freeman and Paul, 1978). There is a strong vibrational mode at ~2000 cm^{-1} that is commonly accepted as due to stretching vibrations of Si–H bonds for isolated (monohydride) hydrogen atoms bonded to four-coordinated silicon atoms. Some of these sites correspond to the hydrogen in the narrow NMR line, but the correspondence does not appear to be one-to-one.

There is a second infrared absorption feature at ~2080 cm^{-1} that is also due to hydrogen and attributed in part to SiH_2 sites (dihydride), where two hydrogens are bonded to the same tetrahedrally coordinated silicon. However, some of the intensity in this absorption may also be due to Si–H sites (Paul, 1980). For this reason, there appears to be no one-to-one correspondence between the broad line in NMR and the 2080 cm^{-1} absorption peak in the infrared absorption.

The NMR experiments are hindered because they can only place limits on the average H–H separations, but these experiments can give the total hydrogen contribution to each line rather accurately. On the other hand, the infrared experiments can, in principle, give detailed information concerning the local silicon–hydrogen bonding, but estimates of absolute intensities are very difficult to make because they involve unknown matrix elements. A coherent and quantitative comparison between the infrared and NMR results would greatly improve our understanding of both the local structural order and the inhomogeneities in a-Si:H films.

3. Hydrogen Diffusion and Evolution

In well-annealed films deposited on high temperature substrates (~200–300°C) at reasonable deposition rates, there is no evidence for hydrogen diffusion or evolution at temperatures up to ~300°C (Reimer *et al.,* 1981b, 1981e; Carlos and Taylor, 1981). When hydrogen evolution does occur in these well-annealed films, the clustered phase is observed to evolve first with little change in either the intensity or the width of the narrow line with temperature until the highest temperatures (≥500°C). No motional narrowing of the ^{1}H NMR lines is observed in these films (Carlos and Taylor, 1981). An increase in the ESR occurs as the hydrogen goes off (see Section 6).

However, in films with high hydrogen content (~20 at. %), which are in general less stable, the ^{1}H NMR provides direct evidence for local internal diffusion before major evolution of hydrogen occurs (Reimer *et al.*, 1981e). In Fig. 9 the hydrogen concentration of the broad and narrow lines is shown as a function of annealing for 20 min at the indicated temperatures. Just as in the more stable films with lower hydrogen content, the hydrogen that constitutes the broad component evolves first, and only at the highest temperature (≥500°C) does appreciable hydrogen from the narrow line evolve. However, in these films the linewidth of the narrow line actually decreases dramatically between 200 and 400°C, a fact which indicates that the hydrogen atoms in this phase are being redistributed into a more uniform (less clustered) configuration over this temperature range. The redistribution is accompanied by a decrease in the ESR spin signal, which implies that the hydrogen diffuses

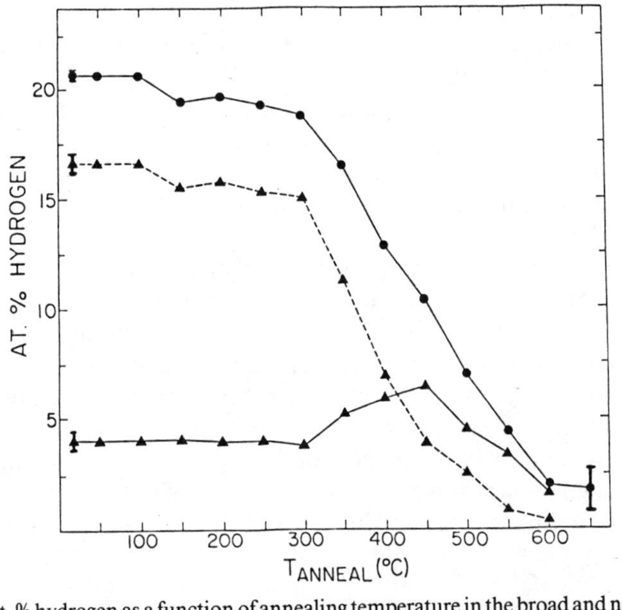

FIG. 9. At. % hydrogen as a function of annealing temperature in the broad and narrow NMR lines. The a-Si:H sample was deposited at a power of 18 W on a room-temperature substrate. —●—, total; —▲—, narrow component; --▲--, broad component. [Reprinted with permission from *Solid State Communications*, Vol. 37, J. A. Reimer, R. W. Vaughan, and J. C. Knights, Proton NMR studies of annealed plasma-deposited amorphous Si:H films. Copyright 1981, Pergamon Press, Ltd.]

locally to existing defects. There is even some evidence for an increase in the total hydrogen concentration in the narrow line between 300 and 450°C. In films of high unpaired spin density ($> 10^{19}$ cm^{-3}) and high hydrogen content, Reimer *et al.* (1981b) have found that internal diffusion of the hydrogen occurs at room temperature on a time scale of months. As mentioned previously, most of these internal diffusion processes are absent in the more stable, low hydrogen-content films, although optically induced metastabilities persist as discussed in Sections 7 and 8.

4. Spin–Lattice Relaxation and Defects

Nuclear spin–lattice relaxation is the process by which the magnetization of an ensemble of nuclei approaches thermal equilibrium after excitation by an appropriate pulse of rf energy. The relaxation process is usually, although by no means always, an exponential decay of rate T_1^{-1}. For nuclei of spin $I > \frac{1}{2}$ the relaxation often results from the modulation of the electronic field gradient [Eq. (6)] by a Raman process involving either phonons or some other lattice excitations. For nuclei of spin $I = \frac{1}{2}$ the relaxation may involve a modulation of the dipolar [Eq. (3)] or the chemical shift [Eq. (4)] interactions, usually by a Raman process. Sometimes paramagnetic states in the gap or impurities control the spin–lattice relaxation in a process that involves direct relaxation of nuclei near the paramagnetic centers and nuclear spin diffusion of the magnetic excitation to the neighborhood of the paramagnetic centers. (For a recent review, see Taylor, 1983.)

a. ^{1}H *NMR*

In the first extended discussion of ^{1}H spin–lattice relaxation in a-Si:H, Carlos and Taylor (1980) described a characteristic minimun in the temperature dependence of T_1 near 40°K, which occurs in all glow-discharge deposited samples studied to date. Typical data (Taylor and Carlos 1980; Carlos and Taylor, 1982b) are shown in Fig. 10 for relatively pure samples (top curves) and samples containing ~0.5–2 at. % oxygen (bottom curves). The decays at any given temperature are exponential over one or two orders of magnitude.

At temperatures above the T_1 minimum the data are independent of the frequency at which the measurement is taken, but there exists a frequency dependence at low temperature as shown in Fig. 11. The simplest models, to be discussed, predict that the frequency dependence should go as ν^2 (solid line in Fig. 11), but the actual dependence is somewhat weaker.

The first interpretation of these data was in terms of relaxation via tunneling or disorder modes (Carlos and Taylor, 1980; Reimer *et al.*, 1981c; Jeffrey *et al.*, 1981a), which were originally invoked to explain anomalies in the low temperature thermal properties of amorphous solids. Movaghar and

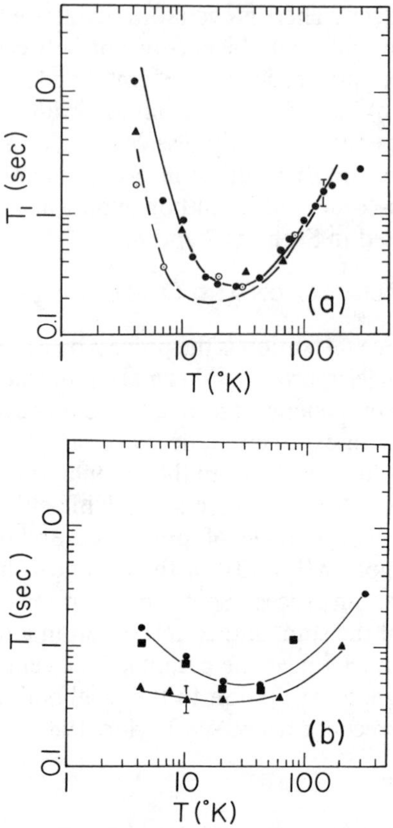

FIG. 10. (a) T_1 as a function of temperature for two samples of a-Si : H at 42.3 MHz (solid line and solid points) and 12.3 MHz (dashed line and open circles). The curves are fits obtained from the model of Conradi and Norberg (1981). ▲, BNL 94; ●, RCA; ⊙, RCA (12.3 mHz). (b) T_1 as a function of temperature for samples containing oxygen impurities. Solid curves are an aid to the eye. ■, NRL; ●, CHI; ▲, BNL95. [From Carlos and Taylor (1982b).]

Schweitzer (1981) pointed out, however, that this mechanism would require that ~ 10% of the hydrogen atoms be involved in the disorder modes and this large fraction should have been directly observable in the NMR experiments. Ngai (1981) has suggested that the relaxation is controlled by "low frequency fluctuations" in the disordered matrix.

The explanation presently accepted is that of Conradi and Norberg (1981), who suggested that the relaxation proceeds via molecular hydrogen (H_2) that is somehow trapped in the amorphous silicon matrix. In this model a small number (~ 1%) of the hydrogen atoms exists as molecular hydrogen, which act as relaxation centers with most of the bonded hydrogen connected to

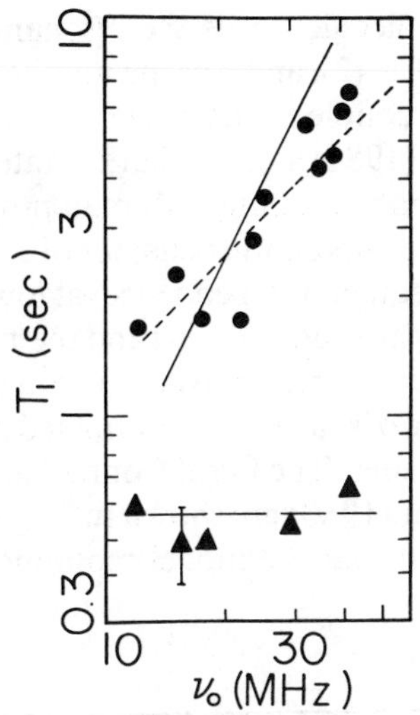

FIG. 11. T_1 as a function of frequency at 5 (●) and 77 (▲) °K. The solid and dashed lines are, respectively, quadratic and linear fits to the data as explained in the text. [From Carlos and Taylor (1982b).]

these centers via nuclear spin diffusion. The direct relaxation of the molecular hydrogen to the lattice then produces the minimum in T_1 and the frequency dependence at low temperatures in a manner directly analogous to the mechanism proposed by Fedders (1979) for relaxation of molecular hydrogen trapped in rare-gas solids (Conradi *et al.*, 1979).

For the bonded hydrogen, T_1 is given as

$$T_1^{\mathrm{H}} = \tfrac{3}{8}\, T_1^0(n_{\mathrm{H}}/n_0) + T_1^{\mathrm{SD}}, \tag{11}$$

where T_1^0 is the spin–lattice relaxation time for molecular hydrogen that is in the ortho ($I = 1$) configuration (parahydrogen has zero nuclear spin), n_{H} the total concentration of hydrogen in the sample, n_0 the concentration of orthohydrogen, and T_1^{SD} a spin-diffusion limited term that can produce a spin-diffusion bottleneck.

The orthohydrogen molecules are relaxed rapidly by modulation of their dipole–dipole interaction via rotational motion of the molecule. The rotational transitions modulate the dipolar interaction within the molecule and lead to a relaxation rate for the orthohydrogen of the form

$$\frac{1}{T_1^0} = \frac{3}{5} n_{\mathrm{r}} \omega_{\mathrm{d}}^2 \left(\frac{\tau}{1 + \omega_0^2 \tau^2} + \frac{4\tau}{1 + 4\omega_0^2 \tau^2} \right), \tag{12}$$

where ω_{d} is the dipolar coupling between protons in an H_2 molecule ($\omega_{\mathrm{d}} = 3.6 \times 10^5$ Hz), τ^{-1} is the rate at which the lattice induces rotational transitions in the H_2 molecule, ω_0 is the resonant angular frequency obtained by diagonalizing Eq. (2), and n_r is the number of the rotational modes of the H_2 molecule that couple to the nuclear spin system ($2 \leq n_r \leq 8$).

Conradi and Norberg (1981) assume that the rate τ^{-1} is that for a quadrupolar nucleus relaxing via two-phonon Raman processes. This assumption affects primarily the low temperature behavior of T_1^H, but not the value of T_1 at the minimum or its frequency dependences at low and high temperatures, which are determined by the ratio (n_H/n_0) and the τ_0 dependence of Eq. (12) for $\omega_0\tau \ll 1$ and $\omega_0\tau \gg 1$, respectively. The term T_1^{SD} provides a rate below which the relaxation is no longer dominated by rapid spin diffusion to the molecular hydrogen sites. The fits of Conradi and Norberg (1981) to the data of Carlos and Taylor (1980) are shown in Fig. 12 for $\omega_0/2\pi = 42.3$ and 12.3 MHz. It is apparent that the model reproduces the temperature and

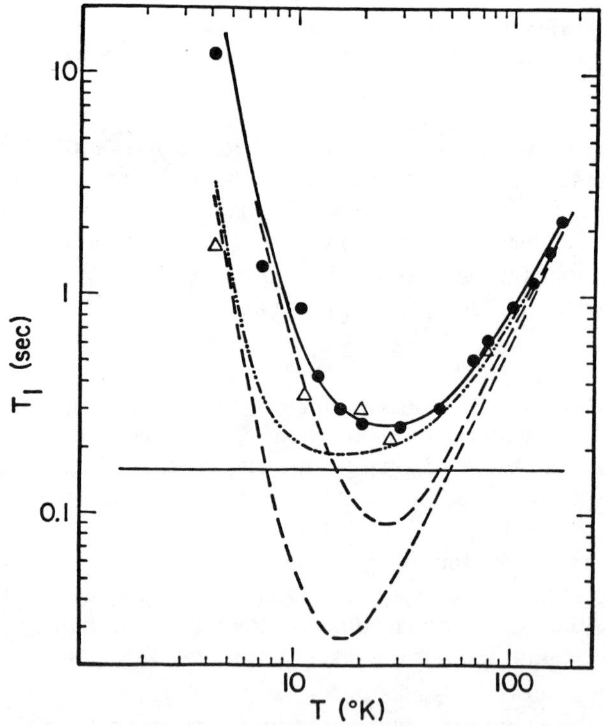

FIG. 12. The data for T_1 as a function of T of Carlos and Taylor (1980, 1982b) at 42.3 MHz (solids points) and 12.3 MHz (open triangles). The solid and dot-dashed lines are fits to the Conradi and Norberg (1981) model including the spin diffusion term of Eq. (11). The dashed lines are the curves without the spin diffusion term. [From Conradi and Norberg (1981).]

frequency dependences fairly accurately if one assumes ~ 1% of the hydrogen atoms in a-Si:H exist as trapped molecular hydrogen.

The most important test of this model is to look for the conversion of orthohydrogen to parahydrogen (the ground state), which is a slow bimolecular process at low temperatures because the conversion is driven by the interaction between protons on adjacent H_2 molecules. At 300°K the ortho/para ratio is $\frac{3}{4}$, and because of the slow conversion rates this ratio is essentially frozen in at low temperatures during the normal running time of an experiment (~ 1 – 10 hr). On a larger time scale, however, the conversion should take place and the relaxation rate will decrease as n_0 decreases.

Carlos and Taylor (1982a) have shown that the expected increase in T_1 does indeed occur on a time scale of months. The inset to Fig. 13 is a plot of the increase in T_1 at the minimum (at ~ 40°K in temperature) as a function of the time that the sample is "annealed" at 4.2°K. The linear increase in T_1 confirms the bimolecular nature of the process, and the rate is consistent with that which is observed for H_2 in rare-gas solids (Conradi *et al.*, 1979). The much faster return to the high-temperature equilibrium of the minimum value of T_1 after annealing at 300°K is also shown in Fig. 13. Although some minor difficulties remain (Carlos and Taylor, 1982b), these low temperature annealing experiments provide firm evidence for the presence of H_2 molecules in a-Si:H films made by the glow-discharge technique.

A similar minimum is sometimes observed in sputtered films (Carlos *et*

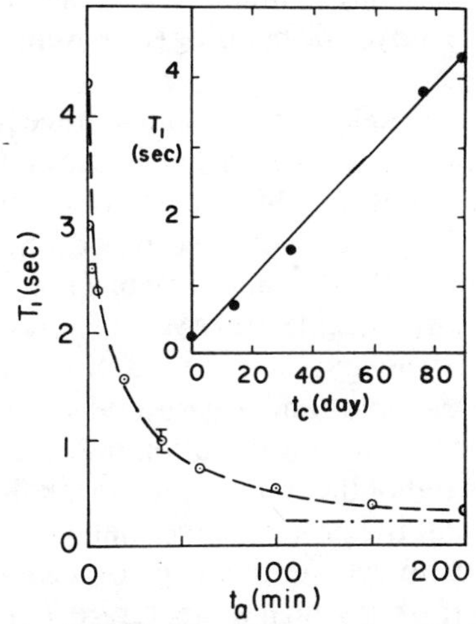

FIG. 13. The inset shows the change in the minimum value of T_1 (at approximately 30°K) as a function of the time held at 4.2°K. The main figure displays the return of the minimum value of T_1 to its high temperatuare (> 200°K) equilibrium value as a function of annealing time at 300°K. [From Carlos and Taylor (1982a).]

al., 1981; Jeffrey *et al.*, 1981a) depending on the hydrogen concentration and the deposition conditions. In some sputtered films much longer relaxation times are observed, which decrease monotonically with increasing temperature. The magnitude of these rates in comparison to the ESR spin densities, as well as the temperature dependences, suggest that these relaxation rates are caused by silicon dangling bonds (Lowry *et al.*, 1981a; Carlos *et al.*, 1981).

In most samples the ^1H spin–lattice relaxation rates are exponential and identical for both hydrogen sites that are observed in the free induction decay. This result is probably a consequence of the rapid nuclear spin diffusion process of the ^1H nuclear magnetization to the neighborhood of the H_2 sites. Reimer *et al.* (1981c) have measured spin–lattice relaxation rates by a technique that suppresses the nuclear spin diffusion processes (the so-called T_{1y} technique). In these measurements one observes a nonexponential decay, as would be expected because those hydrogen atoms closest to an H_2 molecule will relax faster in the absence of rapid spin diffusion. These T_{1y} experiments also indicate that the narrow ^1H NMR line has a relatively faster relaxation rate, which indicates the presence of a greater density of H_2 molecules in this phase than in the clustered (broad NMR line) phase.

In some sputtered samples, double-exponential spin–lattice relaxation rates have been observed (Lowry *et al.*, 1981b). In these samples it has been suggested that faster rates occur at dihydride (SiH_2) bonding sites and slower rates at monohydride sites. This two-component behavior for T_1 has not been reported in glow-discharge samples even when the infrared absorption measurements indicate dihydride bonding sites (Reimer *et al.*, 1981c; Carlos and Taylor, 1980).

At temperatures above 300°K the ^1H spin–lattice relaxation is no longer dominated by H_2 molecules or paramagnetic states. For samples made on high temperature substrates, in which there is essentially no internal rearrangement of the hydrogen between the broad and narrow lines before evolution (see Section 3), Carlos and Taylor (1981) find no measurable hydrogen is driven off until heating to above ~673°K (~400°C). At elevated temperatures (≥300°K) the spin–lattice relaxation rates again increase with increasing temperature. Carlos and Taylor (1981, 1982b) suggest that this increase in rates is due to some local motion of the hydrogen precursory to macroscopic diffusion out of the samples. Although the data are not reliable enough to be certain that the process is thermally activated, if one analyzes the data in this fashion an activation energy of 0.2 eV is obtained.

In the process of performing high temperature T_1 measurements, Carlos and Taylor (1981, 1982) reexamined the low temperature behavior of T_1 after each excursion (~6 hr) to an elevated temperature. As shown in Fig. 14 there is little change in the low temperature T_1 minimum up to an annealing

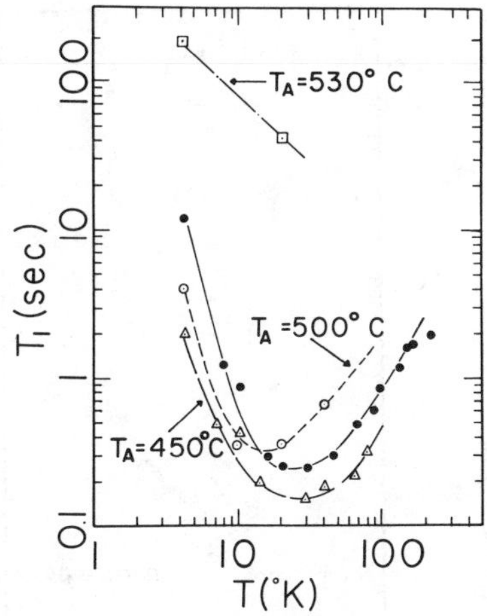

FIG. 14. Low temperature dependence of T_1 for ^1H NMR after annealing at elevated temperatures. The curves are an aid to the eye. Solid circles represent data prior to annealing. [From Carlos and Taylor (1981).]

temperature of ~773°K (~500°C) even though about two thirds of the hydrogen has evolved from the film by that time. After annealing at ~803°K (~530°C) the minimum has abruptly disappeared. This fact indicates that the H_2 relaxation centers must all be trapped in quite similar environments to evolve over such a narrow temperature range. In addition the trapping sites must be small enough, such as microvoids, to trap such a small molecule in an arrangement that is stable up to 500°C. The exact nature of this trapping site remains one of the many unsolved mysteries concerning the defect structure of a-Si:H films (see Part III).

b. ^{2}D *NMR*

When deuterium is employed to displace partially the hydrogen in a-Si:H films, there is evidence that a fraction of the deuterium also gets trapped in the films as D_2 (Leopold *et al.*, 1982). The sample examined by Leopold *et al.* (1982) was made by plasma deposition on a room temperature substrate from 5% silane and 95% D_2 gas. The total hydrogen and deuterium concentrations were 7 and 24 at. %, respectively. Although the results on this sample are very informative, one must be a little careful in making comparisons with films made on high temperature substrates that have lower hydrogen (or deuterium) concentrations.

Figure 15 summarizes the ^{2}D measurements of Leopold *et al.* (1982). The narrow deuterium line described in Section 1b exhibits a T_1 minimum similar to that observed for the two hydrogen sites. Data are shown at 30 and

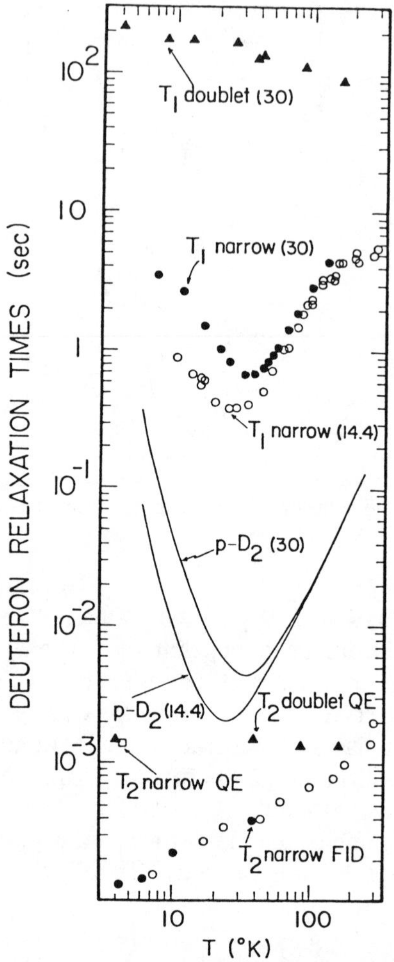

FIG. 15. T_1 for deuterium NMR in a-Si:H. From the top down, the curves display T_1 values for the doublet component at 30 MHz as described in the test (filled triangles), the narrow NMR component at 30 MHz (filled circles) and 14.4 MHz (open circles), and calculations for deuterium in isolated paradeuterium molecules in a solid nonmagnetic host at 14.4 and 30 MHz (solid curves). The bottom portion of the figure gives spin–spin relaxation times T_2 from the doublet quadrupolar echo envelopes at 14.4 MHz (filled triangles) and 30 MHz (open square) and from narrow component measurements which include 14.4. MHz FIDs (open circles) and 30 MHz FIDs (filled circles). [From Leopold *et al.* (1982).]

14.4 MHz in Fig. 15 for this line. These sites are thus relaxed by D_2 molecular impurities (~700 ppm) trapped in the film. The solid lines in Fig. 15 are calculated relaxation rates for the deuterons in paradeuterium molecules. Leopold *et al.* (1982) also find evidence for a spin-diffusion bottleneck, which limits the relaxation rates near the maximum (especially at lower frequencies). There is, in addition, evidence for a contribution to T_1 at low and high temperatures that is probably due to paramagnetic impurities.

The larger fraction of the deuterium (~21 at. %) exhibits a broad doublet line shape (Section 1b), and the values of T_1 for this line are considerably larger as shown at the top of Fig. 15. Leopold *et al.* (1982) speculate that the long T_1 occurs because the quadrupolar splitting greatly reduces the deuterium spin diffusion rate, thus effectively eliminating the paradeuterium molecules as relaxation centers.

III. Electron Spin Resonance

Electron spin resonance (ESR) has been studied in amorphous silicon for many years, long before the deliberate (or not so) introduction of hydrogen to reduce the spin signal. Since the pioneering studies of Brodsky and Title (1969), numerous measurements on ESR in a-Si have been reported. Typically, a-Si exhibits ESR spin densities of $\geq 10^{18}$ cm^{-3}, which have usually been attributed to Si dangling bonds. Excellent reviews of these early measurements on a-Si are available (Stuke, 1976, 1977; Solomon, 1979; Bourgoin, 1981).

When hydrogen is added to the films the spin densities go down dramatically to the point that currently "good quality" films of a-Si:H contain spin densities of $\leq 10^{15}$ cm^{-3}. The ESR measurements in these films have been reviewed by several authors (Voget-Grote and Stuke, 1979; Solomon, 1979; Biegelsen, 1980, 1981; Paul and Anderson, 1981; Bourgoin, 1981,; Yonezawa, 1982).

Although the ESR signal is greatly suppressed in a-Si:H, there appear several strong (10^{16}–10^{17} cm^{-3}) ESR responses that are excited by the application of band-gap light (Knights *et al.*, 1977; Pawlik and Paul, 1977). The properties of these so-called transient light-induced ESR signals have been reviewed by Knights (1977) and Solomon (1979). From studies of these transient light-induced effects, particularly in doped films, one can learn a great deal about the electronically active traps in a-Si:H films. For a review see Street (1981).

Metastable, light-induced ESR responses have been observed (Street *et al.*, 1981; Dersch *et al.*, 1981a–c; Pontuschka *et al.*, 1982) in a-Si:H after irradiation with band-gap light at greater intensities (≥ 100 mW cm^{-2}). Some of these resonances are due to unpaired spins associated with the silicon atoms and some are associated with impurities.

There have also been several novel combinations of ESR with other experimental techniques. Such studies as spin-dependent photoluminescence, spin-dependent photoconductivity, and spin-dependent transient transport provide important information concerning the influence of spin statistics on various electronic processes. Some of these topics are the subject of a separate chapter in this book.

As in the case of NMR, ESR measurements often provide a detailed, local probe of bonding at paramagnetic sites such as defects, impurities, or band tails in amorphous semiconductors. The important terms in the Hamiltonian for most situations of interest in ESR experiments in a-Si and a-Si:H are

$$H = H_{ez} + H_{hfs} + H_{fs}. \qquad (13)$$

In Eq. (13) the first term is the electronic Zeeman interaction, which is of the form

$$H_{ez} = \beta \mathbf{S} \cdot \mathbf{g} \cdot \mathbf{H}, \qquad (14)$$

where $\beta = e\hbar/2m_e c$ is the Bohr magneton, **S** the electronic spin operator, **H** an externally applied magnetic field, and **g** the gyromagnetic or g tensor. The g tensor for an unpaired electron, which is localized to the extent that tight binding calculations apply, can be expressed to second order in perturbation theory as

$$\mathbf{g} = g_e + 2\lambda \sum_n \frac{\langle \psi_0 | L_i | \psi_n \rangle \langle \psi_n | L_i | \psi_0 \rangle}{E_0 - E_n}, \qquad (15)$$

where $g_e \simeq 2.0023$ is the free electron value, λ the spin–orbit coupling constant, which depends on the atom on which the spin resides, ψ_0 the ground state molecular orbital for the unpaired spin, ψ_n an excited state, which is admixed by the ith component of the orbital angular momentum operator L_i, E_0 the ground state energy, and E_n the appropriate excited state energy.

The second term in Eq. (13) is the hyperfine interaction, which couples the electronic spin with surrounding nuclear spins. This interaction can be expressed in the form

$$H_{hfs} = \mathbf{S} \cdot \mathbf{A} \cdot \mathbf{I}, \qquad (16)$$

where **A** is the hyperfine tensor. The elements of **A** depend on the wave function of the unpaired spin and on atomic parameters that essentially determine the coupling of an electron in a given atomic orbital to the nucleus of that atom. In simple tight-binding situations this interaction can be of great help in determining the wave function of the paramagnetic state.

The third term in Eq. (13) is the fine structure interaction that occurs when the effective spin of the paramagnetic center is $S > \frac{1}{2}$. To this extent it is analogous to the quadrupolar interaction discussed in Part II. In parameterized form the fine structure Hamiltonian becomes

$$H_{fs} = \mathbf{S} \cdot \mathbf{D} \cdot \mathbf{S}, \qquad (17)$$

where **D** is the fine structure tensor. The elements of **D** depend upon the symmetries of the wave functions of the "unpaired" electrons and the surrounding electric fields generated by the crystalline symmetry. For example, in the transition metal elements, in which inner shell valance electrons tend to align, the fine-structure term is an important contribution to the total Hamiltonian.

As in the case of NMR, the ESR interactions also depend in general on the orientation of the principal axes of the paramagnetic site with respect to the applied magentic field. Thus in an amorphous solid, powder patterns are also important in ESR, and because of the complexities in forming these averages, perturbation theory is often used to derive approximate solutions.

5. ESR IN a-Si

Essentially all films of amorphous silicon, regardless of the method of preparation, possess more than 10^{18} spins cm^{-3} when examined using the ESR technique. The line shape is a symmetric Lorentzian at the higher spin densities (approaching 10^{20} cm^{-3}) with a characteristic *g*-value of 2.0055 (Brodsky and Title, 1969). At densities below $\sim 10^{19}$ cm^{-3} the line shape possesses some Gaussian character and an asymmetric tail at lower fields (higher *g*-values) as shown in Fig. 16 (Title *et al.*, 1977). When silicon is rendered amorphous by ion implantation, the ESR signal is also anisotropic at low densities (Göetz *et al.*, 1978).

Early experiments determined that the spin signal was a bulk and not a surface effect (Brodsky and Title, 1969), although there are some surface contributions (Suzuki *et al.*, 1980). In measurements of very pure films made under ultrahigh vacuum (UHV) conditions, Thomas *et al.* (1974) showed that the characteristic ESR signal is an intrinsic property of the films and not due to the presence of impurities.

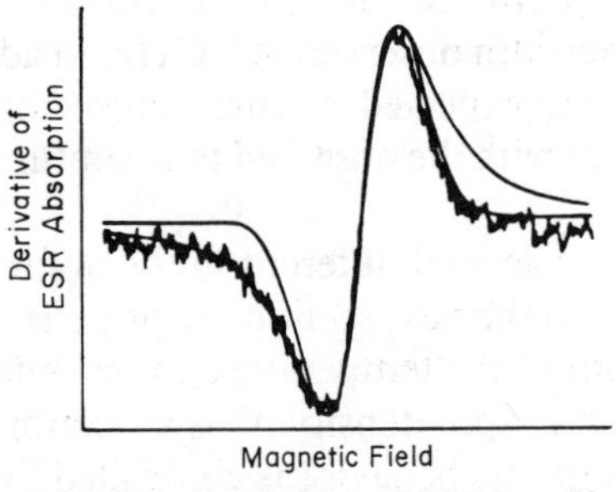

FIG. 16. The derivative of the ESR absorption at 25°K for a-Si with $N_S = 10^{19}$ spins cm^{-3}. The outer and inner solid curves are, respectively, Lorentzian and Gaussian fits to the peak-to-peak derivative width. The noisy curve represents the experimental trace. [From Title *et al.* (1977).]

The Lorentzian linewidth at higher spin densities is commonly attributed to exchange effects between paramagnetic spins (see, for example, Bachus *et al.*, 1979). Such effects are enhanced by hopping from site to site (Brodsky *et al.*, 1970; Bahl and Bhagat, 1975; Voget-Grote *et al.*, 1976; Movaghar and Schweitzer, 1977; Thomas and Flachet, 1981). The linewidth at these spin densities can be decomposed into two components—a low temperature component that is temperature independent and a temperature-dependent component that becomes important at higher temperatures. The temperature-dependent component can be directly related to the hopping conductivity. In addition, the density of ESR sites can be related to the observed conductivity (Title *et al.*, 1970; Hasegawa and Yazaki, 1977; Voget-Grote *et al.*, 1976; Voget-Grote and Stuke, 1979). This hopping conductivity also enhances spin relaxation rates as will be discussed in Section 10.

The temperature-independent component of the line shape, which depends on spin density, is predominantly due to variations in the g-values at different paramagnetic sites and to dipolar interactions between spins (Voget-Grote *et al.*, 1976; Thomas *et al.*, 1978). Estimates of the dipolar contribution, which yield a Lorentzian line shape for "dilute" spin concentrations ($< 10^{21}$ cm^{-3}), can be obtained from Eq. (10) of Section 1a in which γ is the gyromagnetic ratio of the electron and $n = N_S$ is the concentration of paramagnetic spins. For $N_S \sim 10^{20}$ Thomas *et al.* (1978) calculate $\sigma \sim 5$ G, which is consistent with what is commonly observed for the temperature-independent component of σ.

However, estimates of the contributions to the temperature-independent component of σ due to g-tensor anisotropy and variations of the g-values at different (distorted) sites also yield linewidths consistent with those observed (Voget-Grote *et al.*, 1976; Thomas *et al.*, 1978). First, the g-tensor broadening should be linearly dependent on the frequency as can be obtained from Eq. (14). Measurements at ~20 GHz (Voget-Grote, 1976) and ~100 GHz (Thomas *et al.*, 1978) both indicate an increase in σ consistent with that predicted from the linewidth observed at 9 GHz. In addition, Thomas *et al.* (1978) estimate that the expected g-tensor anisotropy for dangling-bond defects is also consistent with the observed temperature-independent contributions to σ.

From experiments on several different kinds of films, including films of a-Si:H to be discussed in the next section, Thomas *et al.* (1978) suggest that the qualitative behavior of the temperature-independent component of the linewidth as a function of spin density N_S is as exhibited in Fig. 17. At low spin densities the asymptotic behavior is controlled by g-tensor effects. The narrowing with increasing N_S is thought to be caused by dipolar spin-flip interactions or exchange effects, and the increase at high N_S is attributed to the "normal" dipolar interaction as expressed in Eq. (3).

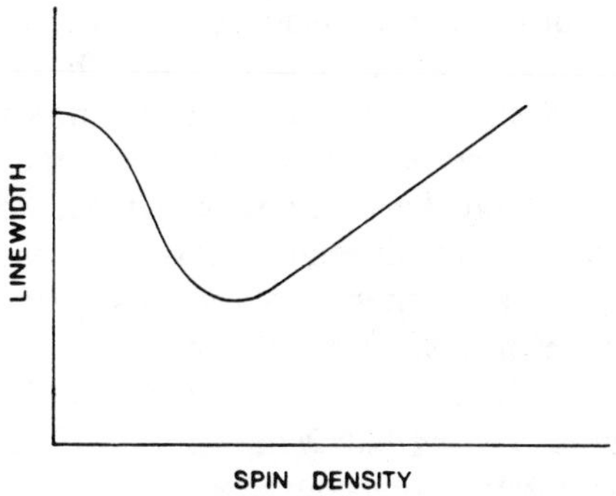

FIG. 17. A schematic representation of the ESR linewidth in a-Si as a function of spin density on a linear scale. [From Thomas *et al.* (1978).]

There is very little detail in the observed ESR line shapes on which to base an unambiguous identification of the microscopic properties of the paramagnetic site; however, a reliable, albeit qualitative, picture has emerged. In their seminal paper, Brodsky and Title (1969) suggested that the ESR signal observed in a-Si is the analog of those spins seen on clean (111) surfaces of crystalline Si by Haneman (1968). In the crystalline experiments the signal has been attributed (Haneman, 1968) to dangling bonds on the surface, whose wave functions have an enhanced s-type character compared with the bulk sp^3 bonding states. Perhaps the most direct, although by no means definitive, evidence that the ESR signal is attributable to silicon dangling bonds comes from comparisons with ESR *g*-values and line shapes observed in single crystal silicon (Title *et al.*, 1970). The general description suggested for a-Si (Brodsky and Title, 1969) was that the dangling bonds reside on internal surfaces of voids or small multivacancy complexes that are incorporated in the films during growth.

Several more recent experiments have cast some doubts on the assertion that the paramagnetic spins reside on internal surfaces, but the qualitative description of these states as Si dangling bonds remains essentially unchallenged. Using evaporation at oblique angles of incidence, Thomas and Kaplan (1976) drastically changed the porosity of the films without any significant change in the spin density. As this procedure greatly alters the internal surface area, these authors (Thomas and Kaplan, 1976; Thomas *et al.*, 1978) concluded that the spins are distributed in the bulk. This conclusion is predicated on the assumption that the spin density does not depend on the film morphology in some complicated fashion, an assumption that may not be entirely accurate. Early studies of Brodsky *et al.* (1972) on a series of evaporated films demonstrated that the ESR spin density is proportional to the film density for ESR spin densities in the range $10^{19}-10^{20}$ spins cm^{-3}, and that an extrapolation to the zero-spin concentration yields the density of crystalline Si. However, recent measurements of Shimizu *et al.* (1980b) on films sputtered in argon show the opposite trend. There is also evidence that the interface structure, or perhaps more accurately the substrate morphology, also affects the ESR spin density and line shape in a-Si (Ohdomari *et al.*, 1981). Clearly, the spin density depends in some complicated fashion on the preparation conditions, a fact that makes generalization difficult.

There are changes in the ESR spin density N_S as a function of impurities such as oxygen (Bahl and Bhagat, 1975; Kubler *et al.*, 1979), and annealing temperature (Thomas *et al.*, 1978; Yonehara *et al.*, 1980; Suzuki *et al.*, 1982), but the variations are rarely over more than an order of magnitude in N_S. For example, oxygen contamination rapidly reduces N_S to $\sim 3 \times 10^{18}$ spins cm^{-3}, but the spin density saturates at this value as the oxygen content is increased further (Kubler *et al.*, 1979). There are two exceptions to this statement. The first is in films made by the decomposition of silane where N_S can be decreased below 10^{17} cm^{-3} (LeComber *et al.*, 1974), but these films are now known to contain substantial hydrogen that itself reduces the spin density. (See Section 6.) The second exception appears to be the case of Mn impurities in which paramagnetic Mn^{2+} is created, which effectively transfers electrons to the Si dangling bonds resulting in values of N_S that approach 10^{17} spins cm^{-3} (Kumeda *et al.*, 1977b).

Annealing at elevated temperatures also reduces N_S (to $\sim 10^{18}$ spins cm^{-3}), but the exact dependence on annealing temperature depends on the purity of the samples (Thomas *et al.*, 1978; Yonehara *et al.*, 1980). In the purest samples N_S appears to be a monotonically decreasing function of annealing temperture.

There is also evidence for clustering of spins in some samples, although the degree of clustering varies considerably from sample to sample. A useful, qualitative measure of the clustering is the departure from Curie susceptibility at low temperatures. Several experiments have indicated that the magnetic susceptibility follows a Curie–Weiss form $[\chi = C/(T - \theta)]$ at low temperatures, but the ordering parameter θ varies considerably from sample to sample. Fritzsche and Hudgens (1975) found $\theta \sim 5\,°K$ in a-Si prepared by evaporation or sputtering or in a-Si:H prepared by the decomposition of silane. In independent measurements, Pawlik *et al.* (1976) found $\theta \sim 1\,°K$ in sputtered a-Si, and Brodsky and Title (1976) determined that $\theta \sim 1\,°K$ in a-Si:H made from silane. In a-Si–Ge alloys Hasegawa *et al.* (1977) have observed much higher values of θ (~ 40–$135\,°K$). On the other hand, in pure UHV evaporated a-Si Thomas *et al.* (1978) found no evidence for any departure from strict Curie-law behavior within their experimental accuracy ($\theta < 1\,°K$). Clearly the degree of spin clustering, and hence of spin ordering, depends sensitively on the method of preparation and perhaps also indirectly on the impurities that are present.

6. ESR IN a-Si:H

When hydrogen is added to a-Si the ESR spin density can be drastically reduced, but the g-value (2.0055) and line shape ($\sigma \sim 7.5$ G) are essentially unchanged. As one varies the deposition conditions and the hydrogen content, the spin density N_S varies over several orders of magnitude, but the shape and position of the ESR response remain remarkably unchanged. In fact, most of the conclusions of the previous section on a-Si hold for a-Si:H at comparable values of N_S. This is perhaps a surprising result when we consider that some of these alloys contain more than 10 or 15 at. % hydrogen, which should affect the ESR line shape through the hyperfine term [Eq. (16)]. This insensitivity is reasonable only if the paramagnetic centers are isolated one from another (at the lower spin densities) and highly localized on Si atoms (Biegelsen, 1981).

Some of the deviations from sample-to-sample that plagued the a-Si results also occur in the a-Si:H system. For example, some authors have observed, on material made by the glow-discharge technique, a Curie–Weiss ordering temperature $\theta \sim 1$ °K for many samples in which N_S varies over three orders of magnitude (Brodsky and Title, 1976). This result implies significant clustering of the paramagnetic centers at all spin densities. However, other groups find less evidence for spin clustering in glow-discharge films (Voget-Grote and Stuke, 1979; Biegelsen *et al.*, 1979a). There have also been reported higher densities of ESR centers in bulk regions near the film surface (Hasegawa and Imai, 1928b). At the lowest spin densities a surface contribution to N_S is observed (Knights *et al.*, 1977).

a. The Case for a Dangling Bond

As mentioned in the previous section, the commonly accepted microscopic interpretation of the ESR response in a-Si and a-Si:H is in terms of a highly localized silicon dangling bond. It is important at this point to examine the evidence for this interpretation. We mentioned in Section 5 the comparison between an ESR response observed on single crystal Si surfaces and the ESR in a-Si. A second comparison of interest is the ESR from Si–SiO_2 interfaces (Caplan *et al.*, 1979), where an axially symmetric response is observed ($g_{\parallel} \simeq 2.001$, $g_{\perp} \simeq 2.008$), which is most probably due to a silicon dangling bond. The powder spectrum of this response is indicated for several values of (isotropic) Gaussian broadening by the broken lines in Fig. 18 (Biegelsen, 1981). The solid curve in Fig. 18 is the observed ESR absorp-

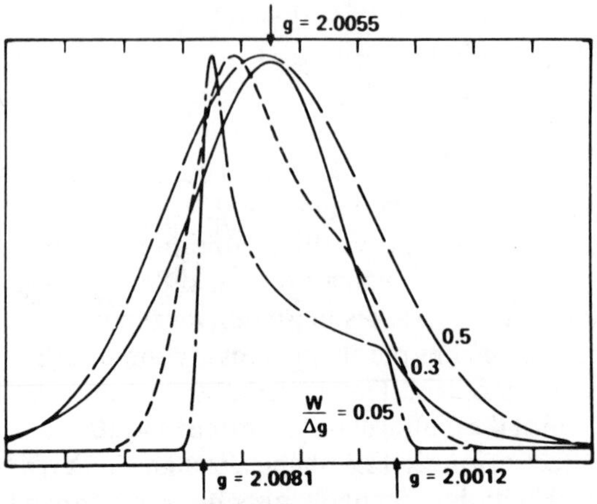

FIG. 18. The dashed curves represent the axially-symmetric powder pattern for the dangling bond observed at $Si-SiO_2$ interfaces (Caplan *et al.*, 1979) for several values of isotropic broadening W. The solid curve is the ESR spectrum in a-Si:H. [Reprinted by permission of the publisher from Electron spin resonance studies of amorphous silicon, by D.K. Biegelsen, *Proceedings of the Electron Resonance Society Symposium,* Vol. 3, pp. 85–94. Copyright 1981 by Elsevier Science Publishing Co., Inc.]

tion in a-Si:H. Because the center of gravity and the width of the ESR response in a-Si:H are similar to those observed for the signal from the $Si-SiO_2$ interface, it is reasonable to assume that the two signals arise from the same source. One would not necessarily expect exact agreement between the two cases because the relaxation at a crystalline interface may be different from that which occurs in the hydrogenated amorphous solid.

These two comparisons provide good evidence for the silicon dangling-bond model, but several minor, although persisting, difficulties remain. The most important of these is the lack of observation of either ^1H or ^{29}Si hyperfine structure. Even if the ESR is due to an electron highly localized on a single Si atom, which might explain the lack of any broadening attributable to H, one would still expect a strong hyperfine broadening for ~ 5% of the ESR centers which reside on ^{29}Si nuclei just as one observes for the silicon E′ centers on SiO_2 (Griscom *et al.*, 1974). Unpublished measurements (Carlos and Taylor, 1982c), which employed a technique designed to enhance broader lines (high power dispersion), have observed a broad underlying response consistent with either ^1H or ^{29}Si broadening, but no sharp hyperfine structure is apparent. Experiments using electron nuclear double resonance (ENDOR) are also inconclusive (Yamasaki *et al.*, 1983; Kerns and Biegelsen, 1980). These experiments confirm the obvious presence of ^{29}Si and ^1H in the samples, but so far they have yielded no definitive details concerning the wave function of the ESR centers. In addition, attempts to calculate the *g*-values (Ishii *et al.*, 1981) have merely shown consistency with the basic suggestions of a Si dangling bond. The situation in a-Ge is even more confusing because the spins are less localized in that material (Stutzmann and Stuke, 1983).

b. Comparisons with Other Experiments

There are some interesting and informative parallels of the ESR in a-Si:H with the results of other experimental techniques. Perhaps the most striking comparison is between the ESR and the photoluminescence (PL), where an inverse relationship exists between the ESR spin density N_S and the PL efficiency (Street *et al.*, 1978; Biegelsen *et al.*, 1979a). Figure 19 shows the relationship between the PL efficiency and N_S for values of N_S ranging from $\sim 10^{15}$ to 10^{18} spins cm^{-3}. It does not appear to matter what deposition parameters are involved—rf power, silane concentration, dilutant atom (He, Ne, etc.) (Knights *et al.*, 1983)—the greater the ESR intensity the smaller the PL efficiency. This correlation holds also when ion bombardment is used to decrease the PL efficiency and increase N_S (Street *et al.*, 1979). It has been suggested (Street *et al.*, 1978; Biegelsen *et al.*, 1979a) that this behavior is due to the fact that the ESR centers form a nonradiative channel into which the trapped, photoexcited carriers can tunnel.

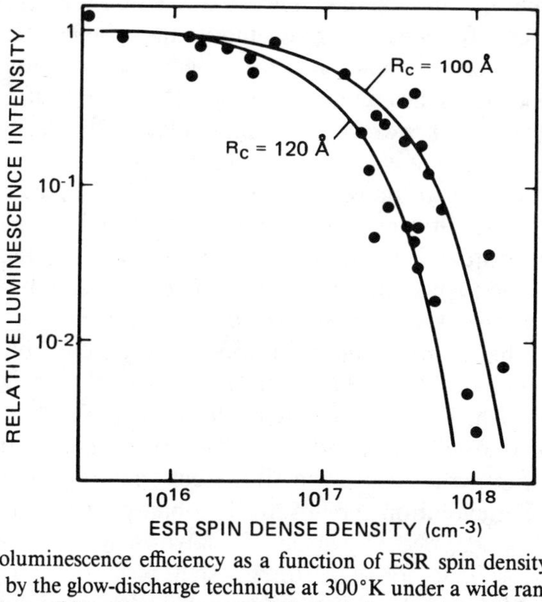

FIG. 19. Photoluminescence efficiency as a function of ESR spin density for samples of a-Si:H prepared by the glow-discharge technique at 300°K under a wide range of deposition conditions. The two solid curves represent model calculations, which give bounds on the data for as-deposited samples. [From Street *et al.* (1978).]

There also appears to be a correlation in a-Si : H between the ESR intensity and both below-gap absorption (Yamasaki *et al.*, 1981; Amer and Shapiro, 1983) and various trap densities derived from transport experiments (Street, 1982a,b). A relatively flat below-gap absorption extending from ~ 1.5 eV toward midgap has been observed. The intensity of this absorption appears to scale with N_S in many samples. In addition to these experiments, the presence of the ESR centers in a-Si : H has been investigated by modulation of the width of the depletion region in *n*-type diode junctions (Cohen *et al.*, 1982). The diode experiments suggest that the ESR state lies slightly above midgap.

c. Hydrogen Diffusion and Evolution

When glow-discharge deposited or sputtered films of a-Si : H are made on room-temperature substrates, the ESR spin density (typically 10^{17} cm^{-3}) is greater than it is for higher temperature substrates (at 300°C N_S is typically less than 10^{16} spins cm^{-3}). When films made at room temperature are annealed, N_S decreases for annealing temperatures up to ~ 300°C (Biegelsen *et al.*, 1980; Kumeda and Shimizu, 1980). Above these temperatures N_S begins to increase again as a substantial quantity of hydrogen evolves from the samples. For the films deposited on higher temperature substrates, at for example 300°C, N_S is constant up to 300–400°C, after which it starts to increase (Biegelsen *et al.*, 1980; Hasegawa and Imai, 1982a). Although the spin densities are lower at any given temperature, these results are very similar to those observed in a-Si and discussed in the previous section.

In most experiments the increase in the spin density is found to be proportional to the amount of hydrogen that has evolved (Biegelsen *et al.*, 1979b) but independent of the type of bonding site as identified by infrared absorption measurements. At the lower temperatures (< 500°C) the evolution process is controlled by diffusion (Biegelsen *et al.*, 1979b) and ESR experiments suggest similar diffusion constants for similar materials (Zellama *et al.*, 1981a,b). There is even some evidence that hydrogen can diffuse into these films at similar rates (Dvurechenskii *et al.*, 1982b,c) at these temperatures. Above 500°C the hydrogen effusion appears to be controlled by simple first-order kinetics with $\Delta E \sim 3.6$ eV, the Si–H bond energy (Zellama *et al.*, 1980, 1981a,b). The low temperature evolution is probably from clustered hydrogen near internal voids and most likely involves a reconstruction of Si–Si bonds after hydrogen is released (Biegelsen *et al.*, 1979b; Zellama *et al.*, 1981a,b). The evolution above 500°C probably involves the breaking of isolated Si–H bonds.

d. Doped Samples

In doped samples additional ESR responses are observed. These resonances, which were first discovered in optically excited ESR experiments

(Knights *et al.*, 1977), are shown in Fig. 20 (Street *et al.*, 1981) for both boron and phosphorus doping. These are not commonly interpreted as signals from un-ionized dopants as one would expect in a crystalline semiconductor at low temperatures, but rather are attributed to singly occupied states in the band tails that are presumably formed from weak or strained bonds (Knights *et al.*, 1977). Similar results are observed in the a-Ge : H system (Stutzmann *et al.*, 1983).

A second interpretation of these resonances is that they are due to the charged states of twofold coordinated silicon defects in a-Si : H films (Adler and Frye, 1981; Adler 1982; Adler and Shapiro, 1983). In this interpretation the ESR that appears on boron doping is due to the positively charged state of these defects and that which appears on phosphorus doping to the negatively charged state. This interpretation depends critically upon the placement of the three states (positively charged, neutral, and negatively charged twofold-coordinated silicon) in the energy gap. These energies are at present controversial (Adler, 1982; Ishii *et al.*, 1983).

In heavily phosphorus-doped a-Si : H samples, the ESR line is narrow and centered at $g = 2.004$, whereas in heavily boron-doped samples the line is much broader and occurs at $g = 2.013$ (Stuke, 1977; Hasegawa *et al.*, 1979, 1980, 1981a; Street *et al.*, 1981; Dersch *et al.*, 1981a; Magarino *et al.*, 1982). Presumably the ESR from un-ionized B or P impurities is not observed because the electrons or holes are trapped by much lower-lying gap states,

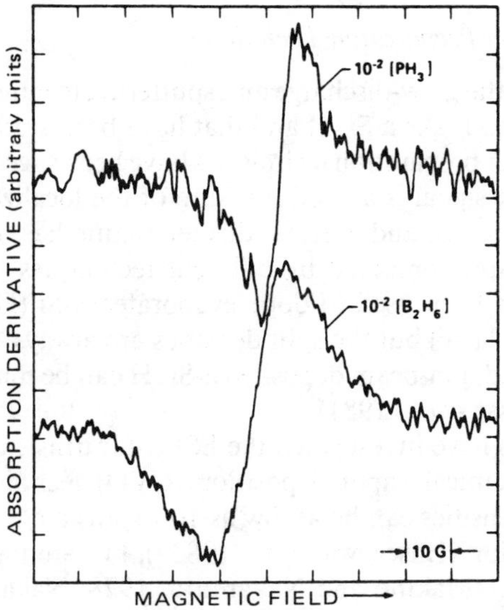

FIG. 20. Equilibrium ESR signals for heavily P- and B-doped a-Si : H at 30°K. [From Street *et al.* (1981).]

which are always present in a-Si:H (Biegelsen *et al.*, 1981). This autocompensation mechanism is consistent with the result that the intitial effect of the doping is to compensate the silicon dangling bond states, which are deep in the gap (Hasegawa *et al.*, 1981a). The spin densities for the band-tail (or doubly coordinated silicon) resonance in these heavily doped samples are greater than 10^{17} cm^{-3}. The identification of these two lines is complicated by the fact that it is difficult to resolve the dangling bond line at $g = 2.0055$ from them. This difficulty is particularly pronounced in the phosphorus-doped samples. Qualitative comparisons of the line shapes suggest that the holes trapped near the band are more highly localized than the electrons trapped near the conduction band (Dersch *et al.*, 1981a). The temperature dependences of the ESR linewidths for the B- and P-doped samples can be used as probes of the conduction properties of trapped electrons and holes (Dersch *et al.*, 1981b). When samples are doped and compensated the ESR signals are greatly reduced, although strong optically excited ESR signals remain, as discussed in the next section (Street *et al.*, 1981). These various features of the ESR results in doped films provide important inputs for determinations of the electronic densities of states (Overhof and Beyer, 1981; Overhof, 1982). In particular, there is much evidence to suggest that doping introduces extra states in the gap, primarily near the valence-band edge (see for example, Biegelsen *et al.*, 1981; Street *et al.*, 1981; or Fischer *et al.*, 1980). The details of these doping experiments have yet to be accounted for in careful model calculations, but the main trends have been reproduced in tight-binding calculations (Robertson, 1983).

e. Dependence on Preparation Technique

In addition to the glow-discharge and sputtering techniques that are most often employed to make a-Si:H and that have been studied most often by ESR, several other preparation techniques have been successfully employed. Because the ESR signal is a good measure of the localized, paramagnetic, deep-gap states in a-Si and a-Si:H, this technique has often been used to characterize samples prepared by different techniques. For example, ion implantation has been used to dope evaporated a-Si (Dvurechenskii and Ryazantsev, 1982a–c) but the spin densities are always $\geq 10^{17}$ spins cm^{-3}. On the other hand, ion-beam deposited a-Si:H can be made with $N_S < 10^{17}$ spins cm^{-3} (Ceasar *et al.*, 1981).

Several groups have investigated the ESR of intrinsic and doped a-Si:H deposited by chemical vapor deposition (CVD) techniques. In undoped films the ESR densities can be as low as 10^{17} spins cm^{-3} in thin films after post-hydrogenation (Hasegawa *et al.*, 1983a), but are typically in the 10^{18} spins cm^{-3} range (Makino and Nakamura, 1978; Nakashita *et al.*, 1981; Hasegawa *et al.*, 1983a) because the films are deposited at high temperatures

(~600 °C) at which little hydrogen is incorporated. In doped films the same three ESR responses are observed in CVD a-Si:H as in the sputtered or glow-discharge deposited films (Magarino *et al.*, 1981, 1982; Hasegawa *et al.*, 1979, 1980, 1981b).

Recently, both polycrystalline and microcrystalline (~200 Å crystallite size) Si films have been investigated by ESR (Hasegawa *et al.*, 1981b, 1983b). A dangling bond resonance in microcrystalline silicon films is observed at $g = 2.0049$. The spin density of this resonance is typically between 10^{17} and 10^{18} spins cm^{-3}. In P-doped samples a conduction-electron spin-resonance signal is observed at $g = 1.997$.

f. Multicomponent Alloys

It is not the primary purpose of this chapter to consider more complicated alloy systems, but we shall summarize some of the more important observations because there has been considerable interest recently in amorphous alloys of Si with such elements as carbon, nitrogen, germanium, and oxygen. In the silicon–carbon alloy system an ESR response at $g = 2.003$ is observed at high carbon content and has been attributed to carbon dangling bonds (Ishii *et al.*, 1982a). This ESR signal is very strong ($N_S \sim 10^{20}$ spins cm^{-3}) in films that contain no hydrogen (Shimizu *et al.*, 1981a), but it is slightly weaker in a-Si_xC_{1-x}:H films (Morimoto *et al.*, 1982a,b; Watanabe *et al.*, 1982). Unlike the a-Si:H system, annealing appears only to increase the spin density in these films (Shimizu *et al.*, 1981b).

The a-Si_xGe_{1-x} alloy system has also been studied in some detail. Without the presence of hydrogen this sytem also possesses a large spin density ($> 10^{19}$ spins cm^{-3}) with a g-value for the ESR line near 2.02 at high Ge concentrations (Shimizu *et al.*, 1981a; Ishii *et al.*, 1982b; Hasegawa *et al.*, 1978; Kumeda *et al.*, 1977a). The ESR intensity exhibits a Curie–Weiss behavior similar to that found in a-Si (Hasewaga *et al.*, 1977; Hasegawa and Yazaki, 1977), which may be due to clustering (Hasegawa *et al.*, 1977) or to a pseudo-Orbach multiphonon process (Dumas, 1978). When hydrogen is added to this alloy system the spin density is reduced to ~10^{17} spins cm^{-3} in analogy with the other alloy systems (Morimoto *et al.*, 1981).

Other alloy systems in which ESR has been an effective experimental tool include silicon nitride (Shimizu *et al.*, 1982b; Ishii *et al.*, 1982c), silicon oxide (Holzenkämpfer *et al.*, 1979; Shimizu *et al.*, 1980a; Dvurechenskii *et al.*, 1982a), and silicon oxynitride (Kubler *et al.*, 1983). In the silicon nitride system the ^{14}N hyperfine interaction provides an additional contribution to the observed g-values and linewidths, which indicates that the spins on the silicon atoms are influenced by nearby nitrogen atoms. This situation contrasts with that observed in a-Si:H or a-Si:F in which no strong influence of the H or F atoms on the Si dangling bond resonance is observed (Ishii *et al.*,

1982b). In the Si–O system one obtains measurable densities of defects that are common in SiO_2, such as the Si E′ center, with increasing oxygen content.

Finally, alloys of Si with various metals or semimetals have been studied. These amorphous systems include Si–Al (Suzuki *et al.*, 1979), Si–Au (Kishimoto *et al.*, 1977, 1981), Si–In (Mell *et al.*, 1980) Si–As (Nemanich and Knights, 1980), and Si–B (Tsai, 1979). The ESR spin densities in most of these systems are high ($\geq 10^{18}$ spins cm^{-3}), except for the Si_xAs_{1-x}:H system in which densities of $N_S \sim 10^{17}$ spins cm^{-3} are observed. At low arsenic concentrations (≤ 5 at. % As) the Si_xAs_{1-x}:H system behaves in a manner similar to the P-doped a-Si:H material previously described.

7. Transient, Optically Induced ESR

Optical excitation with light near or above the band gap creates a nonequilibrium situation that gives rise to several ESR signals (Knights *et al.*, 1977; Pawlik and Paul, 1977; Friederich and Kaplan, 1979). In undoped samples the optically induced signal consists primarily of contributions from the localized electronic states near the conduction-band edge ($g = 2.004$) and from the localized hole states near the valence-band edge ($g = 2.013$) (Friederich and Kaplan, 1979; Street and Biegelsen, 1980a). A typical optically induced ESR spectrum in undoped material is shown at the top of Fig. 21.

In heavily B-doped samples the hole resonance at $g = 2.013$ and the silicon dangling bond resonance at $g = 2.0055$ are observed during optical excitation as shown in the middle trace of Fig. 21. The bottom trace in Fig. 21 shows the light induced ESR from heavily P-doped a-Si:H in which only the $g = 2.0055$ response is observed. In all three samples of Fig. 21 no equilibrium (dark) ESR at $g = 2.0055$ is observed (Street and Biegelsen, 1980b). Because of the featureless nature of these three ESR responses and their mutual overlap, it is difficult in many cases to be certain that the identification is unambiguous. Nonetheless, the trends exhibited in Fig. 21 are well documented. The magnitude of the silicon dangling bond, optically induced response increases with either P- or B-doping. The optically induced response due to holes trapped below the valence band occurs only in undoped or B-doped samples, whereas the response due to electrons trapped below the conduction band is only apparent in undoped samples. The decay of the optically induced ESR at 30°K occurs over a very wide range of times from the shortest time measurable (0.1 sec) to $> 10^3$ sec (Biegelsen *et al.*, 1983).

In CVD or glow-discharge samples that have been annealed to remove some of the hydrogen, the line at $g = 2.0055$ is observed before optical excitation ($N_S \sim 10^{17}-10^{19}$ spins cm^{-3}). In these samples Friederich and Kaplan (1980a,b) find an optically induced decrease in this resonance for

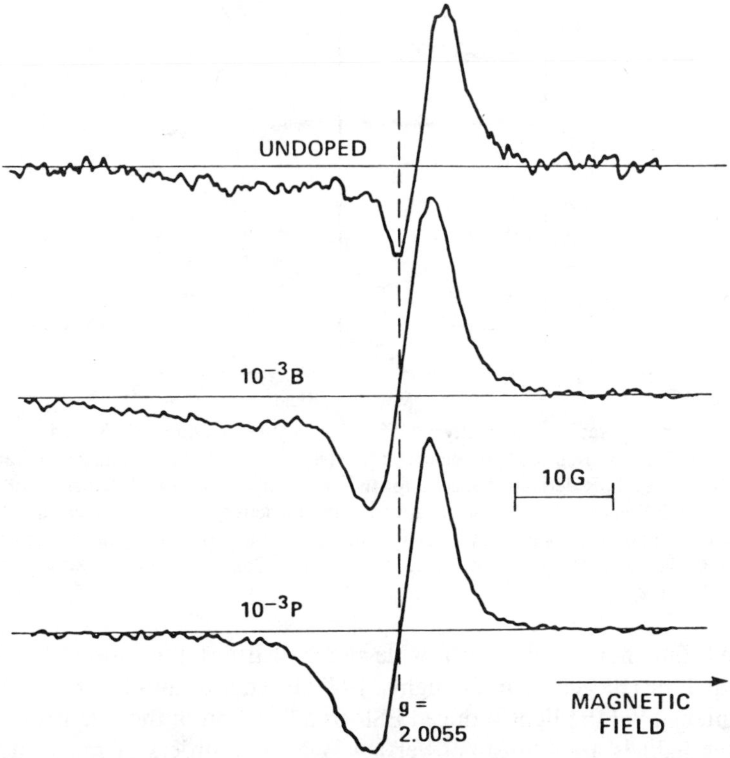

FIG. 21. Transient optically induced ESR in undoped and doped a-Si:H at 30°K, with illumination at 1.915 eV of 100 mW cm^{-2}. Three components have been identified in these spectra as detailed in the text. [After Street and Biegelsen (1980a). Copyright North-Holland Publ. Co., Amsterdam, 1980.]

undoped or lightly doped samples ($<3 \times 10^{-4}$ phosphine or diborane to silane ratio) and no contribution from the resonances at $g = 2.013$ and 2.004 as shown in Fig. 22. This figure shows the comparison between the behavior in glow-discharge films in which no equilibrium ESR is observed and in which a significant ESR at $g = 2.0055$ is observed. Note the difference in phases of the optically induced ESR for these two cases. At high phosphorus doping levels these authors observe the $g = 2.004$ resonance due to electrons trapped near the conduction-band edge and an optically induced increase centered at $g = 2.0055$. This result is identical to that observed in Fig. 21 for the doped samples with no equilibrium response at $g = 2.0055$.

The temperature dependence of the light-induced ESR is relatively weak, especially below ~100°K where the intensity is essentially constant (Biegelsen and Knights, 1977). The ESR signals increase rapidly after the light is

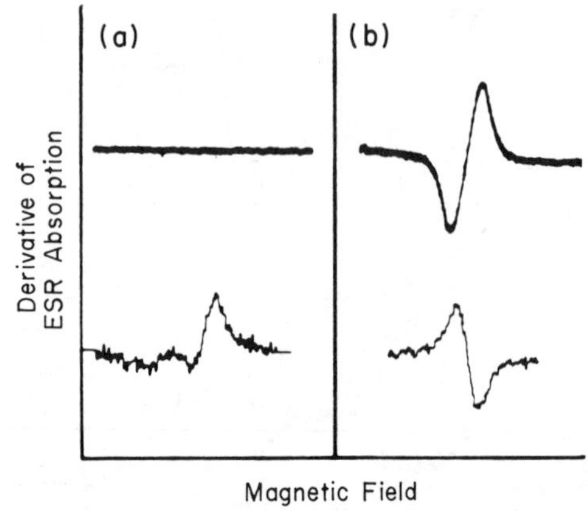

FIG. 22. The top traces are derivatives of ESR absorption in a sample of a-Si : H as deposited (left side) and after substantial hydrogen evolution (right side). The bottom traces are transient optically induced ESR derivative spectra for the same films as-deposited (left side) and after hydrogen evolution (right side). Note the *decrease* in spin density under optical excitation after hydrogen evolution as evidenced by the change in phase of the derivative signal. [From Friederich and Kaplan (1980a). Copyright North-Holland Publ. Co., Amsterdam, 1980.]

applied, but they decay over a wide range of times, the longest of which exceeds 1 sec (Biegelsen and Knights, 1977; Friederich and Kaplan, 1980a). The intensity of the light-induced ESR as a function of the intensity of the exciting light is a sublinear power law over nine orders of magnitude in power density (Street and Biegelsen, 1982; Boulitrop and Dunstan, 1982). Although the original interpretation of this result was in terms of optical saturation of longer-lived geminate pairs (monomolecular kinetics) of trapped electrons and holes (Biegelsen *et al.,* 1978; Street and Biegelsen, 1982), recent experiments over a wide range of power densities suggest that distant pairs (bimolecular kinetics) of trapped electrons and holes may be responsible for the power dependence. This question is still controversial with regard to both optically induced ESR and PL.

Most of the experiments on optically induced ESR in a-Si : H have been performed with white light, sometimes with a filter to block the infrared below-gap light. One study has reported an excitation spectrum for the optically induced ESR (Pawlik and Paul, 1977), but these experiments were performed at high microwave powers and high light intensities that complicate the interpretation. The ESR intensity is, of course, not an absolute measure of the number of centers because the observed signal represents a dynamical balance between the excitation and decay mechanisms. One can also observe an artificial increase in the ESR during optical excitation because the excited electrons and holes allow the equilibrium ESR to be relaxed more rapidly (due to removal of microwave saturation). If the light heats the samples then one can observe an artificial decrease in the equilibrium ESR due to a change in the effective spin temperature (Boltzmann factor). If all of these potential difficulties can be ignored, the excitation spectrum for the optically induced ESR in a-Si : H appears to extend well below the band edge to about 0.8 eV (Pawlik and Paul, 1977).

As with the case of the equilibrium ESR described in the previous section, there are two possible explanations of the light-induced ESR in a-Si : H. The first, and perhaps more commonly accepted, explanation is that the resonances at $g = 2.013$ and 2.004 are due to holes trapped at strained or weak bonds in the valence-band tail and electrons trapped in similar states in the conduction-band tail, respectively (Street and Biegelsen, 1980a; Friederich and Kaplan, 1980b). If the negatively charged (two electron) dangling-bond state lies within the gap as shown in Fig. 23, then the general trends of the optically induced ESR with doping are explained by the position of the Fermi energy with respect to the various defect levels. As mentioned above, whether or not the band-tail electrons and holes are geminate pairs is still a controversial point. The optically induced decrease in the ESR at $g = 2.0055$ in undoped samples is attributed to trapping of electrons or holes at the dangling-bond sites.

In the second approach the light-induced ESR is explained in terms of postively or negatively charged states of two-coordinated silicon atoms (T_2^+ and T_2^-), which are assumed to be a positive-U system where the ground states are charged. The ESR responses at $g = 2.013$ and 2.004 are attributed

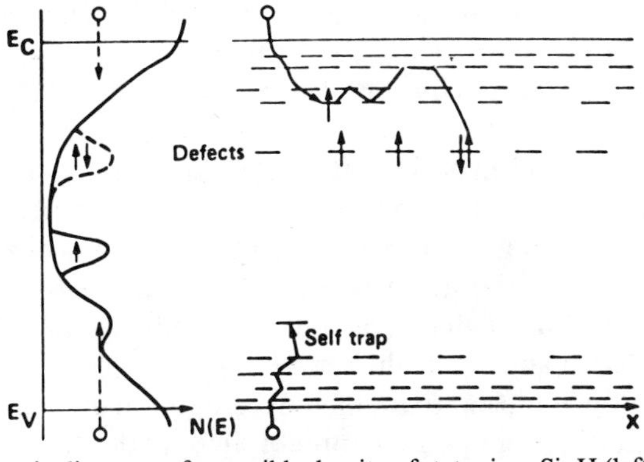

FIG. 23. Schematic diagrams of a possible density of states in a-Si : H (left-hand side) and a schematic representation of the spatial variation of localized electronic states in a-Si : H (right-hand side). [Reprinted by permission of the publisher from Electron spin resonance studies of amorphous silicon, by D. K. Biegelsen, *Proceedings of the Electron Resonance Society Symposium,* Vol. 3, pp. 85–94. Copyright 1981 by Elsevier Science Publishing Co., Inc.]

in this model to holes trapped at T_2^0 sites and electrons trapped at T_3^0 sites, respectively (Adler, 1982).

8. Metastable, Optically Induced ESR

The first observation of a metastable component to the optically induced ESR in a-Si : H was reported by Street *et al.*, (1981) in heavily P- and B-doped samples and in doped and compensated samples. [For a recent review of metastable, optically induced effects in a-Si : H see Pankove (1982).] As near as can be determined, this metastable signal is a fraction (~20%) of the transient optically induced response, and the dependences of the line shapes on doping appear to be similar for the two effects. These ESR centers can be generated with light intensities of ~150 mW cm^{-2}, and they anneal in a few minutes at 80°K. It is particularly surprising that a subset of the photoexcited electrons are trapped in band-tail states or in a T_2^- configuration in *p*-type material in which there exists a large density of holes to facilitate rapid recombination.

Similar results in undoped films have been observed by Morigaki *et al.* (1982) and Sano *et al.*, (1982) at 2°K using optically detected magnetic resonance (ODMR) techniques. In these experiments the metastable, optically induced increase in the ESR at $g = 2.0055$ is accompanied by a fatigue of the PL at ~1.2 eV. These measurements suggest that the PL fatigue results from the creation of additional dangling-bond states that serve as competing nonradiative centers. These metastable centers cannot be bleached by infrared light of comparable intensities (~100–300 mW cm^{-2}), unlike the behavior that is commonly observed in the chalcogenide glasses. At higher exciting light intensities (500 W from a mercury lamp or 2.4 W cm^{-2} at 514.5 nm) metastable ESR centers can be generated at 77°K, which anneal only above about 300°K (Hirabayashi *et al.*, 1981, 1982).

Dersch *et al.* (1981a–c) have employed intense optical excitation (> 100 mW cm^{-2}) for several hours to create a metastable increase in the ESR response at $g = 2.0055$ in a-Si : H at 300°K. Similar results have also been reported following x irradiation (Pontuschka *et al.*, 1982). Taylor and Ohlsen (1983) have demonstrated that these ESR centers, which are created at 300°K, decay at temperatures between ~375 and 475°K in a manner that parallels the annealing of an optically induced enhancement in the PL band at 0.9 eV (Pankove and Berkeyheiser, 1980).

Experimentally, it is clear that different centers are metastable at different temperatures depending upon the initial ESR spin densities and the type of sample, the temperature of irradiation, the power density of the exciting light, and perhaps several other variables. To complicate this situation further, there is once again disagreement as to the interpretation of these metastable, optically induced effects. One model suggests that there is an increase in the density of silicon dangling bonds by optically breaking weak bonds near the valence-band edge (Dersch *et al.* 1981c; Street, 1983). A second model suggests that a change in the hydridization, but not in the total number, of existing dangling bonds accounts for the changes in the PL (Wautelet *et al.*, 1981a,b), although it is difficult to see how such a model can explain the ESR. A third approach suggests that the metastable changes result from an optical rearrangement of electrons and holes in existing diamagnetic defects (Adler and Frye, 1981; Adler and Shapiro, 1983; Adler, 1982). It remains to be seen which of these models will ultimately prove to be correct.

9. ESR OF IMPURITY SPECIES

Although most films of a-Si:H are known to contain substantial amounts of impurities such as oxygen, nitrogen, and carbon, there is remarkably little direct evidence from ESR measurements for the presence of impurity species. Miller and Haneman (1978) have investigated the effect on a-Si of the diffusion of O_2 gas through the film. They find that paramagnetic O_2 readily diffuses into the films and dramatically reduces the spin–lattice relaxation times for the ESR at $g = 2.0055$. These results indicate that the spins are located primarily on internal surfaces that are accessible to oxygen diffusion.

In films that are hydrogenated the situation is slightly different. Street and Knights (1981) find that in a-Si:H films prolonged exposure to air causes an actual increase in the dangling-bond ESR intensity. This increase is roughly proportional to the initial spin density before exposure. For example, films that initially exhibit $N_S \sim 8 \times 10^{16}$ spins cm^{-3} show a factor of five increase in N_S after two months in air, whereas films with $N_S \sim 3 \times 10^{15}$ spins cm^{-3} exhibit only a factor of 2.5 increase in N_S after eight months. Although these results differ qualitatively from those in the unhydrogenated films, the general conclusions are the same. Oxygen can diffuse into internal surfaces in the films (or adhere to external surfaces) and change the local bonding arrangements. These changes produce an increase in the Si dangling bonds on the internal surfaces of voids in the films.

In the studies discussed so far no new ESR centers have been observed, but Pontuschka *et al.* (1982) have discovered several impurity related ESR centers in films purposely doped with oxygen to ~2 at. %. After x irradiation at 77°K these authors observe atomic hydrogen that is trapped in oxygen-rich cages. This atomic hydrogen does not occur in films without significant amounts of oxygen, and the ESR signal decays above ~300°K. In addition to atomic hydrogen these films also exhibit ESR signals that are attributed to Si–O hole centers (holes trapped on singly coordinated oxygen atoms) and silicon E′ centers (electrons trapped on sp^3-hybridized silicon dangling-bond orbitals at sites where the silicon atoms are bonded to three oxygen atoms). The oxygen-related hole center is also observed after optical excitation, albeit at smaller intensity. Elementary considerations that involve the energies of isolated Si–O configurations and the statistically expected probability of SiO_3 structural units in a-Si:H suggest that oxygen is clustered in these films, probably on the walls of internal surfaces. This conclusion is consistent with that deduced from the oxygen diffusion studies previously mentioned (Miller and Haneman, 1978; Street and Knights, 1981).

Pontuschka *et al.* (1982) have also observed ESR from trapped NO_2 molecules after x irradiation at 77°K in some undoped a-Si:H films, but there appears to be no predictability as to which films will exhibit this response. The NO_2 molecules are presumably trapped as NO_2^- radicals in the films. The ESR signal from NO_2 decays rapidly above ~90°K.

In films intentionally doped with transition metals Shimizu *et al.* (1980c, 1981c) have observed ESR responses from Mn, Ni, and Fe. In all cases only a small fraction of the transition elements incorporated into the films contribute to the observed ESR, and details concerning the bonding of these elements in the films are presently very sketchy.

10. Spin–Lattice Relaxation

Spin–lattice relaxation in electron spin resonance is the process by which the ensemble of paramagnetic spins transfers energy to the lattice after microwave excitation in a magnetic field. The relaxation process is often exponential and hence can be characterized by a rate T_1^{-1}, where T_1 is the spin–lattice or longitudinal relaxation time. Measurements of T_1 as a function of temperature, magnetic field, and spin density often provide important insights into the mechanisms that govern the coupling of paramagnetic spins to the amorphous network.

In a-Si films that contain no hydrogen the spin densities are almost always rather high ($N_S \geq 10^{18}$ spins cm^{-3}). For most of these films the spin–lattice relaxation of paramagnetic electrons, like the electrical conductivity, is governed by hopping processes at temperatures ≥ 150°K (Thomas and Flachet, 1981; Gourdon *et al.*, 1981; Movaghar *et al.*, 1977). Good correlations exist between the magnitude and temperature dependences of T_1 and the hopping conductivity in these films.

At lower temperatures ($T < 100$°K) the relaxation rates are proportional to the temperatuare ($T_1^{-1} \propto T$) (Thomas and Flachet, 1981; Gourdon *et al.*, 1981). This temperature dependence is the signature of a relaxation process that involves the excitation of a low frequency phonon (~9 GHz) and the deexcitation of the paramagnetic spin (direct relaxation process). Similar processes are observed in many crystalline solids at low temperatuares.

Some films of a-Si, for which the ESR line shapes are completely determined by motional or exchange narrowing, exhibit spin–lattice relaxation rates above ~100°K that are roughly proportional to the square of the temperatuare ($T_1^{-1} \propto T^\alpha$, where $\alpha \simeq 1.9$) (Hasegawa and Yazaki, 1978). In these films the dominant high temperature relaxation mechanism is not due to hopping but rather to some other process that one can argue probably involves highly anharmonic tunneling or "disorder" modes (Stutzmann and Biegelsen, 1983). A similar temperature dependence is observed for most a-Si:H films in which the values of N_S are usually much smaller (Stutzmann and Biegelsen, 1983). The spin-lattice relaxation rates of the $g = 2.0055$ resonance in a-Si:H thus provide a probe, albeit a rather indirect one, of tunneling modes in these amorphous solids.

Oxygen impurities have a strong effect on the observed values of T_1 in a-Si and a-Si:H. In a-Si:H films doped with oxygen the values of T_1 at a given temperature increase dramatically (Kubler *et al.*, 1979; Pontuschka *et al.*, 1982). The reason for this increase is not well understood. If the oxygen is not bonded to the a-Si network but rather is diffused into a-Si films as O_2, then the values of T_1 at a given temperature are greatly reduced (Miller and Haneman, 1978). This reduction is almost certainly a result of the dipolar interaction of paramagnetic O_2 with the ESR centers.

IV. Summary

The ^1H NMR experiments on a-Si:H have clearly demonstrated the presence of two sites for the hydrogen that is bonded to silicon. One hydrogen site is dilute and may in fact be distributed randomly in some films. This site represents isolated Si–H bonds. The second site is definitely clustered, probably on the internal surfaces of voids. These conlusions are confirmed by ^2D experiments.

The correlations between the two ^1H NMR sites and infrared absorption measurements, which probe Si–H vibrational modes, are still somewhat ambiguous. Several types bonding site appear to contribute to the clustered NMR phase.

Studies of doped samples using ^{11}B and ^{31}P NMR have confirmed that most of these "dopants" are incorporated into the films in threefold coordination and hence are probably ineffective in doping. This situation is particularly true in the case of boron in which no tetrahedral sites have been observed within the accuracy of the experiments.

Spin–lattice relaxation measurements of ^1H and ^2D have uncovered the existence of trapped molecular hydrogen (H_2) in many films of a-Si:H. This molecular hydrogen is trapped at sites where it is more stable during annealing experiments than most of the bonded hydrogen. One important question that remains unanswered is the nature of the trapping site.

The ESR experiments in a-Si and a-Si:H have unambiguously documented the existence of unpaired spins (typically $N_S > 10^{18}$ spins cm^{-3} in a-Si but as low as $N_S \sim 10^{15}$ spins cm^{-3} in a-Si:H) in these films. The most reasonable interpretation of this paramagnetism is in terms of dangling bonds on silicon atoms, but several nagging questions remain concerning the details of the wave function for these unpaired spins. It is clear that, among other effects, hydrogen removes these ESR centers from amorphous silicon. Annealing at higher temperatures ($\gtrsim 300°C$) drives off hydrogen and increases the ESR signal.

There are several important correlations between the ESR results and those of PL and below-gap optical absorption. As N_S increases beyond approximately 10^{18} spins cm^{-3} the PL intensity at 1.2–1.4 eV drops, whereas the PL intensity near 0.8–0.9 eV appears to be approximately proportional to N_S. An absorption, which extends well below the gap, also scales with the ESR intensity.

In doped samples three different ESR responses are observed: the usual silicon dangling-bond resonance, a broad resonance attributed to holes trapped in states near the valence-band edge, and a narrow resonance attributed to electrons trapped in states near the conduction-band edge. Two interpretations have been proposed for the presence of these two additional signals—holes or electrons trapped at weak bonds in the band tails and holes or electrons trapped at twofold-coordinated silicon defect sites.

Both transient and metastable changes in the ESR are observed with optical excitation by light with energy greater than the band gap. Explanations analogous to those proposed to explain the ESR in doped samples have also been suggested to account for the optically induced changes.

Some impurity species, notably oxygen and nitrogen, have been studied by ESR in a-Si:H films. Defects associated with both O and N have been identified and oxygen contamination of internal surfaces has been documented.

Spin–lattice relaxation measurements at higher temperatures in a-Si films often correlate well with hopping conductivity results. In a-Si:H films and in some films of a-Si in which hopping conductivity does not dominate, the spin–lattice relaxation is probably controlled by the presence of tunneling or disorder modes.

Acknowledgment

The author gratefully acknowledges support from the ARCO Solar Corporation for some of the research presented in this chapter.

References

Adler, D. (1982). *Kinam* **C 4,** 225–243.

Adler, D., and Frye, R. C. (1981). *AIP Conf. Proc.* **73,** 146–150.

Adler, D., and Shapiro, F. R. (1983). *Physica* **117B** & **118B,** 932–934.
Amer, N. M., and Skumanich, A. (1983). *Physica* **117B** & **118B,** 897–898.
Bachus, R., Movaghar, B., Schweitzer, L., and Voget-Grote, U. (1979). *Philos. Mag. B* **39,** 27–37.
Bahl, S. K., and Bhagat, S. M. (1975). *J. Non-Cryst. Solids* **17,** 409–427.
Biegelsen, D. K. (1980). *Solar Cells* **2,** 421–430.
Biegelsen, D. K. (1981). *Proc. Electron Resonance Soc. Symp.* **3,** 85–94.
Biegelsen, D. K., and Knights, J. C. (1977). *In* "Amorphous and Liquid Semiconductors" (W. E. Spear, ed.), pp. 429–432. Univ. of Edinburgh, Edinburgh, Scotland.
Biegelsen, D. K., Knights, J. C., Street, R. A., Tsang, C., and White, R. M. (1978). *Philos. Mag. B* **37,** 477–488.
Biegelsen, D. K., Lucovsky, G., Knights, J. C., and Nemanich, R. J. (1979a). *Conf. Ser. Inst. Phys.* **43,** 1143–1146.
Biegelsen, D. K., Street, R. A., Tsai, C. C., and Knights, J. C. (1979b). *Phys. Rev. B* **20,** 4839–4846.
Biegelsen, D. K., Street, R. A., Tsai, C. C., and Knights, J. C. (1980). *J. Non-Cryst. Solids* **35** & **36,** 285–290.
Biegelsen, D. K., Street, R. A., and Knights, J. C. (1981). *AIP Conf. Proc.* **73,** 166–170.
Biegelsen, D. K., Street, R. A., and Jackson, W. B. (1983). *Physica* **117B** & **118B,** 899–901.
Boulitrop, F., and Dunstan, D. J. (1982). *Solid State Commun.* **44,** 841–844.
Bourgoin, J. C. (1981). *In "Electronic Structure of Crystal Defects and of Disordered Systems"* (F. Gautier, M. Gerl, and P. Guyot, eds.), pp. 405–434. Les Editons de Physique, Les Ulis, France.
Brodsky, M. H., and Title, R. S. (1969). *Phys. Rev. Lett.* **23,** 581–585.
Brodsky, M. H., and Title, R. S. (1976). *AIP Conference Proceedings* **31,** 97–101.
Brodsky, M. H., Title, R. S., Weiser, K., and Pettit, G. D. (1970). *Phys. Rev. B* **1,** 2632–2641.
Brodsky, M. H., Kaplan, D., and Ziegler, J. F. (1972). *Appl. Phys. Lett.* **21,** 305–307.
Brodsky, M. H., Cardona, M., and Cuomo, J. J. (1977). *Phys. Rev. B* **16,** 3556–3571.
Caplan, P. J., Poindexter, E. H., Deal, B. E., and Razouk, R. R. (1979). *J. Appl. Phys.* **50,** 5847–5854.
Carlos, W. E., and Taylor, P. C. (1980). *Phys. Rev. Lett.* **45,** 358–362.
Carlos, W. E., and Taylor, P. C. (1981). *J. Phys. Colloq. Orsay, Fr.* **42,** C4-725–727.
Carlos, W. E., and Taylor, P. C. (1982a). *Phys. Rev. B* **25,** 1435–1438.
Carlos, W. E., and Taylor, P. C. (1982b). *Phys. Rev. B* **26,** 3605–3616.
Carlos, W. E., and Taylor, P. C. (1982c). Unpublished data, Naval Research Laboratory, Washington, D. C.
Carlos, W. E., Taylor, P. C., Oguz, S., and Paul, W. (1981). *AIP Conf. Proc.* **73,** 67–72.
Ceasar, G. P., Okumura, K., and Grimshaw, S. F. (1981). *J. Phys. Colloq. Orsay, Fr.* **42,** C4-627–630.
Cohen, J. D., Harbison, J. P., and Wecht, K. W. (1982). *Phys. Rev. Lett.* **48,** 109–112.
Conradi, M. S., and Norberg, R. E. (1981). *Phys. Rev. B* **24,** 2285–2288.
Conradi, M. S., Luszczynski, K., and Norberg, R. E. (1979). *Phys. Rev. B* **19,** 20–31.
Dersch, H., Stuke, J., and Beichler, J. (1981a). *Phys. Status Solidi B* **105,** 265–274.
Dersch, H., Stuke, J., and Beichler, J. (1981b). *Phys. Status Solidi B* **107,** 307–317.
Dersch, H., Stuke, J., and Beichler, J. (1981c). *Appl. Phys. Lett.* **38,** 456–458.
Dumas, J. (1978). *Phys. Status Solidi B* **86,** K75–K77.
Dvurechenskii, A. V., and Ryazantsev, I. A. (1982a). *Sov. Phys. Semicond. Engl. Transl.* **16,** 888–891.
Dvurechenskii, A. V., and Ryazantsev, I. A. (1982b). *Phys. Status Solidi* **A69,** K117–K120.
Dvurechenskii, A. V., and Ryazantsev, I. A. (1982c). *Fiz. Tekh. Poluprovodn. Leningrad* **16,** 1384–1389.

Dvurechenskii, A. V., Edel'man, F. L., and Ryazantsev, I. A. (1982a). *Thin Solid Films* **91,** L55–L57.
Dvurechenskii, A. V., Ryazantsev, I. A., and Smirnov, L. S. (1982b). *Sov. Phys. Semicond. Engl. Trans.* **16,** 400–403.
Dvurechneskii, A. V., Ryazantsev, I. A., and Smirnov, L. S. (1982c). *Fiz. Tekh. Poluprovodn. Leningrad* **16,** 621–624.
Fedders, P. A. (1979). *Phys. Rev. B***20,** 2588–2593.
Fischer, R., Rehm, W., Stuke, J., and Voget-Grote, U. (1980). *J. Non-Cryst. Solids* **35 & 36,** 687–692.
Freeman, E. C., and Paul, W. (1978). *Phys. Rev. B* **18,** 4288–4300.
Friederich, A., and Kaplan, D. (1979). *J. Electron. Mater.* **8,** 79–85.
Friederich, A., and Kaplan, D. (1980a). *J. Non-Cryst. Solids* **35 & 36,** 657–662.
Friederich, A., and Kaplan, D. (1980b). *J. Phys. Soc. Jpn.* **49** *Suppl. A.,* 1237–1240.
Fritzsche, H., and Hudgens, S. J. (1975). *In* "Electronic Phenomena in Non-Crystalline Semiconductors" (B. T. Kolomiets, ed.), pp. 6–15. Nauka, Leningrad, USSR.
Göetz, G., Karthe, W., Schnabel, B., and Sobolev, N. A. (1978). *Phys. Status Solidi A* **50,** K209–K212.
Gourdon, J. C., Fretier, P., and Pescia, J. (1981). *J. Phys. Lett. Orsay, Fr.* **42,** 21–24.
Greenbaum, S. G., Carlos, W. E., and Taylor, P. C. (1982). *Solid State Commun.* **43,** 663–666.
Greenbaum, S. G., Carlos, W. E., and Taylor, P. C. (1983). *Physica* **117B & 118B,** 886–888.
Griscom, D. L., Friebele, E. J., and Sigel, G. H. (1974). *Solid State Commun.* **15,** 479–483.
Haneman, D. (1968). *Phys. Rev.* **170,** 705–718.
Hasegawa, S., and Imai, Y. (1982a). *Philos. Mag. B* **45,** 347–360.
Hasegawa, S., and Imai, Y. (1982b). *Philos. Mag. B* **46,** 239–251.
Hasegawa, S., and Yazaki, S. (1977). *Solid State Commun.* **23,** 41–44.
Hasegawa, S., and Yazaki, S. (1978). *Thin Solid Films* **55,** 15–24.
Hasegawa, S., Yazaki, S., and Shimizu, T. (1977). *Solid State Commun.* **23,** 901–903.
Hasegawa, S., Yazaki, S., and Shimizu, T. (1978). *J. Non-Cryst. Solids* **27,** 215–224.
Hasegawa, S., Kasajima, T., and Shimizu, T. (1979). *Solid State Commun.* **29,** 13–16.
Hasegawa, S., Shimizu, T., and Hirose, M. (1980). *J. Phys. Soc. Jpn.* **49,** Suppl. A., 1237–1240.
Hasegawa, S., Kasajima, T., and Shimizu, T. (1981a). *Philos. Mag. A* **43,** 149–156.
Hasegawa, S., Kurata, Y., Imai, Y., and Narikawa, S. (1981b). *J. Phys. Colloq.* **42,** C4-675–678.
Hasegawa, S., Ando, D., Kurata, Y., and Shimizu, T. (1983a). *Philos. Mag. B* **47,** 139–149.
Hasegawa, S., Narikawa, S., and Kurata, Y. (1983b). *Physica* **117B & 118B,** 914–916.
Hirabayashi, I., Morigaki, K., and Nitta, S. (1980). *Jpn. J. Appl. Phys.* **19,** L357–L360.
Hirabayashi, I., Morigaki, K., and Yoshida, M. (1982). *Solar Energy Mat.* **8,** 153–158.
Holzenkämpfer, E., Richter, F. W., Stuke, J., and Voget-Grote, U. (1979). *J. Non-Cryst. Solids* **32,** 327–338.
Ishii, N., Kumeda, M., and Shimizu, T. (1981). *Jpn. J. Appl. Phys.* **20,** L673–L676.
Ishii, N., Kumeda, M., and Shimizu, T. (1982a). *Solid State Commun.* **41,** 143–146.
Ishii, N., Kumeda, M., and Shimizu, T. (1982b). *Jpn. J. Appl. Phys.* **21,** L92–L94.
Ishii, N., Oozora, S., Kumeda, M., and Shimizu, T. (1982c). *Phys. Status Solidi B* **114,** K111–K114.
Ishii, N., Kumeda, M., and Shimizu, T. (1983). *Phys. Status Solidi B* **116,** 91–100.
Jeffrey, F. R., and Lowry, M. E. (1981). *J. Appl. Phys.* **52,** 5529–5533.
Jeffrey, F. R., Lowry, M. E., Garcia, M. L. S., Barnes, R. G., and Torgeson, D. E. (1981a). *AIP Conf. Proc.* **73,** 83–88.
Jeffrey, F. R., Murphy, P. D., and Gerstein, B. C. (1981b). *Phys. Rev. B* **23,** 2099–2101.
Kerns, R., and Biegelsen, D. K. (1980). Xerox Corporation, Palo Alto Research Center, Palo Alto, Calif., and Solar Energy Research Institute, Golden, Colorado. Unpublished.

Kishimoto, N., Morigaki, K., Murakami, K., Shimizu, A., and Hiraki, A. (1977). *Phys. Status Solidi B* **80,** K113-K115.

Kishimoto, N., Morigaki, K., and Murakami, K. (1981). *J. Phys. Soc. Jpn.* **50,** 1970-1977.

Knights, J. C. (1977). *In* "Amorphous and Liquid Semiconductors" (W. E. Spear, ed.), pp. 433-436. Univ. of Edinburgh, Edinburgh, Scotland.

Knights, J. C., Biegelsen, D. K., and Solomon, I. (1977). *Solid State Commun.* **22,** 133-137.

Knights, J. C., Lujan, R. A., Rosenblum, M. P., Street, R. A., Biegelsen, D. K., and Reimer, J. A. (1983).

Kubler, L., Jaegle, A., and Koulmann, J. J. (1979). *Phys. Status Solidi B* **95,** 307-315.

Kubler, L., Haug, R., Ringeisen, F., and Jaegle, A. (1983). *J. Non-Cryst. Solids* **54,** 27-42.

Kumeda, M., and Shimizu, T. (1980). *Jpn. J. Appl. Phys.* **19,** L197-L200.

Kumeda, M., Jinno, Y., and Shimizu, T. (1977a). *Phys. Status Solidi B* **81,** K71-K74.

Kumeda, M., Jinno, Y., Watanabe, I., and Shimizu, T. (1977b). *Solid State Commun.* **23,** 833-835.

Kumeda, M., Yonezawa, Y., Nakazawa, K., Ueda, S., and Shimizu, T., (1983). *Jpn. J. Appl. Phys.* **22,** L194-L196.

Lamotte, B., Rousseau, A., and Chenevas-Paule, A. (1981a). *Recent Dev. Condens. Matter Phys.* **2,** 247-251.

Lamotte, B., Rousseau, A., and Chenevas-Paule, A. (1981b). *J. Phys. Colloq. Orsay, Fr.* **42,** C4-839-841.

LeComber, P. G., Loveland, R. J., Spear, W. E., and Vaughan, R. A. (1974). *Proc. Int. Conf. on Amorphous and Liquid Semicond., 5th,* (J. Stuke and W. Brenig, eds.), pp. 245-50. Taylor and Francis, London.

Leopold, D. J., Boyce, J. B., Fedders, P. A., and Norberg, R. E. (1982). *Phys. Rev. B* **26,** 6053-6066.

Lowry, M. E., Barnes, R. G., Torgeson, D. R., and Jeffrey, F. R. (1981a). *Proc. Symp. Mater. Res. Soc.* **3,** 341-346.

Lowry, M. E., Jeffrey, F. R., Barnes, R. G., and Torgeson, D. R. (1981b). *Solid State Commun.* **38,** 113-116.

Lucovsky, G., Nemanich, R. J., and Knights, J. C. (1979). *Phys. Rev. B* **19,** 2064-2073.

Mackinnon, A. (1980). *J. Phys. Soc. Jpn.* **49** *Suppl. A.,* 1185-1188.

Mackinnon, A., and Kramer, B. (1981). *In* "Recent Developments in Condensed Matter Physics" (J. T. Devreese, ed.), Vol. 2, pp. 207-215. Plenum, New York.

Magarino, J., Friederich, A., Kaplan, D., and Deneuville, A. (1981). *J. Phys. Colloq. Orsay, Fr.* **42,** C4-737-740.

Magarino, J., Kaplan, D., Friederich, A., and Deneuville, A. (1982). *Philos. Mag. B* **45,** 285-306.

Makino, T., and Nakamura, H. (1978). *Jpn. J. Appl. Phys.* **17,** 1897-1898.

Mell, H., Schweitzer, L., and Voget-Grote, U. (1980). *J. Non-Cryst. Solids* **35** & **36,** 639-644.

Miller, D. J., and Haneman, D. (1978). *Solid State Commun.* **27,** 91-94.

Morigaki, K., Sano, T., and Hirabayashi, I. (1982). *J. Phys. Soc. Jpn.* **51,** 147-152.

Morimoto, A., Miura, T., Kumeda, M., and Shimizu, T. (1981). *Jpn. J. Appl. Phys.* **20,** L833-L836.

Morimoto, A., Miura, T., Kumeda, M., and Shimizu, T. (1982a). *Jpn. J. Appl. Phys.* **21,** L119-L121.

Morimoto, A., Miura, T., Kumeda, M., and Shimizu, T. (1982b). *J. Appl. Phys.* **53,** 7299-7305.

Movaghar, B., and Schweitzer, L. (1977). *Phys. Status Solidi B* **80,** 491-498.

Movaghar, B., and Schweitzer, L. (1981). *AIP Conf. Proc.* **73,** 73-77.

Movaghar, B., Overhof, H., and Schweitzer, L. (1977). *In* "Amorphous and Liquid Semiconductors" (W. E. Spear, ed.), pp. 419-423. Univ. of Edinburgh, Edinburgh, Scotland.

Nakashita, T., Hirose, M., and Osaka, Y. (1981). *Jpn. J. Appl. Phys.* **20,** 471–475.
Nakazawa, K., Ueda, S., Kumeda, M., Morimoto, A., and Shimizu, T. (1982). *Jpn. J. Appl. Phys.* **21,** L176–L178.
Nemanich, R. J., and Knights, J. C. (1980). *J. Non-Cryst. Solids* **35** & **36,** 243–248.
Ngai, K. L. (1981). *J. Phys. Colloq. Orsay, Fr.* **42,** C4-835–838.
Ohdomari, I., Kakumu, M., Sugahara, H., Hori, M., Saito, T., Yonehara, T., and Hajimoto, Y. (1981). *J. Appl. Phys.* **52,** 6617–6622.
Overhof, H. (1982). *Phys. Status Solidi B* **110,** 521–530.
Overhof, H., and Beyer, W. (1981). *Phys. Status Solidi B* **107,** 207–213.
Pankove, J. I. (1982). *Solar Energy Mat.* **8,** 141–151.
Pankove, J. I., and Berkeyheiser, J. E. (1980). *Appl. Phys. Lett.* **37,** 705–706.
Paul, W. (1980). *Solid State Commun.* **34,** 283–285.
Paul, W., and Anderson, D. A. (1981). *Solar Energy Mat.* **5,** 229–316.
Pawlik, J. R., and Paul, W. (1977). *In* "Amorphous and Liquid Semiconductors" (W. E. Spear, ed.), pp. 437–441. Univ. of Edinburgh, Edinburgh, Scotland.
Pawlik, J. R., Connell, G. A. N., and Prober, D. (1976). *In* "Electronic Phenomena in Non-Crystalline Semiconductors" (B. T. Kolomiets, ed.), pp. 304–309. Nauka, Leningrad, USSR.
Pontuschka, W. M., Carlos, W. E., and Taylor, P. C. (1982). *Phys. Rev. B* **25,** 4362–4376.
Reimer, J. A. (1981). *J. Phys. Colloq. Orsay, Fr.* **42,** C4-715–724.
Reimer, J. A., and Duncan, T. M. (1983). *Phys. Rev. B* **27,** 4895–4901.
Reimer, J. A., and Knights, J. C. (1981). *AIP Conf. Proc.* **73,** 78–82.
Reimer, J. A., Vaughan, R. W., and Knights, J. C. (1980). *Phys. Rev. Lett.* **44,** 193–196.
Reimer, J. A., Murphy, P. D., Gerstein, B. C., and Knights, J. C. (1981a). *J. Chem. Phys.* **74,** 1501–1503.
Reimer, J. A., Vaughan, R. W., and Knights, J. C. (1981b). *Phys. Rev. B* **24,** 3360–3370.
Reimer, J. A., Vaughan, R. W., and Knights, J. C. (1981c). *Phys. Rev. B* **23,** 2567–2575.
Reimer, J. A., Vaughan, R. W., Knights, J. C., and Lujan, R. A. (1981d). *J. Vac. Sci. Technol.* **19,** 53–56.
Reimer, J. A., Vaughan, R. W., and Knights, J. C. (1981e). *Solid State Commun.* **37,** 161–164.
Robertson, J. (1983). *Phys. Rev. B* **28,** 4666–4670.
Sano, Y., Morigaki, K., and Hirabayashi, I. (1982). *Solid State Commun.* **43,** 439–442.
Shanks, H. R., Jeffrey, F. R., and Lowry, M. E. (1981). *J. Phys. Colloq. Orsay, Fr.* **42,** C4-773–777.
Shen, S. C., and Cardona, M. (1981). *Phys. Rev. B* **23,** 5322–5328.
Shimizu, T., Kumeda, M., Watanabe, I., and Kamono, K. (1980a). *J. Non-Cryst. Solids* **35** & **36,** 303–308.
Shimizu, T., Kumeda, M., Watanabe, I., and Kiriyama, Y. (1980b). *Jpn. J. Appl. Phys.* **19,** L235–L238.
Shimizu, T., Kumeda, M., Watanabe, I., and Noumi, Y. (1980c). *J. Non-Cryst. Solids* **35** & **36,** 645–650.
Shimizu, T., Kumeda, M., and Kiriyama, Y. (1981a). *Solid State Commun.* **37,** 699–703.
Shimizu, T., Kumeda, M., and Kiriyama, Y. (1981b). *AIP Conf. Proc.* **73,** 171–175.
Shimizu, T., Kumeda, M., Watanabe, I., and Noumi, Y. (1981c). *Philos. Mag. B* **44,** 159–174.
Shimizu, T., Nakazawa, K., Kumeda, M., and Ueda, S. (1982a). *Jpn. J. Appl. Phys.* **21,** L351–L353.
Shimizu, T., Oozora, S., Morimoto, A., Kumeda, M., and Ishii, N. (1982b). *Solar Energy Mat.* **8,** 311–317.
Shimizu, T., Nakazawa, K., Kumeda, M., and Ueda, S. (1983). *Physica* **117B** & **118B,** 926–928.

Solomon, I. (1979). *In* "Amorphous Semiconductors" (M. H. Brodsky, ed.), pp. 189–213. Springer-Verlag, Berlin and New York.
Street, R. A. (1981). *J. Phys. Colloq. Orsay, Fr.* **42,** C4-283–291.
Street, R. A. (1982a). *Philos. Mag. B* **46,** 273–278.
Street, R. A. (1982b). *Appl. Phys. Lett.* **41,** 1060–1062.
Street, R. A. (1983). *Appl. Phys. Lett.* **42,** 507–509.
Street, R. A., and Biegelsen, D. K. (1980a). *J. Non-Cryst. Solids* **35 & 36,** 651–656.
Street, R. A., and Biegelsen, D. K. (1980b). *Solid State Commun.* **33,** 1159–1162.
Street, R. A., and Biegelsen, D. K. (1982). *Solid State Commun.* **44,** 501–505.
Street, R. A., and Knights, J. C. (1981). *Philos. Mag. B* **43,** 1091–1098.
Street, R. A., Knights, J. C., and Biegelsen, D. K. (1978). *Phys. Rev. B* **18,** 1880–1891.
Street, R. A., Biegelsen, D. K., and Stuke, J. (1979). *Philos. Mag. B* **40,** 451–464.
Street, R. A., Biegelsen, D. K., and Knights, J. C. (1981). *Phys. Rev. B* **24,** 969–984.
Stuke, J. (1976). *In* "Electronic Phenomena in Non-Crystalline Semiconductors" (B. T. Kolomiets, ed.), pp. 193–202. Nauka, Leningrad, USSR.
Stuke, J. (1977). *In* "Amorphous and Liquid Semiconductors" (W. E. Spear, ed.), pp. 406–418. Univ. of Edinburgh, Edinburgh, Scotland.
Stutzmann, M., and Biegelsen, D. K. (1983). *Phys. Rev. B* **28,** 6256–6261.
Stutzmann, M., and Stuke, J. (1983). *Solid State Commun.* **47,** 635–639.
Stutzmann, M., Stuke, J., and Dersch, H. (1983). *Phys. Status Solidi B* **115,** 141–151.
Suzuki, M., Maruyama, K., Kumeda, M., and Shimizu, T. (1979). *Phys. Status Solidi A* **56,** K49–K52.
Suzuki, M., Maekawa, T., Nakao, A., Kumeda, M., and Shimizu, T. (1980). *Solid State Commun.* **36,** 393–396.
Suzuki, M., Suzuki, M., Kanada, M., and Kakimoto, Y. (1982). *Jpn. J. Appl. Phys.* **21,** L89–L91.
Taylor, P. C. (1983). *J. Non-Cryst. Solids* **59/60,** 109–116.
Taylor, P. C., and Carlos, W. E. (1980). *J. Phys. Soc. Jpn.* **49,** *Suppl. A.,* 1193–1196.
Taylor, P. C., and Ohlsen, W. D. (1983). *Solar Cells* **9,** 113–118.
Taylor, P. C., Baugher, J. F., and Kriz, H. M. (1975). *Chem. Rev.* **75,** 203–240.
Thomas, P., and Flachet, J. C. (1981). *J. Phys. Colloq. Orsay, Fr.* **42,** C4-151–154.
Thomas, P. A., and Kaplan, D. (1976). *AIP Conf. Proc.* **31,** 85–89.
Thomas, P. A., Lepine, D., and Kaplan, D. (1974). *AIP Conf. Proc.* **20,** 47–52.
Thomas, P. A., Brodsky, M. H., Kaplan, D., and Lepine, D. (1978). *Phys. Rev. B* **18,** 3059–3073.
Title, R. S., Brodsky, M. H., and Crowder, B. L. (1970). *Proc. Int. Conf. Phys. Semicond.* **10th,** (S. P. Keller, Ed.), pp. 794–8. Nat. Tech. Inf. Serv., Springfield, Virginia.
Title, R. S., Brodsky, M. H., and Cuomo, J. J. (1977). *In* "Amorphous Liquid Semiconductors" (W. E. Spear, ed.), pp. 424–428. Univ. of Edinburgh, Edinburgh, Scotland.
Tsai, C. C. (1979). *Phys. Rev. B* **19,** 2041–2055.
Ueda, S., Kumeda, M., and Shimizu, T. (1981a). *Jpn. J. Appl. Phys.* **20,** L399–L402.
Ueda, S., Kumeda, M., and Shimizu, T. (1981b). *J. Phys. Colloq. Orsay, Fr.* **42,** C4-729–732.
Ueda, S., Nakazawa, K., Kumeda, M., and Shimizu, T. (1982). *Solid State Commun.* **42,** 261–266.
Voget-Grote, U., and Stuke, J. (1979). *J. Electron. Mater.* **8,** 749–761.
Voget-Grote, U., Stuke, J., and Wagner, H. (1976). *AIP Conf. Proc.* **31,** 91–95.
Watanabe, I., Hata, Y., Morimoto, A., and Shimizu, T. (1982). *Jpn. J. Appl. Phys.* **21,** L613–L615.
Wautelet, M., Laude, L. D., and Failly-Lovato, M. (1981a). *Solid State Commun.* **39,** 979–981.
Wautelet, M., Andrew, R., Failly-Lovato, M., and Laude, L. D. (1981b). *J. Phys. Colloq. Orsay, Fr.* **42,** C4-395–398.

Weaire, D., Higgins, N., Moore, P., and Marshall, I. (1979). *Philos. Mag. B* **40,** 243–245.
Yamasaki, S., Hata, N., Yoshida, T., Oheda, H., Matsuda, A., Okushi, H., and Tanaka, K. (1981). *J. Phys. Colloq. Orsay, Fr.* **42,** C4-297–300.
Yamasaki, S., Kuroda, S., and Tanaka, K. (1983). *J. Non-Cryst. Solids* **59/60,** 141–144.
Yonehara, T., Saitoh, T., Kawarada, H., Hirata, T., Kakumu, M., and Ohdomari, I. (1980). *Phys. Lett. A* **78,** 192–194.
Yonezawa, F. (1982). *In* "Amorphous Semiconductor Technologies and Devices" (Y. Hamakawa, ed.), pp. 9–31. North-Holland Publ., Amsterdam.
Zellama, K., Germain, P., Squelard, S., Monse, J., and Ligeon, E. (1980). *J. Non-Cryst. Solids* **35/36,** 225–230.
Zellama, K., Germain, P., Picard, C., and Bourbon, B. (1981a). *J. Phys. Colloq. Orsay, Fr.* **42,** C4-815–818.
Zellama, K., Germain, P., Squelard, S., Bourbon, B., Fontenille, J., and Danielou, R. (1981b). *Phys. Rev. B* **23,** 6648–6667.

CHAPTER 4

Optically Detected Magnetic Resonance

K. Morigaki

INSTITUTE FOR SOLID STATE PHYSICS
UNIVERSITY OF TOKYO
ROPPONGI, TOKYO, JAPAN

I. Introduction

Electron spin resonance (ESR) techniques have been widely used to investigate the microscopic nature of magnetic centers, including defects and impurities in solids. Optically detected magnetic resonance (ODMR) provides a powerful tool in a twofold way for those investigations; besides the above purpose, ODMR links ESR and optical phenomena, e.g., luminescence, by monitoring the change in luminescence intensity at resonance. Thus, the recombination processes involved in the luminescence and the nature of recombination centers can be investigated using the ODMR technique [for reviews see Cavenett (1981), Morigaki (1983)]. In this chapter we describe briefly the principle of ODMR and its technique and review the results of ODMR that have been obtained so far in a-Si : H as well as those of time-resolved ODMR (TRODMR), electron nuclear double resonance (ENDOR), and optically detected ENDOR (ODENDOR). Very recently, the conventional ODMR technique has been extended to investigate spin-dependent photoinduced absorption, i.e., to monitor the intensity of pho-

ISBN 0-12-752149-6

toinduced absorption. This new technique is also reviewed as it applies to a-Si:H.

II. Principle of Optically Detected Magnetic Resonance

In the conventional ODMR techniques, one monitors the luminescence intensity to detect ESR signals in terms of changes in luminescence at resonance. These changes are a consequence of spin-dependent recombination processes. For trapped electron–hole pair recombination, illustrated in Fig. 1, the spin-dependent nature of the radiative recombination can be easily understood by considering that radiative recombination is allowed when spins of an electron and a hole are antiparallel and is forbidden, in principle, when they are parallel.

For simplicity, we assume that the generation rate is the same for four Zeeman levels of an electron–hole pair, i.e., $G[N-(n_1+n_2+n_3+n_4)]$, where G and N are the generation rate and the total number of pairs of electron and hole traps, respectively, and n_1, n_2, n_3, and n_4 designate the population of each Zeeman level, as shown in Fig. 1. When the recombination rate R of antiparallel spin states is greater than the recombination rate

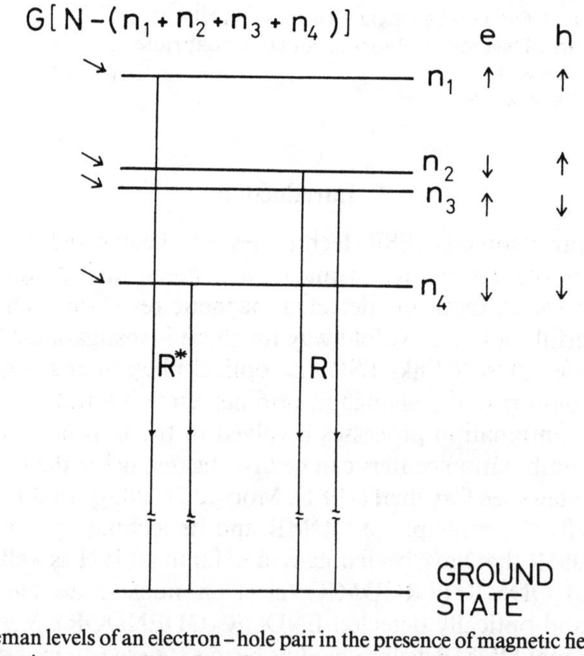

FIG. 1. Zeeman levels of an electron–hole pair in the presence of magnetic field. G, N, and n_i designate generation rate, total number of pairs of electron and hole traps, and population of ith level, respectively. [From Morigaki (1983).]

R^* of parallel spin states and each Zeeman level is unthermalized, the populations of the antiparallel spin states become smaller than those of the parallel spin states. Associated with ESR transitions of either electrons or holes, the population of the antiparallel spin states becomes large. As a result, the luminescence intensity I increases compared with before the occurrence of the ESR transition. The quantitative calculation of $(\Delta I/I)_{\mathrm{ESR}}$, the relative change in I at resonance, has been done by Kaplan *et al.* (1978), Dunstan and Davies (1979), Movaghar *et al.* (1980) and Morigaki (1981). Here, we present the calculated results, using the rate equations (Morigaki, 1981)

$$
\begin{aligned}
\frac{dn_1}{dt} &= G[N-(n_1+n_2+n_3+n_4)] - R^*n_1 - \frac{1}{T_1}n_1 \\
&\quad + \frac{(n_1+n_2+n_3+n_4)\,e^{-E_1/kT}}{T_1(e^{-E_1/kT}+e^{-E_2/kT}+e^{E_2/kT}+e^{E_1/kT})} \\
&\quad + W_{\mathrm{h}}(n_3-n_1) + W_{\mathrm{e}}(n_2-n_1), \\
\frac{dn_2}{dt} &= G[N-(n_1+n_2+n_3+n_4)] - Rn_2 - \frac{1}{T_1}n_2 \\
&\quad + \frac{(n_1+n_2+n_3+n_4)\,e^{-E_2/kT}}{T_1(e^{-E_1/kT}+e^{-E_2/kT}+e^{E_2/kT}+e^{E_1/kT})} \\
&\quad + W_{\mathrm{h}}(n_4-n_2) + W_{\mathrm{e}}(n_1-n_2), \\
\frac{dn_3}{dt} &= G[N-(n_1+n_2+n_3+n_4)] - Rn_3 - \frac{1}{T_1}n_3 \\
&\quad + \frac{(n_1+n_2+n_3+n_4)\,e^{E_2/kT}}{T_1(e^{-E_1/kT}+e^{-E_2/kT}+e^{E_2/kT}+e^{E_1/kT})} \\
&\quad + W_{\mathrm{h}}(n_1-n_3) + W_{\mathrm{e}}(n_4-n_3), \\
\frac{dn_4}{dt} &= G[N-(n_1+n_2+n_3+n_4)] - R^*n_4 - \frac{1}{T_1}n_4 \\
&\quad + \frac{(n_1+n_2+n_3+n_4)\,e^{E_1/kT}}{T_1(e^{-E_1/kT}+e^{-E_2/kT}+e^{E_2/kT}+e^{E_1/kT})} \\
&\quad + W_{\mathrm{h}}(n_2-n_4) + W_{\mathrm{e}}(n_3-n_4),
\end{aligned}
\tag{1}
$$

which describe the time-variation of n_1, n_2, n_3, and n_4 with equal generation rates, as shown in Fig. 1. In Eqs. (1) T_1 is the spin–lattice relaxation time of each Zeeman level assuming all have the same relaxation time; E_1 and E_2 are given by $[(g_{\mathrm{h}}+g_{\mathrm{e}})/2]\mu_{\mathrm{B}}H$ and $[(g_{\mathrm{h}}-g_{\mathrm{e}})/2]\mu_{\mathrm{B}}H$, respectively, using the g-values of the trapped electron g_{e} and trapped hole g_{h}, Bohr magneton μ_{B}, and magnetic field H; and W_e and W_h are the ESR transition rates of trapped electrons and holes, respectively. In these equations, the terms on the right-hand side of each equation are the generation, recombination, relaxation, thermalization, and hole and electron ESR transition terms, respectively. Since the luminescence intensity is proportional to $(n_2 + n_3)$, its relative change at resonance is easily obtained from Eqs. (1) in the steady-state condition as

$$(\Delta I/I)_{\mathrm{ESR}} = \frac{(R^* + 4G)\{(R - R^*) + (2/T_1)[(1 - x_e)(1 - x_h)/(1 + x_e)(1 + x_h)]\}}{(R + R^* + 8G)\{R^* + (2/T_1)[(x_e + x_h)/(1 + x_e)(1 + x_h)]\}} \times \frac{2T_1^* W}{1 + 2T_1^* W}, \tag{2}$$

$$1/T_1^* = 2\{RR^* + 2G[R + R^* + (2/T_1)] + (1/T_1)[1/(1 + x_e)(1 + x_h)] \times [R^*(1 + x_e x_h) + R(x_e + x_h)]\}/(R + R^* + 8G), \tag{3}$$

$$x_e = e^{-g_e \mu_B H/kT}, \qquad x_h = e^{-g_h \mu_B H/kT}, \tag{4}$$

where T_1^* is the effective spin–lattice relaxation time, and W is either W_e or W_h, depending on whether we are considering trapped electrons or holes. We can approximate $(\Delta I/I)_{\mathrm{ESR}}$ in the two limiting cases, i.e., unthermalized and thermalized, as follows.

(1) For the unthermalized case, i.e., R, $R^* \gg 1/T_1$, where T_1 is the spin–lattice relaxation time of either electrons or holes:

When G is small,

$$\left(\frac{\Delta I}{I}\right)_{\mathrm{ESR}} = \frac{R - R^*}{R + R^*}\frac{2T_1^* W}{1 + 2T_1^* W}, \tag{5}$$

$$\frac{1}{T_1^*} = 2\left[R^* + \frac{1}{T_1}\frac{x_e + x_h}{(1 + x_e)(1 + x_h)} + 2G\right], \tag{6}$$

$$x_e = e^{-g_e \mu_B H/kT}, \tag{7}$$

$$x_h = e^{-g_h \mu_B H/kT}. \tag{8}$$

The factor of $2T_1^* W/(1 + 2T_1^* W)$ in Eqs. (2) and (5) is known as the saturation factor in magnetic resonance. For the saturation factor equal to one (strong microwave power case), the magnitude of $(\Delta I/I)_{\mathrm{ESR}}$ reaches 100% when R is greater than R^*.

When G is large,

$$\left(\frac{\Delta I}{I}\right)_{\rm ESR} = \frac{R - R^*}{2R^*} \frac{2T_1^* W}{1 + 2T_1^* W}, \tag{9}$$

$$1/T_1^* = R/2. \tag{10}$$

In this case, one can also expect a large magnitude of $(\Delta I/I)_{\rm ESR}$ when the saturation factor is equal to one.

(2) For the thermalized case, i.e., R, $R^* \ll 1/T_1$,

$$\left(\frac{\Delta I}{I}\right)_{\rm ESR} = \frac{R^* + 4G}{R + R^* + 8G} \frac{(1 - x_{\rm e})(1 - x_{\rm h})}{x_{\rm e} + x_{\rm h}} \frac{2T_1^* W}{1 + 2T_1^* W}. \tag{11}$$

In this case, the magnitude of $(\Delta I/I)_{\rm ESR}$ becomes smaller than that for the unthermalized case, because $x_{\rm e}$ and $x_{\rm h}$ are nearly equal to one except at very low temperatures at which the Zeeman levels are generally unthermalized, so that the thermalization factor $(1 - x_{\rm e})(1 - x_{\rm h})/(x_{\rm e} + x_{\rm h})$ becomes less than one. This result can be easily understood by considering that thermalization among the four Zeeman levels tends to eliminate a greater difference in their populations between antiparallel spin states and parallel spin states after the rapid radiative recombination.

In contrast to the ODMR of radiative centers, the ESR of nonradiative centers gives rise to a decrease in the luminescence intensity in the following way: For example, we consider the dangling-bond center in a-Si:H as a nonradiative center. As shown in Fig. 2, nonradiative recombination occurs

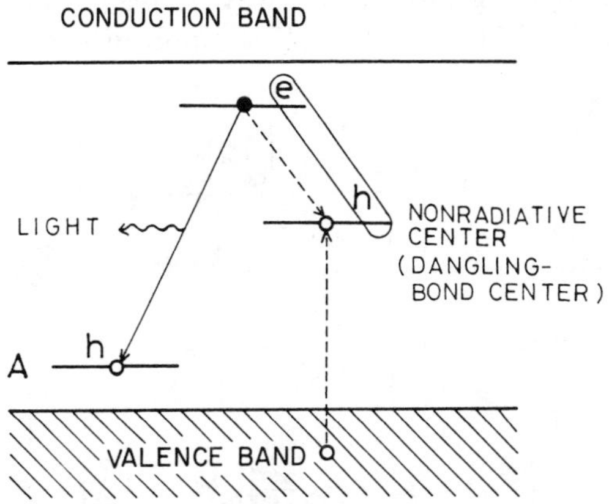

FIG. 2. Schematic diagram of radiative and nonradiative recombination processes. Solid and dashed arrows show radiative and nonradiative recombination, respectively. A radiative hole center is the A center. [From Morigaki (1981).]

at the dangling-bond center, so that we treat a pair of a radiative electron and a nonradiative hole (dangling-bond center) as an electron–hole pair. Considerations similar to those for the radiative electron–hole pair can be applied to the radiative-electron–nonradiative-hole pair. Thus when either radiative electron resonance or nonradiative hole resonance (dangling-bond-center resonance) occurs, the nonradiative recombination is enhanced and eventually the luminescence intensity decreases. The relative change in the luminescence intensity at resonance of the nonradiative centers is given by (Morigaki, 1981).

$$\left(\frac{\Delta I}{I}\right)_{\mathrm{ESR}} = -\frac{R_{\mathrm{ns}}^2}{(2R_1 + R_{\mathrm{ns}})(8G + 2R_1 + R_{\mathrm{ns}})}\frac{2T_1^* W}{1 + 2T_1^* W}, \tag{12}$$

$$\frac{1}{T_1^*} = \frac{4G(2R_1 + R_{\mathrm{ns}}) + 2R_1(R_1 + R_{\mathrm{ns}})}{8G + 2R_1 + R_{\mathrm{ns}}}, \tag{13}$$

where R_1 and R_{ns} designate the radiative recombination rate and the nonradiative recombination rate (antiparallel spin states of an electron–hole pair), shown in Fig. 2. We neglect the nonradiative recombination rate of an electron–hole pair of parallel spin states. As shown in Eq. (12), we obtain a decrease in the luminescence intensity at resonance.

III. Experimental Apparatus

A schematic diagram of the apparatus for ODMR measurements is shown in Fig. 3. Microwave power is chopped at a low frequency, e.g., 1 kHz, using a *p–i–n* diode or applying a rectangular wave to the reflector voltage for the klystron. Emitted light coming through a monochromator from a sample

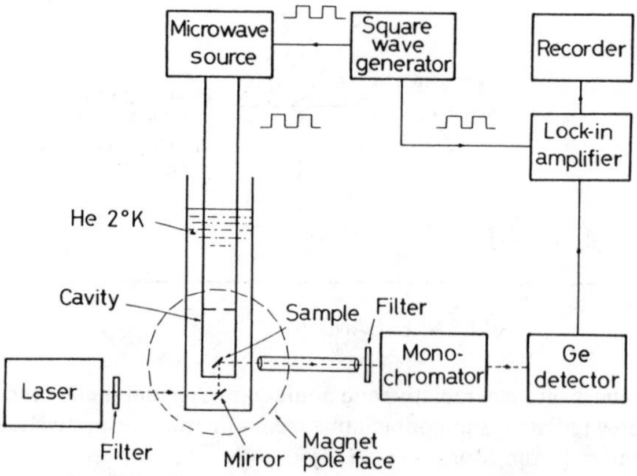

FIG. 3. Block diagram for ODMR measurements.

excited by laser light is detected by either a Ge detector or a photomultiplier. The ODMR signal is lock-in detected and recorded with a scan of the magnetic field.

The detail of our experimental apparatus used in the steady state ODMR experiment is given in Morigaki *et al.* (1983).

IV. Steady State ODMR

Various glow-discharge samples of a-Si : H prepared at different substrate temperatures ranging from 75 °C to 300 °C, sputtered a-Si : H samples, and glow-discharge a-Si : H,F samples have been subjected to the ODMR measurements in our laboratory (Morigaki *et al.*, 1981; 1982a,b, 1983; Sano *et al.*, 1982a,b). The characteristics of these samples and their preparation conditions are listed in Table I.

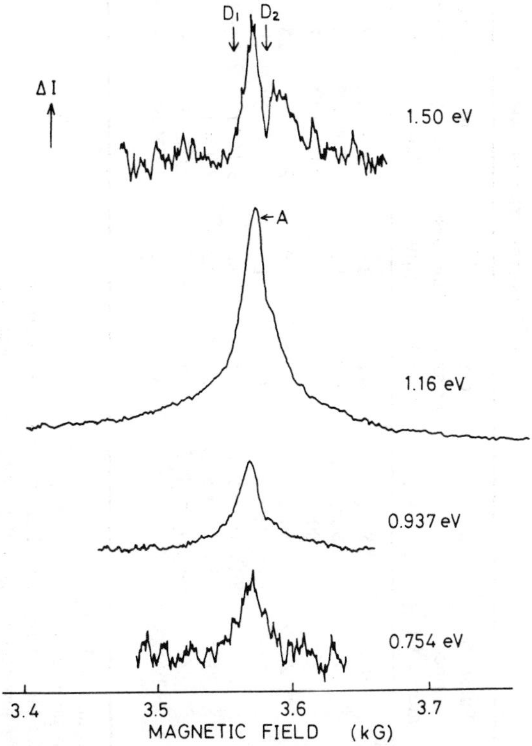

FIG. 4. ODMR spectra observed at 10 GHz and 2°K in a-Si:H No. 519 (T_s = 300°C), monitoring the intensity of emitted light whose photon energy is indicated in the figure, under excitation by unfocused argon ion laser light of 25 mW at 514.5 nm. [From Morigaki *et al.* (1982a).]

TABLE I

CHARACTERISTICS OF SAMPLES[a]

Sample	Substrate temper-ature (°C)	Gas	Gas pressure (Torr)	Thickness (μm)	Spin density (cm^{-3})	H content (at. %)	F content (at. %)	E_{og} (eV)[b]	E_{HL} (eV)[c]	E_{LL} (eV)[d]	E_A (eV)[e]	E_D^- (eV)[f]
a-Si:H, No. 519	300	SiH_4	0.2	2	7.0×10^{15}	7		1.80	1.69	0.80	0.25	0.6
a-Si:H, No. 524	200	SiH_4	0.2	1.25	2.7×10^{16}	18		1.85	1.84	0.72	0.15	0.7
a-Si:H, No. 540	75	SiH_4	0.2	2	2.9×10^{18}	33		~1.9	1.92	~0.6		
a-Si:H, No. 541	120	SiH_4	0.2	1.9	2.5×10^{17}	28		~1.9	1.87	0.66		
a-Si:H, No. 12	300	$SiH_4:Ar = 1:9$	1	1	1.3×10^{16}			1.80	1.75	0.70	0.25	0.6
a-Si:H,F, No. 813	250	$SiF_4:H_2 = 5:1$	1.5	0.2	9.1×10^{16}	15–17	<1	~1.8	1.83	0.65	~0	0.68
a-Si:H,F, No. 814	300	$SiF_4:H_2 = 5:1$	1.5	0.2	7.9×10^{16}			~1.8	1.75	0.60	~−0.05	0.65
a-Si:H,F, No. 995	300	$SiF_4:H_2 = 10:5$	1.5	0.2	2.0×10^{17}	30–40	≳4		1.72	~0.6		
a-Si, No. H6	~100		Ar: 0.3	1.1	4.1×10^{17}	4		1.90	1.90	0.65	~0	0.65

[a] After Morigaki *et al.* (1983).
[b] E_{og}: optical band-gap energy.
[c] H_{HL}: energy of the high energy edge of the luminescence spectrum.
[d] H_{LL}: energy of the low energy edge of the luminescence spectrum.
[e] E_A: energy of the A center above the valence band.
[f] E_D^-: energy of the doubly occupied dangling-bond center below the conduction band.

The ODMR spectra are dependent on various experimental conditions, i.e., microwave power, excitation intensity (laser power), luminescence energy, microwave frequency, etc. Our experimental results on the ODMR spectra are summarized as follows: For low-defect-density samples, the ODMR spectra consist of three components that are designated as the D_1, D_2, and A lines. The D_1 and D_2 lines are the quenching signals for the luminescence, with *g*-values of 2.013 and 2.006, respectively. Here we use the nomenclature of the D_1 and D_2 centers used by Morigaki *et al.* (1978a,b), but the *g*-value of the D_1 center differs from the original value, because our original deconvolution was done for the quenching line overlapped with the enhancing line.

According to the theoretical considerations given in Part II, those two lines (D_1 and D_2) are identified as being due to nonradiative centers. Furthermore, the *g*-value of the D_2 line coincides with that of dangling-bond centers in a-Si (the dark ESR signal), i.e., 2.0055 (Brodsky and Title, 1969), so that the D_2 center is attributed to a dangling-bond center. The A line is observed as an

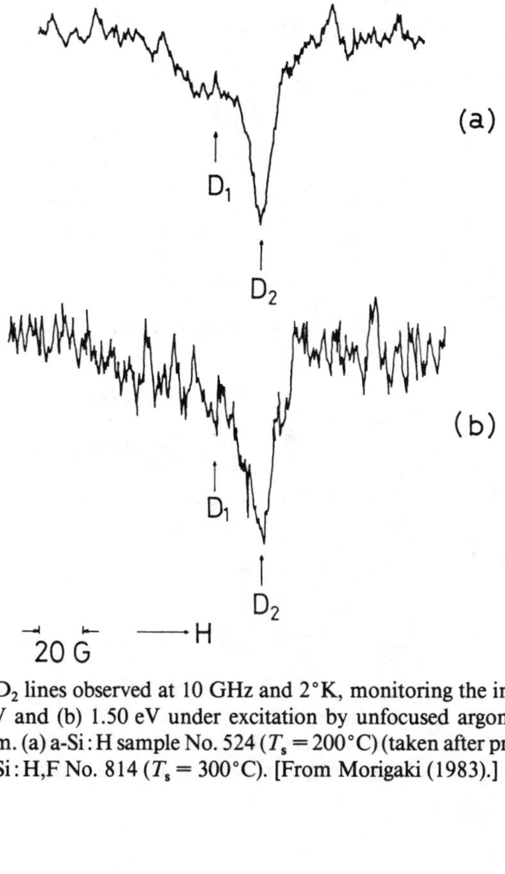

FIG. 5. D_1 and D_2 lines observed at 10 GHz and 2°K, monitoring the intensity of emitted light at (a) 1.54 eV and (b) 1.50 eV under excitation by unfocused argon ion laser light of 25 mW at 514.5 nm. (a) a-Si:H sample No. 524 ($T_s = 200$°C) (taken after prolonged laser light irradiation), (b) a-Si:H,F No. 814 ($T_s = 300$°C). [From Morigaki (1983).]

enhancing signal for the luminescence and thus is attributed to radiative centers, weakly coupled with trapped electrons in the tail or gap states, according to the theoretical considerations given in Part II (Morigaki *et al.*, 1981). The *g*-value of A centers ranges between 2.00 and 2.012, depending on luminescence energy and microwave power.

Typical examples of the ODMR spectra observed in a low-defect-density sample are shown in Fig. 4 for various luminescence energies, fixing the laser power and the microwave power. The D_1 and D_2 lines are not generally resolved because they overlap the A line. Figure 5 shows the D_1 and D_2 lines; here the A line contribution appears to be almost negligible.

The intensity of ODMR signals is proportional to a change in the luminescence intensity ΔI at resonance. The quantity ΔI divided by the luminescence intensity I is used to characterize the spectral dependence of the

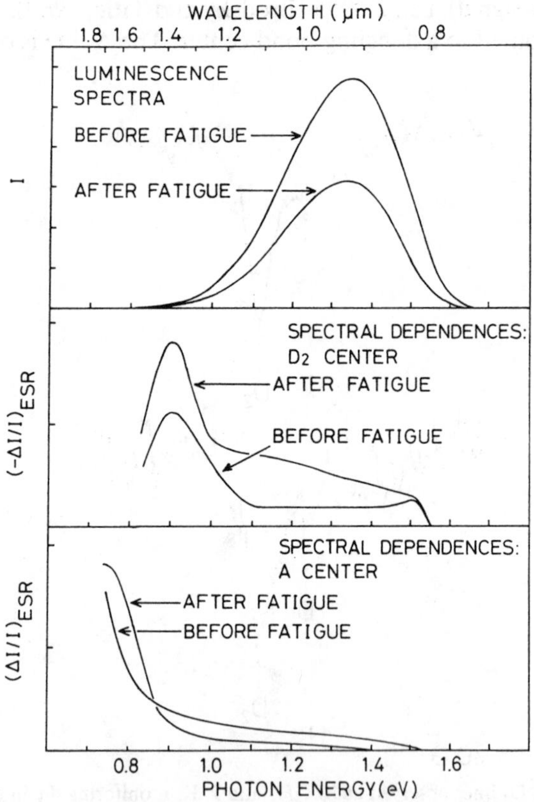

FIG. 6. Luminescence spectrum and spectral dependence of $(\Delta I/I)_{ESR}$ at resonances of the D_2 and A centers at 2°K under excitation by argon ion laser light of 25 mW in a-Si:H No. 519 (T_s = 300°C). Each value in the ordinate is plotted in arbitrary units. [From Morigaki *et al.* (1982a).]

ODMR signal, thus cancelling out the spectral dependence of the detection system of both ΔI and I. In Fig. 6, such a spectral dependence of $(\Delta I/I)_{ESR}$, i.e., relative change in the luminescence intensity at resonance, is shown as a function of luminescence energy, for a low-defect-density sample. In this figure, the luminescence spectrum is also shown. Two curves for the luminescence spectra and the spectral dependencies of $(\Delta I/I)_{ESR}$ for D_2 and A centers are referred to as those for before and after fatigue due to light exposure. Fatigue effects on the luminescence and the ODMR spectra will be discussed in Part V.

The magnitude of $(\Delta I/I)_{ESR}$ has been measured by observing a transient response of the ODMR signal to pulsed microwaves (Morigaki *et al.*, 1983). Typical magnitudes of $(\Delta I/I)_{ESR}$ for the A center resonance were 0.1–2%. Also the transient wave form of the ODMR signal is quite different for resonances of radiative and nonradiative centers. Hence, the transient wave form has been used to distinguish the nature of recombination centers, either radiative or nonradiative (Dunstan and Davies, 1979; Depinna *et al.*, 1982c; Street, 1982).

The ODMR spectra have been observed also at 34 GHz in a-Si:H. A typical trace of the ODMR spectrum (A line) monitoring emitted light of 1.16 eV is shown in Fig. 7. The full width at half amplitude of the A line is 92 G at 34 GHz, whereas it is 38 G at 10 GHz. The line broadening at 34 GHz is significant. This result shows that the broadening of the A line is inhomogeneous. However, the spectral dependence of $(\Delta I/I)_{ESR}$ for the D_2 and A centers at 34 GHz is almost similar to that measured at 10 GHz.

The D_2 and A lines are observed for all samples listed in Table I. However, the D_1 line is observed for samples containing few dangling bonds such as

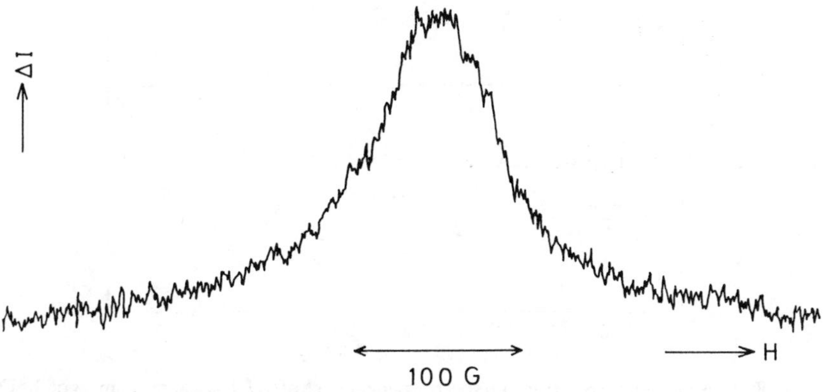

FIG. 7. ODMR spectrum observed at 34.36 GHz and 2°K in a-Si:H No. 12 ($T_s = 300$°C), monitoring emitted light of 1.16 eV under excitation by argon ion laser light of 200 mW at 514.5 nm. [From Morigaki *et al.* (1983).]

a-Si:H No. 519. The spectral dependence of the ODMR signals also depends on preparation conditions of samples, in particular for the D_2 center. A few examples of the spectral dependences of $(\Delta I/I)_{ESR}$ are shown in Figs. 8–11 (a-Si:H Nos. 541, 540; a-Si:H,F No. 813, and sputtered a-Si:H No. H6). Samples of a-Si:H,F No. 995 exhibit different luminescence and ODMR properties from other samples. The reasons are given by Sano *et al.* (1982a).

The spectral dependence of $(\Delta I/I)_{ESR}$ at the D_2 center resonance has a peak at low photon energy, as seen in Fig. 6. The position of this peak and its full width at half maximum (FWHM) are plotted against the photon energy of the high energy edge of the luminescence spectra E_{HL} as shown in Fig. 12.

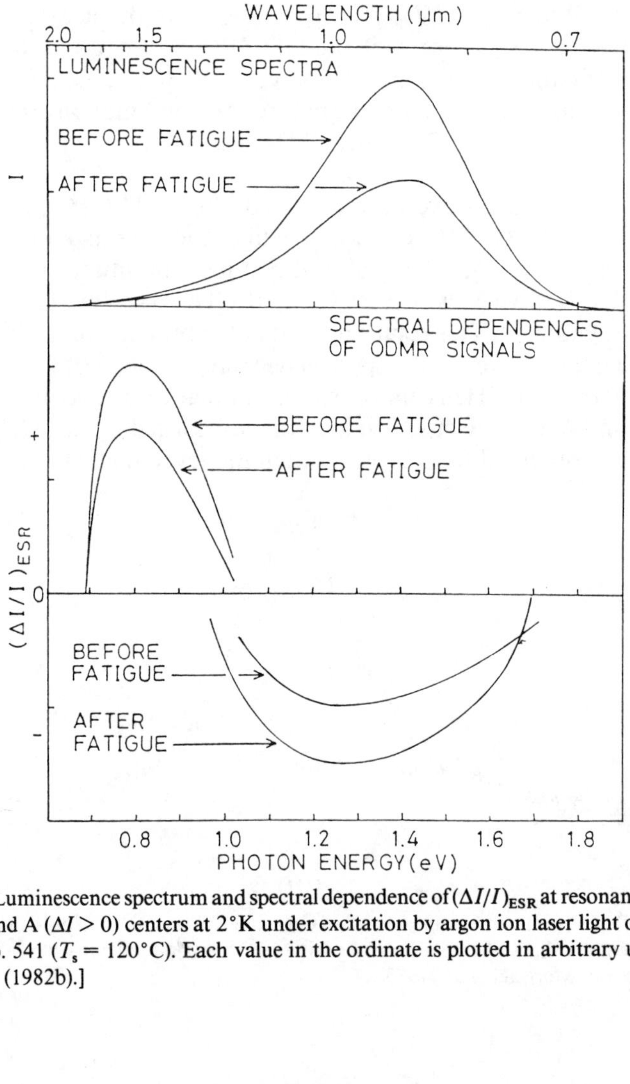

FIG. 8. Luminescence spectrum and spectral dependence of $(\Delta I/I)_{ESR}$ at resonances of the D_2 ($\Delta I < 0$) and A ($\Delta I > 0$) centers at 2°K under excitation by argon ion laser light of 25 mW in a-Si:H No. 541 ($T_s = 120$°C). Each value in the ordinate is plotted in arbitrary units. [From Sano *et al.* (1982b).]

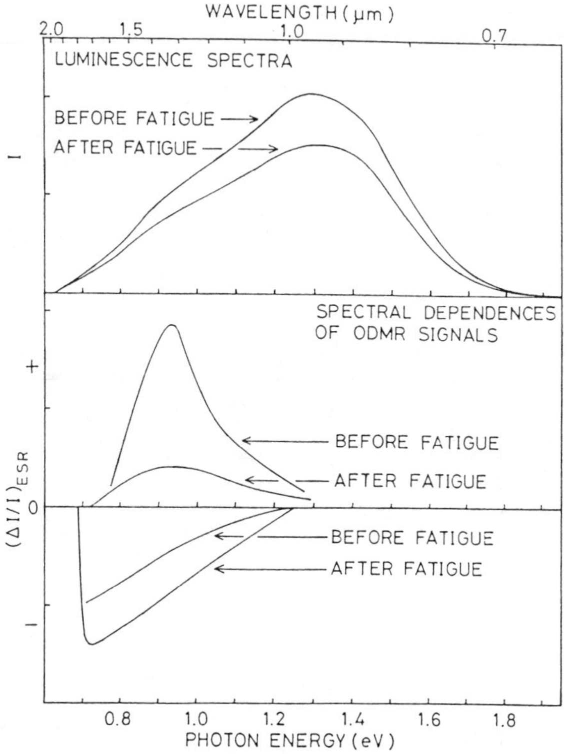

FIG. 9. Luminescence spectrum and spectral dependence of $(\Delta I/I)_{ESR}$ at resonances of the D_2 ($\Delta I < 0$) and A ($\Delta I > 0$) centers at 2°K under excitation by argon ion laser light of 200 mW in a-Si : H No. 540 ($T_s = 75$°C). Each value in the ordinate is plotted in arbitrary units. [From Sano *et al.* (1982b).]

Variations of the peak energy and the FWHM with E_{HL} are seen in this figure.

In the following, we present a model for interpreting our ODMR results: The luminescence arises from radiative recombination of trapped electron–hole pairs, in which trapped electrons lie in the band-tail and gap regions and trapped holes are localized at the A centers.

Nonradiative recombination occurs at the D_2 center (dangling-bond center) through activated transition or tunneling transition of trapped electrons (Morigaki *et al.*, 1981). Figure 13 shows a schematic diagram of recombination processes and levels of the D_2 [or dangling bonds (DB)] and A centers in a-Si : H.

The spectral dependences of $(\Delta I/I)_{ESR}$ at the D_2 and A center resonances are interpreted in terms of the preceding model as follows: A peak in the low

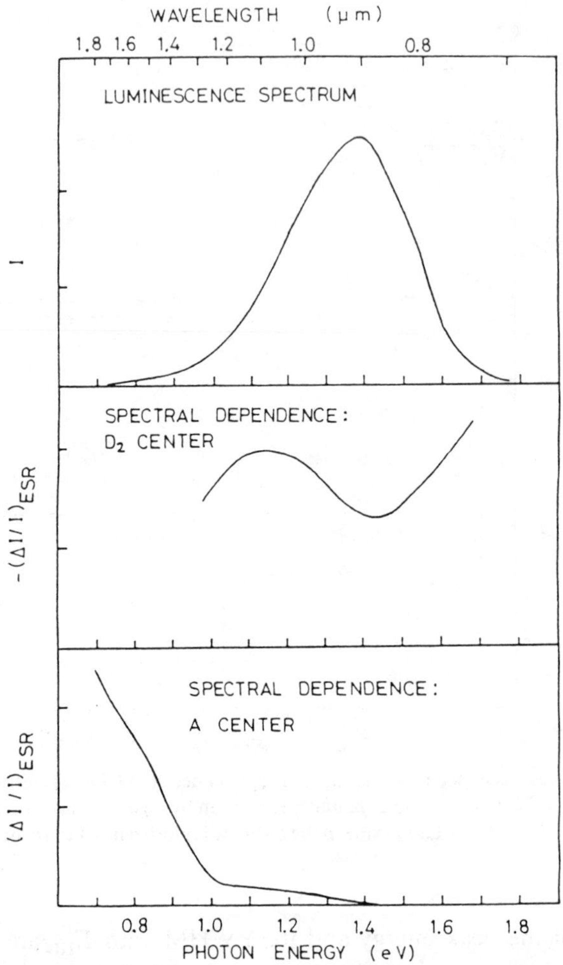

FIG. 10. Luminescence spectrum and spectral dependence of $(\Delta I/I)_{ESR}$ at resonances of the D_2 and A centers at 2°K in a-Si:H,F No. 813 ($T_s = 250$°C). Each value in the ordinate is plotted in arbitrary units. [Reprinted with permission from *Solid State Communications,* Vol. 43, K. Morigaki, Y. Sano, I. Hirabayashi, M. Konagai, and M. Suzuki, Level of dangling-bond centers and its broadening due to disorder in amorphous silicon as elucidated by optically detected magnetic resonance measurements. Copyright 1982, Pergamon Press, Ltd.]

photon energy region of the spectral dependence of the D_2 center may be due to the tunneling transition of a trapped electron at a radiative center into a dangling-bond center. The higher energy side of the spectral dependence may be accounted for in terms of activated nonradiative transition of a trapped electron into a dangling-bond center. On the other hand, the A center resonance enhances the luminescence intensity. The magnitude of $(\Delta I/I)_{ESR}$ at the A center resonance increases with decreasing photon energy. This is accounted for as follows: The luminescence at the lower energy arises

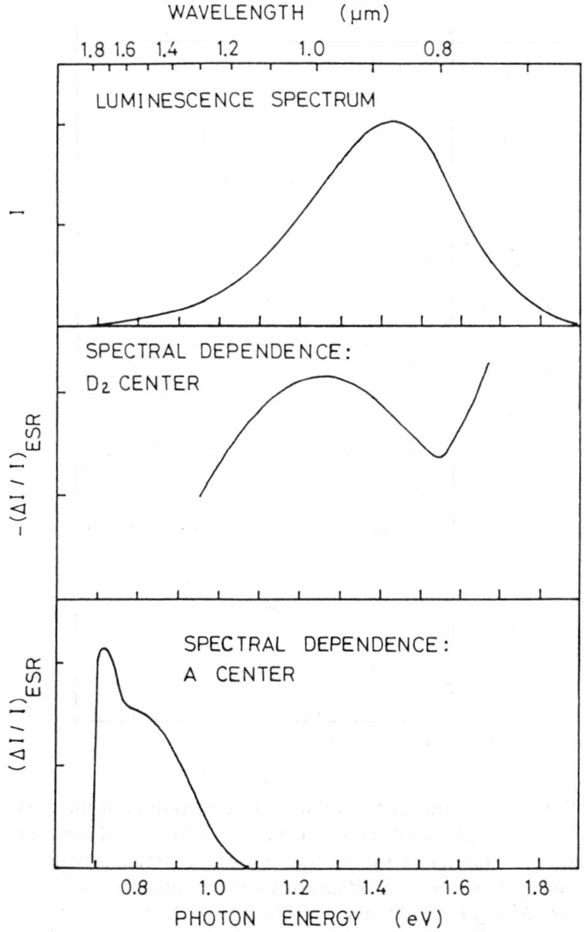

FIG. 11. Luminescence spectrum and spectral dependence of $(\Delta I/I)_{ESR}$ at resonances of the D_2 and A centers at 2°K in sputtered a-Si No. H6. [Reprinted with permission from *Solid State Communications,* Vol. 43, K. Morigaki, Y. Sano, I. Hirabayashi, M. Konagai, and M. Suzuki, Level of dangling-bond centers and its broadening due to disorder in amorphous silicon as elucidated by optically detected magnetic resonance measurements. Copyright 1982, Pergamon Press, Ltd.]

from radiative recombination of deeply trapped electrons with holes trapped at the A centers; the nonradiative recombination of these electrons via dangling-bond centers is weak at low temperatures. This results in an enhancement of $(\Delta I/I)_{ESR}$ at the A center resonance as can be derived from Eq. (5) for the case of weak microwave power regime:

$$(\Delta I/I)_{ESR} \cong [(R - R^*)/(R + R^*)]\,(W/R^*). \tag{14}$$

An analysis of the spectral dependence of ODMR signals allows us to estimate the positions of the levels of the A centers and the D^- doubly occupied

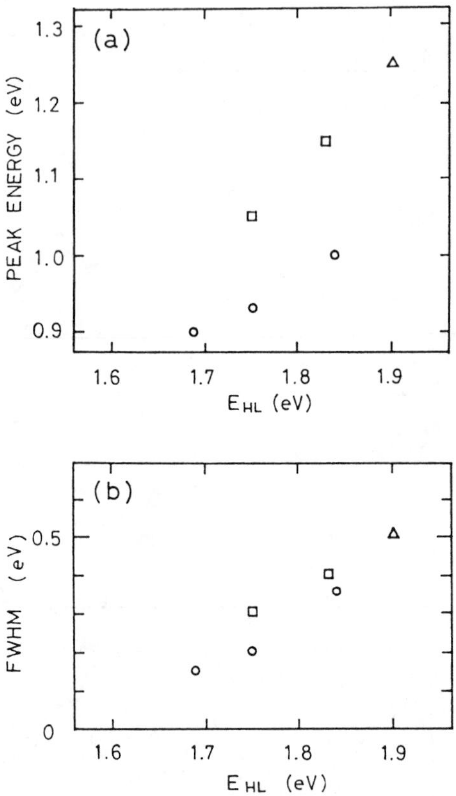

FIG. 12. (a) Peak energy and (b) full width at half maximum of the peak in the spectral dependence of the D_2 dangling-bond center resonance as a function of photon energy E_{HL} at the high energy edge of the luminescence spectra for various samples: triangles, sputtered a-Si; circles, a-Si : H; and squares, a-Si : H,F. [Reprinted with permission from *Solid State Communications,* Vol. 43, K. Morigaki, Y. Sano, I. Hirabayashi, M. Konagai, and M. Suzuki, Level of dangling-bond centers and its broadening due to disorder in amorphous silicon as elucidated by optically detected magnetic resonance measurements. Copyright 1982, Pergamon Press, Ltd.]

dangling-bond centers with respect to the edge of the conduction band. A detailed account of this analysis is given by Morigaki *et al.* (1982b). The depths of the A center and D^- doubly occupied dangling-bond centers, i.e., E_A and E_{D^-}, are listed in Table I.

The depth of the doubly occupied dangling-bond centers, i.e., the ionization energy of the second electron at the dangling-bond center, is also estimated from other experiments. The value of E_{D^-} of 0.6 eV for a-Si : H sample No. 519 agrees with that estimated from isothermal capacitance transient spectroscopy (ICTS) measurements by Okushi *et al.* (1982), i.e., 0.56 eV. The depth of singly occupied neutral dangling-bond centers cannot be estimated from the ODMR measurement. This level lies lower than the D^- level

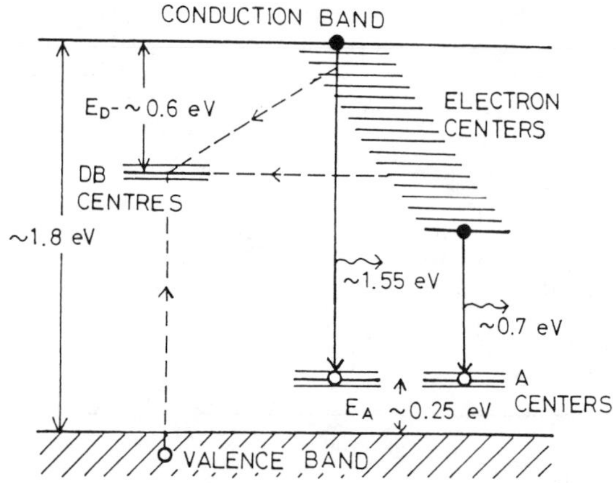

FIG. 13. Schematic diagram of radiative and nonradiative recombination processes, which are indicated by solid line and dashed line, respectively. Energies relevant to the positions of the localized levels are shown for a-Si:H No. 12. DB centers are doubly occupied dangling-bond centers D^-. [From Morigaki *et al.* (1982a).]

with a separation of the correlation energy U, which is estimated to be 0.4 eV by Dersch *et al.* (1981a) and 0.5 eV by Schweitzer *et al.* (1981) for a-Si:H.

It is worth noting that there is a correlation (see Fig. 12) between FWHM of the peak in the spectral dependence of $(\Delta I/I)_{ESR}$ at the D_2 center and E_{HL}, which roughly approximates the optical band-gap energy. The FWHM gives a measure of the broadening of the D^- level, which is enhanced by disorder. Cody *et al.* (1981) have pointed out that structural disorder is the fundamental determining factor of the optical band gap. The observed increase of the FWHM with E_{HL}, as shown in Fig. 12, may be accounted for by disorder resulting from the randomness of the Si network and also from the presence of foreign atoms such as H and F.

Experiments using ODMR have also been reported by Depinna *et al.* (1981, 1982a–d), Street *et al.* (1982), and Street (1982). Great care is required for comparing their results, because, as already mentioned, the ODMR results depend on experimental conditions. It seems to us that most of the experimental results are consistent with each other. However, the interpretation of the results is different. This comes, in part, from a difference in how the ODMR spectrum is deconvoluted.

According to Depinna *et al.* (1982c), two types of trapped hole centers, i.e., A_1 and A_2 centers, are responsible for the luminescence at 1.25 eV and 0.9 eV, respectively. The dangling-bond center with $g = 2.006$ acts as a radiative center for the luminescence at 0.9 eV and also possibly for those at 1.4 eV and 1.25 eV, although the tail states are equally considered as radia-

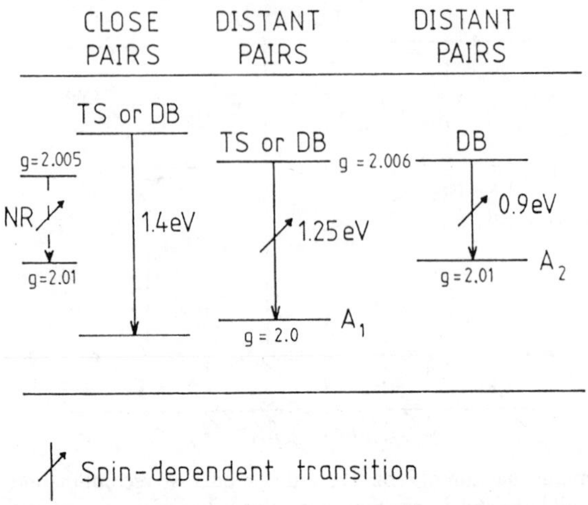

FIG. 14. Recombination model for a-Si : H deduced from the ODMR results. The 0.9 and 1.25 eV emissions are spin-dependent pair processes, whereas the 1.4 eV emission is linked to a nonradiative pair process. DB, dangling bond ($g = 2.0055$); TS, electron tail state ($g = 2.0045$). [From Depinna *et al.* (1982d).]

tive centers for these luminescences, as shown in Fig. 14. This conclusion has been deduced from their deconvolution of the enhancing lines as shown in Fig. 15.

Figure 16 shows the ODMR spectra observed by Street (1982) for two different types of a-Si : H samples, i.e., high-defect-density samples and low-defect-density samples. He identified a quenching line Q_1 observed in high-defect-density samples as being a coupled resonance of the dangling bond

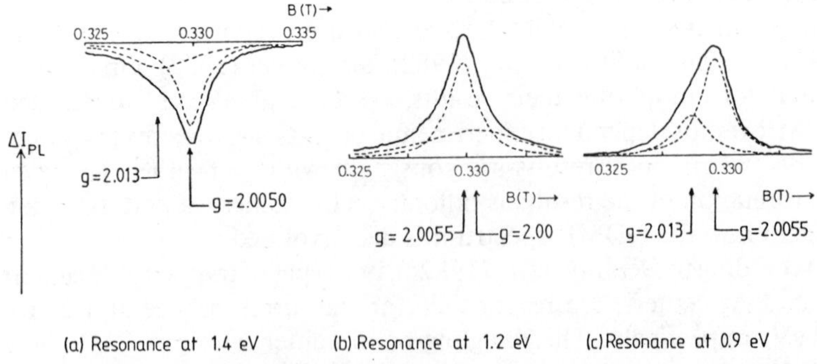

FIG. 15. ODMR signals taken at 9.36 GHz and 2°K. [From Depinna *et al.* (1982d).]

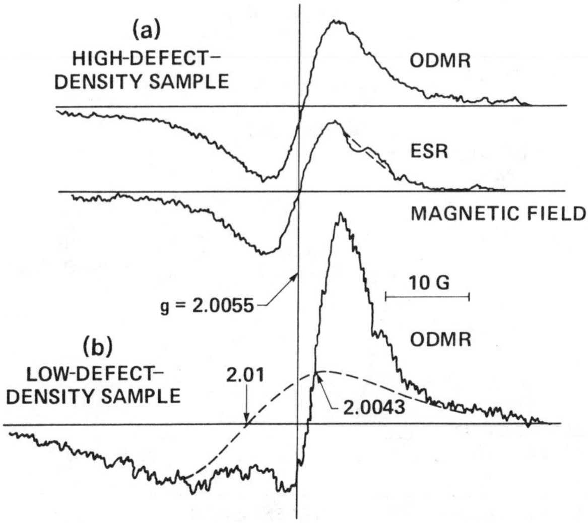

FIG. 16. ODMR quenching signals (derivatives) for the high-defect-density and low-defect-density samples at 85°K. The ESR signal is also shown for the high-defect-density sample. [From Street (1982).]

and the conduction-band-tail electron. For low-defect-density samples besides the Q_1 line with $g = 2.0055$ observed at low excitation level, he attributed a pair of quenching lines Q_2 with $g = 2.0045$ and $g = 2.011$, which have been observed at high excitation level, to electrons and holes in band-tail states. The band-tail electrons and holes are considered to contribute to the band-edge luminescence (see Fig. 17), hence he thought that the fact that those electron and hole resonances quench the luminescence is accounted for by nonradiative Auger processes. The enhancing line is associated with a radiative transition and is also due to the band-tail electrons and holes coupled with each other by an exchange interaction, because, according to Street (1982), the g-value of the enhancing line coincides with the average of the g-values of those electrons ($g = 2.0045$) and holes ($g = 2.011$).

Such different identifications of the centers responsible for ODMR signals lead to different models for radiative and nonradiative recombination processes. The g-values of paramagnetic centers that have been identified by Morigaki *et al.* (1981), Depinna *et al.* (1982d), and Street (1982) are summarized in Table II together with tentative models for paramagnetic centers. The models for recombination processes by Depinna *et al.* (1982c) and Street (1982) are shown in Figs. 14 and 17.

Now we shall comment on the identification of paramagnetic centers by Depinna *et al.* (1982d), Morigaki *et al.* (1981, 1982a), and Street (1982). Depinna *et al.* deconvolute the enhancing line into two components,

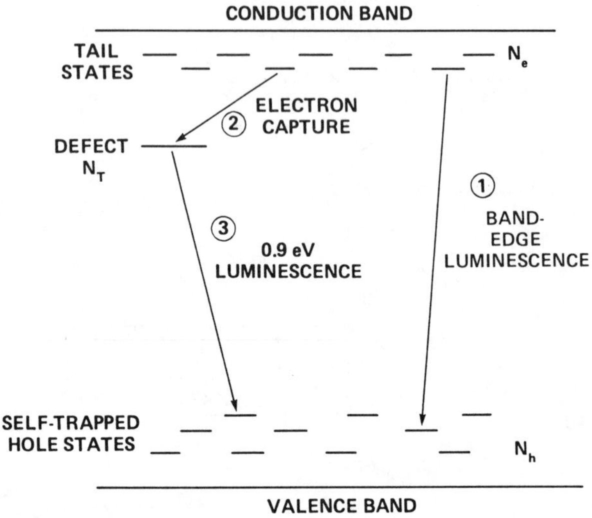

FIG. 17. Schematic recombination diagram for defect and band-edge luminescence. [From Street *et al.* (1979).]

whereas Morigaki *et al.* (1981, 1982a) identify this line as being due to one center, i.e., the A center, which is a trapped hole presumably weakly coupled with trapped electrons. How the A center is coupled with the trapped electrons depends on the luminescence energy. The A line appears to have an asymmetric shape that may be due to an anisotropic nature of the electronic state of the A center. The shape and broadening of the A line are partly due to the strength of the exchange interaction between the trapped electrons and the A center hole and its distribution, and also they depend on microwave power and excitation intensity. In this regard, we should emphasize that the broadening of the A line is inhomogeneous, as has already been mentioned. Thus the variation of the shape and broadening of the A line with photon energy is directly or indirectly related to the properties of the states occupied by electron–hole pairs participating in the luminescence one monitors during the ODMR measurement. Therefore, it seems to us that the model of Depinna *et al.* in which dangling-bond centers act as radiative centers may be due to an inappropriate deconvolution of the enhancing line for both the band-edge luminescence and the low energy luminescence. Our conclusion that dangling-bond centers act as nonradiative centers for both the band-edge luminescence and the low energy luminescence is based on the observation that the dangling-bond center resonance obviously quenches those luminescences and also that prolonged light irradiation creates more dangling-bond centers, as will be described in Part V. One of the quenching pair

TABLE II

g-VALUES AND NATURE OF RECOMBINATION CENTERS RESPONSIBLE FOR ODMR[a]

Model	Sample type	Center	*g*-Value	Recombination involving the center[b]	Assignments
Morigaki *et al.* (1981)	High-defect-density	D_2	2.005	NR	Dangling-bond center
		A	2.009	R	Trapped-hole center
	Low-defect-density	D_1	2.013	NR	Trapped hole at T_3^- (tentative)
		D_2	2.006	NR	Dangling-bond center
				NR	Trapped electron at T_3^+ (tentative)
		A	2.0–2.01	R	Trapped-hole center
Street (1982)	High-defect-density	Q_1	2.0052	NR	Dangling-bond center
	Low-defect-density	Q_1	2.0055	NR	Dangling-bond center
		Q_2	2.0045	R	Band-tail electron
			2.011	R	Band-tail hole
		E_1	2.0085	R	Coupled pair of band-tail electron and band-tail hole
Depinna *et al.* (1982d)		E_1	2.006	R	Dangling-bond center for 0.9 eV luminescence
					Dangling-bond center or band-tail electron for 1.4 and 1.25 eV luminescence
		A_1	2.0	R	Band-tail hole
		A_2	2.01	R	Band-tail hole
		Q_1	2.01	NR	Band-tail hole
		Q_2	2.005	NR	Band-tail electron

[a] Data from Morigaki *et al.* (1981), Street (1982), and Depinna *et al.* (1982d).
[b] R and NR designate radiative and nonradiative recombination, respectively.

lines, Q_2, with $g = 2.011$, seems to correspond to the D_1 line. The other line Q_2 with $g = 2.0045$ may overlap with the D_2 line in our measurements. However, our interpretation of these two lines is completely different from Street's and more straightforward: We have tentatively identified them as being nonradiative centers that have unpaired spins, i.e., a trapped electron at a positively charged threefold coordinated center T_3^+, whose wave function is formed from an sp^3 orbital and a trapped hole at a negatively charged threefold coordinated center T_3^-, whose wave function is formed from a p orbital. Their nature is also discussed in Part VI. It is worth noting that these signals have been observed as light-induced ESR signals. In our measurements, we could not separate two lines with $g = 2.0045$ and $g = 2.0055$ because of the line broadening, so that the D_2 line observed in low-defect-density samples appears to have two components, one due to light-induced centers consisting of trapped electrons at T_3^+ and the other due to dangling-bond centers. In our model, any ODMR signal due to trapped electrons in the conduction-band-tail states are assumed to be inobservable because the spread of the energy levels would broaden the ODMR signal (spread of *g*-value distribution) beyond detection. Furthermore, the resonance of trapped electrons would result in two opposing effects on the luminescence, i.e., enhancement and quenching through radiative and nonradiative recombination, respectively, so that the ODMR signal would become too weak to be observed.

V. Light-Induced Effect in ODMR

Prolonged exposure of a-Si:H samples to light whose energy exceeds the optical band gap induces a fatigue effect in the luminescence at low temperatures (Morigaki *et al.*, 1980b; Pankove and Berkeyheiser, 1980; Hirabayashi *et al.*, 1981). This effect is attributed to the creation of dangling bonds by prolonged exposure to light. The dangling bonds act as nonradiative centers, and as a result, the luminescence intensity decreases compared with that before exposure. Such defect creation has been confirmed by ESR experiments (Hirabayashi *et al.*, 1980; Dersch *et al.*, 1981b; Hirabayashi *et al.*, 1982) and also by ODMR experiments (Morigaki *et al.*, 1982a). Figure 18 shows ODMR spectra, monitoring the intensity of emitted light at 1.16 eV, taken at 2°K before and after exposure to light (total photon number of 2.4×10^{22} cm^{-2} from an argon ion laser operating at 514.5 nm) and also after annealing the sample at room temperature using a-Si:H sample No. 519. The peak intensity of luminescence decreases to 57% of its initial value after exposure to light. As seen in Fig. 18, the D_2 line increases in intensity, whereas the A line decreases from the initial value before exposure to light. Such an increase in the D_2 line intensity clearly suggests that dangling-bond centers are created by prolonged exposure to light at low temperatures. Annealing the sample at room temperature causes a partial recovery of the ODMR spectrum and a similar recovery of the luminescence. The spectral dependence of $(\Delta I/I)_{ESR}$ at the D_2 and A center resonances before and after exposure to light is shown in Fig. 6. As seen in Fig. 6, the dangling-bond-center resonance reduces the luminescence over almost the whole spectral region. The nature of such a light-created dangling-bond center is not clear,

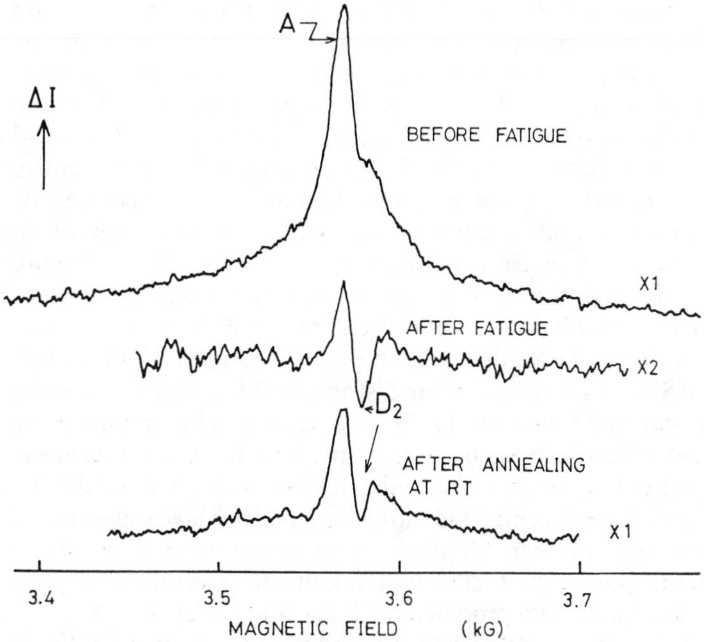

FIG. 18. ODMR spectra taken at 2°K before and after fatigue and also after annealing the sample at room temperature, monitoring the intensity of emitted light at 1.16 eV from a-Si:H No. 519 (T_s = 300°C). [From Morigaki *et al.* (1982a).]

but its model is briefly discussed in Part VI. The magnitude of $(\Delta I/I)_{ESR}$ at the A center resonance decreases after exposure to light, as seen in Fig. 6, except for the low energy luminescence. This result can be interpreted by considering that, in Eq. (5), R^* is enhanced as a result of the creation of dangling-bond centers by prolonged exposure to light. An increase in $(\Delta I/I)_{ESR}$ in the low energy luminescence after exposure to light is attributed to radiative centers (trapped electron centers) created by prolonged exposure to light. This is consistent with the observation by Pankove and Berkeyheiser (1980) that the low energy luminescence is enhanced after exposure to light.

VI. Nature of the Recombination Centers in a-Si:H

As was mentioned in Part IV, the D_2 center with $g = 2.006$ is the dangling-bond center whose dark ESR has been observed with $g = 2.0055$ in a-Si and a-Si:H samples. However, it is difficult to uniquely identify which type of dangling bonds, i.e., dangling bonds in the voids or singly occupied dangling bonds whose electronic configuration is designated by T_3^0, contribute to the D_2 line. On the other hand, in our tentative model, two light-induced ESR signals (Street and Biegelsen, 1980) are attributed to an electron trapped at T_3^+ (sp^3 orbital) and a hole trapped at T_3^- (p orbital), respectively, as was mentioned in Part IV. Such charged threefold coordinated centers have been discussed by Adler (1978, 1981), Adler and Shapiro (1983), and Elliott (1978). (See also Chapter 14 by Adler in Volume 21A.) The light-induced centers may differ from the simple T_3^0 centers (dangling-bond centers) in that they are created by optical excitation and thus appear to cause a distortion among the surrounding atoms in order to find a local minimum energy state, even if the unpaired electron lies in the sp^3 orbital (T_3^+ + electron).

The nature of the conduction-band-tail states that are responsible for radiative recombination is not clear. Potential fluctuations occur at the conduction-band edge due to the disorder associated with the random network of Si atoms and the random distribution of foreign atoms such as H and F (Dunstan and Boulitrop, 1981). As a result, local minimum energy states are created, in which electrons are captured after band-to-band excitation. These states may be possible candidates for conduction-band-tail states.

In Part IV, the A center is identified as a trapped hole state, but its detailed nature is also not clear. Although a three-center bond model has been proposed (Morigaki *et al.*, 1980a), we think that the A center corresponds to the self-trapped hole state proposed by Tsang and Street (1979).

According to a model of light-induced centers proposed by Hirabayashi *et al.* (1980), a dangling bond is created by breaking a Si–Si bond near a Si–H bond. However, experimental evidence for this model has not been obtained.

The creation of dangling bonds stabilizes the atomic configuration by using the energy released by the nonradiative recombination (Morigaki *et al.*, 1980b).

VII. Remark on Recombination Processes

The luminescence of a-Si:H has, in general, two emission bands, i.e., the band-edge luminescence, which peaks at 1.3–1.4 eV, and the low energy luminescence, which peaks at 0.8 eV [for reviews see Street (1981) and Chapter 7 by Street in Volume 21B].

The low energy luminescence is considered by Street *et al.* (1979) and Street (1980) to arise from radiative recombination of an electron at the doubly occupied dangling-bond center with a valence-band-tail hole, as shown in Fig. 17. Since there is a correlation between the intensity of low energy luminescence and the dangling-bond density, a similar conclusion has been proposed by Voget-Grote *et al.* (1980). However, this conclusion is inconsistent with our recent ODMR experiments on hydrogen-effused a-Si:H samples (Yoshida and Morigaki, 1983). Figure 19 shows the lumines-

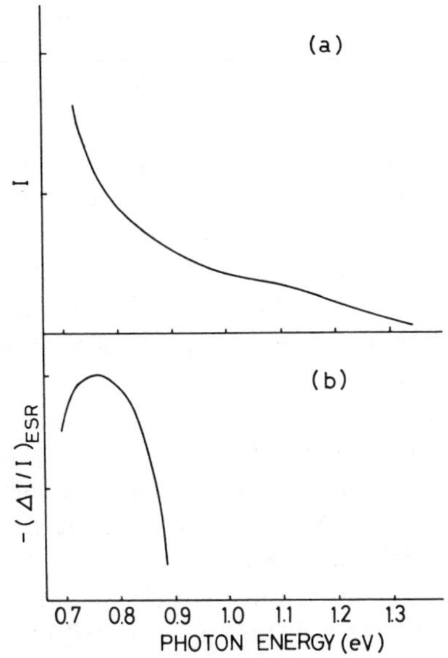

FIG. 19. Luminescence spectrum and spectral dependence of $(\Delta I/I)_{ESR}$ at the D_2 center resonance at 2°K in a-Si:H sample No. 519 annealed at 600°C in vacuum. Each value in the ordinate is plotted in arbitrary units. The luminescence spectrum was measured using a cooled Ge detector and corrected for the spectral response of the detection system. [From Yoshida and Morigaki (1983). Copyright North-Holland Publ. Co., Amsterdam, 1983.]

cence spectrum and the spectral dependence of $(\Delta I/I)_{ESR}$ at the dangling-bond-center resonance taken at 2°K in a-Si:H sample No. 519 annealed in vacuum at 600°C for 30 min. The luminescence spectrum consists of the two components described above, but the low energy luminescence is dominant. The ODMR spectrum consists of only the quenching line due to dangling-bond centers with $g = 2.005$. Thus this result clearly indicates that dangling-bond centers act as nonradiative centers for the low energy luminescence in contrast with the model of Street (1980), which is illustrated in Fig. 17. No enhancing ODMR signal has been observed in this annealed sample. Thus, even if the deconvolution of an enhancing line into a dangling-bond-center resonance line and other lines as by Depinna *et al.* (1982d) would be reasonable, this deconvolution would not be applicable for this annealed sample. So, the result of Fig. 19 is direct evidence against the models by Street and Depinna *et al.*

VIII. Time-Resolved ODMR

The time-resolved ODMR (TRODMR) measurements on a-Si:H have been carried out by Morigaki *et al.* (1978b), Depinna and Cavenett (1982),

Street (1982), and Boulitrop (1982). These measurements have previously been reviewed (Morigaki, 1983).

The main results are summarized as follows: Morigaki *et al.* (1978b) observed TRODMR spectra in a low-defect-density sample, monitoring the total emitted light using a GaAs photomultiplier (detection wavelength <900 nm). A microwave pulse of 40 μsec duration is applied 40 μsec after a pulse from a nitrogen laser is switched off, and the delay time of the sampling gate for observation of ODMR signals is changed from 40 μsec to 140 μsec after the laser pulse. The GaAs photomultiplier detects mainly the band-edge luminescence, so that the quenching signal due to the dangling-bond center is observed, as shown in Fig. 20, in which the steady state ODMR spectrum is also shown for comparison. The width of the quenching line decreases with increasing delay time of the sampling gate. Just after the laser pulse is switched off, close electron–hole pairs [radiative-electron-center–nonradiative-hole-center (dangling-bond-center) pairs] contribute to TRODMR. Thus the exchange interaction between an electron and a hole causes ODMR line broadening. On the other hand, with increased delay time of the sampling gate, more distant pairs contribute to TRODMR. Since the exchange interaction is weak for distant pairs, the line broadening is less than for close pairs. Such behavior of the TRODMR with delay time has also been observed for the enhancing signal in sputtered a-Si:H (Boulitrop,

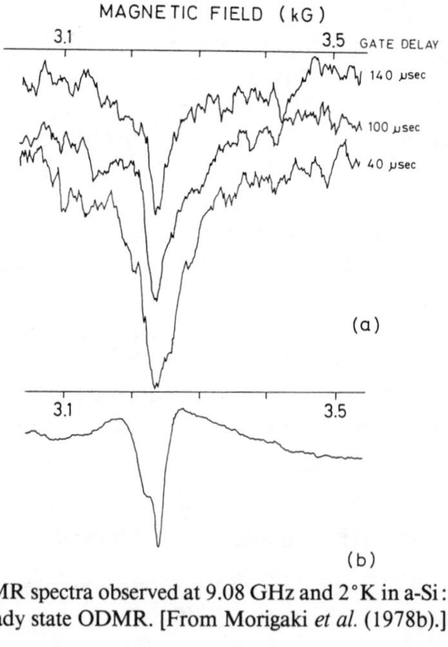

FIG. 20. (a) TRODMR spectra observed at 9.08 GHz and 2°K in a-Si:H No. 12 for various gate delays and (b) steady state ODMR. [From Morigaki *et al.* (1978b).]

1982) and for donor–acceptor recombination in crystalline semiconductors (Block and Cox, 1981).

Depinna and Cavenett (1982) and Street (1982) used a Ge detector (detection wavelength $<1.9\ \mu m$) that detects both the band-edge luminescence and the low energy luminescence. In monitoring the total emitted light, Depinna and Cavenett have observed TRODMR signals as a function of emission lifetime, using frequency response spectroscopy (FRS) (Dunstan *et al.*, 1982) to avoid the contribution of longer lived states, as shown in Fig. 21. Street has also observed TRODMR signals as a function of delay time of the sampling gate after an excitation pulse, as shown in Fig. 22. However, the results are different between these two groups in the appearances of enhancing and quenching lines. The enhancing line seems to be due to the A center.

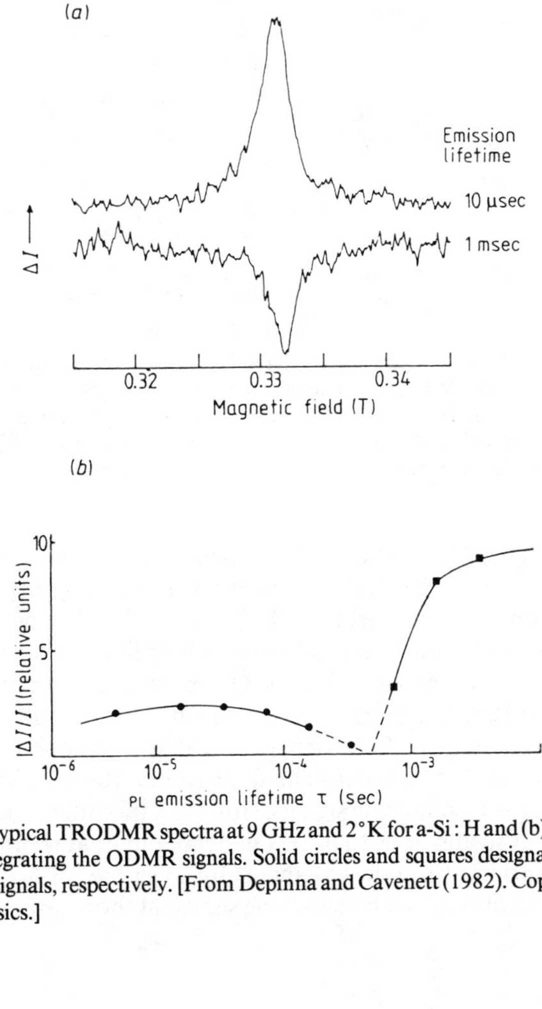

FIG. 21. (a) Typical TRODMR spectra at 9 GHz and 2°K for a-Si : H and (b) plot of $(\Delta I/I)_{ESR}$ obtained by integrating the ODMR signals. Solid circles and squares designate the enhancing and quenching signals, respectively. [From Depinna and Cavenett (1982). Copyright 1982, The Institute of Physics.]

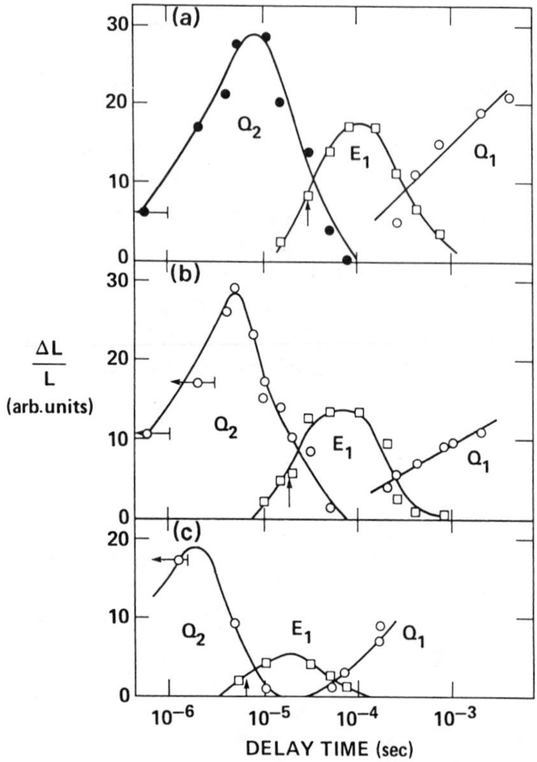

FIG. 22. TRODMR $(\Delta I/I)_{ESR}$ obtained from peak to peak height of the derivative resonances for a-Si : H at $P_{mW} = 100$ mW under excitation by argon ion laser light of 5145 Å. Temperatures are (a) 14°K, (b) 25°K, and (c) 50°K. The vertical arrows indicate the time at which E_1 reaches half its maximum intensity and is taken as the measure of T_1. For the detail of Q_2 and E_1 signals, see Table II. [From Street (1982).]

Thus, the results by Depinna and Cavenett are accounted for by considering the difference in T_1 (spin–lattice relaxation time) between the D_2 and A centers for unthermalized electron–hole pairs. The Q_2 lines observed by Street at shorter delay times appear as quenching lines with $g = 2.0045$ and $g = 2.011$. This result indicates that the Q_2 centers are nonradiative centers, as discussed in Part IV. Street has concluded that the enhancing line E_1 occurs at a geminate pair (Biegelsen *et al.,* 1978) when one spin is thermalized and the other is not, because he thought that the hole line appears in both quenching and enhancing signals. However, this interpretation appears unreasonable. It seems to us that the quenching lines at shorter delay time have different origins from the enhancing line, i.e., the A line. Depinna and Cavenett did not observe such quenching signals at short emission lifetime in their TRODMR (FRS) experiments perhaps because of different preparation conditions for the Depinna *et al.* samples compared with those of Street. Therefore, further study of the TRODMR properties of a-Si : H is required.

IX. Temperature Dependence of ODMR

The ODMR signals have been observed at higher temperatures (Depinna *et al.*, 1982b; Biegelsen *et al.*, 1978), up to room temperature for the quenching signal and up to 88°K for the enhancing signal. The temperature dependence of those ODMR signals indicates that the centers involved in the enhancing resonances and in the quenching resonances are unthermalized up to 80°K and 300°K, respectively. This result confirms an assumption in Part IV that the ODMR results on the A and D_2 centers at 2°K can be interpreted in terms of the unthermalized electron–hole pair model. The spectral dependences of the D_2 and A center resonances have been measured at 77°K for a-Si : H sample No. 541 (Yoshida *et al.*, 1983). The result is similar to those measured at 2°K except for a slight shift of the peak toward low energy. The spin–lattice relaxation times of the dangling-bond center and the hole center have been measured as a function of temperature using the TRODMR method, as well as the conventional saturation method (Street, 1982; Street *et al.* 1982). From these measurements it has been found that the dangling bonds relax through a Raman process at temperatures between 50°K and 120°K, whereas the trapped holes (A centers) relax by a single phonon direct process (Street, 1982). However, no theoretical study on the spin–lattice relaxation rate has been reported in a-Si : H. This would be valuable for the understanding of the nature of recombination centers in a-Si : H. Time-resolved ODMR provides a novel technique for measuring the spin–lattice relaxation time of recombination centers.

X. ENDOR and Optically Detected ENDOR

Electron nuclear double resonance is a powerful tool for obtaining information about the electronic structure of defects and impurities in solids. In spite of the usefulness of ENDOR, few attempts have been made in using this technique on a-Si : H.

In ENDOR experiments, one monitors the ESR signal under the application of those rf powers that excite the NMR transitions of nuclear spins neighboring the magnetic center or of the nuclear spin of the center itself. The rf frequency is usually scanned to find such NMR transitions.

The conventional ENDOR experiments have been independently carried out by Biegelsen (1980), Sano *et al.* (1983), Boulitrop (1982), and Yamasaki *et al.* (1983) by monitoring the dangling-bond ESR. Figures 23 and 24 show the ENDOR spectrum observed for a-Si : H samples with low defect density ($N_s = 10^{16}$ cm^{-3}). Figure 23 clearly shows the distant ENDOR signals due to nuclear spins of ^{29}Si and ^{1}H whose frequencies are 2.8 and 14 MHz, respectively. As shown in Fig. 24, besides the peak corresponding to the distant ENDOR signal, subsidiary peaks are observed around 14 MHz (Boulitrop, 1982; Yamasaki *et al.*, 1983). Similar appearances of subsidiary peaks are observed by Sano *et al.* around the distant ENDOR signal at 2.8 MHz for the high-defect-density sample No. 540. The broadening of the distant ENDOR lines is attributed to spatial fluctuations of the hyperfine interaction of dangling-bond spins with nuclear spins, i.e., either ^{29}Si or ^{1}H, and also to a fluctuation of the extent of the wave function of dangling-bond spins. The subsidiary peaks such as those seen in Fig. 24 are due to hyperfine interaction of the dangling-bond spin with a nuclear spin of ^{29}Si at a specific site. When the microwave power becomes high, the splitting of two subsidiary peaks becomes great and the line is broadened as a whole. This result at high microwave power can be explained as a saturation of the strong coupling between dangling-bond spins and nuclear spins within their Zeeman levels. This also explains a very broad ENDOR line at 14 MHz observed by Sano *et al.* (1983). Until now, no ENDOR signal associated with nearest-

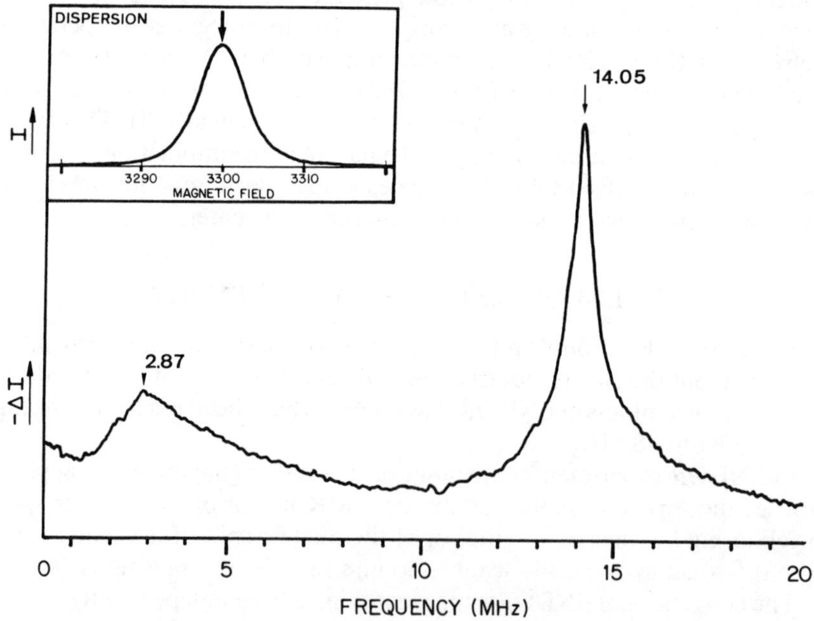

FIG. 23. ENDOR spectra at 30°K for a low-defect-density sample of sputtered a-Si : H with $P_{MO} = 2\ \mu$W and $P_{rf} = 40$ W ($N_s = 10^{16}$ cm^{-3}). [From Boulitrop (1982).]

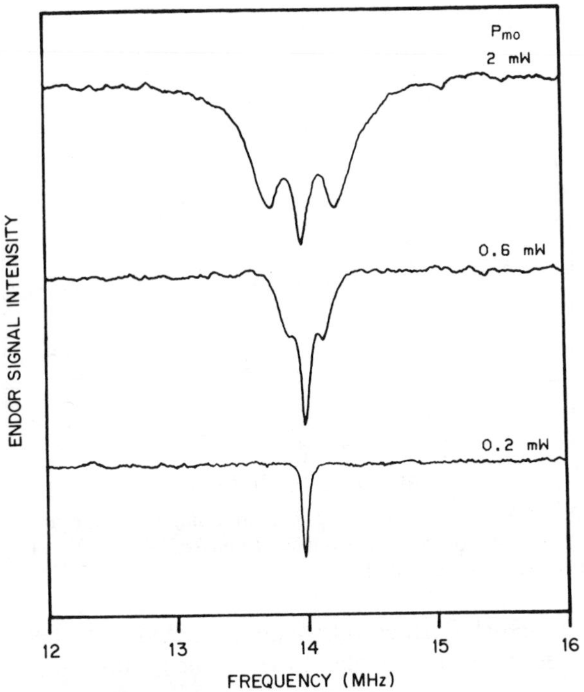

FIG. 24. ENDOR absorption spectra associated with ^1H at 77°K around 14 MHz for various microwave powers in sputtered a-Si : H. (a) 2 mW, (b) 0.6 mW, and (c) 0.2 mW. [From Boulitrop (1982).]

neighbor nuclear spins of ^{29}Si and ^1H has been observed for dangling-bond centers, whose resonance frequencies appear to be beyond 20 MHz (Ishii *et al.*, 1982, 1983).

Optically detected ENDOR (ODENDOR) experiments are of particular interest from the point of view that photoinduced magnetic centers such as the A center can be investigated in greater detail than is possible with the ODMR experiment. The first ODENDOR experiment was done by Sano *et al.* (1983). In this experiment, a broad ENDOR spectrum is observed in the rf frequency range of 1–20 MHz by monitoring the intensity of the total emitted light and setting the magnetic field at the peaks of ODMR signals for the A and D_2 centers (see Fig. 25). As in the case of ODMR, ODENDOR yields enhancement and quenching of luminescence for radiative centers such as the A centers and nonradiative centers such as the D_2 centers, respectively. Figure 25 shows the ODENDOR spectrum associated with the A center resonance. The broadening of ODENDOR lines associated with ^{29}Si nuclear spins is accounted for in a way similar to that for the conventional

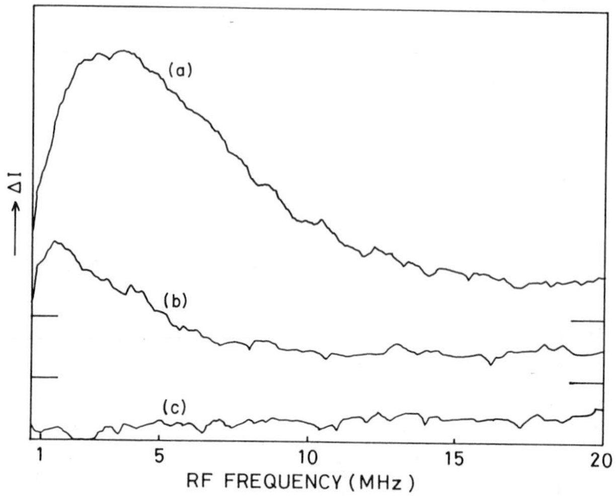

FIG. 25. ODENDOR spectra observed at 2°K, monitoring the A center at 514.5 nm. Microwave powers: (a) 640 mW, (b) 35 mW, (c) 0 mW. [From Sano *et al.* (1983). Copyright North-Holland Publ. Co., Amsterdam, 1983.]

ENDOR experiments. Further investigations including higher rf frequency ENDOR are required to understand the detailed nature of the A and D_2 centers.

XI. Spin-Dependent Photoinduced Absorption

Photoinduced absorption (PA) measurements provide useful information about relaxation processes of photocarriers and tail and gap states in a-Si : H [Tauc, (1982); see also Chapter 9 by Tauc in Volume 21B]. Photoinduced absorption also depends on the recombination processes of those trapped electrons and holes that are created under optical excitation. Thus it is expected that PA will be spin-dependent, as is luminescence. Hirabayashi and Morigaki (1983a,b) have observed, for the first time, spin-dependent PA at 2°K in a-Si : H, monitoring the PA intensity while irradiating the sample with microwaves at 9.6 GHz. This is called photoinduced absorption-detected ESR (PADESR), which provides complementary information to the conventional ODMR (luminescence-detected ESR). However, since the PADESR technique can be applied to nonluminescent materials, it may provide a more general means for highly sensitive detection of ESR than conventional ODMR.

Figure 26 shows the PADESR spectra observed for a-Si : H sample No. 541, as well as the conventional ODMR spectra for comparison. The microwave power is chopped at 1 kHz while the excitation is by argon ion laser light of 3 W cm^{-2} at 514.5 nm; the PADESR signals are observed by moni-

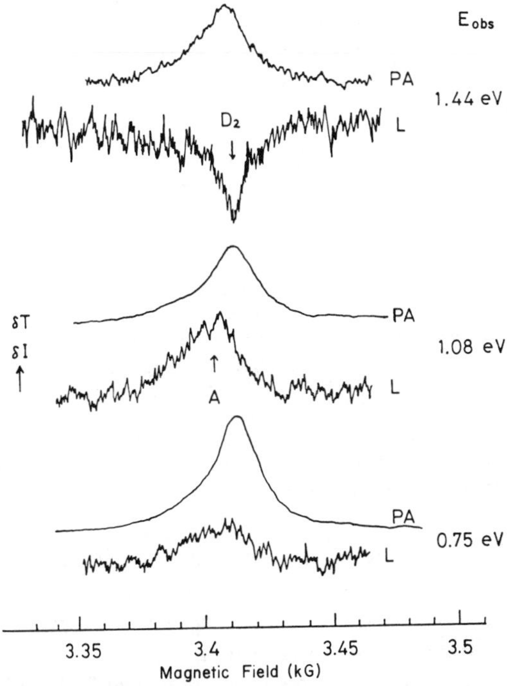

FIG. 26. PADESR and ODMR spectra at 2°K in a-Si:H No. 541, monitoring the PA intensity or luminescence (L) intensity at various photon energies. [From Hirabayashi and Morigaki (1983b). Copyright North-Holland Publ. Co., Amsterdam, 1983.]

toring the intensity of transmitted probe light of various photon energies. In contrast to ODMR spectra, PADESR is observed with the same sign of change in transmitted light intensity ΔT for both the A and D_2 centers, i.e., $\delta\,\Delta T > 0$ (decrease in the PA intensity). Here ΔT is the PA contribution to the intensity of the transmitted light.

The mechanism for PADESR is as follows: As shown in Fig. 27, holes are excited into the valence band by the probing light, as suggested by O'Connor and Tauc (1982). From our PADESR results, these holes appear to be mainly trapped at the A centers. From a simple consideration based on the trapped electron–hole pair model given in Part II, the ESR of the A centers shows a decrease in the number of trapped holes at the A centers. The change in the number of trapped holes n_h is given by

$$\frac{\Delta n_h}{n_h} = -\frac{R_1^2}{(2R_n + R_1)(8G + 2R_n + R_1)} \frac{2T_1^* W}{1 + 2T_1^* W}, \qquad \frac{1}{T_1^*} = \frac{4G(2R_n + R_1) + 2R_n(R_n + R_1)}{8G + 2R_n + R_1}, \tag{15}$$

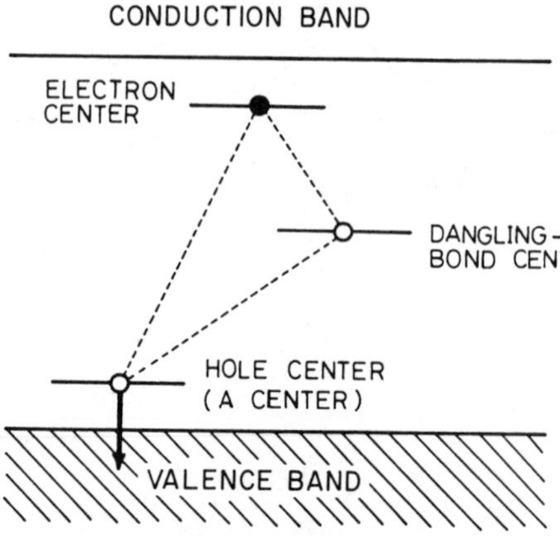

FIG. 27. Schematic diagram of PA and recombination processes involved in PADESR. The arrow indicates the excitation of holes from the A centers into the valence band. Dashed lines show recombination of electrons with holes.

where R_1 and R_n designate the radiative and nonradiative recombination rates of the electron–hole pairs, and other notations are the same as for Eq. (1). The PA intensity is proportional to n_h, so that $\Delta n_h/n_h$ corresponds to PADESR signal intensity in relative units.

For the dangling-bond-center resonance D_2, the PADESR mechanism is as follows: Trapped electrons and trapped holes recombine at the dangling-bond centers, as shown in Fig. 27, whereas the dangling-bond-center resonance enhances such a recombination in a way similar to that of conventional ODMR, and the number of trapped holes is also decreased at resonance. The change in n_h with the ESR of the D_2 center is also easily derived.

A detailed discussion of the PADESR results for various a-Si : H samples is given in Hirabayashi and Morigaki (1983a,b). Here, it is only mentioned that the excitation of trapped electrons at the dangling-bond centers D^- contributes to PA in a-Si : H sample No. 540 according to the PA and PADESR measurements.

XII. Summary

The ODMR technique we have reviewed in this chapter uses spin-dependent properties of semiconductors. Therefore, this technique can be extended to other optical phenomena than luminescence and photoinduced absorption such as treated here, e.g., electronic Raman scattering (Romestain *et al.*, 1974; Geschwind, 1978), optical absorption (Mollenauer *et al.*, 1969) etc. Since the pioneering work by Lepine (1972) on crystalline silicon, spin-dependent recombination has been investigated through photoconductivity during microwave irradiation (Solomon *et al.*, 1977; Solomon, 1980). The measurements of spin-dependent photoconductivity in a-Si:H provide information about recombination processes and tail and gap states that is complementary to ODMR measurements (Dersch *et al.*, 1983).

As far as we are concerned, with ODMR detailed TRODMR measurements are required, including the use of short pulses of optical excitation and measurements of the spectral dependence of TRODMR signals. These measurements should be done on a-Si : H samples prepared under various conditions. No evidence for localized exciton formation has been obtained in a-Si:H. However, it might be worthwhile to search for such evidence by measuring TRODMR under short pulse excitation.

Acknowledgments

I wish to thank Mr. Y. Sano, Ms. M. Yoshida, and Dr. I. Hirabayashi for their collaboration on the experiments reported in this chapter. I am also grateful to Dr. B. C. Cavenett, Dr. S. Depinna, and Professor S. Nitta for stimulating discussions on the topics in this chapter.

References

Adler, D. (1978). *Phys. Rev. Lett.* **41,** 1755.

Adler, D. (1981). *J. Phys. Colloq. Orsay, Fr.* **42,** C4-3.

Adler, D., and Shapiro, F. R. (1983). *Physica Amsterdam* **117B** & **118B,** 932.

Biegelsen, D. K. (1980). *Solar Cells* **2,** 421.

Biegelsen, D. K., Knights, J. C., Street, R. A., Tsang, C., and White, R. M. (1978). *Philos. Mag. B* **37,** 477.

Block, D., and Cox, R. T. (1981). *J. Luminescence* **24/25,** 167.

Boulitrop, F. (1982). Thesis L'Universitè Scientifique et Mèdicale et L'Institut National Polytechnique de Grenoble.

Brodsky, M. H., and Title, R. S. (1969). *Phys. Rev. Lett.* **23,** 581.

Cavenett, B. C. (1981). *Adv. Phys.* **30,** 475.

Cody, G. D., Tiedje, T., Abeles, B., Brooks, B., and Goldstein, Y. (1981). *Phys. Rev. Lett.* **47,** 1480.

Depinna, S. P., and Cavenett, B. C. (1982). *J. Phys. C* **15,** L489.

Depinna, S., Cavenett, B. C., Austin, I. G., and Searle, T. M. (1981). *J. Phys. Colloq. Orsay, Fr.* **42,** C4-323.

Depinna, S., Cavenett, B. C., Austin, I. G., and Searle, T. M. (1982a). *Solid State Commun.* **41,** 263.

Depinna, S., Cavenett, B. C., Austin, I. G., and Searle, T. M. (1982b). *Solid State Commun.* **43,** 79.

Depinna, S., Cavenett, B. C., Austin, I. G., and Searle, T. M. (1982c). *Philos. Mag. B* **46,** 501.

Depinna, S., Cavenett, B. C., Austin, I. G., Searle, T. M., Thompson, M. J., and Allison, J. (1982d). *Philos. Mag. B* **46,** 473.

Dersch, H., Stuke, J., and Beichler, J. (1981a). *Phys. Status Solidi B* **105,** 265.

Dersch, H., Stuke, J., and Beichler, J. (1981b). *Appl. Phys. Lett.* **38,** 456.

Dersch, H., Schweitzer, L., and Stuke, J. (1983). *Phys. Rev. B* **28,** 4678.
Dunstan, D. J., and Boulitrop, F. (1981). *J. Phys. Colloq. Orsay, Fr.* **42,** C4-331.
Dunstan, D. J., and Davies, J. J. (1979). *J. Phys. C* **12,** 2927.
Dunstan, D. J., Depinna, S. P., and Cavenett, B. C. (1982). *J. Phys. C* **14,** L425.
Elliott, S. R. (1978). *Philos. Mag. B* **38,** 325.
Geschwind, S. (1978). *Proc. Int. Conf. Phys. Semicond., 14th, 1978 Edinburgh* (B. L. H. Wilson, ed.), p. 1013. Inst. Phys., London.
Hirabayashi, I., and Morigaki, K. (1983a). *Solid State Commun.* **47,** 469.
Hirabayashi, I., and Morigaki, K. (1983b). *J. Non-Cryst. Solids* **59/60,** 133.
Hirabayashi, I., Morigaki, K., and Nitta, S. (1980). *Jpn. J. Appl. Phys.* **19,** L357.
Hirabayashi, I., Morigaki, K., and Nitta, S. (1981). *J. Phys. Soc. Jpn.* **50,** 2961.
Hirabayashi, I., Morigaki, K., and Yoshida, M. (1982). *Solar Energy Mat.* **8,** 153.
Ishii, N., Kumeda, M., and Shimizu, T. (1982). *Jpn J. Appl. Phys.* **21,** L92.
Ishii, N., Kumeda, M., and Shimizu, T. (1983). *Phys. Status Solidi B,* **116,** 91.
Kaplan, D., Solomon, I., and Mott, N. F. (1978). *J. Phys. Paris* **19,** L51.
Lepine, D. (1972). *Phys. Rev. B* **6,** 436.
Mollenauer, L. F., Pan, S., and Yngvesson, S. Y. (1969). *Phys. Rev. Lett.* **23,** 683.
Morigaki, K. (1981). *J. Phys. Soc. Jpn.* **50,** 2279.
Morigaki, K. (1983). *Jpn. J. Appl. Phys.* **22,** 375.
Morigaki, K., Dunstan, D. J., Cavenett, B. C., Dawson, P., Nicholls, J. E., Nitta, S., and Shimakawa, K. (1978a). *Solid State Commun.* **26,** 981.
Morigaki, K., Dawson, P., Cavenett, B. C., Dunstan, D. J., Nitta, S., and Shimakawa, K. (1978b). *Proc. Int. Conf. Phys. Semicond., 14th, Edinburgh, 1978* (B. L. H. Wilson, ed.), p. 1163. Inst. Phys., London.
Morigaki, K., Cavenett, B. C., Dawson, P., Nitta, S., and Shimakawa, K. (1980a). *J. Non-Cryst. Solids* **35/36,** 633.
Morigaki, K., Hirabayashi, I., Nakayama, M., Nitta, S., and Shimakawa, K. (1980b). *Solid State Commun.* **33,** 851.
Morigaki, K., Sano, Y., and Hirabayashi, I. (1981). *Solid State Commun.* **39,** 947.
Morigaki, K., Sano, Y., and Hirabayashi, I. (1982a). *J. Phys. Soc. Jpn.* **51,** 147.
Morigaki, K., Sano, Y., Hirabayashi, I., Konagai, M., and Suzuki, M. (1982b). *Solid State Commun.* **43,** 751.
Morigaki, K., Sano, Y., and Hirabayashi, I. (1983). *In* "Amorphous Semiconductor Technologies and Devices—1983" (Y. Hamakawa, ed.). Ohmsha, Tokyo, and North-Holland, Amsterdam.
Movaghar, B., Ries, B., and Schweitzer, L. (1980). *Philos. Mag. B* **41,** 141.
O'Connor, P., and Tauc, J. (1982). *Phys. Rev. B* **25,** 2748.
Okushi, H., Tokumaru, Y., Yamasaki, S., Oheda, H., and Tanaka, K. (1982). *Phys. Rev. B* **25,** 4313.
Pankove, J. I., and Berkeyheiser, J. E. (1980). *Appl. Phys. Lett.* **37,** 705.
Romestain, R., Geschwind, S., Devlin, G. E., and Wolff, P. A. (1974). *Phys. Rev. Lett.* **33,** 10.
Sano, Y., Morigaki, K., Hirabayashi, I., and Konagai, M. (1982a). *Jpn. J. Appl. Phys.* **21,** L291.
Sano, Y., Morigaki, K., and Hirabayashi, I. (1982b). *Solid State Commun.* **43,** 439.
Sano, Y., Morigaki, K., and Hirabayashi, I. (1983). *Physica Amsterdam* **117B & 118B,** 923.
Schweitzer, L., Grünewald, M., and Dersch, H. (1981). *J. Phys. Colloq. Orsay Fr.* **42,** C4-827.
Solomon, I. (1980). *J. Non-Cryst. Solids* **35 & 36,** 625.
Solomon, I., Biegelsen, D. K., and Knights, J. C. (1977). *Solid State Commun.* **22,** 505.
Street, R. A. (1980). *Phys. Rev. B* **21,** 5775.
Street, R. A. (1981). *Adv. Phys.* **30,** 593.
Street, R. A. (1982). *Phys. Rev. B* **26,** 3588.

Street, R. A., and Biegelsen, D. K. (1980). *Solid State Commun.* **33,** 1159.
Street, R. A., Biegelsen, D. K., and Stuke, J. (1979). *Philos. Mag. B* **40,** 451.
Street, R. A., Biegelsen, D. K., and Zesch, J. (1982). *Phys. Rev. B* **25,** 4334.
Tauc, J. (1982). *Solar Energy Mat.* **8,** 259.
Tsang, C., and Street, R. A. (1979). *Phys. Rev. B* **19,** 3027.
Voget-Grote, U., Kümmerle, W., Fischer, R., and Stuke, J. (1980). *Philos. Mag. B* **41,** 127.
Yamasaki, S., Kuroda, S., and Tanaka, K. (1983). *J. Non-Cryst. Solids* **59/60,** 141.
Yoshida, M., Hirabayashi, I., and Morigaki, K. (1983). To be published.
Yoshida, M., and Morigaki, K. (1983). *J. Non-Cryst. Solids* **59/60,** 357.

CHAPTER 5

Carrier Mobility in a-Si:H

J. Dresner

RCA/DAVID SARNOFF RESEARCH CENTER
PRINCETON, NEW JERSEY

I. General Considerations

1. INTRODUCTION

The mobilities of majority and minority carriers are some of the most important quantities one wishes to measure for a semiconductor. The magnitudes of the mobilities μ_n and μ_p for electrons and holes, respectively, play a large role in determining the performance of a device, whether transistor, photocell, electrophotographic receptor, or vidicon camera tube. In a conventional semiconductor the temperature dependence of μ is the primary source of information about the scattering and trapping processes that control carrier transport. In combination with other magnetogalvanic effects, the structure of the various conduction and valence bands in **k** space can often be elucidated. These methods are discussed in many good textbooks on semiconductors; surveys of the Hall and other magnetogalvanic effects can be found in the texts by Putley (1960) and Smith (1959).

In hydrogenated amorphous silicon (a-Si:H) mobility measurements are much more difficult to perform and to interpret for two reasons: (1) the material has a disordered structure, and (2) undoped a-Si:H is a photoconducting insulator whose transport properties have much in common with those of dielectrics. In many instances, even doped a-Si:H may be very resistive compared with conventional semiconductors. Nevertheless, the Hall mobility can still yield valuable information about the bands involved in electron and hole transport. In this chapter some of the ideas that make the

ISBN 0-12-752149-6

interpretation of the Hall effect different in a-Si : H will be sketched. It should be stated at the outset that this interpretation depends on the results of recondite quantum mechanical calculations. To date, only a few results are generally accepted and the field is still under study. The experimental modifications necessary to measure the Hall effect in very resistive materials will then be discussed. Finally, some experimental results will be described, particularly cases that shed light on the states involved in electron and hole conduction.

2. Microscopic, Drift, and Hall Mobilities

The mobility μ is defined as carrier velocity per unit electric field and is usually given in units of $cm^2\ V^{-1}\ sec^{-1}$. There are three somewhat different quantities called μ, which are not always clearly distinguished.

The microscopic or conductivity mobility μ_c. This is the mobility of carriers during the time they are energetically located in a particular conducting state. It is the quantity appearing in the expression for conductivity $\sigma = (n, p)e\mu$ It is often misnamed drift mobility in the literature.

The Hall mobility μ_H. In conventional semiconductors μ_H differs from μ_c by a factor of order unity that depends on the detailed band structure of the material. This factor a also appears in the Hall constant $R_H = \mu_H/\sigma = a/(n, p)e$.

Drift mobility or "time-of-flight" mobility μ_d. This is the measurement most often performed on insulators. Although this method is discussed in the following chapter by T. Tiedje, some comments are needed here. In the most common version of the method, a layer of the material is placed between two noninjecting contacts, at least one of which is transparent. A sheet of charge is then generated at one contact by a short pulse of strongly absorbed radiation and a voltage is applied simultaneously to the electrodes. The length of the radiation pulse must be short compared with the transit time of the carriers across the sample t_T. By measuring the shape of the current transients one can determine $\mu_d = d^2/t_T V$, where d is the sample thickness; both electrons and holes can be measured by using the proper polarity for the applied voltage. It is also required that $t_T < \varepsilon/\sigma$, where ε/σ is the dielectric relaxation time of the material. This condition is that which is used to distinguish operationally a semiconductor from an insulator. If it is not met, the injected sheet of charge will be immediately screened by the mobile carriers in the sample. A common alternative scheme is to use an ohmic injecting contact and apply a potential pulse. The shape of the space-charge-limited current transient can then be related to μ_d (Many and Rakavy, 1962). The first use of the time-of-flight method in its modern form is generally credited to Spear (1957). A summary of this method can be found in the review article by Martini, Mayer, and Zanio (1972).

When both μ_d and μ_H can be measured, it is nearly always observed that $\mu_d \ll \mu_H$. Furthermore, even when the temperature dependence of μ_H is weak (the most common case), μ_d may show an exponentially activated dependence on temperature. The explanation is derived from ideas first advanced by Rose (1951). During their transit across the layer, the carriers may be captured by traps in which they remain until they are thermally released; most of the transit time is spent in these traps. In the simplest case, seldom encountered, of one well-defined trap depth E, one expects $\mu_d \propto \exp(-E/kt)$. However, during the last decade it has been understood that the trapping kinetics can be considerably more complex. The first observations of this were made on the photoconductive chalcogenide glasses used in xerography. This led to the introduction of stochastic or "dispersion" models in which not all carriers have the same μ_d or in which μ_d can change with time (Pfister and Scher, 1978). The dispersion originates from carriers falling into traps that have very different release times. Studies of dispersive transport can yield much useful data about the distribution of defect states in amorphous semiconductors, in contrast to μ_H, which gives information about the states actually involved in conduction. Note that μ_d is the quantity actually involved in determining the performance of devices such as transistors and photoconductors, where the transit time across a region of the device is important. In a photovoltaic solar cell, both μ_d and μ_H may play an important role, depending on whether there are field-free regions in the structure; in that case, carrier transport is by diffusion, which is directly related to the magnitude of μ_H (or μ_c).

3. Hall Effect in a Disordered Semiconductor

From standard transport theory we have the equations

$$l = (D\tau)^{1/2}, \qquad D = \mu kT/e, \qquad \mu = (e/kT)\langle v^2\tau\rangle, \tag{1}$$

where v is the thermal velocity of the carriers and τ the mean time between scattering events. Combining these equations we can relate the mean free path to the mobility

$$L\,(\text{cm}) = 3.8 \times 10^{-9}(m^*/m)^{1/2}\mu\,(\text{cm}^2\ \text{V}^{-1}\ \text{sec}^{-1}), \tag{2}$$

where m is the free electron mass and m^* the carrier effective mass. The classical case is defined by these equations together with the relations

$$R_H = \pm(1/ne)(\mu_H/\mu_c), \qquad \mu_H = R_H\sigma, \qquad \mu_c = \sigma/ne, \tag{3}$$

where R_H is the Hall constant and n the carrier concentration. Equation (2) shows that if $\mu_c < 5$ cm^2 V^{-1} sec^{-1}, the mean free path will be smaller than 2 Å (for $m^* \sim m$). When L becomes shorter than the mean distance between atomic sites in the semiconductor, one cannot assign a classical velocity to the charge carriers; i.e., one cannot analyze the Hall effect by calculating the Lorentz force as though the electron were delocalized. Instead, one must calculate quantum mechanically the jump probabilities between localized states in the lattice. The mode of transport will depend on the energy of the charge carriers relative to the mobility edges; those edges are the energies that separate localized states from extended states in disordered semiconductors (Cohen *et al.*, 1969; Cohen, 1970). Electrons well above or holes well below their respective mobility edges will be delocalized and can be treated in a classical way. If the carriers move in extended states very close to the mobility edge, the situation is complicated by the long-range disorder in the semiconductor; the energies of those states will be spatially modulated by the random potential fluctuations. For deeper states, the carriers move between localized states by multiphonon-assisted hopping or by tunneling.

The last two regimes have been studied theoretically. The random phase approximation (RPA) has been applied to carriers moving immediately above the mobility edge (Friedman 1971, 1978; Overhof and Beyer, 1981). The hopping case has been studied by Emin (1977), Friedman (1978), Friedman and Pollack (1978), and Movaghar *et al.* (1981). For highly localized carriers, a somewhat different treatment based on chemical bond orbital theory has also been used (Grünewald *et al.*, 1981). The calculations predict that in the RPA model μ_H would be smaller than μ_c, with $\mu_H/\mu_c \sim 0.1$, and that μ_H would be only weakly temperature dependent. For hopping transport, magnitudes in the range $10^{-4} < \mu_H < 10^{-2}$ cm^2 V^{-1} sec^{-1} have been estimated at room temperature, and μ_H is strongly temperature dependent.

A more striking result of these calculations is that the sign of the Hall effect is found to be dependent on the geometry of the semiconductor lattice. Sign anomalies have been predicted for both electrons and holes (and have been found experimentally). The sign of μ_H is defined as normal if it agrees with the sign of the thermoelectric power. Whether the sign is normal or reversed has been predicted to depend on whether conduction takes place through odd- or even-membered rings, or, more precisely, whether the transition between the initial and final states comprises an odd or even number of steps. In the chemical bond approach, one obtains the result that the sign depends on whether the sites involved are threefold or fourfold coordinated. The sign of the Hall constant has been given by Emin (1977) as

$$\operatorname{sgn} R_H = \operatorname{sgn}\left[\varepsilon^{n+1} \prod_i^n J_{i,i+1}\right], \tag{4}$$

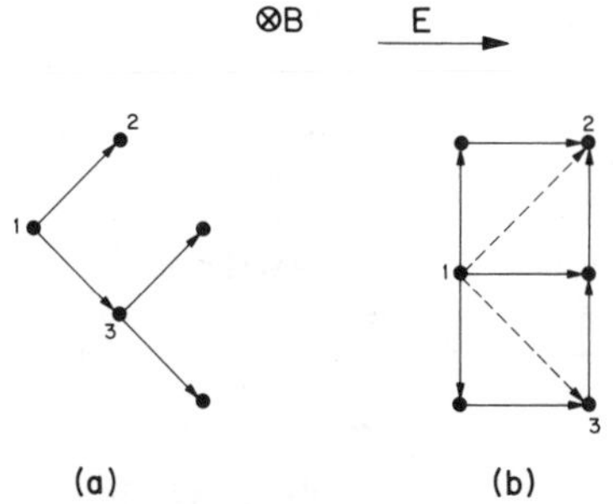

FIG. 1. Transitions involved in the Hall effect for (a) triangular and (b) square lattices.

where $\varepsilon = -1$ for electrons and $\varepsilon = +1$ for holes. The number of steps in the transition is n, and J is the electronic transfer integral for each step. The latter can be written as the matrix element $\langle a_1|H|a_2\rangle$, where the Hamiltonian H is given by $H = (\mathbf{P} + e\mathbf{A}/c)^2/2m + V$, $\mathbf{A}$ being the vector potential, V the total one-electron potential, and $\mathbf{P}$ the electron momentum. The question facing the theorist is the proper selection of the state functions a_i and the actual computation of the matrix element. The geometry is illustrated in Fig. 1. For a triangular lattice (a), an electron located initially at position 1 can move in a direction with components along the electric field $\mathbf{E}$ and normal to the magnetic field $\mathbf{B}$ in a single step; it is the difference in the jump probabilities to sites 2 and 3 that constitutes the Hall effect. Only transitions between nearest-neighbor sites are considered in calculating the J_i, since those to more distant sites are much less probable. For a square lattice (b), the dashed transitions are then not allowed and two jumps are needed in order to reach sites 2 or 3. The actual situation is more complex than illustrated in Fig. 1. Some calculations of the transition probabilities predict that the minimum number of jumps needed to generate a Hall effect may be three (Friedman, 1978) or even five (Emin, 1971). The geometries are often described by theorists as the cases in which the smallest rings in the continuously random network contain an odd or even number of bonds. One should also keep in mind that in a covalent semiconductor like a-Si:H the carriers will be localized on p-orbitals (antibonding for electrons, bonding for holes) and therefore the jumps take place between regions located midway between the atomic sites; this does not change the geometry of Fig. 1. For antibonding orbitals (electrons), the product of the transfer integrals in Eq. (4) is positive for both the three-site and four-site geometry. For bonding orbitals (holes), the product of the transfer integrals is negative for the three-site geometry and positive for the four-site geometry. The signs of the Hall effect predicted by these various calculations are summarized in Table I.

TABLE I

PREDICTED SIGN OF HALL EFFECT FOR a-Si:H

Even-numbered rings	Sign normal for electrons and holes
Odd-numbered rings	Sign reversed for electrons and holes
Tetrahedral bonding	Sign reversed for electrons and holes
Trihedral bonding	*p*-type for electrons and holes

II. Experimental Method

In measuring μ_H in a-Si:H some difficulties arise that are not usually encountered in conventional semiconductors. These are the following.

1. The value of μ_H is small. Measured values have ranged from 5×10^{-3} to 0.5 $cm^2\ V^{-1}\ sec^{-1}$. For a typical value, $\mu_H \sim 10^{-2}\ cm^2\ V^{-1}\ sec^{-1}$, one must measure a Hall voltage V_H of the order of 10 μV.
2. The noise level of V_H is often high, particularly if the current contacts to the sample are not ohmic. An inhomogeneous sample exhibiting a columnar or granular structure will also have a higher noise level. The noise can be reduced considerably by preparing the contacts on the substrate prior to depositing the a-Si layer. Satisfactory results have been obtained with Cr and Cr + Sb (~ 100 Å) for undoped and *n*-type samples and with Cr + Au (~ 100 Å) for *p*-type samples. For contacts applied to the outer surface of the sample, evaporated Al has been successfully used for *p*-type material and baked painted Au (from liquid bright gold solution) for *n*-type samples (LeComber *et al.*, 1977). Although Au diffuses rapidly and reacts with Si, particularly during the Si deposition, the effect can usually be localized to the contact area. For Al the diffusion coefficient is much smaller and doping is localized to the contact area.
3. Because the samples are films of thickness seldom exceeding 1 μm, the sample resistance may be high. For undoped material, for which the resistance between contacts exceeds $10^{11}\ \Omega$ in the dark, the capacitance of the measuring circuits must be taken into consideration.
4. The thermoelectric power for a-Si:H is typically in the range 100–1000 $\mu V\ {}^{\circ}K^{-1}$ for doped material and may be larger if the material is undoped. Thus small temperature gradients can generate signals much greater than V_H. It is important to ensure the stability and uniformity of the sample temperature.
5. The major source of drift is the large density of deep states that can act as electron and hole traps. The occupancy of these states, whose energies extend throughout the energy gap, can vary spatially with time, with temperature gradients, and on the application of the magnetic field. The magnetic

field acts on trap occupancy through the generation of the Hall field and also through its influence on the photoconductivity of a-Si (Mell *et al.*, 1978). It may take as long as 8 hr to stabilize sufficiently the potential at the Hall electrodes after changing the sample temperature or reversing the current direction. On application or reversal of the magnetic field, the response time of V_H may range from 10 to 200 sec.

6. Spatial inhomogeneities in the structure or in the electrical properties of the specimen can cause magnetoresistance effects, yielding offset signals that do not reverse with magnetic field direction and that can be much larger than the Hall voltage (Herring, 1960). For example, these can be induced by fine scratches on the substrate or by local variations of the mobility. It is essential to reverse the current direction as well as the magnetic field for each data point.

The circuit shown in Fig. 2 alleviates some of these problems and has been used to measure μ_H in specimens of various materials with a resistance between contacts of up to 10^{12} Ω. The balanced electrometer arrangement reduces to a second-order effect the slow potential drift at the Hall electrodes caused by trap filling near the current contacts and also greatly reduces the noise originating at those contacts. It also eliminates any effect of the capacitance to ground of the current source V_x; in the usual configuration this source would be floating. The low input capacitance of the feedback-driven shields permits even high-resistance samples to be remotely located in a Dewar for accurate temperature control. It is also important to be able to apply compensating voltages to the Hall electrodes without compromising their resistance to ground; with the vibrating capacitor electrometer this can be done easily. Finally, the use of digital techniques makes it much easier to extract signals from the noise when necessary.

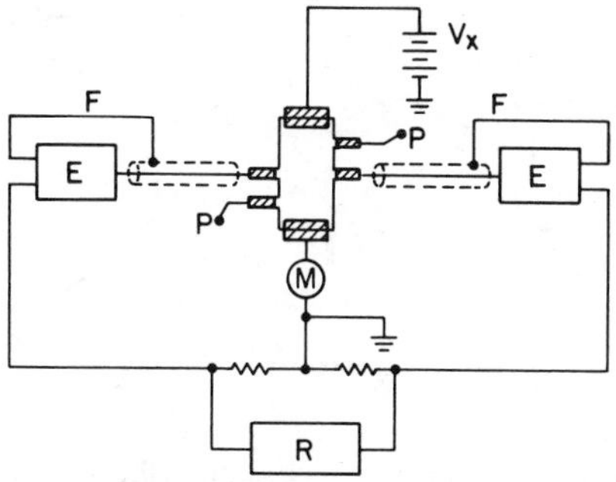

FIG. 2. Experimental arrangement for measuring μ_H in semi-insulating materials. The sample is located in a Dewar between the magnet poles. E, vibrating reed electrometer; R, recorder and digital data processor; F, feedback-driven shields; P, potential probes; M, current meter.

III. Experimental Results

At this date there is not a universally accepted model for electron and hole conduction in a-Si:H. The theory is still not adequate to unequivocally fit the experimental Hall-effect data to the various models that have been proposed. Some of the theoretical difficulties that remain have been discussed by Mott (1978). It should be pointed out that anomalous behavior of the Hall effect, both in sign and in temperature dependence, is not necessarily limited to noncrystalline semiconductors but can occur whenever there is strong localization of the carriers, for example in impurity bands and inversion layers at low temperature ($T < 10°K$). Despite these complications, the measurements of μ_H combined with other transport parameters have shed much light on the states involved in the conduction process.

The first measurements of μ_H on materials doped with phosphorus and boron were performed by LeComber *et al.* (1977) and are shown in Fig. 3. The most important result is that the sign of μ_H is reversed for both electrons and holes. For the *n*-type samples, as the ratio PH_3/SiH_4 increases, μ_H decreases in magnitude and becomes more strongly exponentially activated. The doping concentrations given in Fig. 3 are by volume in the discharge ambient. The phosphorus concentration in the deposited layers will be nearly the same. For *p*-type samples, however, the incorporation ratio for boron may range between 0.65 and 4.4, depending on details of the deposition method. Thus the concentration of B in the samples may be higher than the volume fractions given (Dresner, 1983). The data, together with conductivity and drift mobility measurements, were analyzed in terms of a model with two conduction paths (LeComber *et al.*, 1972). At least for the lightly

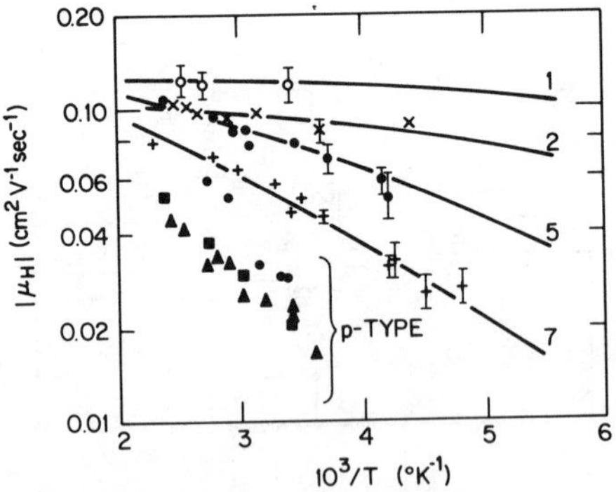

FIG. 3. Mobility for doped a-Si. For the *n*-type samples the volume gas ratios PH_3/SiH_4 are (1) 98 ppm; (2) 304 ppm; (5) 2000 ppm; (7) 3×10^4 ppm. The three *p*-type samples were prepared with B_2H_6/SiH_4 ratios: ■ and ▲, 5×10^4 ppm; ●, 2.3×10^4 ppm. [From LeComber *et al.* (1977).]

doped n-type samples, conduction above 250°K is through extended states, whereas below this temperature it takes place by hopping between localized states that in doped material would be impurity levels. As the donor concentration increases above 5×10^{18} cm^{-3}, impurity conduction becomes the dominant transport mechanism even at room temperature. For impurity-band conduction one can also relate the mobility to the average separation between impurities.

Another group of investigators (Beyer *et al.*, 1977, 1979a,b) from an analysis of thermoelectric power and conductivity as function of temperature in n-type materials inferred that there is only one mode of conduction between 200°K and 600°K with a thermally activated mobility. For samples prepared with PH_3/SiH_4 ratios in the range 1 to 10^4 ppm, their measured values of μ_H are shown in Fig. 4. It was proposed that the single current path is located in extended states above the mobility edge, whose energy is spatially modulated by the electrostatic potential of charged centers (Overhof and Beyer, 1981; Overhof, 1981). The solid curves in Fig. 4 show the results of a calculation of μ_H for this model for different densities of charged centers. These were obtained by computer simulation, with the material divided into small cells for which the energy of the mobility edge is taken to be the average value of the fluctuating potential.

Another question on which measurements of μ_H have yielded valuable information is that of which states are involved in electron conduction for undoped a-Si:H. Based on an analysis of conduction and drift mobility in

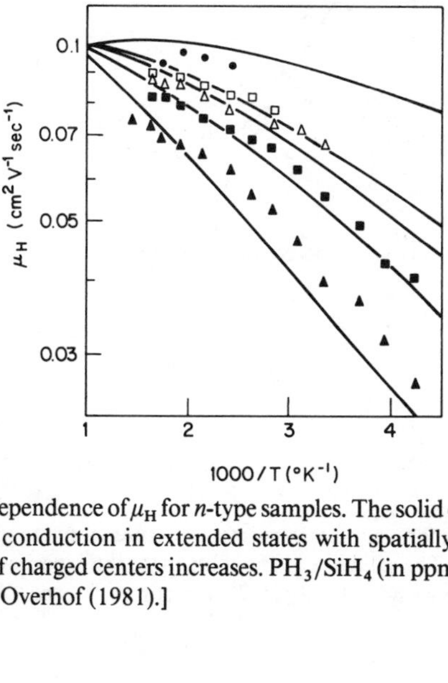

FIG. 4. Temperature dependence of μ_H for n-type samples. The solid curves are calculated for a model of one-channel conduction in extended states with spatially modulated energy. μ_H decreases as the density of charged centers increases. PH_3/SiH_4 (in ppm): ●, 1; □, 3; △, 250; ■, 1000; ▲, 10,000. [From Overhof (1981).]

the framework of the Cohen–Fritsche–Ovshinsky model (Cohen *et al.*, 1969, 1970), LeComber *et al.* (1972, 1977) concluded that conduction above 250°K took place in extended states. The Hall mobility was measured in undoped a-Si:H under illumination and also in the dark (Dresner, 1980), and the results are shown in Fig. 5. There are two regions: below 360°K, $\mu_H \sim 0.1$ cm^2 V^{-1} sec^{-1} and is temperature independent; above 360°K, μ_H is exponentially activated. The best fit for data obtained under illumination gives $\mu_H = 7.9 \exp(-0.13/kT)$. The signals were always *p*-type, although it is known from other transport experiments that the majority carriers are electrons. The existence of the exponential region above 360°K indicates that in undoped material conduction at room temperature cannot be through extended states. The temperature independent μ_H is compatible with tunneling through barriers. For the high-temperature regime, the activation energy $E = 0.13$ eV is characteristic of phonon-assisted hopping. One might visualize a two-channel model in which μ_H reflects mixed conduction: in extended and in localized states with the exponential region reflecting changes in the

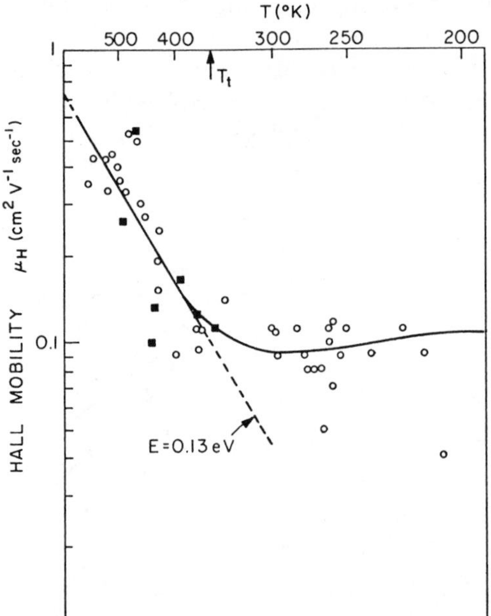

FIG. 5. Hall mobility for electrons versus temperature in undoped a-Si:H. The signals were always reversed in sign. The solid line is a fit of Eq. (6) to the experimental points. ■, dark; ○, 40 mW cm^{-2} illumination with $0.59 < \lambda < 0.79$ μm. [From Dresner (1980).]

relative occupancy of the two sets of states. In that case,

$$\mu_H = (n_1\mu_1^2 + n_2\mu_2^2)/(n_1\mu_1 + n_2\mu_2), \tag{5}$$

but the μ_H versus T data could not be fitted to such a model. However, if one assumes that there is only one conducting state with two modes of transport, Eq. (5) becomes

$$\mu_H = (\mu_1^2 + \mu_2^2)/(\mu_1 + \mu_2) \tag{6}$$

and a good fit to the experimental data is obtained by taking $\mu_1 = 0.11$ and $\mu_2 = 7.9 \exp(-0.13/kT)$ $cm^2\ V^{-1}\ sec^{-1}$. By combining the $\mu_H(T)$ and $\sigma(T)$ data, one can infer that the conducting state lies close to the mobility edge. One is led to a picture of conduction that resembles that of Overhof and Beyer (1981). Electron conduction takes place in a set of states spatially modulated in energy with a mean barrier height of 0.13 eV. At low temperature, conduction is by tunneling through these barriers; at high temperature, it is by thermally activated hopping over them. The prefactor $\mu_0 = 7.9$ $cm^2\ V^{-1}\ sec^{-1}$ then corresponds to a mobility in the extended states. This value is within the range 3–16 $cm^2\ V^{-1}\ sec^{-1}$ deduced by LeComber *et al.* (1972) for the extended states by using the time-of-flight technique.

Another example of the type of information obtainable from μ_H in a-Si:H is taken from the work of Dresner (1983). Figure 6 shows μ_H for two samples prepared with 600 ppm B_2H_6/SiH_4 (compare with Fig. 3) containing 4400 ppm of incorporated boron. In this doping range, μ_H changes sign with temperature. It is *p*-type below ~380°K and *n*-type above that temperature, although the sign of the thermoelectric power remains *p*-type throughout. Hence conduction is always by holes; the sign change is not due to mixing with electron conduction. From the discussion preceding Table I, this implies that at least two sets of states participate in hole conduction and that the sites forming these states must be coordinated in different ways. If the boron concentration is increased tenfold, the sign reversal of μ_H with temperature is not seen; μ_H is always *n*-type and the magnitude and temperature dependence of μ_H are close to those reported by LeComber *et al.* (1977). From this work a two-channel model for hole conduction in B-doped a-Si:H has emerged. The deep channel, dominant at low temperature, may be associated with three-fold coordinated boron and would always yield a *p*-type signal; only a small fraction of the incorporated boron (~10^{-3}) forms acceptors in fourfold coordinated sites. Studies of NMR in boron-doped samples have shown that nearly all the boron is threefold coordinated and that the boron is clustered rather than uniformly distributed (Greenbaum *et al.*, 1982). The valence-band extended states become dominant at high temperature, and in those one expects the sign of μ_H to be negative. The change of sign

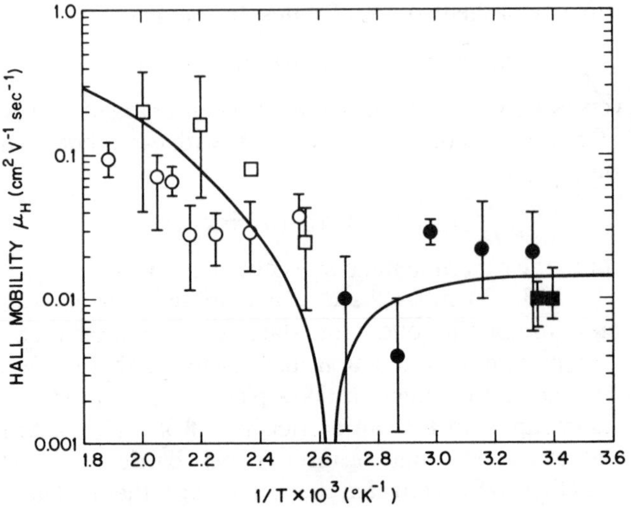

FIG. 6. μ_H versus $1000/T$ for two specimens containing 4400 ppm of incorporated boron. The sign of μ_H changes with temperature while the sign of the thermoelectric power remains *p*-type. The curve is calculated for a two-channel model. Open symbols: *n*-type, filled symbols: *p*-type. [From Dresner (1983). Copyright North-Holland Physics Publishing, Amsterdam, 1983.]

with doping level at low temperature was qualitatively related to the motion of the Fermi level through those states.

IV. Concluding Comments

We see from the preceeding examples that at the present time the information obtained from μ_H measurements in a-Si:H consists mostly of zero-order effects, such as changes in the sign of μ_H or whether μ_H is temperature dependent in a certain range. Nevertheless, the method is useful for testing general ideas about the conducting states and may contribute to the development of better models. At the present time, it is not yet capable of elucidating the detailed processes that control the magnitude of the carrier mobility and lifetime. This is only partly due to the experimental difficulties attendant to making accurate measurements; these difficulties can be overcome. A more severe problem is the need for a better theoretical model for interpreting the experimental data. Until this happens, we can expect that most mobility measurements will be of μ_d and will use some form of the time-of-flight method. As pointed out in this chapter, this method generally gives information about the defect states rather than about the conducting states.

REFERENCES

Beyer, W., Mell, A., and Overhof, H. (1977). *In Proc. Int. Conf. Amorphous and Liquid Semicond., 7th* (W. E. Spear, ed.), p. 328. Edinburgh University, Edinburgh.

Beyer, W., Fischer, R., and Overhof, H. (1979a). *Philos. Mag. B* **39,** 205.

Beyer, W., Fischer, R., and Wagner, H. (1979b). *J. Electron. Mater.* **8,** 127.

Cohen, M. H. (1970). *J. Non-Cryst. Solids* **2,** 432.

Cohen, M. H., Fritsche, H., and Ovshinsky, S. R. (1969). *Phys. Rev. Lett.* **22,** 1065.

Dresner, J. (1980). *Appl. Phys. Lett.* **37,** 742.

Dresner, J. (1983). *J. Non-Cryst Solids* **58,** 353.

Emin, D. (1971). *Ann. Phys.* **64,** 336.

Emin, D. (1977). *Philos. Mag.* **35,** 1189.

Friedman, L. (1971). *J. Non-Cryst. Solids* **6,** 329.

Friedman, L. (1978). *Philos. Mag.* **38,** 467.

Friedman, L., and Pollack, M. (1978). *Philos. Mag. B* **38,** 173.

Friedman, L., and Pollack, M. (1981). *Philos. Mag. B* **44,** 487.

Greenbaum, S. G., Carlos, W. E., and Taylor, P. C. (1982). *Solid State Commun.* **43,** 663.

Grünewald, M., Thomas, P., and Wurtz, D. (1981). *J. Phys. C.* **14,** 4083.

Herring, C. (1960). *J. Appl. Phys.* **31,** 1939.

LeComber, P. G., Madan, A., and Spear, W. E. (1972). *J. Non-Cryst. Solids* **11,** 219.

LeComber, P. G., Jones, D. I., and Spear, W. E. (1977). *Philos. Mag.* **35,** 1173.

Many, A., and Rakavy, G. (1962). *Phys. Rev.* **126,** 1980.

Martini, M., Mayer, J. W., and Zanio, K. R. (1972). *Appl. Solid State Sci.* **3,** 182.

Mell, H., Movaghar, B., and Schweitzer, L. (1978). *Phys. Status Solidi* **88,** 531.

Mott, N. F. (1978). *Philos. Mag. B* **38,** 549.

Movaghar, B., Pohlmann, B., and Wurtz, D. (1981). *J. Phys. C.* **14,** 5127.

Overhof, H. (1981). *Philos. Mag. B* **44,** 317.

Overhof, H., and Beyer, W. (1981). *Philos. Mag. B* **43,** 433.

Pfister, G., and Scher, H. (1978). *Adv. Phys.* **27,** 747.

Putley, E. H. (1960). "The Hall Effect and Related Phenomena." Butterworth, London.

Rose, A. (1951). *RCA Rev.* **12,** 362.

Smith, R. A. (1959). "Semiconductors," Chap. 5. Cambridge Univ. Press, London and New York.

Spear, W. E. (1957). *Proc. Phys. Soc. London Sect. B* **70,** 669.

CHAPTER 6

Information about Band-Tail States from Time-of-Flight Experiments

T. Tiedje

EXXON RESEARCH AND ENGINEERING COMPANY
CORPORATE RESEARCH SCIENCE LABORATORIES
ANNANDALE, NEW JERSEY

I. Introduction

In an amorphous semiconductor with a low defect density such as hydrogenated amorphous silicon (a-Si:H), charge transport takes place in the electronic states in the vicinity of the conduction- and valence-band edges. However, no complete theory of the electronic structure near the band edges in a-Si:H or any amorphous semiconductor has yet been devised. The problem appears to be extraordinarily complex. The disorder generates localized states near the band edges that are not present in crystalline material.

ISBN 0-12-752149-6

According to deep theoretical arguments, these states are separated from the extended states by a mobility edge, which is sharp at zero temperature, somewhat analogous to a thermodynamic phase transition [see, for example, Mott (1967), Mott (1978), Mott and Davis (1978)]. Thermal effects broaden the mobility edge at finite temperature. These localization effects are unique to the amorphous state and cannot be derived by perturbation techniques from conventional theories of electronic structure in crystalline semiconductors.

An additional complication arises from the fact that the probability of an electron (or hole) being self-trapped due to the electron–phonon interaction increases strongly as the electronic wave function shrinks in size to the order of atomic dimensions (Emin, 1982). A consequence of this is that electrons in disorder-induced localized states are believed to be more susceptible to small polaron formation and self-trapping than are ordinary extended-state electrons (Emin, 1984; Cohen *et al.,* 1983). Thus, not only does the disordered structure of amorphous semiconductors introduce new physical phenomena, namely, the mobility edge, but also the effect of known phenomena, such as the electron–phonon interaction, can be qualitatively different.

Certain aspects of the electronic structure of amorphous semiconductors are generally accepted: band tails and mobility edges fall into this category. Beyond this, at least as far as transport properties are concerned, the agreement is not universal. In fact, there are three fairly distinct pictures of the dominant electronic phenomena involved in the charge transport, which we will call the Mott–CFO model (Mott, 1967; Cohen, Fritzsche, and Ovshinsky, 1969), the small polaron model (Emin, 1982, 1984), and the Adler–Silver model (Adler, 1982; Silver *et al.,* 1981). In the Mott–CFO model, transport proceeds via a band of relatively high mobility extended states interrupted by zero-mobility localized states at lower energy. The Adler–Silver model is an elaboration of the Mott–CFO model in which the mobility in the extended-band states is essentially the same as in the corresponding crystalline material, and the band tails are Gaussian and very sharp. In this model the localized states, which control carrier drift transport, are due to well-defined structural defects not related to the structural disorder that gives rise to the band-tail states. In general, the structural defect states will be separated from the band edge to leave a gap in the density of states just below the band edge (see Adler, 1982).

In the small polaron picture, on the other hand, the electron–phonon interaction, compounded by the localization effects introduced by the disorder, leads to the formation of small polarons. The polaron binding energy is then the largest energy in the problem, and charge transport involves multiphonon-assisted hopping of small polarons (Emin, 1984).

In this chapter we show that time-resolved charge transport experiments are a fertile source of information about the charge transport mechanisms in a-Si:H. In some ways, these experiments give less ambiguous results than do more conventional steady-state experiments, which of necessity are always long-time averages. The time-of-flight experiment, which is probably the simplest time-resolved transport experiment, will be the main subject of our discussion. Although it is not possible to rule out other explanations, we show that virtually all of the results of the time-of-flight experiments can be explained in a straightforward way by a Mott–CFO model in which the band-tail states are distributed exponentially in energy.

II. Time-of-Flight Experiment

1. Electrostatics

In the simplest embodiment of the time-of-flight experiment, electron–hole pairs are injected at one face of a dielectric material at $t = 0$ with a short flash of strongly absorbed light. The electrons (or holes depending on the sign of the applied field) are drawn across the material by an externally applied field, as shown in Fig. 1. The two-dimensional sheet of electrons, in the plane perpendicular to the plane of the figure, drifts across the sample with velocity $v_d = \mu_d F$, where μ_d is the drift mobility and F the applied field. For a fixed

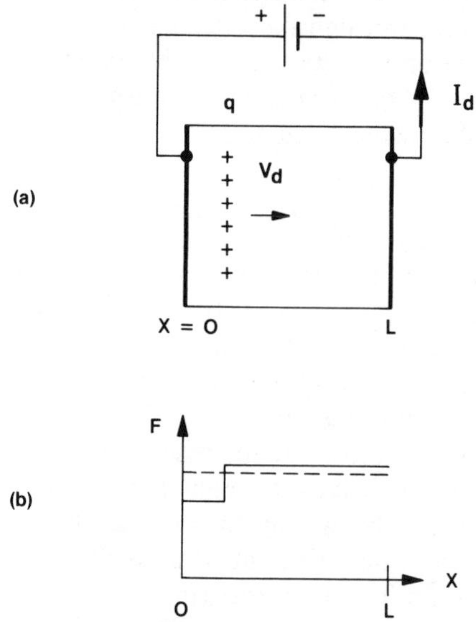

FIG. 1. Schematic of the time-of-flight experiment illustrating (a) the electrode configuration and the electrical bias and (b) the step in the electric field inside the sample due to the photoinjected charge packet.

potential difference V across the sample, simple electrostatic considerations show that for small injected charge density q, the electric field to the left of the drifting charge sheet decreases to $V/L - v_d qt/L\varepsilon$ and the electric field to the right increases to $(V/L) + (q/\varepsilon)(1 - v_d t/L)$, where L is sample thickness, V applied bias, and t time [see, for example, Spear (1969), Spear and Steemers (1983a), or Lampert and Mark (1970)]. The net current $J = q(x)\mu_d F(x) + \varepsilon(dF(x)/dt)$ is the sum of the drift current, which is zero everywhere except at the position of the charge sheet, and a displacement current, which is nonzero everywhere else inside the material.

At low frequencies, that is, for frequencies for which the wavelength of light is much larger than the largest dimension of the circuit, the current is the same everywhere in the circuit. In this case, the displacement current inside the material must be matched by an identical current in the external circuit used to maintain the potential difference across the sample. Thus one can observe the drift motion of the electrons inside the sample by monitoring the current induced in the external circuit. In our example, the current is qv_d/L, or the product of the injected charge density with its average drift velocity normalized to the sample thickness.

Two important conclusions should be self-evident at this point: First, an injected charge inside the material contributes to the integrated charge flowing in the external circuit in proportion to how far it moves through the sample. That is, if one electron moves halfway across the sample, one-half an electron charge will flow through the external circuit. Conversely, for constant applied bias, the only way a current can be induced in the external circuit is by the motion of charge inside the sample. These points should be kept in mind in the following discussion.

2. Experimental Technique

Most of the time-of-flight experiments on a-Si : H that have been reported in the literature were performed on small (2-mm^2 diode structures, either metal Schottky diodes or p–i–n-type structures in the range of 1–10 μm thick. The most common excitation source is a N_2 laser-pumped dye laser with a 1–10-nsec pulse in the blue or green part of the spectrum that is absorbed in the top few hundred angstroms of the film. The optical pulse is timed to arrive at the sample a few microseconds after the turn on of the electrical bias pulse. (A typical experimental apparatus is illustrated in the block diagram in Fig. 2.) This timing ensures that the applied field remains uniform until after the transit-time measurement is completed. After the applied bias has been on for some time (the dielectric relaxation time is a lower limit), the space charge in the vicinity of the blocking contact will redistribute itself in response to the applied potential, and the applied field will no longer be uniform (Tiedje *et al.*, 1980b).

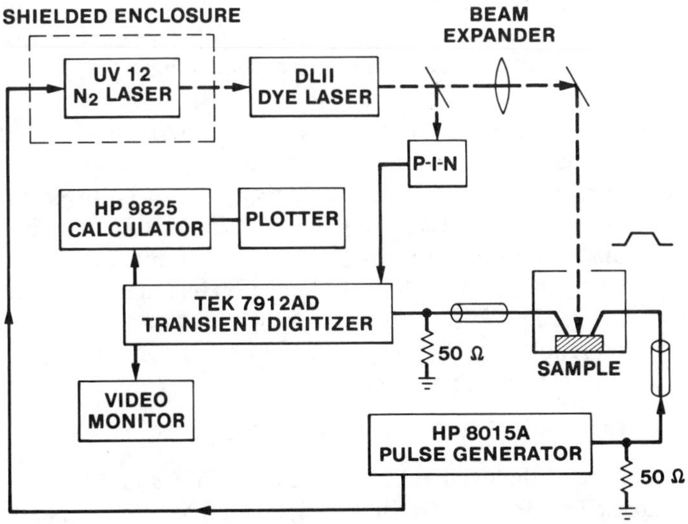

FIG. 2. Experimental time-of-flight apparatus used to apply the electrical bias and record the current induced by the drifting charges in the time-of-flight experiment. [From Tiedje *et al.* (1980b).]

The motion of the drifting charge inside the sample is normally observed through the current in a 50-Ω series resistor, monitored with a high-gain wide-bandwidth oscilloscope or transient digitizer (see Fig. 2). The experiments are usually performed in the space-charge-free regime, where the intensity of the incident light pulse is low enough that the integrated charge q collected in the external circuit is small compared to $Q = CV$, the charge due to the bias voltage on the sample treated as a capacitor. Very little work has been done on the space-charge-limited transients in amorphous semiconductors, probably because of the complexity of the interpretation problem (see, for example, Lampert and Mark, 1970).

The observation of a carrier transit time t_T is normally the first objective of the time-of-flight experiment. The transit time can only be observed if the carrier lifetime τ against deep trapping is longer than the transit time (Spear and Steemers, 1983b). Mathematically the requirement is that the drift length $\mu_0 \tau E$ be greater than the sample thickness, where μ_0 is the band mobility and τ the lifetime of an extended-state electron (or hole) against deep trapping. In the best solar-cell material, the electron and hole $\mu_0 \tau$ products are about equal and of order 10^{-7} cm^2 V^{-1} (Moore, 1983). It is by now well known that the electron (or hole) lifetimes are highly sensitive to doping, increasing dramatically in the case of electrons with n-type doping (or p-type doping in the case of holes). At the same time the lifetime of the opposite sign minority carrier decreases dramatically (see, for example, Oda *et al.*, 1982). This characteristic of the material means that time-of-flight measurements on electrons require material that is slightly *n* type, and conversely measurements on holes require slightly *p*-type material. Both carriers can be observed in the same sample only if the material is of high quality.

This characteristic of the material can cause complications in the vicinity of blocking contacts where the conductivity type can be inverted due to band-bending effects. Thus the carrier lifetime is likely to be anomalous near blocking contacts, an effect that is particularly significant in low-quality material and in the samples in which the depletion region is a large fraction of the total thickness. This effect may be responsible for some of the difficulties that have been reported with electron drift mobility measurements on thin ($\lesssim 1$-μm) samples (Datta and Silver, 1981).

3. "True" Electron Drift Mobility

Electron time-of-flight current transients for a 3.5-μm film and three different applied fields are shown in Fig. 3, taken from Tiedje *et al.* (1980b). Taking the half-widths of the current decay curves in Fig. 3, or nearly equivalently the inflection points, as transit times, one can infer an electron drift mobility of 0.8 $cm^2\ V^{-1}\ sec^{-1}$. Silver and his collaborators (Silver *et al.*, 1982a,b,c, 1983a,b) have rejected this interpretation. In their interpretation the electron transit takes place during the rise time of the laser flash, and the observed current transients are actually due to an optically induced rearrangement of the space charge in the depletion layer. They claim by certain indirect arguments that the electron drift mobility is actually $\sim 1000\ cm^2\ V^{-1}\ sec^{-1}$ (Silver *et al.*, 1982b).

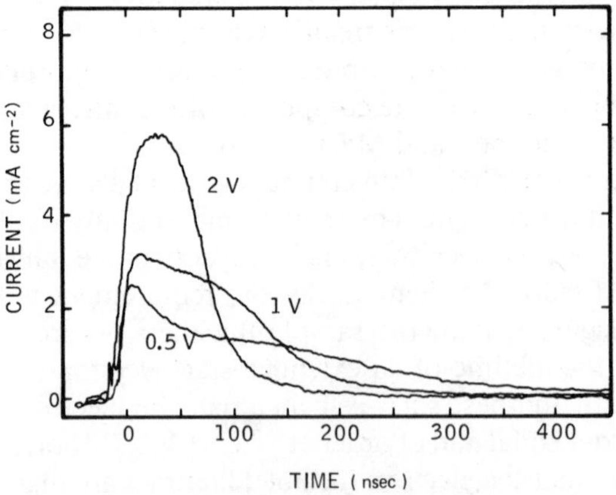

Fig. 3. Electron photocurrent measured at room temperature in the time-of-flight experiment for three different applied biases for a 3.8-μm-thick film. [From Tiedje *et al.* (1980b).]

First we note that the integrated charge collected in the external circuit for the transients in Fig. 3 is equal to the number of absorbed photons, as accurately as the number of absorbed photons can be measured, or $\pm 30\%$ (Tiedje *et al.,* 1980a; Spear and Steemers, 1983a). Thus, if there were a fast component to the current that traversed the sample during the on time of the laser pulse, there should also be a large spike coincident with the laser, with an integrated area similar to the observed signal. A signal of this magnitude is not observed. Why the relaxation of the space-charge layer should give rise to an integrated current equal to the photogenerated charge is not clear. If a charge equal to the photogenerated charge moved one depletion width ($\sim 0.3\ \mu m$ for the sample in Fig. 2), the current would be an order of magnitude smaller. Somehow, each photon must cause on the order of ten electrons to move across the depletion width.

Silver and his collaborators have also tested their hypothesis by reversing the bias field in the middle of the current transient (Silver *et al.,* 1982a,b, 1983a,b). If the electrons are really traversing the sample, the current should reverse; however, no reverse current was observed. Spear and Steemers (1983b) have repeated the experiment independently and in their case showed that a reverse current does exist. Furthermore, they claim that the slow rise time of the bias pulses used by Silver *et al.* (1982b, 1983a,b) and other peculiarities of the experiment made it unlikely that they would, in fact, have been able to observe a reverse current if it did exist.

In my opinion, the simplest explanation of the transient phtotcurrents, namely, that they represent charges drifting from one side of the sample, is also the correct one. This conclusion is supported by Street (1982) and Kirby and Paul (1982), in addition to Spear and Steemers (1983a,b), who have investigated the question in some detail. In the remainder of this chapter we shall assume that the conventional interpretation of the electron time-of-flight experiment is the correct one and that the electron drift mobility is on the order of 1 $cm^2\ V^{-1}\ sec^{-1}$ at room temperature.

III. Experimental Results

4. Temperature Dependence of Electron and Hole Transport

The character of both the electron and hole current transients changes dramatically at low temperatures. At room temperature and above, the electron transit time is inversely proportional to the applied bias; on a log scale the current is relatively constant up to the transit time, at which point it drops rapidly to zero (Tiedje *et al.,* 1980b; Kirby and Paul, 1982). At low temperatures all of the classic features characteristic of dispersive transport are evident. After the excitation pulse the current decay follows a power law in time with a gradual change to a faster power law after the transit time; in addition, the transit time is a superlinear function of electric field, so that the corresponding drift mobility is field dependent. These properties of dispersive transport were first explained by Scher and Montroll (1975) as being due to a distribution of localization times of the form $\psi(t) \propto t^{-\alpha-1}$ with no characteristic long-time cutoff. Dispersive transport has been studied extensively both experimentally and theoretically in connection with the amorphous chalcogenides and various organic semiconductors, and more recently in a-Si:H. See Marshall (1983) for a review. Before discussing the physical origin of dispersive transport in a-Si:H, we first describe the experimental results, beginning with the electron transport.

Figure 4 shows electron current transients plotted on a linear scale for

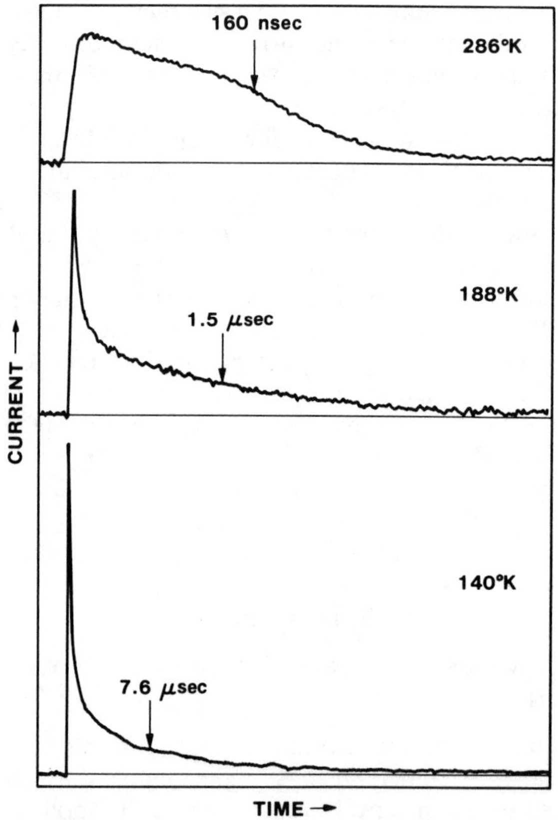

FIG. 4. Electron photocurrent transients measured at three different temperatures. The transit times indicated by the arrows were determined from the log plots shown in Fig. 5. [From Tiedje (1984).]

three different temperatures: 300°K, 220°K, and 170°K. Notice that the current transients become progressively more dispersive and the transit times progressively longer at low temperatures. The transit times are determined from the break in the slope of the current decays when they are plotted on a log scale as shown in Fig. 5, in which the same data shown in Fig. 4 are replotted as log I versus log t. Notice that the current decays are approximately linear on the logarithmic scale, indicative of power-law behavior. However, the data in Fig. 5 are clearly not a convincing demonstration of a power-law time dependence in itself because of the limited time range.

More extensive measurements at 160°K illustrated in Fig. 6 do show power-law behavior over the entire 10-nsec–3-msec time range that was experimentally accessible. The small deviation from linearity at short times can be related to the nonuniform field close to the blocking contact due to

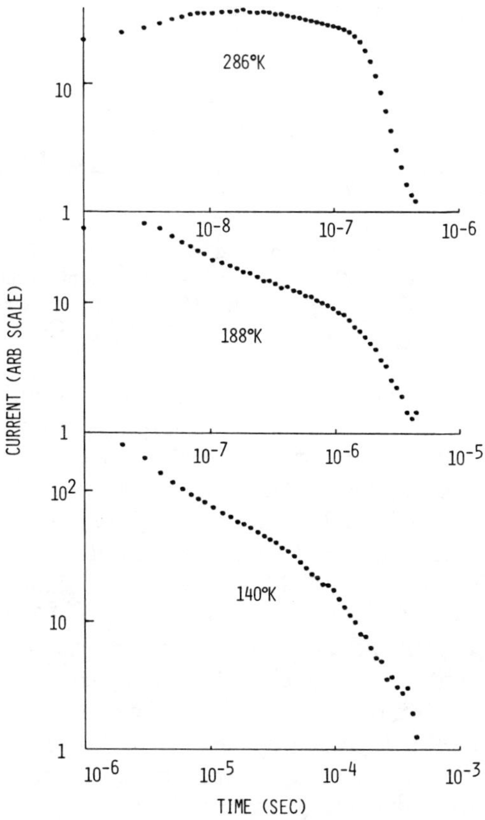

FIG. 5. The same electron photocurrent data as plotted in Fig. 4, replotted on a logarithmic scale.

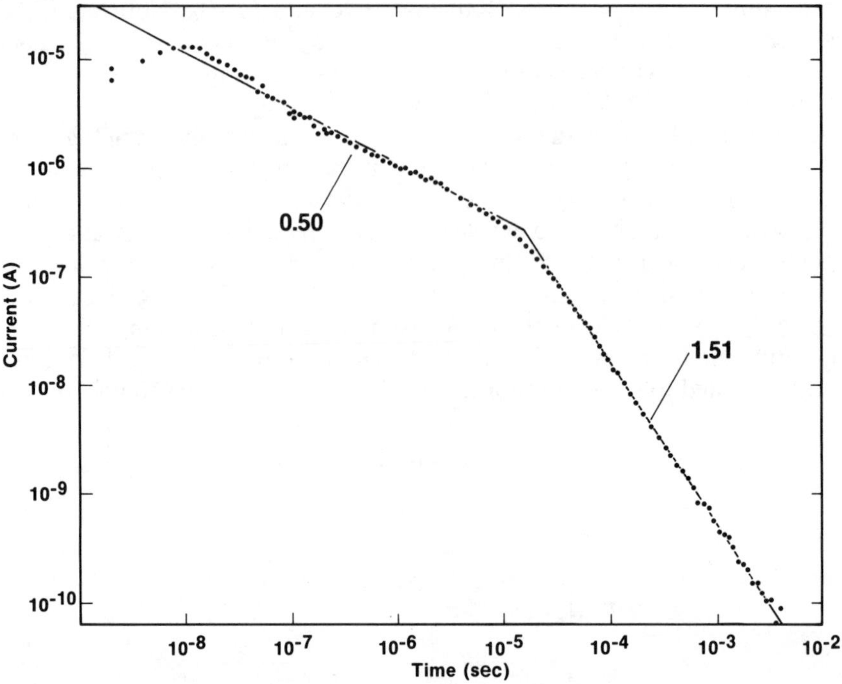

FIG. 6. Electron photocurrent decay obtained by fitting together the current decays measured with three different amplifier bandwidths over three different overlapping time intervals. The sample was 3.8 μm thick, the applied bias was 16 V, and the temperature was 160°K. The numbers indicate the slope of the curve. [From Tiedje (1984).]

the space-charge layer. Otherwise, any deviation is within the experimental error. The data in Fig. 6 were obtained by joining current decays measured over three successive time intervals, each 2.7 orders of magnitude in time, along with progressively higher gains and lower bandwidths.

The transit time, which is clearly evident as a knee in Fig. 6, was measured as a function of applied bias V. The resulting drift mobility defined by $\mu_d = L^2/Vt_\tau$, has a power-law field dependence, as shown in Fig. 7. In the theory of dispersive transport worked out by Scher and Montroll (1975) and others, the field dependence of the transit time is related to the time dependence of the current decay through a dispersion parameter α. In the theory, the current decay at short times ($t < t_T$) has the form $t^{\alpha-1}$ and at long times ($t > t_T$) the form $t^{-\alpha-1}$. Similarly the transit time t_T is proprotional to $(L/F)^{1/\alpha}$. Note that the data in Figs. 6 and 7 are consistent with these predictions of the theory with $\alpha = 0.51$ at 160°K.

In the theory of dispersive transport, all the information about the physical properties of the material is contained in the two parameters α and t_T. Both α

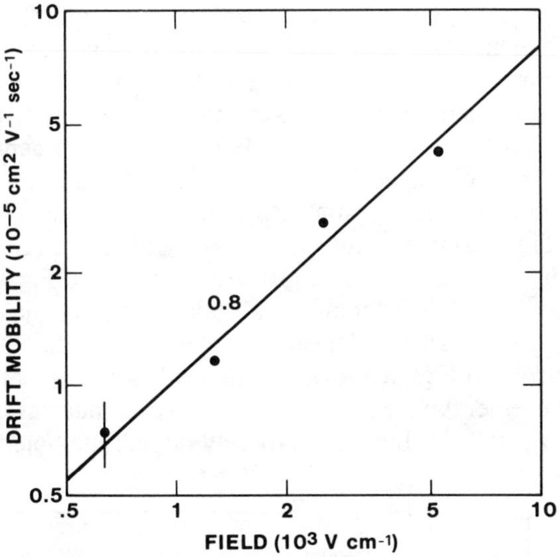

FIG. 7. Field dependence of the electron drift mobility. The solid line is a fit to the data with $\alpha = 0.56$ in Eq. (5).

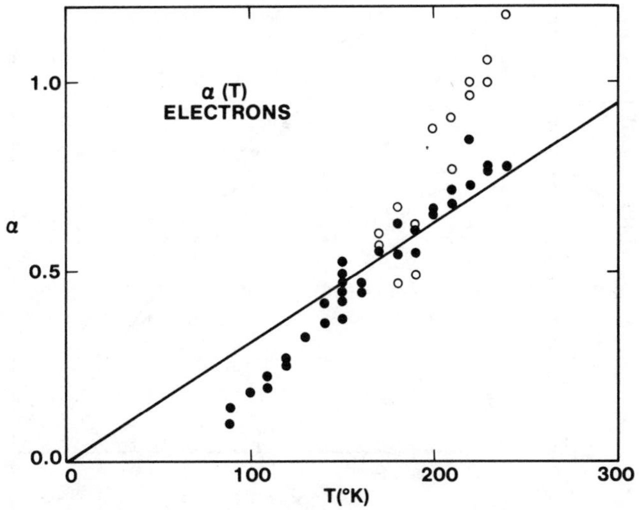

FIG. 8. Temperature dependence of the dispersion parameter α for electrons determined from the slope of the first branch of the photocurrent decay (solid circles) and the second branch (open circles).

and t_T are temperature dependent for the electrons in a-Si : H, as shown in Figs. 8 and 9, in which α and the drift mobility, respectively, are plotted as functions of temperature (Tiedje *et al.*, 1981a).

Time-of-flight measurements on holes show similar dispersive transport phenomena, except that the hole drift mobility at room temperature is about two orders of magnitude lower than for the electrons and the dispersion is greater with a correspondingly smaller α (Allan, 1978; Tiedje *et al.*, 1981a; Street, 1982). Time-of-flight results for holes are illustrated in Fig. 10 (current decays), Fig. 11 (temperature dependence of α), and Fig. 12 (temperature dependence of the drift mobility), all taken from Tiedje *et al.* (1983). Note that the α's are smaller and the drift mobility lower than the analogous results for electrons in Figs. 8 and 9. Another difference is that the electron transport measurements were performed on material that was slightly *n* type (activation energy 0.7 eV) but not purposely doped; the hole transport ex-

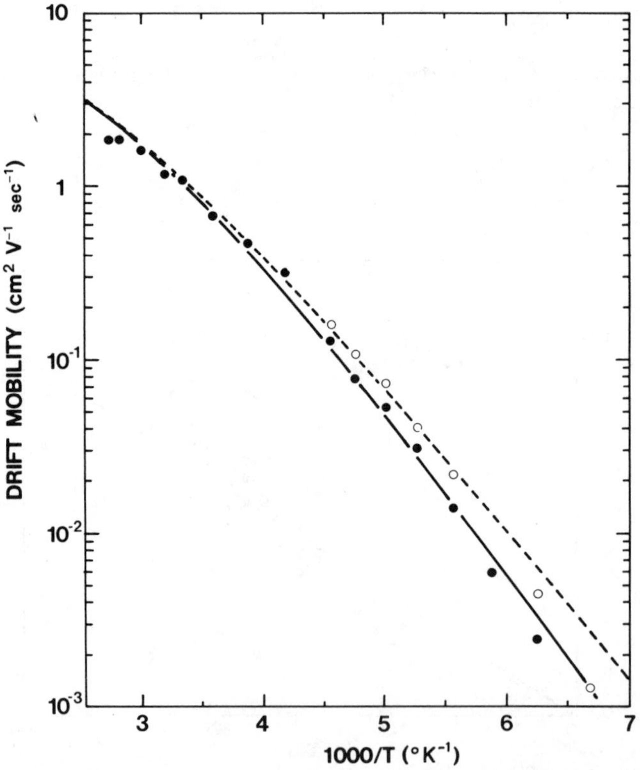

FIG. 9. Temperature dependence of the electron drift mobility for a 3.8-μm-thick sample measured with an applied field of 10^4 V cm^{-1} (solid circles) and 2×10^4 V cm^{-1} (open circles). The lines are fits to the data, as discussed in the text. [After Tiedje *et al.* (1981a).]

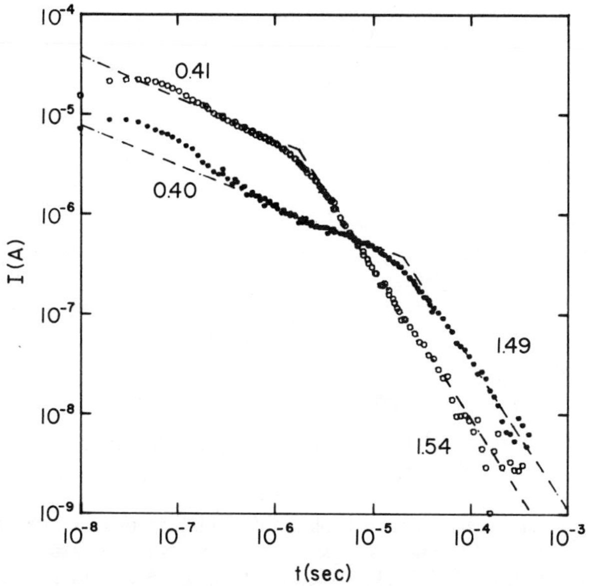

FIG. 10. Hole photocurrent transients measured with 2- and 8-V applied bias (solid and open circles, respectively) at 299°K in the time-of-flight experiment. [Reprinted with permission from *Solid State Communications,* Vol. 47, T. Tiedje, B. Abeles, and J. M. Cebulka, Urbach edge and the density of states of hydrogenated amorphous silicon, Copyright 1983, Pergamon Press, Ltd.]

periments were also performed on material produced by plasma-assisted chemical vapor deposition (CVD) of SiH_4. However, this material was *p* type doped with 100 ppm B_2H_6 and had a conductivity activation energy of ~0.8 eV (Tiedje *et al.,* 1983).

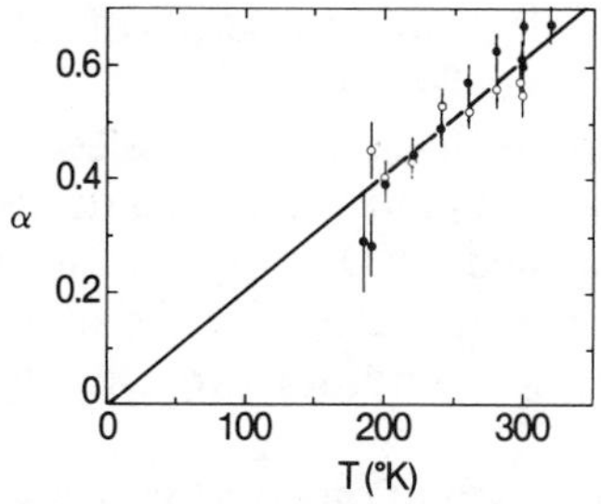

FIG. 11. Temperature dependence of the dispersion parameter α for holes for the sample of Fig. 10 determined from the field dependence of the transit time (open circles) and the slope of the photocurrent decays (solid circles). [Reprinted with permission from *Solid State Communications,* Vol. 47, T. Tiedje, B. Abeles, and J. M. Cebulka, Urbach edge and the density of states of hydrogenated amorphous silicon, Copyright 1983, Pergamon Press, Ltd.]

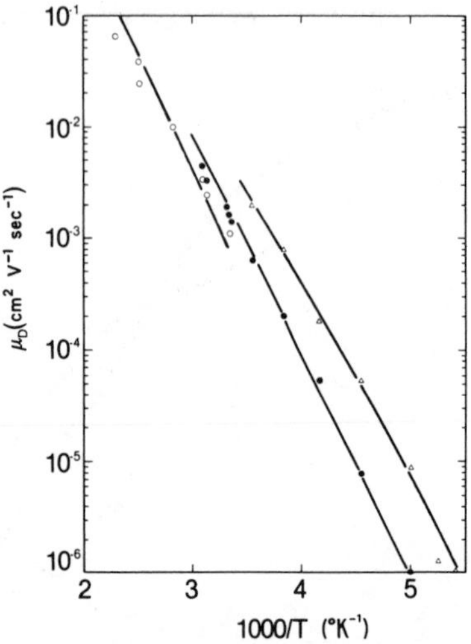

FIG. 12. Temperature dependence of the hole drift mobility. Solid lines are theoretical fits as discussed in the text. △, 17 V; ●, 4 V; ○, 1 V. [Reprinted with permission from *Solid State Communications,* Vol. 47, T. Tiedje, B. Abeles, and J. M. Cebulka, Urbach edge and the density of states of hydrogenated amorphous silicon, Copyright 1983, Pergamon Press, Ltd.]

5. Influence of Preparation Conditions

The extent to which the carrier mobilities depend on preparation conditions is not well understood. One reason for this is that, unlike optical measurements, for example, the time-of-flight measurements can only be made on material for which the carrier lifetimes against deep trapping are sufficiently long for the charges to traverse the sample with a reasonable field, as pointed out earlier. This feature of the time-of-flight experiment has been exploited by Street (1982) and Steemers *et al.* (1983), who have used the technique to explore the electronic properties of defects and the influence of preparation conditions on defect density.

The experimental results discussed in this section were obtained from measurements on "solar-cell-grade" material deposited from low-power rf (13.56-MHz) plasma-assisted CVD of pure silane at 30 mTorr, with a high flow rate, on substrates held at ~250°C on the anode of a capacitive-geometry reactor. These deposition conditions are fairly typical for preparation of high-quality films.

The effect of hydrogen content on electron transport has been studied by Tiedje *et al.* (1981b) in reactively sputtered material. A series of sputtered films prepared with different H contents, varying from 10 to 20%, exhibited no systematic trend in the electron drift mobility with H content, as is illustrated in Fig. 13. The conclusion drawn from this experiment was that

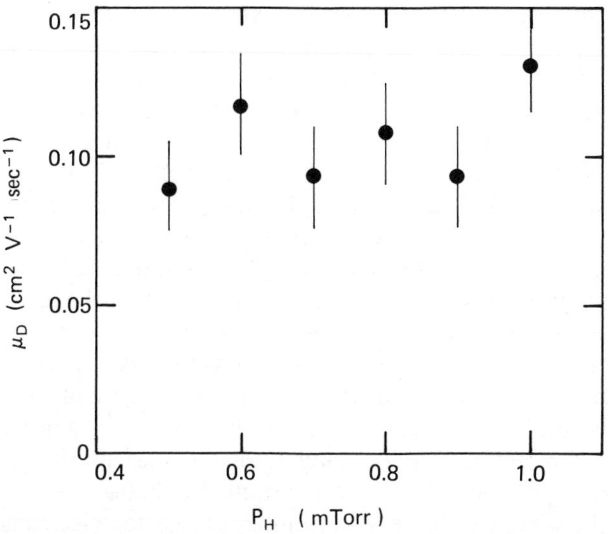

FIG. 13. Electron drift mobility at room temperature as a function of the partial pressure of H_2 in the sputtering plasma for a series of sputtered a-Si : H films. The atomic percent hydrogen in the films shown in the figure increased monotonically with H_2 partial pressure from 14 to 19.5%. [From Tiedje *et al.* (1981b).]

the states at the conduction-band edge in a-Si : H are not sensitive to Si–H bonds; this is consistent with theoretical estimates for the positions of the energy levels associated with Si–H in a-Si : H. In general, the electron drift mobility is lower and more variable in reactively sputtered material (Tiedje *et al.*, 1980a; Kirby and Paul, 1982) than in plasma-assisted CVD material; this is consistent with the larger deposition-parameter space accessible by reactive sputtering.

IV. Transport Model

6. Electron Thermalization in Localized States

What is the origin of the charge transport phenomena and what do these experimental observations tell us about the material? We show that the Mott–CFO model can answer these questions at least to first order, with the additional assumption that the density of localized band-tail states falls off exponentially away from the mobility edges. In this picture, the time-dependent charge transport is dominated by the statistical process associated with the progressive thermalization of electrons (or holes) into the band-tail states. We confine the discussion to electrons and assume that it can be generalized to holes trivially.

At the beginning of the time-of-flight experiment, electrons are injected into the conduction band by a short flash of light. In a time of order 10^{-12} sec, the electrons drop down through the extended-band states to the bottom of the conduction band, where they are captured by the localized states of the band tail (Auston *et al.*, 1980; see also Chapter 9 by Tauc in Volume 21B). Provided that the localized states all have the same capture cross section, each localized state has the same probability of capturing an electron. Thus the initial energy distribution of the electrons trapped in the band-tail states simply parallels the density of tail states, as shown in Fig. 14. For reasons that will become clear later, we assume that the band-tail states decrease in density exponentially away from the mobility edge with the form $N_t(E) = N_0 \exp(-E/kT_c)$, where kT_c is the characteristic width of the conduction-band tail. (We adopt the convention that the zero of energy is at the mobility edge and that positive energy points toward midgap.) Thus, right after the electrons are trapped in the tail states, their distribution has the form $fN_0 \exp(-E/kT_c)$, where $f < 1$ is a constant occupation factor set by the volume of the sample and the total number of injected electrons.

At some time ω_0^{-1} later, where ω_0 is of the order of a phonon frequency,

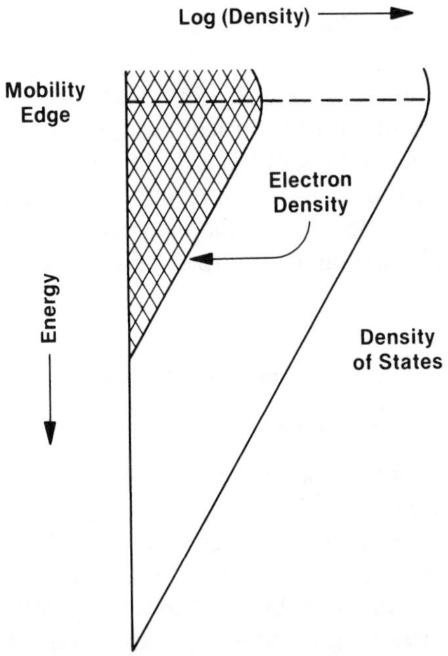

FIG. 14. The shaded area shows the energy distribution of the trapped electrons immediately after the photoinjected electrons have thermalized to the bottom of the conduction band.

the electrons in the shallow localized states are thermally reexcited back into the band states above the mobility edge, where they can diffuse or drift in response to an electric field. Meanwhile, the electrons that happened to fall into deep states for which $E > kT$ must stay in these states for an exponentially longer time of order $\omega_0^{-1} \exp(E/kT)$ before the probability of thermal emission is significant (Rose, 1978). Each time an electron is thermally excited above the mobility edge, it has a new chance to sample the distribution of localized states, and each time it has a small but nonzero chance to fall into a deep state from which reemission takes a very long time. The net result is that the relatively numerous shallow states with the high thermal emission rates are sampled many times for each thermal emission event from a deep state. It follows that the electrons tend to accumulate in the deep states.

The physics of this multiple-trapping process is clear, but how do we quantify it? The multiple trapping model has been solved exactly for a number of special distributions of localized states by several authors, including Noolandi (1977) and Rudenko and Arkhipov (1982). The approximate analysis that we give here is due to Tiedje and Rose (1981) and Orenstein and Kastner (1981).

We begin by noting that at some time t, all the deep states for which $E > E^* = kT \ln \omega_0 t$ have been accumulating electrons without reemitting them, since their mean time for thermal emission $\tau = \omega_0^{-1} \exp(E/kT)$ is much longer than t. It follows that the distribution of electrons in these states must parallel the density of states. Similarly, the electrons in the shallow states for which $E < E^*$ have on average experienced many thermal-emission–recapture events. The end result of this statistical scrambling for the shallow states ($E < E^*$) is well known in solid-state physics—a Boltzmann distribution results. Thus the distribution function for the electrons in the shallow states must have the Boltzmann form $\sim \exp(E/kT)N_t(E)$. The proportionality constant can be calculated from the requirement that the distribution of electrons in the shallow localized states be continuous with the distribution in the deep states. If the (constant) fractional occupancy of the deep states is f, then for continuity the population in the shallow states ($E < E^*$) must be $f \exp[(E - E^*)/kT]\, N_t(E)$.

The interesting thing to note here is that as long as $T < T_c$, the trapped charge distribution is strongly peaked at E^*, as shown in Fig. 15, decreasing toward the mobility edge because of the Boltzmann factor and decreasing toward midgap because of the declining density of states. Furthermore, the energy at which the charge is concentrated moves deeper into the localized state distribution, logarithmically with time according to $E^* = kT \ln \omega_0 t$. Physically E^* is the deepest level from which an electron can be thermally emitted in a time t. Meanwhile, the fractional occupancy f of the deep states

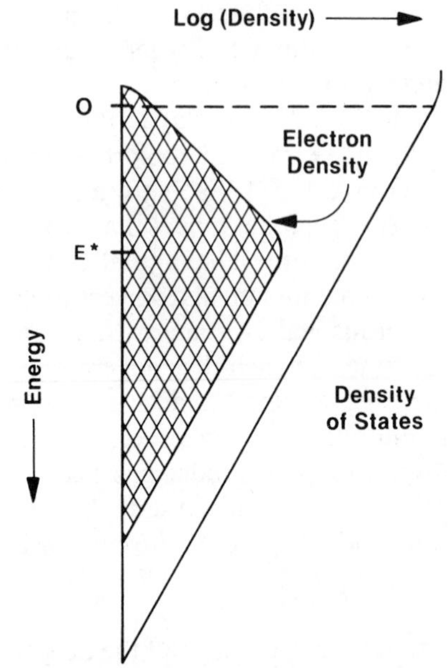

FIG. 15. Trapped electron distribution some time $t \gg \omega_0^{-1}$ after the initial excitation pulse.

progressively increases until the deep states for which $E > E^*$ are all filled or until E^* approaches the Fermi level (see Schiff, 1981).

7. Time-Dependent Drift Mobility

The information in the preceding subsection about the electron thermalization process can be used to calculate the time-dependent transport properties. As far as the transport properties are concerned, the crucial quantity is the number of electrons in the extended states above the mobility edge, since these are the only electrons that can move according to the model. In practice, the model is not strictly correct here since the finite-temperature mobility edge is not perfectly sharp; however, as long as the mobility edge is the sharpest cut-off energy in the problem, neither the broadening of the mobility edge nor phonon-assisted hopping will materially affect the results.

Since the electron density is strongly peaked at E^*, it is possible to treat the drift mobility problem as a two-level problem: a conducting level near the mobility edge with an effective density of states N_c and a single trap level at E^*. This trap level has the unusual feature that its energy position E^* and number density are both time dependent.

For a single trap level, the drift mobility is by definition (Spear, 1969)

$$\mu_d = \mu_0 \frac{n_{\text{free}}}{n_{\text{free}} + n_{\text{trap}}}, \tag{1}$$

where μ_0 is the band-edge mobility and n_{free} and n_{trap} are the number densities of free and trapped electrons, respectively. A straightforward calculation of the free and trapped charge densities following the arguments outlined above leads to

$$\mu_d \simeq \mu_0 \frac{\alpha(1-\alpha)N_c}{\alpha(1-\alpha)N_c + N_0 \exp[E^*(1-\alpha)/kT]}, \tag{2}$$

where $\alpha \equiv T/T_c$. This expression is approximate in the sense that we have treated the trapped charge density as a double-sided exponential at E^* and neglected rounding of the peak. Equation (2) can be simplified if we substitute $kT \ln \omega_0 t$ for E^* and assume that most of the electrons are trapped at times of interest in the time-of-flight experiment. The result

$$\mu_d \simeq \mu_0 (N_c/N_0)\alpha(1-\alpha)(\omega_0 t)^{\alpha-1} \tag{3}$$

shows that the fraction of the injected electrons that are in the transport states above the mobility edge decreases with time according to the power law $t^{\alpha-1}$, where α is a temperature-dependent constant. This result reproduces the first branch ($t < t_T$) of the photocurrent decay observed in the experiments discussed in Part III and in more sophisticated theoretical treatments of dispersive transport.

The power-law decay in Eq. (3) can be regarded as a decay of free carrier density or as a decay of the drift mobility. The latter interpretation and the usual definition of the transit time t_T lead to

$$\int_0^{t_T} \mu_d F \, dt = L, \tag{4}$$

which can be readily solved for t_T. By substituting Eq. (3) into Eq. (4) we find for $\alpha < 1$

$$t_T = \frac{1}{\omega_0} \left(\frac{1}{1-\alpha} \frac{N_0}{N_c} \frac{L}{\mu_0 F} \right)^{1/\alpha}. \tag{5}$$

Note that this expression has the nonlinear sample thickness L and electric field dependences characteristic of dispersive transport.

8. Current Decay after the Transit Time

After the transit time, the nature of the photocurrent decay changes because on the average an electron that is thermally emitted from a trap near E^* after t_T will be extracted at the back contact without being retrapped below E^*. As a result, the photocurrent for $t > t_T$ is controlled by the rate of thermal emission of trapped electrons near E^*. For every factor of e increase in time, another slice of charge kT wide boils off the top of the trapped charge distribution and is collected. It follows that the current is

$$I(t > t_T) = \frac{\Delta Q}{\Delta t} = ef_T \frac{kTN_0}{t} \exp\left(-\frac{E^*}{kT_c}\right), \tag{6}$$

where f_T is the fractional occupancy of the deep traps frozen in at the transit time and e the electron charge.

This factor can be estimated from the assumption that all of the photoinjected electrons are trapped at $t = t_T$ in a distribution that is centered at $E^*(t_T)$ and decays exponentially away from $E^*(t_T)$, as discussed earlier. If the effective density of traps at $E^*(t_T)$ is N_T, then

$$f_T = \frac{g}{LN_T} = \frac{g(1-\alpha)}{LN_0 kT} \exp\left(\frac{E^*(t_T)}{kT_c}\right). \tag{7}$$

Replacing E^* by $kT \ln \omega_0 t$ as before and substituting Eq. (7) into Eq. (6), we find

$$I(t > t_T) = eg(\mu_0 N_c / LN_0)E\alpha(1-\alpha)(\omega_0 t_T)^{2\alpha}(\omega_0 t)^{-\alpha-1}, \tag{8}$$

where the transit time is given by Eq. (5). Note that the current decay in Eq. (8) has the $t^{-\alpha-1}$ time-dependence characteristic of the second branch ($t > t_T$) of the current decay for dispersive transport.

Thus we have been able to reproduce all of the principal results of the dispersive transport theory by simple arguments based on the interaction of the electrons with a distribution of localized states. How the predictions of the model compare with experiment will be discussed in Part V.

9. High Temperature Limit

The preceding discussion is based on the hypothesis that $T_c > T$; however, there is no physical reason why this condition must be true. In the opposite limit $T_c < T$, the density of states is a more rapidly varying function of energy than the Boltzmann factor, and the peak in the electron density stays in the vicinity of the mobility edge, as illustrated in Fig. 16, and does not follow E^* down into the tail states as in the dispersive $\alpha < 1$ limit. After an initial logarithmic decay, it can be shown with the help of Fig. 16 and arguments similar to those used above that at long times the drift mobility becomes a time-independent constant given by

$$\mu_d = \mu_0 \frac{(\alpha - 1)N_c}{(\alpha - 1)N_c + N_0}. \tag{9}$$

The relatively weak explicit temperature dependence of the drift mobility in this expression should not be taken too seriously because it is likely that the temperature dependence of the band mobility, which has been neglected, is equally strong. About all that can be said is that at high temperatures

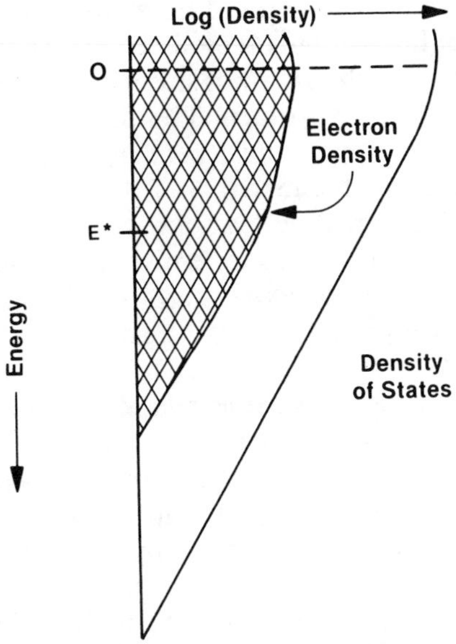

FIG. 16. Trapped electron distribution for $T > T_c$ and at long times ($t \gg \omega_0^{-1}$) when the density of states in the band tail is a more rapidly varying function of energy than the Boltzmann factor.

($T > T_c$), the dispersion goes away at long times and the drift mobility becomes an intensive quantity, independent of time and sample thickness. Even in this regime, the drift mobility is smaller than the free-carrier (band) mobility, and it remains time dependent at short times, as discussed by Silver *et al.* (1982a).

V. Discussion

10. Electron Transport

Our simple model for dispersive transport is in remarkably good agreement with the results of the time-of-flight experiments considering the strength of the starting assumptions. The temperature dependence of the measured dispersion parameter α for electrons can be fit quite well by the form $\alpha = T/T_c$ as predicted by the model, with $T_c \simeq 300°K$ (Tiedje *et al.*, 1981a), as shown in Fig. 8. This value for T_c explains the essentially nondispersive electron transport at room temperature, shown in Fig. 3 ($\alpha \simeq 1$). Taking $T_c = 312°K$, we can fit the temperature dependence of the electron drift mobility in Fig. 9 by using the free-carrier mobility μ_0 and the attempt rate ω_0 as adjustable parameters. For simplicity we assume that $N_c = N_0$. An excellent fit can be obtained (see Fig. 9) with physically reasonable values for the two parameters ($\mu_0 = 13\ cm^2\ V^{-1}\ sec^{-1}$, $\omega_0 = 10^{11}\ sec^{-1}$). In the vicinity of the mathematically singular $\alpha = 1$ point, the fit was calculated from a numerical solution of Eqs. (2) and (4) modified to take into account the fact that for $\alpha \simeq 1$ the trapped charge is not strongly concentrated near E^*.

Some of the details of the experimental results are not consistent with the model. For example, the temperature dependence of α in Fig. 8 is somewhat stronger than predicted by the model, and the α values for the first and second branches of the current decays show systematic differences at high temperatures. More work is needed to resolve these discrepancies.

Although in the model the localized states are distributed exponentially right up to the mobility edge, experimental constraints at both short and long times limit the region of the density of states to which the experiment is sensitive. The experimentally accessible region is defined by the long-time limit of the photocurrent decay at the highest temperature (10^{-6} sec at 300°K) and the short-time limit at the lowest temperature (10^{-8} sec at 140°K). Using these numbers and the attempt rate of 10^{11} sec^{-1} determined from the fit in Fig. 9, we conclude that the time-of-flight experiment shows that the localized state distribution is exponential from 0.08 to 0.30 eV below the mobility edge. The data for 160°K in Fig. 6 alone are sensitive to localized states from 0.1 to 0.27 eV away from the mobility edge. This range over which the density of states is exponential leaves only $\lesssim 0.1$ eV for the "dip" in the density of states near the mobility edge postulated by Adler (1982) and Silver *et al.* (1981).

The picosecond photoconductivity experiments of Auston *et al.* (1980) and Johnson *et al.* (1981) and the photoinduced absorption work of Vardeny *et al.* (1981) can be interpreted by trapping and thermal emission processes that extend to time scales as short as 10 psec. The simplest interpretation of this data is that there is no gap in the density of states on the conduction-band side.

11. Hole Transport

The hole transport data can be interpreted in a similar way. For holes, T_c turns out to be 490°K (see Fig. 11), which is significantly larger than for the electrons. Once again the temperature dependence of the hole drift mobility can be fit with μ_0 and ω_0 as adjustable parameters. The best fit, illustrated in Fig. 12, was obtained with $\mu_0 = 0.5$ cm^2 V^{-1} sec^{-1} and $\omega_0 = 10^{12}$ sec^{-1}. Since the hole transport is dispersive over the entire experimentally accessible temperature range, the exact expression for the transit time derived in the Appendix was used in the fit rather than the approximate expression derived in Part IV. The exact expression could not be used with the fit to the electron data because the exact solution has not been worked out for $\alpha \simeq 1$. It is worth noting that the fit with the approximate expression for the transit time (see Tiedje *et al.*, 1981a) gives μ_0 and ω_0 values that are within a factor of two of the results of the fit to the exact expression.

Once again, the range of sensitivity of the experiments to the density of states can be inferred from E^* at the shortest measurement time at the lowest temperature (10^{-8} sec at 200°K) and the longest measurement time at the highest temperature (10^{-4} sec at 350°K). Using $\omega_0 = 10^{12}$ sec^{-1}, we find that the experiment is sensitive to states between 0.16 and 0.5 eV above the valence-band mobility edge.

The extended-state mobility for holes of 0.5 cm^2 V^{-1} sec^{-1} is low for a true band mobility, since a mobility of unity corresponds to a mean free path on the order of a lattice constant (see, for example, Mott and Davis, 1978). There are several possible explanations. One possibility is that the transport mechanism in the vicinity of the mobility edge is a hopping process. Another possibility is that $N_c < N_0$ since we are actually measuring $\mu_0 N_c/N_0$ and not μ_0 itself. For example, the extrapolated trap density at the mobility edge N_0 could be significantly larger than the effective density of extended states N_c if the energy dependence of the density of localized states flattens out just above the valence-band mobility edge.

Over a limited temperature range, both the electron and hole drift mobilities exhibit a thermally activated temperature dependence with an activation energy near room temperature of 0.15 and 0.38 eV, respectively (see Figs. 9 and 12). One of the conclusions of this chapter is that this activation energy is not necessarily due to a feature (edge) in the distribution of localized states at this energy as has been implicitly assumed in the past (see, for example, Spear *et al.,* 1980). Rather, the activation energy can be regarded as a kind of kinetic limit on how far down into the continuously decreasing density of tail states the injected charge packet can sink before the transit time.

12. Density of States

Clearly, the concept of a mobility edge with the exponential density-of-states tails, illustrated in Fig. 17, can explain the data. In this figure the experimentally accessible part of the density of states is shown by a solid line, where $N_0 = 10^{21}$ cm^{-3} eV^{-1} has been assumed (see Mott and Davis, 1978). How much deviation from a strict exponential distribution would still fit the data is a question that has not been explored yet. Marshall *et al.* (1983a) and Schonherr *et al.* (1981) have shown that Gaussian band tails lead to similar transport phenomena, albeit over more restricted time intervals. Although data such as those shown in Figs. 6 and 10 make it possible to rule out Gaussian band tails, the data are not sufficiently precise to rule out some more subtle difference, such as a functional dependence intermediate between an exponential and a Gaussian or nonexponential trap distributions very close to or very far from the mobility edges.

Furthermore, if the capture cross section of the localized states is allowed to be an exponential function of energy, then the temperature dependence of α in Figs. 8 and 11 should be fit with two parameters rather than one. The

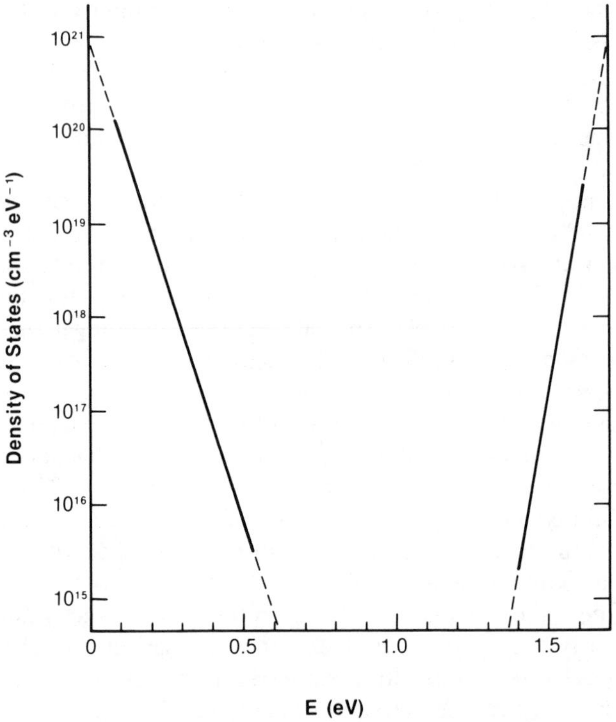

FIG. 17. Solid lines indicate a density of states that is consistent with the experimental time-of-flight data in a-Si:H as discussed in the text. The broken lines indicate regions of the density of states outside the range to which the experiments are directly sensitive.

second parameter describes the energy dependence of the capture cross section (see, for example, Arkhipov *et al.*, 1979). If this extension of the theory is applied to the α data for holes in Fig. 11, then the best fit has a trap capture cross section that increases with depth, with a characteristic temperature analogous to T_c of 2.5×10^3 °K. However, the two-parameter fit is not significantly better than the single-parameter fit; this added refinement is somewhat marginal.

Another fact that has been ignored here is the temperature dependence of the density-of-states parameter T_c itself. Based on the temperature dependence of the optical absorption tail (see, for example, Cody *et al.*, 1981a), this quantity may deviate from its average value by ~10% over the temperature range of interest. Better data and more understanding are needed before this effect can be included in the transport model in a meaningful way.

In addition to being temperature dependent, the width of the optical absorption tail is also dependent on the sample preparation conditions

(Cody *et al.*, 1981b). If T_c for the valence-band tail is related to the exponential tail on the absorption edge, as has been proposed by Tiedje *et al.* (1983), then there should be large variations in the room-temperature hole drift mobility with different preparation conditions since μ_d is an exponential function of T_c. No systematic study of this relationship has yet been undertaken, however.

13. Related Experimental Results

Several completely different experiments support our interpretation of the time-of-flight transport process and the conclusions we have drawn about the distribution of band-tail states. The time-resolved photoinduced absorption experiments of Ray *et al.* (1981) support the view that the photogenerated holes are concentrated in the vicinity of an energy E^*, which moves deeper into the localized state distribution, linearly with temperature and logarithmically with time. Furthermore, the time decay of the photoinduced absorption, which is controlled by the more mobile of the two carriers (electrons), has the $t^{-\alpha}$ form expected from the multiple trapping model (see, for example, Orenstein *et al.*, 1982). The $\alpha = T/300°\mathrm{K}$ temperature dependence for α reported by Tauc (1982) is in excellent agreement with the electron time-of-flight results.

Perhaps the most convincing non-transport-related evidence for exponential density-of-states tails comes from the electron spin resonance work of Dersch *et al.* (1981) on doped material. These workers find a characteristic spin resonance signal, distinct from the dangling-bond resonance, when the Fermi level lies close to either band edge. This signal, which they attribute to unpaired spins in the band-tail states, increases exponentially as the Fermi level approaches either the valence- or conduction-band edge. The characteristic energy for the exponential rise in spin density is 43 meV on the valence-band side and 31 meV on the conduction-band side. These values agree within experimental error with the values obtained from the time-of-flight experiments (42 and 27 meV).

Another interesting comparison is with the optical absorption tail. In principle, the optical absorption coefficient is a convolution of the valence-band density of states with the conduction-band density of states multiplied by a matrix element. However, if the band tails are exponential and one band tail is broader than the other, an elementary mathematical analysis shows that the optical absorption tail has the same energy dependence as the broader band tail, with the energy dependence of the matrix element neglected. In our picture of the electronic structure of a-Si : H, the valence-band tail is broader, and hence the characteristic width of the absorption tail should be compared with the width of the valence-band tail (~42 meV). The optical absorption tail for material prepared under conditions similar to the

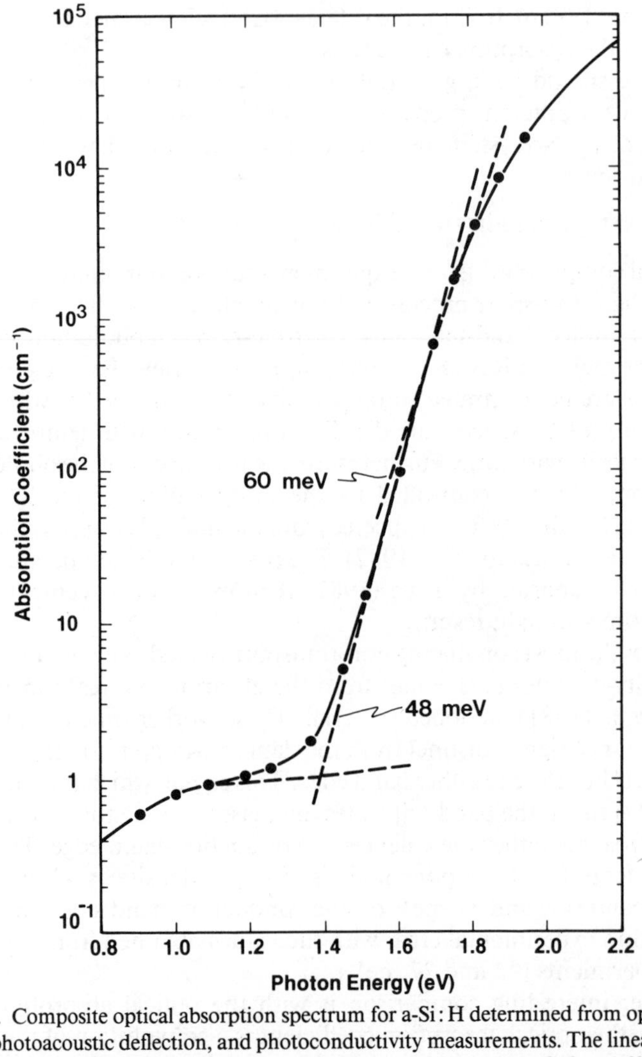

FIG. 18. Composite optical absorption spectrum for a-Si: H determined from optical transmission, photoacoustic deflection, and photoconductivity measurements. The linear fits to the data indicate exponential absorption edges with characteristic widths of 48 and 60 meV. [Reprinted with permission from *Solid State Communications,* C. B. Roxlo, B. Abeles, C. R. Wronski, G. D. Cody, and T. Tiedje, Comment on the optical absorption edge in a-Si: H, Copyright 1983, Pergamon Press, Ltd.]

samples used in the transport experiments was found to have a characteristic width of 48 meV by the photoacoustic deflection measurements of Roxlo *et al.* (1983), illustrated in Fig. 18. The remarkably close agreement between the two different results (48 versus 42 meV) suggests that the Urbach edge in a-Si : H is primarily due to the valence-band density-of-states tail and that the optical matrix element has little if any energy dependence.

14. Practical Implications of Band Tails for Solar Cells

If the band-tail states are a fundamental property of the amorphous state, as seems likely, then one inescapable consequence is that amorphous semiconductors have intrinsic nonradiative recombination centers that cannot be removed, unlike the conventional defects and impurities in crystalline semiconductors. It is well known that nonradiative recombination limits the output voltage in solar cells. It has been shown (Tiedje, 1982) that the band-tail distributions, inferred from the transport experiments described in this chapter, limit the output voltage of amorphous silicon solar cells to 1.0 V for material with an optical (Tauc) gap of 1.7 eV. For comparison, a crystalline semiconductor with the same gap limited only by radiative recombination would have a maximum output voltage of 1.4 eV.

VI. Conclusions

By very simple arguments, we have reproduced all of the principal phenomena of the time-of-flight experiments on a-Si : H from an extension of the Mott–CFO model in which the band tails are exponential. In this model, electron transport occurs only via thermal excitation of localized electrons into the extended bandlike states above the mobility edge. Furthermore, the dispersive transport phenomena that are observed in time-resolved charge-transport experiments are due to the statistical process associated with the thermalization of electrons into a continuous distribution of band-tail states. The widths of the localized state distributions in the band tails, as determined from the transport experiments, are found to be in remarkably good agreement with a variety of other measurements, notably with the electron spin density as a function of doping and with the width of the exponential part of the optical absorption edge.

Many important fundamental issues are unresolved. Virtually no experimental information is available about the energy dependence of the mobility in the vicinity of the mobility edge at some finite temperature. Very little is known about the nature of the electronic structure of the band-tail states or about the strength of the electron–phonon interaction for these states. Finally, it is not clear why the disorder produces exponential rather than Gaussian band tails or even why the valence-band tail is wider than the conduction-band tail, although some theoretical models have been suggested (see, for example, Yonezawa and Cohen, 1981). Much remains to be done.

Appendix

The multiple-trapping model can also be solved analytically for the transit time by the method of Laplace transforms. The one-dimensional transport equations for the free-electron density $n(x, t)$ in a semiconductor with a distribution of discrete trapping levels are

$$\frac{\partial n}{\partial t} + \mu_0 F \frac{\partial n}{\partial x} = \sum_i (k_r^i n_i - k_c^i n) + g\delta(x)\delta(t) \tag{A1}$$

and

$$\frac{\partial n_i}{\partial t} = k_c^i n - k_r^i n_i, \tag{A2}$$

where the δ functions on the right-hand side of the first equation define the initial condition appropriate for the time-of-flight experiment. Diffusion has been neglected because the dispersion in the charge packet due to the interaction with the localized states is expected to dominate diffusion except near $t = 0$ and for very narrow distributions of localized states ($T_c < T$). The rate constants k_r^i and k_c^i define the release rate and capture rate, respectively, for the ith localized state; $n_i(x, t)$ is the electron density in the ith localized state at position x and time t.

The Laplace transform of $n(x, t)$ is defined by

$$\tilde{n}(x, s) = \int_0^\infty e^{-st} n(x, t)\, dt. \tag{A3}$$

It follows that the Laplace transforms of Eqs. (A1) and (A2) can be written as

$$s\tilde{n} + \mu_0 F \frac{\partial \tilde{n}}{\partial x} = s\tilde{n} \sum_i \frac{k_c^i}{s + k_r^i} + g\delta(x) \tag{A4}$$

and

$$\tilde{n}_i = \frac{k_c^i}{s + k_r^i}\tilde{n}, \tag{A5}$$

where $\tilde{n}_i$ has been eliminated from transformed Eq. (A1) with the help of Eq. (A5). Since we are interested in a continuous distribution of localized states, in fact, an exponential distribution, it is convenient to convert the summations over discrete traps to integrals over a continuous distribution.

In the model of interest here, the capture and release rates $k_c(E)$ and $k_r(E)$ are determined only by the energy E of the localized states. We assume that the thermal emission rate $k_r(E)$ is given by

$$k_r(E) = \omega_0 \exp(-E/kT), \tag{A6}$$

as discussed in Part IV. By detailed balance, the corresponding capture rate must be given by

$$k_c(E)\,dE = \omega_0 \exp\left(-\frac{E}{kT_c}\right)\frac{dE}{kT}, \tag{A7}$$

where we have assumed that the density of stages above the mobility edge is a constant given by $N_c\,(\mathrm{cm^{-3}\,eV^{-1}})$ and the density of traps below the mobility edge $(E > 0)$ is

$$N_t(E) = N_c \exp\left(-\frac{E}{kT_c}\right). \tag{A8}$$

The summation over discrete traps reduces to

$$\sum_i \frac{k_c^i}{s + k_r^i} \to \int_0^\infty \frac{\omega_0 \exp(-E/kT_c)}{s + \omega_0 \exp(-E/kT)}\frac{dE}{kT}, \tag{A9}$$

and after a change of variables it becomes

$$\left(\frac{s}{\omega_0}\right)^{\alpha-1} \int_{s/\omega_0}^\infty \frac{u^{-\alpha}\,du}{1+u}, \tag{A10}$$

where $\alpha = T/T_c$.

The Laplace variable s is a frequency parameter that characterizes the time scale. As long as the times of interest are long compared to ω_0^{-1} ($\sim 10^{-12}$ sec), the lower limit of integration in Eq. (A10) can be taken to be zero (in the $\alpha < 1$ limit). In this limit the integral can be evaluated analytically and Eq. (A4) becomes

$$\mu_0 F \frac{\partial \tilde{n}}{\partial x} + \left(s + \omega_0\left(\frac{s}{\omega_0}\right)^\alpha \frac{\pi}{\sin \alpha\pi}\right)\tilde{n} = g\delta(x), \tag{A11}$$

which has the solution

$$\tilde{n}(x, s) = \frac{g}{\mu_0 F}\exp\left[-\frac{x}{\mu_0 F}\left(s + \omega_0\left(\frac{s}{\omega_0}\right)^\alpha \frac{\pi}{\sin \alpha\pi}\right)\right]. \tag{A12}$$

The problem now reduces to finding the inverse transform of Eq. (A12) or an inverse Laplace transform of the form $L^{-1}[\exp(-bs^\alpha)]$. It is convenient to evaluate $L^{-1}[(1/s)\exp(-bs^\alpha)]$ first and then use the property $L^{-1}[s\tilde{f}(s)] = df(t)/dt$ of Laplace transforms. The result is

$$n(x, t) = \frac{g}{\mu_0 F}\frac{1}{\pi t_1}\int_0^\infty \exp(-u - a\cot \alpha\pi u^\alpha)\sin(au^\alpha)\,du, \tag{A13}$$

where $a = \pi(x/t_1^\alpha)(\omega_0^{1-\alpha}/\mu_0 F)$, $t_1 = t - x/\mu_0 F$, and $n(x, t) = 0$ for $x > \mu_0 Ft$.

The total current is given by

$$I(t) = \frac{\mu_0 F}{L} \int_0^L n(x, t)\, dt, \tag{A14}$$

where L is the sample thickness.

In the long-time limit ($t \gg t_T$), where $a_L = \pi(L/t^\alpha)(\omega_0^{1-\alpha}/\mu_0 F) \ll 1$, the free-carrier density $n(x, t) \propto x/t^{1+\alpha}$, and the drift current can easily be shown to be

$$I(t \gg t_T) = g\frac{\mu_0 F}{L}\frac{\Gamma(1+\alpha)}{2}\left(\frac{\omega_0 L}{\mu_0 F}\right)^2 (\omega_0 t)^{-1-\alpha}, \tag{A15}$$

where $\Gamma(1+\alpha)$ is a gamma function. In the opposite (short-time) limit, for which $a_L \gg 1$, the upper limit of integration in Eq. (A14) can be replaced by infinity. The result of the integration when the difference between t_1 and t is neglected is

$$I(t \ll t_T) = g(\mu_0 F/L)\Gamma(1-\alpha)(\sin^2 \alpha\pi/\pi)(\omega_0 t)^{\alpha-1}. \tag{A16}$$

The standard experimental definition of the transit time corresponds to the extrapolated intersection point of the short- and long-time current decays. By solving Eqs. (A15) and (A16) for their intersection point, we find

$$t_T = \frac{1}{\omega_0}\left(\frac{\omega_0 L}{\mu_0 F}\right)^{1/\alpha}\left(\frac{\alpha\Gamma(\alpha)^2}{2\sin\alpha\pi}\right)^{1/2\alpha}. \tag{A17}$$

This expression for the transit time has the same $(L/\mu F)^{1/\alpha}$ dependence as Eq. (5). Also, Eq. (A17) is the same as the expression derived by Rudenko and Arkhipov (1982) in a different way if the difference in initial assumptions is taken into account. They have assumed that the number of states at the mobility edge is N_c (cm^{-3}), whereas we have used kTN_c, where N_c in our case is a density of states (cm^{-3}eV^{-1}). It is interesting to compare numerical factors between the approximate expression [Eq. (5)] and the "exact" expression Eq. (A17); at $\alpha = \frac{1}{2}$, the exact and approximate expressions give $\pi/4$ and 4, respectively.

References

Adler, D. (1982). *Solar Energy Mat.* **8,** 53.

Allan, D. (1978). *Philos. Mag. B* **38,** 381.

Arkhipov, V. I., Iovu, M. S., Rudenko, A. I., and Shutov, S. D. (1979). *Phys. Status Solidi A* **54,** 67.

Auston, D. H., Johnson, A. M., Smith, P. R., and Bean, J. C. (1980). *Appl. Phys. Lett.* **37,** 371.

Cody, G. D., Tiedje, T., Abeles, B., Brooks, B., and Goldstein, Y. (1981a). *Phys. Rev. Lett.* **47,** 1480.

Cody, G. D., Tiedje, T., Abeles, B., Moustakas, T. D., Brooks, B., and Goldstein, Y. (1981b). *J. Phys. Orsay, Colloq. Fr.* **42,** C4-301.

Cohen, M. H., Fritzsche, H., and Ovshinsky, S. R. (1969). *Phys. Rev. Lett.* **22,** 1065.

Cohen, M. H., Economou, E. N., and Soukoulis, C. M. (1983). *Phys. Rev. Lett.* **51,** 1202.

Datta, T., and Silver, M. (1981). *Solid State Commun.* **38,** 1067.

Dersch, H., Stuke, J., and Beichler, J. (1981). *Phys. Status Solidi B* **105,** 265.

Emin, D. (1982). *Phys. Today* **35,** 34.

Emin, D. (1984a). *Comments Solid State Phys.* (in press).

Johnson, A. M., Auston, D. H., Smith, P. R., Bean, T. C., Harbison, J. P., and Adams, A. C. (1981). *Phys. Rev. B* **23,** 6816.

Kirby, P. B., and Paul, W. (1982). *Phys. Rev. B* **25,** 5373.

Lampert, M. A., and Mark, P. (1970). "Current Injection in Solids." Academic Press, New York.

Marshall, J. M., Michiel, H., and Adrianenssens, G. J. (1983). *Philos. Mag. B.* **47,** 211.

Marshall, J. M. (1983). *Rep Prog. Phys.* **46,** 1235.

Moore, A. R. (1983). *J. Appl. Phys.* **54,** 222.

Mott, N. F. (1967). *Adv. Phys.* **16,** 49.

Mott, N. F., and Davis, E. A. (1978). "Electronic Processes in Non-Crystalline Materials," 2nd ed. Oxford Univ. Press, London.

Mott, N. F. (1978). *Rev. Mod. Phys.* **50,** 203.

Noolandi, J. (1977). *Phys. Rev. B* **16,** 4466.

Oda, S., Terazono, S., and Shimizu, I. (1982). *Solar Energy Mat.* **8,** 123.

Orenstein, J., and Kastner, M. (1981). *Phys. Rev. Lett.* **46,** 1421.

Orenstein, J., Kastner, M., and Vaninov, V. (1982). *Philos. Mag. B* **46,** 23.

Paul, W., and Anderson, D. A. (1981). *Solar Energy Mat.* **5,** 229.

Ray, S., Vardeny, Z., and Tauc, J. (1981). *J. Phys. Colloq. Orsay, Fr.* **42,** C4-555.

Rose, A. (1978). "Concepts in Photoconductivity and Allied Problems." R. E. Krieger and Sons, Huntington, New York.

Roxlo, C. B., Abeles, B., Wronski, C. R., Cody, G. D., and Tiedje, T. (1983). *Solid State Commun.* **47,** 985.

Rudenko, A. I., and Arkhipov, V. I. (1982). *Philos. Mag. B* **45,** 209.

Scher, H., and Montroll, E. W. (1975). *Phys. Rev. B* **12,** 2245.

Schiff, E. A. (1981). *Phys. Rev. B* **24,** 6189.

Schonherr, G., Bassler, H., and Silver, M. (1981). *Philos. Mag. B* **44,** 47.

Silver, M., Cohen, L., and Adler, D. (1981). *Phys. Rev. B* **24,** 4855.

Silver, M., Cohen, L., and Adler, D. (1982a). *Appl. Phys. Lett.* **40,** 261.

Silver, M., Giles, N. C., and Snow, E. (1982b). *Solar Energy Mat.* **8,** 303.

Silver, M., Giles, N. C., Snow, E., Shaw, M. P., Cannella, V., and Adler, D. (1982c). *Appl. Phys. Lett.* **41,** 935.

Silver, M., Snow, E., Giles, N. C., Shaw, M. P., Cannella, V., Payson, S., Ross, R., and Hudgens, S. (1983a). *Physica B* **117–118,** 905.

Silver, M., Snow, E., Wright, B., Aiga, M., Moore, L., Cannella, V., Ross, R., Payson, S., Shaw, M. P., and Adler, D. (1983b). *Philos. Mag. B* **47,** L39.

Spear, W. E. (1969). *J. Non-Cryst. Solids* **1,** 197.

Spear, W. E., Allan, D., LeComber, P. G., and Ghaith, A. (1980). *Philos. Mag. B* **41,** 419.

Spear, W. E., and Steemers, H. (1983a). *Philos. Mag. B* **47,** L77.

Spear, W. E., and Steemers, H. L. (1983b). *Philos. Mag. B* **47,** L107

Steemers, H., Spear, W. E., and LeComber, P. G. (1983). *Philos. Mag. B* **47,** L83.

Street, R. A. (1982). *Appl. Phys. Lett.* **41,** 1060.

Tauc, J. (1982). *Solar Energy Mat.* **8,** 269.

Tiedje, T. (1984). "The Physics of Hydrogenated Amorphous Silicon II" (J. Joannopoulos and G. Lucovsky, eds.), Chap. 6, p. 261. Springer-Verlag, Berlin and New York.
Tiedje, T., Abeles, B., Morel, D. L., Moustakas, T. D., and Wronski, C. R. (1980a). *Appl. Phys. Lett.* **36,** 695.
Tiedje, T., Wronski, C. R., Abeles, B., and Cebulka, J. M. (1980b). *Solar Cells* **2,** 301.
Tiedje, T., and Rose, A. (1981). *Solid State Commun.* **37,** 49.
Tiedje, T., Cebulka, J. M., Morel, D. L., and Abeles, B. (1981a). *Phys. Rev. Lett.* **46,** 1425.
Tiedje, T., Moustakas, T. D., and Cebulka, J. M. (1981b). *J. Phys. Colloq. Orsay, Fr.* **42,** C4-155.
Tiedje, T. (1982). *Appl. Phys. Lett.* **40,** 627.
Tiedje, T., Abeles, B., and Cebulka, J. M. (1983). *Solid State Commun.* **47,** 493.
Vardeny, Z., Tauc, J., and Fang, C. J. (1981). *J. Phys. Colloq. Orsay, Fr.* **42,** C4-539.
Yonezawa, F., and Cohen, M. H. (1981). *Solid State Sciences* **25,** 119.

CHAPTER 7

Diffusion Length in Undoped a-Si:H

Arnold R. Moore†

RCA/DAVID SARNOFF RESEARCH CENTER
PRINCETON, NEW JERSEY

I. Introduction

In bipolar devices made from conventional semiconducting materials such as crystalline germanium or silicon, the diffusion length is a vital parameter of the material used. In a p^+ on n-type crystalline silicon solar cell, for example, the photogenerated hole–electron pairs flow only by diffusion from the point of generation back to the p–n junction, where the holes are collected to form the external current. The hole diffusion length is here a measure of the survival of these photogenerated pairs against recombination. It determines the distance from the surface into the semiconductor from which one can hope to obtain useful current. In other configurations, the electron diffusion length may be of primary importance, or sometimes both electron and hole diffusion lengths in various combinations can be important parameters. We shall call the operable quantity in the particular device by the generic term L_D. Its value clearly has a direct effect on the conversion efficiency of a solar cell made from such materials. Typical values of L_D range from a few micrometers in III–V materials to several hundred micrometers in high-quality crystalline silicon (Sze, 1969). From a

† Present address: Institute of Energy Conversion, University of Delaware, Newark, Delaware.

ISBN 0-12-752149-6

theoretical viewpoint, L_D itself depends on the more fundamental quantities μ and τ for both electrons and holes.

In thin film materials such as a-S:H, the L_D is very short, often less than 1 μm. In order to use these materials effectively in devices that depend on carrier collection, designers try to build in electric fields with depletion layers, doping gradients, or external electrodes. Then collection depends on the drift or collection length, a measure of motion in the electric field. Sometimes a combination of diffusion and local field is involved, either spatially superimposed or in contiguous regions. But in any case the carrier collection will again depend on some combination of μ and τ for electrons and holes. Therefore, a measure of the purely diffusion-related quantity L_D may be expected to be of importance in determining material quality regardless of the details of the particular configuration of the device in which it is used. The exact relationship between the collection length and the diffusion length depends on the particular model. We shall return to this point later. This chapter will for the most part be confined to a discussion of the meaning and determination of diffusion length in a-Si:H as derived from experiments in which the diffusive process forms the dominant theme, notwithstanding the fact that L_D may be in principle indirectly derived from the $\mu\tau$ product for electrons and holes obtained from drift-field experiments.

In the bulk of a conventional semiconductor, say of n type, in which a local disturbance raises the concentration of minority carriers from its thermal equilibrium value p_0 to $p(x)$, the steady-state flow of excess carriers $\Delta p = p(x) - p_0$ will be governed, in the absence of electric fields, by a diffusion equation of the form

$$D_p \frac{\partial^2 \Delta p}{\partial x^2} - \frac{\Delta p}{\tau_p} + G(x) = 0. \tag{1}$$

Here, τ_p is the minority carrier (hole) lifetime. The diffusion coefficient for holes D_p is related to the microscopic mobility μ_p by the Einstein relation $D_p = (kT/q)\mu_p$. Here $G(x)$ is a generation function that describes the local disturbance. The solution to this diffusion equation is of the form

$$\Delta p = A \exp(-x/L_p) + B \exp(x/L_p) + H[G(x)], \tag{2}$$

where L_p is the hole diffusion length given by $L_p^2 = D_p\tau_p$. The constants A and B and the functional $H[G(x)]$ depend on the exact form of the generation function and the boundary conditions of the problem.

II. Experimental Determination of L_p in Conventional Semiconductors

Although many methods have been used to determine L_p in normal semiconductors, we wish to consider here only a very successful one, which has proven applicable, in modified form, to a-Si:H as well. We refer to the surface photovoltage method pioneered by A. M. Goodman (1961, 1980). In this method the surface is illuminated with monochromatic light at various wavelengths near the absorption edge. A photovoltage is developed between the surface and the bulk due to the back diffusion of minority carriers to a surface barrier and the subsequent charge separation of the hole-electron pairs in the barrier field. The light intensity at each wavelength is adjusted for a constant photovoltage. It will be instructive to see how application of the diffusion equation [Eq. (1)] leads to a practical method.

We shall assume that our conventional semiconductor is strongly extrinsic, i.e., $n \gg p$, and that there is no electric field and no space charge. Indeed, it is only then that the simple Eq. (1) describes the minority-carrier flow. Yet a strong electric field is required in the surface-barrier region in order to separate the hole-electron pairs and thereby generate a measurable surface photovoltage. Therefore, we also assume that the width W of this space-charge region at the surface is so small that the diffusive flow occurs entirely outside this region, within the sample width d. This amounts to taking $W \ll L_p \ll d$.

When a monochromatic light beam of photon intensity I_0 strikes the surface, the resulting pair generation rate per unit volume $G(x) = I_0\eta(1-R)\alpha\exp(-\alpha x)$. Here α is the absorption coefficient of the light in the semiconductor at that particular wavelength, η the quantum efficiency for the generation process and R the surface reflection coefficient. To simplify the boundary conditions needed to solve the differential equation and also for practical reasons, we further assume that the sample is sufficiently thick (or α is sufficiently large) so that all the light is absorbed within the semiconductor bulk. Yet no light is deemed lost by absorption within the thin surface space-charge region W. These requirements on the optical beam are summarized by the inequality $W \ll 1/\alpha \ll d$. The two sets of inequalities mean that we imagine the sample to be infinitely thick, thus disposing of the boundary at the back. The front boundary condition is given in terms of surface recombination velocity s such that the net rate at which carriers are lost there is proportional to Δp at $x = 0$. Since these carriers arrive at the front surface entirely by diffusion, the diffusion current must balance the recombination so that there is no net accumulation. This leads to the front boundary condition: $s/D_p = (\partial\,\Delta p/\partial x)/\Delta p$.

With the boundary conditions and $G(x)$ specified, Eq. (1) can now be solved. The result for $\Delta p(x)$ is

$$\Delta p(x) = \frac{I_0\eta(1-R)\alpha}{D_p(\alpha^2 - 1/L_p^2)}\left[\left(\frac{1+\alpha D_p/s}{1+D_p/sL_p}\right)\exp(-x/L_p) - \exp(-\alpha x)\right]. \quad (3)$$

Note that Eq. (3) is indeed of the form of Eq. (2) with $B = 0$. The surface photovoltage depends on the minority carrier concentration at the edge of

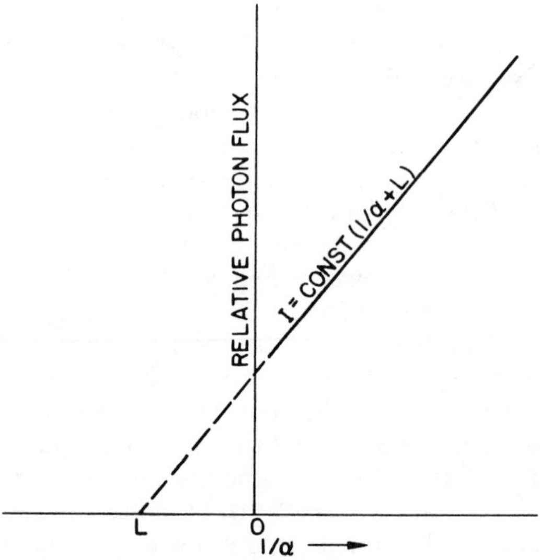

FIG. 1. Plot of Eq. (6), showing how diffusion length is obtained in the SPV method applied to conventional semiconductors. [From Moore (1983)].

the surface space-charge region at $x = 0$. From Eq. (3), $\Delta p(0)$ is given by

$$\Delta p(0) = I_0\eta(1 - R)(D_p/L_p + s)^{-1}[\alpha L_p/(1 + \alpha L_p)]. \tag{4}$$

The measured quantity is the surface photovoltage ΔV. It is not necessary to know the exact form of the dependence of ΔV on $\Delta p(0)$, if the experiment is performed with a constant photovoltage at each wavelength. We need only require that ΔV be some monotonic function of $\Delta p(0)$:

$$\Delta V = f(\Delta p(0)) = \text{const.} \tag{5}$$

If ΔV is fixed by experiment, then so also will be $\Delta p(0)$. If η and R are essentially constant over the range of wavelengths (and of α) used, then Eq. (4) can be written

$$I = \text{const}(1/\alpha + L_p). \tag{6}$$

Thus if the light intensity needed to keep the photovoltage constant at each wavelength is plotted versus $1/\alpha$, the result will be a straight line whose intercept on the negative $1/\alpha$ axis yields L_p. This is illustrated in Fig. 1.

III. Estimates of Diffusion Length in Amorphous Silicon

Amorphous silicon is not a conventional semiconductor in the sense already discussed. Undoped a-Si : H as used in the base layer of an a-Si solar cell is more accurately described as a photoconductive semi-insulator. In the dark it has a resistivity of 10^7–10^9 Ω cm; in sunlight the resistivity is ten thousand times less. And the diffusion length is very short. Early experiments were concerned mainly with secondary photoconductivity and therefore measured essentially the electron $\mu\tau$ product, which was remarkably large for an amorphous material. The fact that successful solar cells could be made from this material, however, demonstrated that both electron and hole motion occur.

The diffusion length itself was first estimated from the properties of solar cells (Wronski, 1977; Konagai *et al.*, 1980), usually from the dependence of short circuit current on reverse voltage at different wavelengths of excitation. Wronski estimated $L_p = 0.2$ μm, whereas the Japanese authors found their results consistent with a L_p of not greater than 0.05 μm.

The first direct estimate was by Staebler (1980), who used a wedge-shaped base layer in a p–i–n cell. He measured the reverse current as a function of position along the wedge when excited by blue (low-penetration) light. His value was not greater than 0.035 μm. In this determination, as well as in others, it must be recognized that diffusion length is a sensitive property of structure and composition. Even if the different methods are presumed sound, consistency is not to be expected while preparations techniques are still being constantly revised unless identical samples are used in the compar-

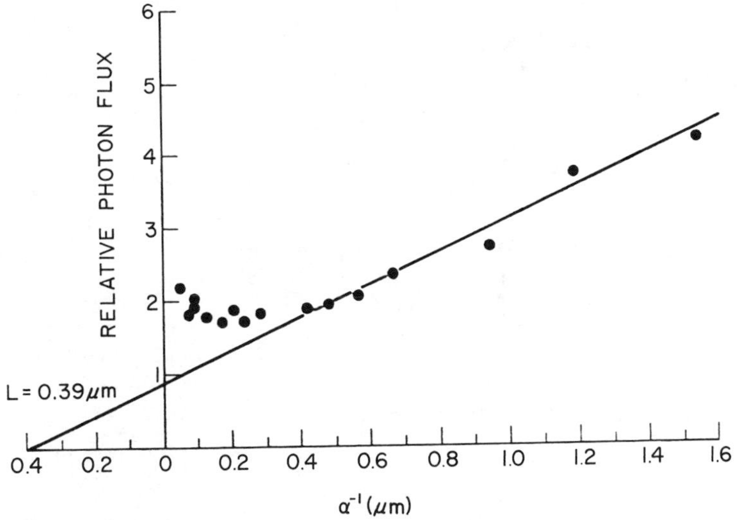

FIG. 2. SPV curve obtained by Dresner *et al.* (1981) on a sample of a-Si:H (sample D-09-04-0 with substrate temperature 240°C and sample width 2.3 μm) using a Kelvin electrostatic probe pickup. [From Dresner *et al.* (1981).]

ison. Indeed, the diffusion length itself can be used as a measure of the quality of the material in bipolar applications.

A direct method that depends on diffusion alone employed the photoelectromagnetic (PEM) effect (Moore, 1980). In this method a diffusive flow of pairs is acted on by a magnetic field perpendicular to the flow. A current mutually perpendicular to both flow and field results. Analysis of the data yielded $L_p = 0.09\ \mu m$.

The first use of the surface photovoltage method to determine minority-carrier diffusion length in a-Si : H was by Dresner *et al.* (1981) and Goldstein *et al.* (1982). They used a vibrating capacitor plate (Kelvin probe) to pick up the surface photovoltage generated by incident light at various wavelengths. The monochromatic light was a steady or dc source; this method avoids the changes of surface-state charge density that in a-Si:H interfere with the chopped light or ac modulation method usually employed with conventional semiconductors. They found that a curve of light intensity versus $1/\alpha$

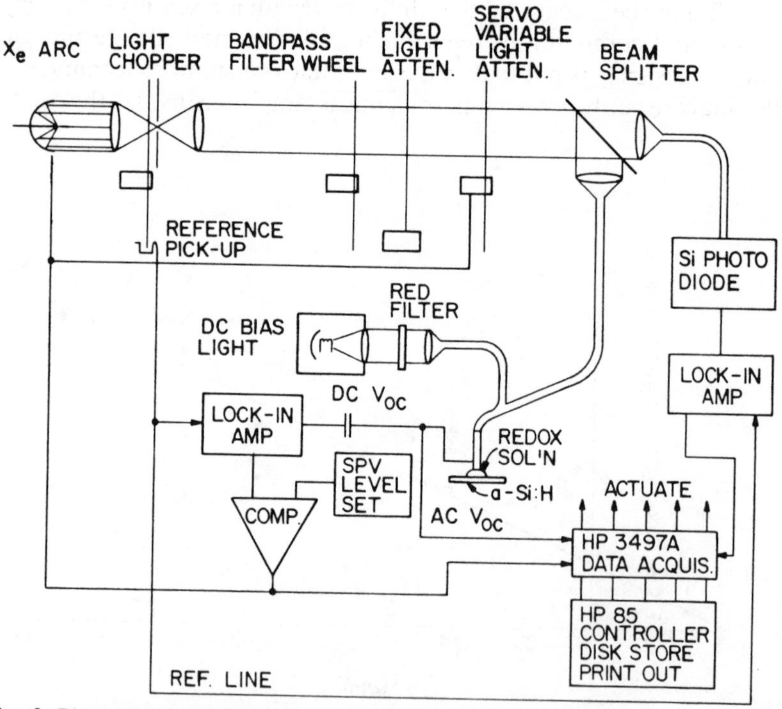

FIG. 3. Block diagram of SPV apparatus using redox electrochemical liquid pickup. Light is conducted to the sample surface by fiber optics. A servo loop adjusts the light intensity to keep the SPV constant at all wavelengths.

could be obtained, as in Fig. 2, similar to that of Fig. 1. Their values for L_p varied from 0.2 to 0.4 μm. Moore (1982) improved the surface photovoltage method by employing an electrochemical oxidation–reduction (redox) system, which acts as a liquid Schottky barrier, to pick up the surface photovoltage signal. This enabled the use of an ac modulation of the incident monochromatic light combined with a dc bias light to fix the pair concentration level at the value commonly used in solar cells (one sun). A feedback system keeps the ac component of the surface photovoltage constant as the wavelength of the monochromatic light is changed. Figure 3 is a simplified block diagram of the system. Moore found L_p values of ~ 0.5 μm, in general agreement with Dresner *et al.* (1981) for similar material prepared in the same way.

IV. Theory of the Surface Photovoltage Method in a-Si:H

Since the SPV method has been intensively used for diffusion length measurement in undoped a-Si:H, we shall discuss the theory of the method in some detail. The approach is to contrast and compare the theory to that already given for conventional semiconductors. The differences arise from two basic facts: (1) Undoped a-Si:H is a photoconductive semi-insulator rather than an extrinsic semiconductor; (2) The thickness of the surface space-charge region (the surface barrier) may be comparable to the diffusion length, whereas in the SPV theory for conventional semiconductors it is assumed that $W \ll L_p$.

As long as the electron concentration (majority carrier) dominates over holes (minority carrier), equations of the form of Eq. (1) will yield proper solutions to diffusion problems in terms of the minority-carrier diffusion length $L_p = \sqrt{D_p \tau_p}$. When $n \sim p$ the diffusion length must be represented by an ambipolar value (Smith, 1959). The ambipolar diffusion length represents the influence of the faster-diffusing electrons on the holes in maintaining overall space-charge neutrality. Because of free-carrier trapping, the steady-state ratio of n to p can vary at high light intensity even when both n and p are much greater than n_0 and p_0. Therefore, for the moment we shall simply state that with the SPV method we obtain an "effective" L_D and postpone until later a discussion of the exact relationship between the measured L_D, the theoretical L_p, and the trapping parameters. Thus the first major difference between a-Si:H and conventional semiconductors is reconciled.

The second difference, the incompatibility of the surface space-charge width and L_D, is resolved by changing the model (Moore, 1983). We assume the sample to be one-dimensional and semi-infinite, divided into the space-charge region $0 < x < W$ and the bulk $W < x < \infty$. The theory assumes that in the surface space-charge region holes and electrons flow in response to the local electric field and to concentration gradients but that there is no recombination. The assumption of no recombination in the space-charge region is reasonable because the transit time in the strong field is normally very short, and it is desirable because it simplifies the mathematical solution to the problem. The electric field in the space-charge region corresponds to the potential $\phi(x)$ set up by the fixed space charge and is uncoupled from the photogenerated carriers. This allows the theory to be compared to experiments like that of Moore (1982) in which a weak ac light modulation is superimposed on a strong penetrating (red) light, which fixes the background charge density. This feature of the model avoids the necessity of solving Poisson's equation simultaneously with the flow equations, a considerable simplification. In the bulk $W < x < \infty$, it is assumed that there is no electric field at all, that carriers flow only by diffusion, and now recombination is included; therefore, holes have an effective diffusion length in the bulk, the ambipolar diffusion length L_D.

The transport equations are now solved, within this model, for the hole and electron currents at $x = 0$. Just as in the case for conventional semiconductors, boundary conditions are required at the surface that take into account the surface recombination. In addition, the boundary $x = W$ between the space-charge region and the bulk must meet the conditions that $n(W) = N$ and $\phi(W) = V_{bi} - V$, where N is the equilibrium electron concentration in the bulk and V_{bi} is the built-in potential. This forces a smooth transition in n and ϕ between the two regions. No such requirement is needed for the holes because, due to the assumption of no recombination, all holes arriving at the depletion edge are collected. Using this approach to obtain the collection efficiency in low-mobility solar cells, Reichman (1981) first gave a solution for the short-circuit electron and hole currents. Moore (1983) gave the corresponding result for the open-circuit voltage and, by applying it to the SPV experiment, gave the light level needed to keep the SPV constant as

$$I = \frac{\text{const}(1/\alpha + L_D)(1 + V_T\mu_n/sF_1)}{(1/\alpha + L_D)(1 + V_T\mu_n/sF_2)F_2 - (1 + V_T\mu_n/sF_1)F_1(1/\alpha)\exp(-\alpha W)}. \tag{7}$$

Here, F_1 and F_2 are certain integrals involving the potential distribution that arise from the integrating factors needed to solve the differential equation. They are given by

$$F_1 = \int_0^W \exp[-\phi(x)/V_T]\,dx = -\frac{V_T}{|E_0|}[\exp(-|E_0|W/V_T) - 1]$$

and

$$F_2 = \int_0^W \exp[-\alpha x - \phi(x)/V_T]\,dx = \frac{V_T}{|E_o| + V_T}[\exp(-\alpha W - |E_0| W/V_T) - 1]. \tag{7a}$$

The potential distribution assumed is that of the classical Schottky barrier with surface field E_0. Thermal voltage kT/q is represented by V_T. In the context of the model, which has been tailored to apply to a-Si:H and similar materials, Eq. (7) takes the place of Eq. (6), which applies only to conventional extrinsic semiconductors with $W \ll L_p$.

The complexity of Eq. (7) obscures its close relationship to the original SPV equation. Some of this obfuscation can be removed by considering an approximation to Eq. (7) obtained by assuming that $1/\alpha$ becomes large. Then Eq. (7) tends toward

$$I = \text{const}(1/\alpha + L_D)[1 + \alpha W^2/2(L + W)]. \tag{8}$$

If W/L and αW are both much less than one, then Eq. (8) reverts to Eq. (6) (except for the definition of L_D). These are just the conditions that make the surface space-charge region negligible with respect to the diffusion length.

The strong influence of the surface depletion region on the interpretation of the results of an SPV experiment on a material with the properites of a-Si:H is illustrated in Fig. 4. This is a plot of Eq. (7). The true L_D is taken as 0.10 μm. The sample has no bias light on it, i.e., the dc open-circuit voltage $V_{oc} = 0$; $V_{bi} = 0.5$ V, $s = \infty$, and the space-charge density is such that $W = 0.50$ μm. The plotted range of $1/\alpha$ corresponds to the normal experimental one, covering the visible light range. A straight line fitted between the two indicated points is an approximation of the treatment of experimental data. It yields an apparent L_D value of 0.369 μm, a substantial difference from the true L_D. Clearly, the surface field is dominating the result.

Another interesting feature that comes out of the plot of Eq. (7) is the departure of the computed curve from the straight line approximation as $1/\alpha$ goes toward zero. This departure is observable in the experimental curve of Fig. 2. In the present model this deviation is caused by electrons photogenerated in the space-charge region flowing back toward the surface counter to the electric field. This is sometimes called back diffusion, because it is in response to the strong concentration gradient created at high α and occurs when the surface recombination velocity is large. The reverse electron flow constitutes a photocurrent that subtracts from the normal hole photocurrent. It therefore requires more light to keep the open-circuit voltage constant as α increases and the curve turns upward.

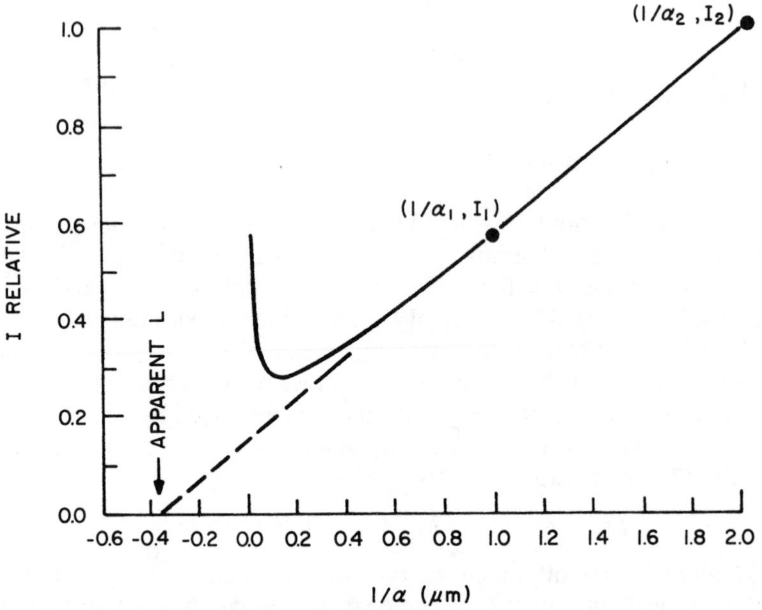

FIG. 4. SPV curve according to Eq. (7). Parameters are $V_{oc} = 0$ (in the dark), $V_{bi} = 0.5$ V, $W = 0.50$ μm, $L_D = 0.10$ μm, $s = \infty$. The apparent L from the $1/\alpha$ intercept is 0.369 μm. [From Moore (1983).]

When the field in the depletion region dominates the transport, the SPV results become insensitive to the true L_D. Then it becomes impossible to extract the diffusion length value from the experimental data by the use of Eq. (7). Fortunately there is a simple way out of this predicament. If a sufficiently strong dc bias light is superimposed on the ac probe light, a dc open-circuit voltage is generated, which approaches the built-in voltage. Accompanying this V_{oc} is a reduction in the space-charge width. The surface band bending is reduced to the same degree as the V_{oc} appearing in the external circuit. With a smaller difference to support, the space-charge width collapses. Experimentally it has been found (Moore, 1983) that intensities near one sun are sufficient to reduce the value of W so that $W \ll L_D$. Capacitance measurements independently confirm this collapse. The reduction is generally greater than can be explained by a simple application of the Schottky-barrier theory. The greater collapse is thought to be due to the capture of photogenerated positive charge by traps in the space-charge region. This too has been observed independently (Williams *et al.*, 1979). The result is an SPV curve of the type shown in Fig. 5. The experimental points were obtained at zero- and one-sun light bias. The solid lines are calculated according to Eq. (7) by using an L_D value of 0.55 μm. This value is a limiting

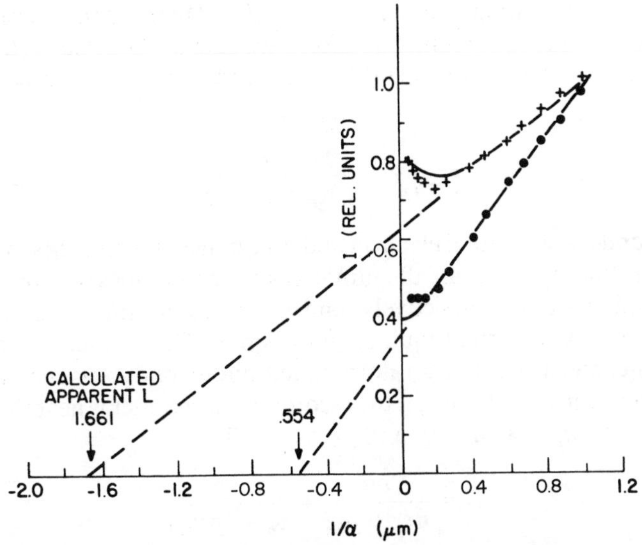

FIG. 5. Fit of Eq. (7) to experimental data at zero- and one-sun bias light intensity. Data were taken with the redox liquid Schottky-barrier method, a red bias light, and on ac probe light. Plus signs refer to zero-sun bias, filled circles to one-sun bias light. [From Moore (1983).]

value as bias light level is increased. Thus strong bias light is an essential component in the SPV experiment to determine the correct value of L_D.

V. Relation of Measured L_D to L_p

It has already been remarked that when the hole concentration approaches the electron concentration as in a near-intrinsic semiconductor or a photoconductive insulator, the diffusion length is no longer the minority-carrier diffusion length $L_p = \sqrt{D_p \tau_p}$, but rather an ambipolar or effective L_D. In this section we wish to show the relationship of the measured L_D to the minority-carrier diffusion length L_p in the context of a material with the photoconductive properties of undoped a-Si:H.

Since we assume free electrons and holes in roughly equal numbers, we must consider the continuity and flow equations for both, combining them simultaneously. In the bulk region of the a-Si:H sample ($W < x < \infty$), the field is taken as zero. Under field-free conditions the resultant diffusion equation is (Rothwarf and Boer, 1975)

$$\frac{kT}{q}\left(\frac{\mu_n \mu_p}{\mu_n n + \mu_p p}\right)\left(p\frac{\partial^2 n}{\partial x^2} + n\frac{\partial^2 p}{\partial x^2}\right) - U + G = 0. \tag{9}$$

This equation is a more general form of which Eq. (1) is a special case. Here,

G is the pair-generating function already defined in that connection, and U is a general recombination function. For U we take the Shockley–Read single-recombination-center formula (Shockley and Read, 1952) in the trap-free case:

$$U = \frac{pn - n_i^2}{(n + n_1)\tau_p + (p + p_1)\tau_n}. \tag{10}$$

Here, n and p are the total electron and hole concentration, n_i is the intrinsic concentration, τ_p and τ_n are the minority-carrier lifetimes in extrinsic n-type and p-type material, respectively, and n_1 and p_1 are the electron and hole concentrations at thermal equilibrium when the Fermi level is at the recombination-center level. For an illuminated photoconductor such as a-Si:H with a band gap of ~ 1.7 eV, if the recombination center is near the center of the gap, then $n, p \gg n_1, p_1$, and $pn \gg n_i^2$. Then

$$U = \frac{pn}{n\tau_p + p\tau_n} = \frac{p}{\tau_p + (p/n)\tau_n}. \tag{11}$$

We can put this value of U into Eq. (9). Since the bulk region is field free, we can also assume space-charge neutrality, $n = p$. Then the applicable diffusion equation becomes

$$\frac{2kT}{q}\left(\frac{\mu_n\mu_p}{\mu_n + \mu_p}\right)\frac{\partial^2 p}{\partial x^2} - \frac{p}{\tau_p + \tau_n} + G = 0. \tag{12}$$

If we consider just the homogeneous part, we can see that this equation is a diffusion equation with a diffusion length given by

$$L_D = \sqrt{\frac{2kT}{q}\left(\frac{\mu_n\mu_p}{\mu_n + \mu_p}\right)(\tau_n + \tau_p)}. \tag{13}$$

Actually Eqs. (12) and (13) would apply as well to conventional semiconductors that were near-intrinsic. We now consider two extra limitations that are specific to our experimental method and to our material.

The SPV diffusion length measurement as developed by Moore uses a weak ac probe light superimposed on a strong dc (red, penetrating) light that sets the background concentration level. We assume a quasi-steady state in which the ac chopping frequency is low enough for all concentrations to follow the fluctuating light intensity. Then the total electron and hole concentrations can be written

$$p = P + \Delta p \qquad \text{and} \qquad n = N + \Delta n. \tag{14}$$

Here, P and N are the dc concentrations and Δp and Δn the ac values. The use of the notations Δp (and Δn) here and in Eq. (1) should not lead to confusion.

In Eq. (14) Δp (and Δn) represents small signal deviations from dc level, whereas in Eq. (1) it represents deviation from thermal equilibrium. The recombination term in Eq. (11) is then

$$U = \frac{P + \Delta p}{\tau_p + [(P + \Delta p)/(N + \Delta n)]\tau_n}. \tag{15}$$

Our approximation now is to replace $P + \Delta p$ and $N + \Delta n$ in the denominator (but not the numerator) by P and N, respectively. The recombination function then splits into dc and ac terms:

$$U = \frac{P}{\tau_p + (P/N)\tau_n} + \frac{\Delta p}{\tau_p + (P/N)\tau_n}. \tag{16}$$

We want a differential equation in the variable Δp analogous to Eq. (12), which is in the variable p. Put Eqs. (14) and (16) back into the original Eq. (9). We can ignore all dc terms (or at least simply treat them as constants) and higher-order Δp and Δn terms. In addition, we can neglect all terms involving gradients of P and N as multiplicative factors, even if these terms are ac terms. This is allowed because these represent ac diffusive flow set up by spatial variations in the dc carrier concentrations. But in our experiment these are created only by the dc red penetrating light that generates carriers almost uniformly through the volume. This emphasizes the importance of the proper color of bias light. The result is

$$\frac{kT}{q}\left(\frac{\mu_n \mu_p}{N\mu_n + P\mu_p}\right)\left(P\frac{\partial^2 \Delta n}{\partial x^2} + N\frac{\partial^2 \Delta p}{\partial x^2}\right) - \frac{\Delta p}{\tau_p + (P/N)\tau_n} + G + \text{dc terms} = 0. \tag{17}$$

Now space-change neutrality requires $N = P$ and $\Delta n = \Delta p$, so that

$$\frac{2kT}{q}\left(\frac{\mu_n \mu_p}{\mu_n + \mu_p}\right)\frac{\partial^2 \Delta p}{\partial x^2} - \frac{\Delta p}{\tau_p + \tau_n} + G + \text{dc terms} = 0. \tag{18}$$

When only the homogeneous equation is considered, this is an equation in Δp with a diffusion length

$$L_{\mathrm{D}} = \sqrt{\frac{2kT}{q}\left(\frac{\mu_p \mu_n}{\mu_n + \mu_p}\right)(\tau_n + \tau_p)}. \tag{19}$$

This is identical to the result formerly obtained. The small-signal and large-signal approaches are therefore equivalent, provided that the basis light is penetrating.

A feature that distinguishes a-Si:H from conventional semiconductors, although it is common enough among large-band-gap insulators, is that during photogeneration a large fraction of the carriers may be trapped in shallow traps in equilibrium with the free carriers (Rose, 1963). The trapped carriers do not contribute directly to photoconduction or diffusion, but they do contribute to the space charge. Indirectly, the trapped carriers modify the mobility, which leads us to introduce the concept of drift mobility. In photoconductivity theory the electron and hole drift mobility are defined as (Bube, 1960)

$$\mu_n^{\mathrm{D}} = \theta_n \mu_n \qquad \text{and} \qquad \mu_p^{\mathrm{D}} = \theta_p \mu_p, \tag{20}$$

where θ_n and θ_p represent the fraction of charge that is free; i.e.,

$$\theta_n = n_{\mathrm{f}}/(n_{\mathrm{f}} + n_{\mathrm{t}}) \qquad \text{and} \qquad \theta_p = p_{\mathrm{f}}/(p_{\mathrm{f}} + p_{\mathrm{t}}). \tag{21}$$

The subscripts f and t stand for free and trapped. Since in photoconductors much of the charge is trapped, θ_n and θ_p are often taken as simply the ratio of free to trapped charge.

To take this new property of our material into account in the diffusion equation, we must introduce the requirement that in imposing space-charge neutrality the space charge is composed of both free and trapped charge. Mathematically this means that $N/\theta_n = P/\theta_p$ and $\Delta n/\theta_n = \Delta p/\theta_p$. Going back to Eq. (17) and introducing this requirement, together with the definitions of Eq. (20), we get

$$\frac{2kT}{q}\frac{\partial^2 \Delta p}{\partial x^2}\left(\frac{\mu_n^{\mathrm{D}}\mu_p}{\mu_n^{\mathrm{D}} + \mu_p^{\mathrm{D}}}\right) - \frac{\Delta p}{\tau_p + (\theta_p/\theta_n)\tau_n} + G + \text{dc terms} = 0. \tag{22}$$

Equation (22) is a diffusion equation with diffusion length

$$L_{\mathrm{D}} = \sqrt{\frac{2kT}{q}\left(\frac{\mu_n^{\mathrm{D}}\mu_p}{\mu_n^{\mathrm{D}} + \mu_p^{\mathrm{D}}}\right)\left(\tau_p + \frac{\theta_p}{\theta_n}\tau_n\right)} = \sqrt{\frac{2kT}{q}\left(\frac{\mu_n^{\mathrm{D}}\mu_p^{\mathrm{D}}}{\mu_n^{\mathrm{D}} + \mu_p^{\mathrm{D}}}\right)\left(\frac{\tau_p}{\theta_p} + \frac{\tau_n}{\theta_n}\right)}. \tag{23}$$

Available evidence suggests (Moore, 1977) that in undoped a-Si:H, $\mu_n^{\mathrm{D}} \gg \mu_p^{\mathrm{D}}$. Then

$$L_{\mathrm{D}} = \sqrt{\frac{2kT}{q}\mu_p\left(\tau_p + \frac{\theta_p}{\theta_n}\tau_n\right)}. \tag{24}$$

If we assume that the microscopic mobilities μ_n and μ_p are of the same order of magnitude, then $\mu_n^{\mathrm{D}} \gg \mu_p^{\mathrm{D}}$ implies $\theta_n \gg \theta_p$. Under these assumptions Eq. (24) becomes

$$L_{\mathrm{D}} = \sqrt{\frac{2kT}{q}\mu_p\tau_p} = \sqrt{2}\sqrt{D_p\tau_p} = \sqrt{2}L_p. \tag{25}$$

Thus except for the factor of $\sqrt{2}$, L_D is a direct measure of the minority-carrier diffusion length L_p, with the assumptions stated (Faughnan *et al.*, 1984).

VI. Application of Diffusion Length Measurements in a-Si:H

Diffusion length measurements have been used as a quality-control technique in the preparation of a-Si:H from glow discharge in silane. The reason such quality-control methods are necessary is that L_D is a very sensitive indicator of contamination by chemical or physical imperfections. Some contaminants are far more effective than others in their influence on diffusion length (and hence on the performance of solar cells), presumably because of chemical interaction that leads to the formation of recombination centers. One of the most pernicious impurities in the gaseous atmosphere during the discharge is oxygen. Figure 6 shows the results of an experiment performed to demonstrate this sensitivity to oxygen contamination of the undoped a-Si:H layer (P. Longeway, 1983, private communication). A

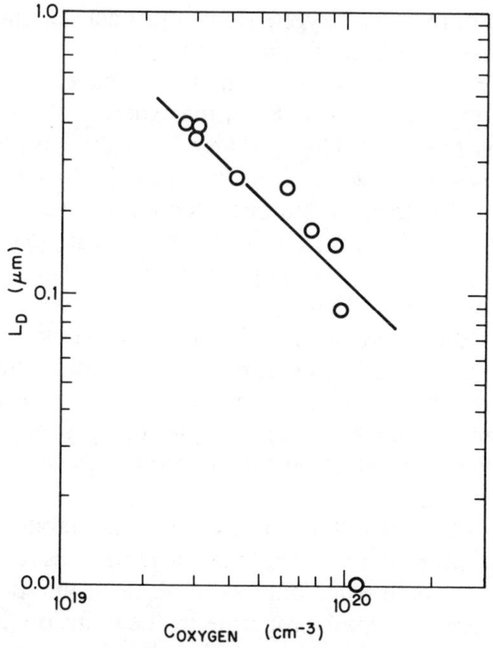

FIG. 6. Effect of oxygen contamination of the a-Si:H film on measured L_D. The oxygen was introduced into the discharge atmosphere during film growth, but its concentration was measured later in the film by SIMS. The line represents a least squares fit to the data, which gives a slope of -0.97 on the log–log plot.

series of samples were produced in a dc discharge system in which the oxygen concentration in the discharge atmosphere was varied by changing the relative flow rates of silane and bleed-in oxygen. Diffusion length was measured on all the samples by the SPV redox pickup method with a bias light of one sun (red light). Then the samples were subjected to a SIMS analysis for oxygen in the deposited films.

The figure shows that over an atomic oxygen concentration range of $\sim 10^{19}-10^{20}$ cm^{-3} in the film, the diffusion length decreased by roughly an order of magnitude. For the highest oxygen concentration the measured diffusion length was 0.01 μm, a number so small that it is considered below the reliable limit. Disregarding this extreme point, a least squares fit to the data gives L_D proportional to $C_{ox}^{-0.97}$. Aside from showing the importance of eliminating oxygen from the discharge, the data tell us something about the nature of the recombination center. The minority-carrier lifetime τ_p can be written as $\tau_p = 1/N_r v\sigma$, where N_r is the density of recombination centers, v the thermal velocity, and σ a recombination cross section. Since $L_D \propto \tau_p^{1/2}$, we would expect to find $L_D \propto 1/C_{ox}^{1/2}$ if each oxygen atom contributed linearly to the recombination-center density. The fact that we actually find $L_D \propto 1/C_{ox}$ implies that two oxygen atoms must act in concert to form the recombination center. The most likely assumption is that the oxygen must be associated in the solid as O–Si–O in order to have its deleterious effect.

One other example will suffice. By fitting experimental I versus $1/\alpha$ data on a given sample to Eq. (7) at low and high bias light, it is possible to extract both L_D and the space-charge width at low light W_0. This can be done rapidly by a computer fit. As already mentioned, for a single type of recombination center we expect $L_D \sim 1/N_r^{1/2}$. In the ordinary Schottky-barrier formula the surface space-charge region W_0 is proportional to $1/\rho^{1/2}$, where ρ is the space-charge density at low light. Now in a material like a-Si:H, the space-charge density at low or zero bias light is composed of positive charge trapped in deep states. If whatever is generating the deep states is linearly related to the recombination center density, then one might expect that since the functional dependences are the same, a plot of L_D versus W_0 would be a straight line as the recombination-center density was changed, other things remaining the same.

One subject under active study at this time is the stability of a-Si:H solar cells under solar illumination. There may be many reasons for cell instability, but we here are concerned only with the electronic properties of the i layer. We shall represent these properties by the diffusion length. A form of degradation can be demonstrated by using the plotting principle just outlined. Figure 7 shows the value of L_D plotted versus the corresponding W_0 obtained on the same sample as the sample was subjected to the indicated number of hours of illumination at an intensity of one sun (equivalent).

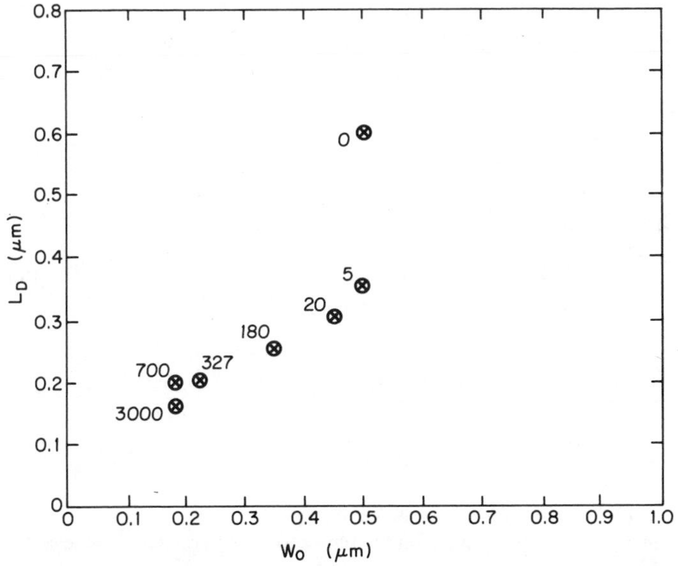

FIG. 7. Stability of a sample of a-Si:H under solar illumination, plotted according to the diffusion length method outlined in the text. The light-aged time (hours at one sun) is shown alongside each point.

Reduction in L_D and W_0 (increased space-charge density) are apparent up to 3000 hr of illumination. Between 5 and 3000 hr the L_D versus W_0 curve is a reasonably straight line, indicating that the light acts, directly or indirectly, to create or allow access to deep traps and recombination centers at the same rate. But note that the early stages of life test between 0 and 5 hr do not fit this pattern. The implication is that there was initially a certain concentration (about 10^{14}) of deep traps not associated with recombination centers and probably of quite different origin. This number was overcome by the light-related variety only after 5 hr of illumination. The subject of light-related defects (Staebler–Wronski effect) is complicated. This illustration is only meant to show how diffusion length measurements can possibly contribute to its understanding.

VII. Conclusion

The diffusion length as a measure of $\mu\tau$ in undoped a-Si:H is a useful parameter to predict the performance of materials made by different methods. The surface photovoltage method has proven to be a rapid and reliable method of determining the diffusion length, provided provision is made to apply bias light to the sample to the level of one sun. This collapses the surface space-charge region so that it no longer dominates the carrier flow and so allows a reliable estimate of the diffusion length. This measured value is simply related to the hole diffusion length if carrier trapping is taking into account.

References

Bube, R. (1960). "Photoconductivity of Solids." Wiley, New York.
Dresner, J., Szostak, D. J., and Goldstein, B. (1981). *Appl. Phys. Lett.* **38,** 998.
Faughnan, B., Moore, A. R., and Crandall, R. S. (1984). *Appl. Phys. Lett.* **44,** 613.
Goldstein, B., Dresner, J., and Szostak, D. J. (1982). *Philos. Mag. B* **46,** 63.
Goodman, A. M. (1961). *J. Appl. Phys.* **32,** 2550.
Goodman, A. M. (1980b). *Dig. Tech. Pap., Int. Electron Device Meet., Washington, 1980.*
Konagai, M., Miyamoto, H., and Takahashi, K. (1980). *Jpn. J. Appl. Phys.* **19,** 1923.
Moore, A. R. (1977). *Appl. Phys. Lett.* **31,** 762.
Moore, A. R. (1980). *Appl. Phys. Lett.* **37,** 327.
Moore, A. R. (1982). *Appl. Phys. Lett.* **40,** 403.
Moore, A. R. (1983). *J. Appl. Phys.* **54,** 322.
Reichman, J. (1981). *Appl. Phys. Lett.* **38,** 251.
Rose, A. (1963). "Concepts in Photoconductivity and Allied Problems." Wiley (Interscience), New York.
Rothwarf, A., and Boer, K. W. (1975). *Prog. Solid State Chem* **10,** Part 2, 84.
Shockley, W., and Read, W. T. (1952). *Phys. Rev.* **87,** 835.
Smith, R. A. (1959). "Semiconductors," p. 25. Cambridge Univ. Press, London and New York.
Staebler, D. (1980). *J. Non-Crys. Solids* **35/36,** 387.
Sze, S. M. (1969). "Physics of Semiconducting Devices," p. 58, Table 2.3. Wiley (Interscience), New York.
Williams, R., and Crandall, R. S. (1979). *RCA Rev.* **40,** 371.
Wronski, C. (1977). *IEEE Trans. Electron Devices* **ED-24,** 351.

SEMICONDUCTORS AND SEMIMETALS, VOL. 21, PART C

CHAPTER 8

Doping Effects in a-Si : H

W. Beyer

INSTITUT FÜR GRENZFLÄCHENFORSCHUNG UND VAKUUMPHYSIK
KERNFORSCHUNGSANLAGE JÜLICH
JÜLICH, FEDERAL REPUBLIC OF GERMANY

and

H. Overhof

FACHBEREICH PHYSIK
UNIVERSITÄT PADERBORN
PADERBORN, FEDERAL REPUBLIC OF GERMANY

ISBN 0-12-752149-6

I. Introduction

In this chapter we discuss the influence of dopants on the electronic transport properties of a-Si : H. In Part I, subsequent to a historical perspective, we try to elucidate why the electronic action of impurities in a-Si : H and the effect of different doping techniques are of interest. We list some experimental techniques yielding data sensitive to doping. In Part II we develop the theoretical background for those experimental techniques on which this chapter mainly relies, namely, the thermoelectric power and the conductivity (dark conductivity). Details of the experimental setup and of the doping methods are given in Part III. The experimental results regarding the doping effect of various impurities in a-Si : H are reviewed in Part IV. Part V gives a discussion of the experimental data in terms of present transport models. A perspective of work necessary in the future is presented in Part VI.

1. Doping of Amorphous Semiconductors

Motivated by the wide range of control of the electrical properties of many crystalline semiconductors by intentional incorporation of impurities, several attempts were made in the early sixties to dope amorphous material, in particular amorphous chalcogenide glasses (Kolomiets, 1964) and amorphous germanium (Grigorovici, 1968). Their failure has been explained by Mott (1967, 1969) suggesting that in amorphous semiconductors the coordination number of an impurity atom adjusts itself so that all electrons are taken up by bonds. Since doping effects in crystalline semiconductors generally rely on atoms having an electron (or hole) not in a bond, doping of amorphous semiconductors appeared to be impossible for physical reasons.

Indications that Mott's rule on the doping insensitivity of amorphous semiconductors may not be strictly obeyed in amorphous silicon could be taken already from the work of Chittick *et al.* (1969). For a-Si films prepared by glow-discharge decomposition of SiH_4 (monosilane) these authors reported a rise in the conductivity by factors of $\sim 10^2$ and $\sim 10^6$ when 200 ppm and 4% PH_3 (phosphine), respectively, were added to the silane. The observed instability of these films, however, sheds doubt about whether this was a true doping effect or rather an effect of defects, which were known to cause even higher conductivity changes (Beyer and Stuke, 1972).

The breakthrough was reached by the Dundee group (Spear and LeComber, 1975, 1976; Madan *et al.,* 1976) showing by field-effect measurements that glow-discharge a-Si (because of its high hydrogen content better termed a-Si : H) can be prepared with a low defect concentration, and demonstrating that by addition of phosphine and diborane (B_2H_6) *n*- and *p*-type material, respectively, can be obtained with room temperature conductivities as high as 10^{-2} $(\Omega\ \mathrm{cm})^{-1}$. The first direct evidence for substitutional doping in a-Si:H by group V elements of the periodic table was presented by Knights and associates (1977). These authors were able to reveal by extended x-ray absorption fine structure (EXAFS) studies that the increase of the conductivity upon arsenic incorporation is correlated with an increase of fourfold coordination of these impurity atoms.

A completely different approach for the doping of a-Si:H has been followed by Beyer and Fischer (1977). They showed that *n*-type doping is also obtained by interstitially incorporated lithium brought into the films by in-diffusion as well as by ion implantation. The possibility of introducing substitutional dopants in a-Si:H by ion implantation has been proved by Müller *et al.* (1977). Subsequently, these authors demonstrated the doping effect of most elements of the groups I, III, and V of the periodic table by ion implantation experiments (Kalbitzer *et al.*, 1980).

Doping effects have also been observed for (almost) hydrogen-free a-Si films prepared by vacuum evaporation (Beyer *et al.*, 1979a) and by pyrolytic chemical vapor deposition (Hirose, 1981), as well as for a-Si films containing oxygen (Beyer, 1979) and fluorine (Matsumura *et al.*, 1980). For other amorphous semiconductors successful doping has also been reported, as, e.g., for Ge:H (Spear and LeComber, 1976), Si:C:H (Tawada *et al.*, 1981), and Si:O:H (Holzenkämpfer *et al.*, 1982).

2. Effect of Doping on the Electronic Properties of a-Si:H

Generally, the aim of doping a semiconductor is to control the electronic properties exclusively by shifting the Fermi energy. In the study of a-Si:H, the question arose early as to whether the incorporation of dopants causes side effects as well. The formation of a phosphorus impurity band ~0.13 eV below the conduction-band mobility edge E_c has been proposed by LeComber *et al.* (1977) from their results of Hall effect experiments. An arsenic impurity level 0.35 eV below E_c and a boron impurity level 0.42 eV above the valence-band mobility edge E_v have been inferred by Jan *et al.* (1979, 1980) from thermoelectric power and conductivity measurements, assuming the validity of a two-path transport model (Mott *et al.*, 1975). Kinks in the temperature dependence of the conductivity of phosphorus-doped films led Beyer *et al.* (1977a) to postulate the buildup of a density-of-states peak close to midgap due to phosphorus doping. Further evidence for a phosphorus-related density-of-states structure near midgap has been obtained in the meantime by Lang *et al.* (1982) using space-charge spectroscopy, by Jackson and Amer (1982) employing photothermal deflection spectrometry, by Tanaka and Yamasaki (1982) from photoacoustic spectroscopy, and by

Wronski *et al.* (1982) from photoconductivity measurements. Using depletion-width modulated ESR, Cohen *et al.* (1982) were able to establish that the midgap structure is due to silicon dangling bonds with $g = 2.0055$. An analysis of the photovoltaic properties suggests that boron creates deep levels in a-Si:H as well (Carlson and Wronski, 1979; Carlson, 1980). The presence of an autocompensation mechanism in a-Si:H has been proposed by Street *et al.* (1981). Street (1982) has suggested that it might be impossible to incorporate dopant atoms in the neutral charge state into fourfold coordinated sites.

The fabrication of high quality doped a-Si:H films is desirable for many technical applications. In particular, amorphous silicon solar cells rely on highly conductive p^+ and n^+ layers with good photovoltaic properties (Carlson and Wronski, 1979). Carrier trapping or carrier recombination effects due to impurities are problems of the utmost importance (Carlson, 1980). It appears necessary, therefore, to study the electrical behavior of any impurity in a-Si:H. Since the doping actions of impurities may differ by the way the impurities have been incorporated into the material, the investigation of different doping methods is also of considerable interest.

Answers to the question of whether doping by a given impurity affects the electronic transport in a-Si:H exclusively via shifting the Fermi level, or whether additional effects occur as well, cannot be obtained from conductivity measurements alone. In crystalline semiconductors, the Hall effect is generally used to obtain a distinction between concentration- and mobility-related effects. The interpretation of the Hall effect in amorphous semiconductors, however, has so far been inhibited by the sign anomaly (LeComber *et al.*, 1977). On the other hand, the thermoelectric power (thermopower, Seebeck coefficient) has turned out to be a useful tool. Thermopower measurements have first been applied to amorphous silicon by Grigorovici *et al.* (1967). The analysis of combined measurements of conductivity and thermopower as a function of temperature has been shown to lead to meaningful results, allowing a separation of effects due to the Fermi-level position in the gap from those related to the carrier transport mechanism (Beyer *et al.*, 1979b,c; Overhof and Beyer, 1981a,b).

Further information on the transport processes in a-Si:H and on the influence of doping can be obtained, e.g., from measurements of the drift mobility (Allan *et al.*, 1977; Moore, 1977), of the photoconductivity (Rehm *et al.*, 1977; Anderson and Spear, 1977), as well as of the magnetic field dependence of the photo- and dark conductivity (Weller *et al.*, 1981). In this chapter, however, we shall confine ourselves mainly to results of conductivity and thermopower measurements. Some results from Hall effect and photoconductivity studies are also discussed.

II. Transport Parameters of a-Si:H

In crystalline semiconductors, the investigation of electronic transport properties (measured as a function of temperature or electric or magnetic field strength) can provide information on the scattering mechanism, carrier concentration and mobility, Fermi-level position, etc. Yet none of these latter quantities is obtained directly from the experiment: Theoretical models for the transport are fitted to the experimental data to obtain transport parameters. The validity of both transport theories and transport parameters is checked by the quality of the fit and by a comparison of parameters obtained from different transport experiments.

In contrast to crystalline semiconductors, our knowledge about the electronic transport mechanism in amorphous materials is still rather limited. It is clear, however, that we are not dealing with transport by free electrons in the sense that the mean free path exceeds considerably the interatomic spacing. In the case of a-Si:H the interpretation of experimental data is severely impeded by the fact that we are dealing with an alloy system of silicon and hydrogen, and that this material can be prepared only as a thin film. In addition, the temperature range accessible to transport experiments usually is rather limited: At low temperatures the material becomes a good insulator, whereas at elevated temperatures the films change their properties due to structural annealing, hydrogen migration, and effusion, and, finally, crystallization.

In an attempt to cope with these difficulties, we shall discuss in this chapter several transport parameters of a-Si:H. Our emphasis lies on parameters related to the conductivity and thermoelectric power, because for amorphous semiconductors these effects appear to be fairly basic and relatively easy to measure, and yet they contain relevant information on the electronic transport mechanism. The Hall effect is touched only marginally. This latter effect is very useful in crystalline semiconductors, in which it yields directly the carrier density. In amorphous semiconductors rather generally a sign anomaly of the Hall effect is observed [LeComber *et al.* (1977); Roilos (1978)]: The sign of the Hall effect is opposite to that of the dominant charge carriers. This sign anomaly is not too astonishing since we are no longer dealing with free electrons subject to a Lorentz force. Both magnitude and sign of the Hall effect are, therefore, determined by details of the electronic wave functions and of the atomic geometry. What is surprising is the universality of the sign anomaly. The reader is referred to Emin (1977a) for a model of how this anomaly could be understood. The present authors do not believe, however, that a proper understanding of this phenomenon has already been achieved.

3. General Expressions for Conductivity and Thermopower

The conductivity σ relates the current density $\mathbf{j}$ to the external electric field $\mathbf{E}$ as

$$\mathbf{j} = \sigma \mathbf{E}. \tag{1}$$

For unipolar conduction and for a single conduction mechanism, i.e., for identical charge carriers,

$$\sigma = n_c |q| \mu \tag{2}$$

determines the carrier mobility μ if n_c is the carrier density and q ($q = +e$ for holes and $-e$ for electrons) is the carrier charge.

The Seebeck coefficient S is defined by the electric field $\mathbf{E}_S$ generated in the sample by a temperature gradient under open circuit conditions

$$\mathbf{E}_S = S \operatorname{grad} T. \tag{3}$$

The temperature gradient results in a gradient of the carrier density, which in turn leads to a diffusion of carriers to the cold end. A space charge and an electric field is thus built up. In crystalline semiconductors S may have a contribution originating from the directed phonon flow from the warmer to the colder end of the sample. If scattering of the carriers is by collisions with phonons, momentum of the directed phonon flow will be transferred to the carrier system enhancing the Seebeck coefficient. This "phonon drag" component can be neglected in amorphous materials (Beyer *et al.*, 1977b; Jones *et al.*, 1977). Because of their negligible mean free path, the carriers lose the extra momentum immediately. The thermopower is related to the Peltier coefficient Π by Onsager's relation

$$\Pi = ST. \tag{4}$$

The Peltier energy $-e\Pi$ has a simple interpretation as the mean energy with respect to the Fermi energy transported per charge carrier.

For a nondegenerate semiconductor with all mobile carriers above (for holes below) some energy E_t the conductivity and thermopower are conveniently written as

$$\sigma = \sigma_0(T) \exp(-|E_t - E_F|/kT), \tag{5}$$

$$S = (|E_t - E_F|/qT) + (k/q)A(T). \tag{6}$$

Here k is Boltzmann's constant and E_F denotes the Fermi energy. The conductivity prefactor σ_0 depends on the carrier mobility and on the density of states at the energy E_t and above. The heat of transport term A of the thermopower is related to the kinetic energy of the charge carriers and is determined by the carrier scattering process.

Both transport parameters σ_0 and A may depend on temperature. Moreover, the Fermi energy and E_t may also be temperature dependent. Therefore, even if plots of $\ln \sigma$ and S as a function of inverse temperature (Arrhenius plots) yield straight lines as is often observed in a-Si:H, the apparent activation energies E_σ^* and E_S^* determined by

$$E_\sigma^* = -kT\frac{d \ln \sigma}{d(1/T)}, \tag{7}$$

$$E_S^* = q\frac{dS}{d(1/T)} \tag{8}$$

may differ considerably from $(E_t - E_F)$. Likewise the intercepts at $1/T = 0$,

$$\sigma_0^* = \sigma \exp(E_\sigma^*/kT) \tag{9}$$

and

$$A^* = (q/k)S - E_S^*/kT, \tag{10}$$

can be different from σ_0 and A, respectively. These complications cannot be ignored in amorphous semiconductors: Rather generally, a difference between E_σ^* and E_S^* is observed (see, e.g., Beyer and Overhof, 1979; Nagels, 1979; Anderson and Paul, 1981, 1982). This difference is one of the main topics in this chapter. The presence of a shift of E_t with temperature is suggested by a corresponding shift of the optical gap (Connell, 1972). A significant shift of E_F with temperature has been deduced for a-Si:H from transport data (Jones *et al.*, 1977; Beyer *et al.*, 1977a,b) and has recently been measured directly by photoemission spectroscopy (Gruntz, 1981).

4. The Influence of a Temperature Dependence of the Fermi Energy on Transport Data

In this section we shall demonstrate that a temperature dependence of the Fermi energy can drastically alter the observed transport parameters like E_σ^* and σ_0^*. We start from a simple model: Let us assume that σ_0 and A in Eqs. (5) and (6), respectively, are constants and that the full line in Fig. 1 gives the position of E_F with respect to E_t as a function of temperature. We shall see in Section 16 that $E_F(T)$ in Fig. 1 is representative of highly doped *n*-type a-Si:H films.

The Fermi energy $E_F(T)$ may be approximated by straight lines as

$$E_F = E_{F,i}^*(0) - \gamma_i T \tag{11}$$

with different slopes γ_i and different intercepts $E_{F,i}^*(0)$ at $T = 0$ for the three

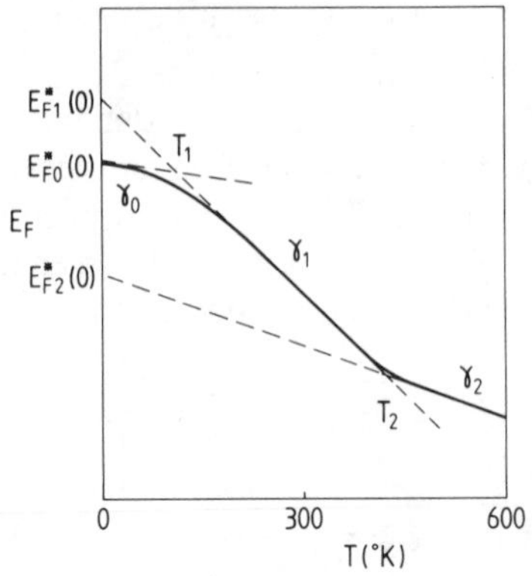

FIG. 1. Schematic representation of the temperature dependence of the Fermi energy E_F. Conductivity and thermopower measurements yield apparent Fermi-level positions $E^*_F(0)$.

temperature ranges: $T < T_1$, $T_1 < T < T_2$, and $T_2 < T$. By inserting Eq. (11) into Eq. (5), we obtain

$$\sigma = \sigma_0 \exp\left(-\frac{\gamma_i}{k}\right) \exp\left[\frac{E_t - E^*_{F,i}(0)}{kT}\right]. \tag{12}$$

Naively,

$$\sigma^*_{0,i} = \sigma_0 \exp(-\gamma_i/k) \tag{13}$$

could be interpreted as the conductivity prefactor and

$$E^*_{\sigma,i} = E_t - E^*_{F,i}(0) \tag{14}$$

as the Fermi-level position. However, a comparison with Fig. 1 shows that E^*_σ yields the actual Fermi-level position only in the low temperature limit, which in most cases is inaccessible to experiment. Generally, E^*_σ can be considerably different from $E_t - E_F$ and σ^*_0 may differ from σ_0 by orders of magnitude.

If the temperature dependence of $E_t - E_F$ can be approximated by different straight lines, an Arrhenius plot of the conductivity must show kinks, at which both the apparent conductivity prefactor and the apparent activation energy change. Yet this discontinuity does not mean automatically that the dominant transport path has changed. In fact, from conductivity measurements alone it is impossible to show conclusively that the dominant transport path is altered. Additional measurements, like the thermopower, are necessary. In doped crystalline semiconductors [see, e.g., Madelung (1957)] it is common that at low temperatures (in the "exhaustion regime") a negative "activation energy" of σ is observed, whereas at high temperatures (in the "intrinsic regime") E^*_σ is positive and equal to approximately half of the gap width. Yet the dominant transport path does not change at all except that ambipolar conduction is present at high temperatures. The effect is entirely due to a large statistical shift of the Fermi level at low temperatures, which is stopped as E_F approaches midgap.

5. The Q-Function

The discussion of the previous section shows that no firm statement about the transport mechanism can be made from conductivity measurements alone unless the influence of temperature shifts of E_F with respect to E_t are eliminated. At the film surface, the position of E_F relative to the valence-band edge can be measured directly by photoemission [Ley (1984); see also Chapter 12 by Ley in Volume 21B]. Thus, σ_0 could be determined according to Eq. (5) if E_t and the valence-band edge are known to coincide and if no band bending at the surface occurs. An alternative way is offered by the fact that both conductivity and thermopower formulas [Eqs. (5) and (6)] contain the term $|E_t - E_F|/T$. Any dependence of E_F and E_t on temperature, therefore, cancels if we combine conductivity and thermopower data to form (Beyer *et al.*, 1979b)

$$Q(T) = \ln(\sigma\ \Omega\ \mathrm{cm}) + (q/k)S = \ln(\sigma_0\ \Omega\ \mathrm{cm}) + A. \tag{15}$$

Thus, although we are unable to obtain the transport parameters σ_0 and A separately from conductivity and thermopower measurements, we can obtain a transport parameter Q containing essentially the same information as σ_0 and A. Accordingly, any change in the dominant transport path must reveal itself in $Q(T)$ as it would in $\sigma_0(T)$ and $A(T)$, except for the unlikely case that σ_0 and A would show a completely complementary temperature dependence. Note, furthermore, that any difference between E_σ^* and E_S^* will show up as a nonzero slope of Q when this quantity is plotted versus reciprocal temperature. Anderson and Paul (1982) have questioned the use of $Q(T)$ as a key transport parameter. They argue that Q obscures rather than reveals structures. What we propose, on the contrary, is that structures visible in $\sigma(T)$ and $S(T)$ but invisible in $Q(T)$ are due to a nonlinear temperature shift of the Fermi level. Otherwise they are caused by $\sigma_0(T)$ and $A(T)$. We must admit, however, that Eq. (15) may not eliminate exactly any temperature dependence of $E_t - E_F$. Emin (1977b) has pointed out that the shrinking of the width of the band gap E_G due to electron–phonon coupling cannot be treated as a temperature dependence of the electronic energies because interaction energies couple the electron system to the phonon system. In spite of objections by Butcher and Friedman (1977) we believe this argument to be correct. Reviewing Emin's paper (Overhof and Beyer, 1983) we have shown that to a first approximation a proper treatment of $E_G(T)$ will lead to a modification of σ_0 according to Eq. (13), whereas the heat of transport term A

is practically unchanged. The influence of $E_G(T)$ on σ_0 can be estimated from the temperature dependence of E_G as known from optical data. According to these data (Fritzsche, 1980),

$$E_G \sim E^*_{G0} - 5kT \tag{16}$$

holds in the temperature range from ~ 200 to $450°K$. Therefore, it seems reasonable to assume a shift $\gamma \sim -2.5k$ for the temperature shift of the states at the band edge with respect to those near midgap. Accordingly, the conductivity prefactor σ_0 differs by

$$\sigma_0 \sim \sigma_{00} e^{2.5}, \tag{17}$$

i.e., by a factor of ~ 10, from the value σ_{00}, which would be observed if the electron–phonon coupling were absent and which has to be compared with theoretical predictions.

6. Transport Processes

In this section we shall discuss different models proposed for the transport mechanism in amorphous silicon, concentrating on the implications for $\sigma_0(T)$, $A(T)$, and $Q(T)$. The framework of our discussion is based on the picture of localized and delocalized states. This goes back to Anderson's pioneering work on localization in a single band. If the disorder potential V_0 is comparable to the bandwidth B, all states of the band will be localized. If $V_0 < B$, localization will set in at the band edges, leaving the states in the center of the band delocalized.

This picture must be modified to include two bands if it is to be applied to a semiconductor (Cohen, *et al.*, 1969; Davis and Mott, 1970). For weak disorder, the valence-band states are delocalized up to E_v near the top of the band. Localized states just above this energy form the valence-band tail, whereas other gap states may originate from defects. The conduction-band states are delocalized above some energy E_c near the bottom of this band, whereas states in the tail below E_c are again localized.

a. Minimum Metallic Conductivity and the Random Phase Model

Mott has shown in 1967 that in an Anderson model for a disordered metal the conductivity cannot be arbitrarily small if E_F is above E_c. The reason is that the position of E_c, the mobility edge, is given by a balance between the electronic overlap J of wave functions centered at adjacent atomic sites and the disorder potential V_0. For a mobility edge the former energy is a definite fraction of the latter and hence the lower limit of the conductivity is always the minimum metallic conductivity $\sigma_{min} \sim 200\ \Omega^{-1}\ cm^{-1}$, the value depending somewhat on assumptions about unknown constants. This concept of a minimum metallic conductivity has initiated a large body of experimental and, in particular, theoretical work. Space does not allow us even to summarize the different controversial papers and the reader is referred to Mott (1981) and to Mott and Kaveh (1983a,b) for a critical and fair review. We note, however, that the concept of minimum metallic conductivity predicts for amorphous semiconductors that the conductivity prefactor equals σ_{min} and that $A \sim 1$. This yields (see Section 5) $\sigma_0 \sim 2000\ \Omega^{-1}\ cm^{-1}$ for a-Si:H and $Q \sim 9$ independent of temperature.

A different approach to the calculation of transport properties above a mobility edge was used by Hindley (1970) and Friedman (1971). Both authors discuss the random phase model (RPM) proposed by Mott (1967, 1972) and Cohen (1970). In the RPM a delocalized state $\Phi_n(\mathbf{r})$ can be written as a linear combination of atomic orbitals ψ centered at $\mathbf{R}_m$:

$$\Phi_n = \frac{1}{\sqrt{N}} \sum_m \exp(i\phi_{nm})\psi(\mathbf{r} - \mathbf{R}_m) \tag{18}$$

with a random phase term $\exp(i\phi_{nm})$ replacing the Bloch factor $\exp(i\mathbf{k} \cdot \mathbf{R}_m)$ in crystals.

The predictions of the RPM for the transport parameters σ_0, A, and Q are similar to what is concluded from the minimum metallic conductivity if the density-of-states distribution near E_c is a slowly varying function of energy. The Hall mobility has been evaluated by Friedman (1971). It is found to be temperature independent and of the order $0.1\ cm^2\ V^{-1}\ sec^{-1}$.

b. Potential Fluctuations

The results of Section 6a are substantially modified if long-range internal potential fluctuations play a role. Fritzsche (1971a) has discussed several reasons why a static fluctuating potential could be present in amorphous semiconductors. Among these are the presence of local density fluctuations, growth inhomogeneities, and electric fields due to charged centers. The latter case has been treated in detail by the present authors (Overhof and Beyer, 1981a). Since in highly doped material charges are distributed at random in space (i.e., their distribution is not strictly uniform) electrostatic internal fields of a typical length scale of 200 Å will arise.

We have treated the effect of these potential fluctuations in a semiclassical way assuming that only the energy of the mobility edges is modulated according to this potential. If the fluctuations do not form regular patternlike planes or linear chains the detailed shape of the fluctuations is of minor importance compared to their magnitude Δ. Since the fluctuations enter the transport equations in terms of $\exp(-\Delta/kT)$, a reduced temperature can be defined as

$$T' = kT/\Delta, \tag{19}$$

and one expects that the influence of the potential fluctuations on Q will depend on T' only. Numerical model calculations (Overhof and Beyer, 1981a) confirm this point. The results of such calculations are shown in Fig. 2. Here

$$\hat{Q} = Q - \ln(\sigma_0\ \Omega\ \mathrm{cm}) - A \tag{20}$$

is plotted as a function of $1/T'$ for Δ ranging from 0.05 to 0.25 eV. For $1/T' > 3$ ($\Delta > 3kT$) the data can be fitted by

$$\hat{Q} = 1.8 - 1.25/T', \tag{21}$$

i.e., we obtain with Eqs. (19) and (20)

$$Q = \ln(\sigma_0\ \Omega\ \mathrm{cm}) + A + 1.8 - 1.25\Delta/kT. \tag{22}$$

Thus, this model predicts a nonzero slope of Q versus $1/T$, i.e., a difference between the apparent conductivity and thermopower activation energies of

$$E_\sigma^* - E_S^* = 1.25\Delta \tag{23}$$

at low temperatures. Likewise, the apparent intercept of Q at $1/T = 0$ becomes

$$Q_0^* = \ln(\sigma_0\ \Omega\ \mathrm{cm}) + A + 1.8. \tag{24}$$

At high temperatures $\hat{Q}$ versus $1/T'$, i.e., Q versus $1/T$, bends over so that E_S^* is expected to approach E_σ^*.

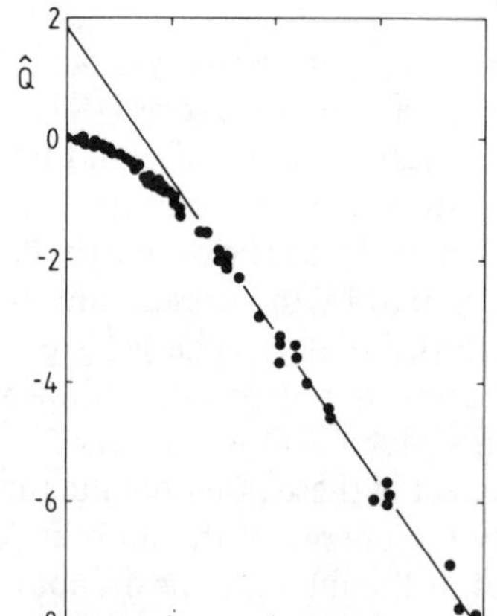

FIG. 2. Computer calculation for potential fluctuation model of $\hat{Q} = Q - \ln(\sigma\ \Omega\ \mathrm{cm}) - A$ as a function of inverse reduced temperature $T' = kT/\Delta$. Δ ranges from 0.05 to 0.25 eV. [From Overhof and Beyer (1983).]

In a similar way the computer calculations yield for σ and S at low temperatures

$$\ln\left(\frac{\sigma}{\sigma_0}\right) = \frac{|E_t - E_F|}{kT} - 0.2 + 0.25\frac{\Delta}{kT}, \tag{25}$$

$$\left(\frac{q}{k}\right)S = \frac{|E_t - E_F|}{kT} + A + 2 - 1.5\frac{\Delta}{kT}, \tag{26}$$

whereas the Hall mobility μ_H is given by (Overhof, 1981)

$$\ln(\mu_H/\mu_{H0}) = 0.25 - 0.42\Delta/kT. \tag{27}$$

In these formulas E_t is the mean energy of the mobility edge; μ_{H0} in Eq. (27) is the Hall mobility that would be observed if no potential fluctuations were present. Note that the conductivity term $\ln(\sigma/\sigma_0)$ is less affected by the fluctuations than the thermopower term $(q/k)S$. In particular, we mention that the Peltier energy $-eST$ given by

$$-eST = E_t - E_F - (e/q)[(A + 2)kT - 1.5\Delta] \tag{28}$$

may be quite different from $E_t - E_F$ even at low temperatures.

For convenience in discussing the experimental data in Part IV and V we estimate Q_{RT}, the room temperature value of Q, for the presence of potential fluctuations. With $\sigma_0 = 2000\ (\Omega\ \text{cm})^{-1}$ and $A = 1$ (see Section 6a), we obtain $-2 < Q_{RT} < 8$ for $0.25\ \text{eV} > \Delta > 0.05\ \text{eV}$.

c. *Kubo–Greenwood Formulas*

If transport is through extended states and no inelastic scattering processes occur, we can assume that contributions to the current from states at different energies are independent of each other. (Note, however, that hopping conduction discussed in Section 6d is a counterexample.) Starting from the Kubo formula, one defines an energy dependent conductivity (Fritzsche, 1971b)

$$\sigma(E) = |q|\mu(E)g(E) \tag{29}$$

as a product of the carrier mobility $\mu(E)$ and the density of states $g(E)$ and calculates the conductivity by integrating with respect to the energy as

$$\sigma(T) = \int \sigma(E) f_F(E_F,T)[1 - f_F(E_F,T)]\, dE. \tag{30}$$

Here $f_F(E_F,T)$ is the Fermi distribution function

$$f_F(E_F,T) = \{\exp[(E - E_F)/kT] + 1\}^{-1}. \tag{31}$$

For the case of nondegenerate statistics Eq. (30) reduces to

$$\sigma(T) = \int \sigma(E) \exp\left(-\frac{|E - E_F|}{kT}\right) dE. \tag{32}$$

In a similar way one obtains for the Peltier energy (see Section 3)

$$-eST = \frac{1}{\sigma(T)} \int (E - E_F)\sigma(E) \exp\left(-\frac{|E - E_F|}{kT}\right) dE. \tag{33}$$

These formulas have been inverted by Döhler (1979) in order to obtain $\sigma(E)$ directly from experimental conductivity and thermopower data. His main emphasis was to explain the difference between E_σ^* and E_S^*. Döhler showed that this can be achieved by a $\sigma(E)$ function that rises gradually from zero to rather high values within some 0.2 eV. It must be noted, however, that the shape of the $Q(T)$ curves depends on the shape of $\sigma(E)$ in a rather sensitive way. Since $\sigma(E)$ can be regarded as the Laplace transform of $\sigma(T)$, the low temperature behavior of $Q(T)$ is extremely sensitive to the shape of $\sigma(E)$ at low energies for which the absolute value of $\sigma(E)$ is small. Accordingly, $Q(T)$ is expected to be strongly dependent on the density-of-states distribution in the gap, i.e., on the preparation conditions of the a-Si:H films.

d. Hopping Conduction

Electrons in localized states can contribute to the conductivity by hopping. This transport mechanism has been observed first in partially compensated crystalline semiconductors at low temperatures: Electrons are able to tunnel from occupied donor sites to empty neighboring sites if the difference in energy can be absorbed from or emitted into the phonon system (note that in compensated material the donor states are not strictly degenerate). For a simple hydrogenlike impurity state characterized by a localization length a, the jump probability p_j from an occupied to an empty site, being separated by R in space and by $\Delta E > 0$ in energy is given by (Miller and Abrahams, 1960)

$$p_j = \nu_0 \exp(-2R/a - \Delta E/kT), \tag{34}$$

where ν_0 is a characteristic phonon frequency. For $\Delta E < 0$ the $\Delta E/kT$ term in Eq. (34) is replaced by zero. If hopping is always to the nearest neighbor in space [nearest-neighbor hopping (NNH)] the properly averaged jump probability yields for conduction in localized states E_{tl}

$$\sigma = \sigma_0 \exp[-(|E_{tl} - E_F| + W)/kT], \tag{35}$$

$$(q/k)S = |E_{tl} - E_F|/kT + A, \tag{36}$$

that is,

$$Q = \ln(\sigma_0\ \Omega\ \mathrm{cm}) + A - W/kT. \tag{37}$$

Here, W is a typical energy difference between nearest-neighbor states. Accordingly, hopping in a donor (or acceptor) band can explain a difference between E_σ^* and E_S^*. However, since the conductivity prefactor σ_0 is expected to depend strongly on the density of impurity states, Q_0^* should be strongly dependent on the doping level as well. With rising impurity concentration it should increase rapidly.

The averaging procedure of the jump probability Eq. (34) is completely altered if the energy factor W/kT becomes comparable to $2R/a$. This situation is present in unhydrogenated a-Si in which a high density of localized states extends throughout the pseudogap. At low temperatures hopping will be by carriers close to the Fermi level with small hopping energies ΔE at the expense of larger hopping distances R. At elevated temperatures the mean hopping distance will decrease since larger energy differences are more easily overcome. This variable range hopping (VRH) has been proposed by Mott (1969) and, for hopping via states near the Fermi level, leads to the characteristic $T^{-1/4}$ law

$$\sigma_H = \sigma_{0H} \exp(-T_0/T)^{-1/4} \tag{38}$$

with $T_0 \sim 10^8\,°K$ for unhydrogenated a-Si. The thermopower is small (Overhof, 1975; Zvyagin, 1973) and is given by

$$S = C\sqrt{T}\,\frac{d}{dE}\ln g(E)_{E=E_F}, \tag{39}$$

where C is a constant. We shall not discuss in detail the unsolved problems of the VRH theory [see, e.g., Movaghar (1981)] but simply estimate Q_{RT} for hopping conduction near the Fermi level. The absolute value of σ_H at 300°K will depend on the density of localized states near E_F. High values for $g(E_F)$ are expected for hydrogen-free amorphous silicon. For such a material $\sigma_H \sim 10^{-3}\ \Omega^{-1}\ cm^{-1}$ has been observed (Beyer and Stuke, 1975), leading to $Q_{RT} \sim -7$ as an upper limit for this type of a transport process.

The NNH formulas [Eqs. (35) and (36)] have been applied by numerous workers also for hopping in tail states. It should be noted that it is not correct to do so because in a tail with a rapidly rising density-of-states function nearest neighbors cannot be defined. In this case, clearly, a VRH model has to be considered. This has been done by Grünewald and Thomas (1979). The authors find for a density-of-states distribution rising exponentially as $g(E) = g_0 \exp(E/\epsilon)$ that a difference between apparent conductivity and thermopower activation energies of $E_\sigma^* - E_S^* = 3.5\epsilon$ will result.

e. Ambipolar Conduction and Accumulation Layer

Conduction by both electrons and holes will arise for a Fermi-level position close to midgap or for the presence of an inversion layer. Considering the first case only, the conductivity is given by

$$\sigma = \sigma_n + \sigma_p \tag{40}$$

and the thermopower follows as

$$S = (\sigma_n/\sigma)S_n + (\sigma_p/\sigma)S_p, \tag{41}$$

where subscripts n and p denote contributions of electrons and holes, respectively. At the point of electronic compensation $S \sim 0$, whereas the RT conductivity in a-Si:H reaches values as low as $10^{-12}\ \Omega^{-1}\ \mathrm{cm}^{-1}$ (Beyer *et al.*, 1981a). Hence Q_{RT} may be as small as -27 for ambipolar transport in extended states.

An accumulation layer will also lead to a decrease of Q_{RT}. In a simplified model in which a conductive layer of thickness d_1 dominates the transport in a sample of thickness d, a conductivity reduced by a factor d_1/d is measured while the thermopower remains virtually unchanged. Hence Q_{RT} will be smaller by an additive constant $\ln(d_1/d)$. For a thin highly conductive layer this effect can be quite sizable.

7. Combination of Different Transport Processes

Following the pioneering work of the Dundee group (LeComber and Spear, 1970; LeComber *et al.*, 1977; Jones *et al.*, 1977; Allan *et al.*, 1977) several researchers (Jan *et al.*, 1979, 1980; Anderson and Paul, 1982) have interpreted their transport data in terms of a two-carrier model: At higher temperatures, transport is by extended states, whereas at lower temperatures NNH in tail states or donor states is pedominant. The transport formulas

$$\begin{aligned}\sigma = \sigma_{\mathrm{x}} + \sigma_{\mathrm{l}} = \sigma_{0\mathrm{x}} &\exp(-|E_{\mathrm{tx}} - E_{\mathrm{F}}|/kT \\ &+ \sigma_{0\mathrm{l}} \exp[-(|E_{\mathrm{tl}} - E_{\mathrm{F}}| + W)/kT]\end{aligned} \tag{42}$$

and

$$\frac{q}{k}S = \frac{\sigma_{\mathrm{x}}}{\sigma}\left(\frac{|E_{\mathrm{tx}} - E_{\mathrm{F}}|}{kT} + A_{\mathrm{x}}\right) + \frac{\sigma_{\mathrm{l}}}{\sigma}\left(\frac{|E_{\mathrm{tl}} - E_{\mathrm{F}}|}{kT} + A_{\mathrm{l}}\right). \tag{43}$$

have been applied. The subscripts x and l stand for extended and localized states, respectively. Note that these NNH formulas are also used throughout in the literature for hopping in tails, although this is incorrect according to Section 6d. In Fig. 3 several curves of Q versus $1/T$ are shown calculated from the parameters used by Jones *et al.* (1977) and Anderson and Paul (1982) to fit conductivity and thermoelectric power of phosphorus-doped a-Si:H, assuming a two-carrier model. The results demonstrate the sensitivity of Q to changes in the transport path.

Anderson and Paul (1980–1982) have tried to correlate kinks in the

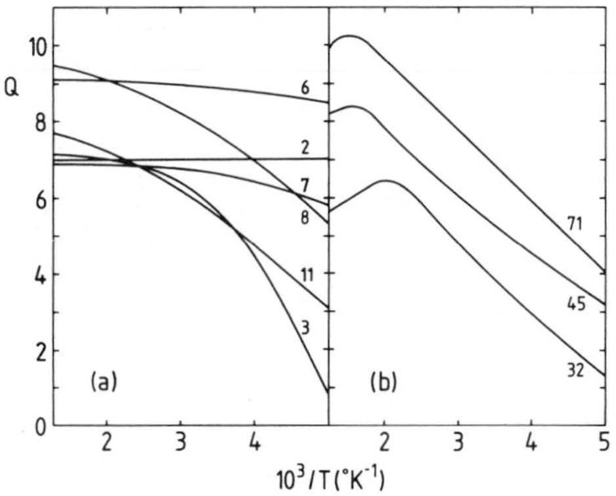

FIG. 3. Calculated curves of Q as a function of $1/T$ for two-carrier model using parameters of (a) Jones *et al.* (1977) and (b) Anderson and Paul (1982). Curve numbers indicate sample numbers in references.

temperature dependence of σ and S with a structural model, suggesting that a-Si:H films consist of two phases: A densely packed phase containing relatively little hydrogen, and a defect- and hydrogen-rich phase. The authors suggest that the latter phase dominates the transport at low temperatures, whereas the former one is important at high temperatures. The same model has been used recently by Yang and Lee (1983). We note, however, that such a change in transport should show up in $Q(T)$. Trying various sets of parameters, we have found none leading to kinks in $\sigma(T)$ and $S(T)$ without revealing structure in $Q(T)$.

Finally, we show in Fig. 4 calculated curves of σ and Q as functions of $1/T$ for extended-states conduction allowing for potential fluctuations (curve 1), for a contribution of Fermi-level hopping (curve 2), and for the presence of a thin highly conductive layer (curve 3). These curves apply for lightly doped a-Si:H ($E_\sigma^* = 0.6$ eV); a nonlinear temperature shift of E_F has been neglected for simplicity. Also plotted is the behavior of a-Si:H with $E_\sigma^* = 0.8$ eV assuming a contribution of ambipolar conduction (curve 4). The results reveal a rather characteristic pattern of Q versus $1/T$ for the different contributions to σ. It is seen that a measurement of the temperature dependence of Q can allow one to distinguish between the different transport processes. Only in the case of a narrow temperature range accessible to the experiment (as is true for undoped or electrically compensated a-Si:H), this may not be possible. Measurements of other transport parameters are necessary in this case for a clear assignment. To distinguish between the contribution of Fermi-level hopping and of an accumulation layer, measurements of the photoconductivity $\Delta\sigma_{ph}$, in particular of its absolute value, can be used. This quantity (at a given illumination level) is known to be rather sensitive to

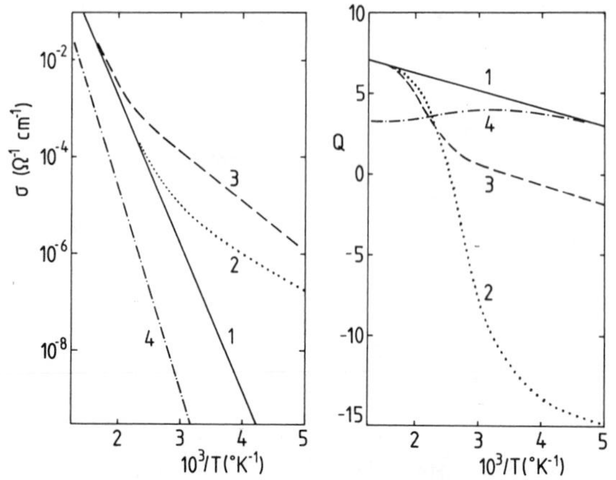

FIG. 4. Calculated curves of σ and Q versus $1/T$. (1) Extended-states transport modulated by potential fluctuations, (2) contribution of hopping conduction near E_F, (3) presence of thin, highly conducting layer, and (4) contribution of ambipolar conduction.

the presence of gap states acting as recombination centers (Jones *et al.*, 1979). Hopping near the Fermi level needs a high density of gap states, i.e., $\Delta\sigma_{ph}$ is expected to be low; the presence of an accumulation layer, on the other hand, generally indicates a low density of gap states, i.e., $\Delta\sigma_{ph}$ may attain rather high values.

III. Experimental Methods

8. Sample Preparation and Doping Techniques

This chapter deals mainly with a-Si:H films prepared by the glow-discharge technique (Sterling and Swann, 1965). The major part of the experimental results is obtained on films prepared in an inductively coupled system at Marburg (M–I) or in a capacitive system at Jülich. The deposition conditions were rather similar in both cases, i.e., undiluted silane at a pressure of ~0.4 Torr and a flow of ~5 standard cm³ per minute (sccm) was employed. An rf power of <5 W was used, leading to deposition rates of 1–3 Å sec⁻¹.

Gas-phase doping is accomplished by adding suitable amounts of dopant gases like phosphine or diborane to the silane (Spear and LeComber, 1975, 1976). Alternatively, gas-phase doping can be achieved by plasma-chemical methods (Bauer and Bilger, 1982). Interaction of the plasma or of the neutral gas with surfaces containing dopants is used to introduce dopant atoms into the gas phase. In the case of doping by ion implantation, dopant ions are accelerated to an energy of typically 10–100 keV and are allowed to impinge on the semiconductor surface. The distribution of the implanted atoms is approximately Gaussian in shape for an amorphous target, the mean penetration depth depending mainly on the ion mass and on the incident energy (Mayer *et al.*, 1970). An approximately constant dopant concentration within a film can be produced by varying the implantation energy. The ion impact leads to implantation damage. In a-Si:H this can give rise to variable range hopping conductivity (Beyer *et al.*, 1975). Therefore, either the implantations are carried out at an elevated temperature ("hot" implantation) so that the implantation damage is annealed during irradiation (Müller *et al.*, 1977) or the samples are annealed subsequent to the implantation process (Beyer and Fischer, 1977). Care has to be taken to avoid the formation of a film of cracked hydrocarbons on the sample surface during the implantation process since such layers can become highly conductive upon annealing. For doping of a-Si:H by in-diffusion of lithium, a thin Li-layer is evaporated onto the film in a high vacuum and the sample is heated subsequently in vacuo to 350–400°C for about one hour to allow the lithium to diffuse into the sample. The remaining lithium is removed afterward by rinsing the sample in HCl and water (Beyer and Fischer, 1977).

9. Electrical Measurements

For the electronic transport measurements reported in this chapter, the experimental setup is fairly standard. The main problems are posed by the high sample resistance, in particular at low temperatures. Therefore, measuring instruments with a high input resistance must be used. Care must be taken, furthermore, to create a high insulation resistance as well as a thorough shielding of all leads. Suitable substrates are fused silica, single crystalline sapphire, and Corning 7059. Some glass substrates can lead to erroneous results, in particular at high temperatures, due to ionic conduction as well as due to alkali doping of the films. The choice of the contact material is most critical when a two-probe technique is applied for dark- and photoconductivity measurements. Ohmic surface contacts have been obtained with evaporated aluminum on *p*-type samples and with evaporated gold–antimony contacts on *n*-type material (Spear and LeComber, 1976). Other techniques to obtain good contacts on *n*-type films involve the deposition of an n^+ layer (Wronski *et al.*, 1976) or the predeposition of molybdenum (Ast and Brodsky, 1980) or antimony-coated chromium (Staebler and Wronski, 1980).

In the following we describe the technique used at Jülich for combined measurements of thermopower and dark- and photoconductivity. A cryostat is employed having two cold fingers that can be cooled independently by insertion of liquid nitrogen or by a flow of cool N_2 or air. Each cold finger bears a ~100-W resistance heater (H1 and H2 in Fig. 5a). The heater mass

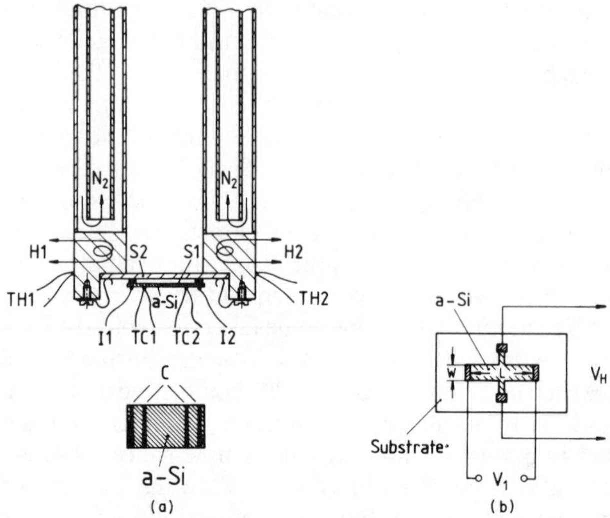

FIG. 5. Diagram illustrating experimental setup for (a) thermopower and (b) Hall mobility: H1, H2, heaters; S1, S2, substrates; C, contacts; TH1, TH2, TC1, TC2, thermocouples; I1, I2, current leads; V_1, applied voltage; V_H, Hall voltage.

has been kept small to ensure a thorough control of the heater temperature, which is monitored by thermocouples (TH1 and TH2) on the heater surface. The electric contact pattern on the amorphous film consists of four parallel metal strips (C) 0.5–1 mm wide. The gap between the inner two strips is >5 mm, between the outer and inner strips 1–2 mm. The sample (S1) of dimensions 0.5 × 8 × 15 mm is attached by silver paint to a 1 × 15 × 25 mm sapphire plate S2 (providing electrical insulation), which itself is brought in good thermal contact with the heaters using clamps and silver paint. Two thermocouples TC1 and TC2 (nickel–chromium/copper–nickel; wire thickness 50 μm) are attached to the inner contact strips using silver paint. Current leads I1 and I2 (thin gold wires) are fixed to the outer strips. Electronic switches, placed outside the cryostat allow the measurement of the thermovoltage ΔV (measured between the NiCr leads of TC1 and TC2), of the temperatures T_1 and T_2 (measured by TC1 and TC2, respectively) as well as of the resistance of the sample by the two- or four-probe technique within less than a second. In practical operation, the measuring time is determined by the RC rise time. The measuring instrument is a Keithley 616 electrometer of $> 10^{14}\ \Omega$ input impedance. All measurements are performed in a turbomolecular-pumped vacuum.

Prior to the thermopower and conductivity measurements the samples usually are heated in vacuo and in darkness to ~150°C to remove adsorbed gas layers (Tanielian *et al.*, 1978) as well as the Staebler–Wronski effect

(Staebler and Wronski, 1977). In order to avoid irreversible annealing effects the measurements are performed going from higher to lower temperatures. For each data point of the thermoelectric power S, about 10 values of the thermovoltage ΔV as a function of $\Delta T = T_1 - T_2$ are taken at a given mean temperature $T = (T_1 + T_2)/2$, and $S_{\text{Si-NiCr}}$ is determined from the slope of ΔV versus ΔT ($\Delta T < 10°\text{K}$) using a least squares fit procedure. Subsequently, the value of the thermopower is corrected allowing for the fact (see Fig. 5a) that the thermovoltage ΔV is measured across a smaller gap than the temperature difference ΔT. The absolute thermopower S is determined from the corrected value of $S_{\text{Si-NiCr}}$ by adding the absolute thermopower of NiCr, which is taken from the literature. Reliable thermopower data are obtained up to an electrical resistance of 10^{10}–10^{11} Ω.

The resistance is measured when T_1 equals T_2. When the two-probe technique is applied, the ohmic characteristics of the contacts are verified by measuring the resistance at different current levels. To obtain the conductivity σ from the resistance measurements the thickness of the amorphous film must be known accurately.

The photoconductivity measurements are usually performed after the dark conductivity–thermopower run. As a light source, a tungsten–iodine lamp is used with an intensity of ~50 mW cm^{-2} of white light at the sample position. The results presented in this chapter refer to room temperature measurements of the photoconductivity only. The conductance under illumination σ_{ill} is determined 15 sec after switching on the light source and the photoconductivity $\Delta\sigma_{\text{ph}}$ is calculated by subtracting the dark conductivity.

For the Hall effect measurements performed at Marburg the films are inserted into a cryostat that allows a magnetic field to be applied perpendicular to the sample plane. A sawtooth-type magnetic field B with an amplitude of 1 T at a frequency of 1.5 Hz is generated by chopping the magnet current. A voltage V_1 is applied to the sample (see Fig. 5b) and the Hall voltage V_H is detected using a lock-in technique. The Hall mobility follows by $\mu_H = (L/W)(V_H/V_1)B^{-1}$. Here, L and W ($L > 4W$) are the length and the width, respectively, of the amorphous layer.

IV. Experimental Results

10. Doping of a-Si:H

Dopants in a semiconductor are characterized by their ability to give or to take electrons from their host material. For crystalline silicon, practically all atoms of a conventional dopant (phosphorus or boron) are located at lattice positions where they are able to dope, i.e., they are fourfold coordinated to silicon. In amorphous semiconductors, Mott's rule on the insensitivity to doping (Mott, 1967) would suggest that none of the potential dopants are electrically active because all electrons are in the bonds. As this rule is apparently not strictly valid in a-Si:H (Spear and LeComber, 1975), the doping efficiency of a given dopant may be defined as

$$\eta = n_D/n_I \tag{44}$$

with n_I the total concentration of incorporated dopant atoms and n_D the concentration of dopant atoms in a doping configuration (i.e., fourfold coordinated (P_4) in the case of phosphorus).

Neither n_I nor n_D is generally known in a-Si:H. Considering gas-phase doping using doping gases, the doping level is usually described by the volume ratio of the doping gas to silane. Since the film growth by plasma deposition is a complicated process, the concentration ratio of dopant atoms to silicon atoms in the film $(n_I/n_{Si})^f$ may be different from that in the gas phase $(n_I/n_{Si})^g$ (Spear and LeComber, 1975). To account for this effect, one introduces the incorporation ratio

$$p = (n_I/n_{Si})^f/(n_I/n_{Si})^g. \tag{45}$$

In the case of doping by ion implantation, n_I can be determined from the ion dose if the range distribution of the implanted ions is known. The concentration n_D of impurity atoms in a doping configuration has been directly measured by EXAFS studies (Knights *et al.*, 1977); indirectly it can be estimated by monitoring the electrical activity of these atoms. However, only ionized dopant atoms (n_D^i) are electrically active, and we may define

$$\beta = n_D^i/n_D. \tag{46}$$

In thermal equilibrium

$$n_D^i = n_D - \int_{-\infty}^{\infty} N_D(E) f_F(E_F, T)\, dE \tag{47}$$

is valid for any energy distribution $N_D(E)$ of donor (or acceptor) levels, so that

$$\beta = 1 - n_D^{-1} \int_{-\infty}^{\infty} N_D(E) f_F(E_F, T)\, dE. \tag{48}$$

Thus β approximately equals unity if the Fermi level lies far below the centroid of the donor distribution. In a crystalline semiconductor with a low density of gap states, donor levels deliver their electrons directly into the conduction band. In an amorphous semiconductor, the situation usually is different (Spear and LeComber, 1975, 1976): Most of the electrons will condense into empty states near the Fermi level. Let Δn_D^i be the concentration of additionally introduced ionized dopant atoms. A shift of the Fermi energy ΔE_F will result that depends on the density of states $g_1(E)$ and $g_2(E)$ prior to and after dopant incorporation. Changes of $g(E)$ due to doping could be induced by energy levels related to dopant atoms not in a doping configuration or by defect states introduced by doping. We may define

$$\Delta n_D^i / \Delta E_F = \alpha. \tag{49}$$

In the simple case of a doping-independent $g(E)$ and for small Δn_D^i

$$\alpha = g(E) \tag{50}$$

is valid in the low temperature approximation. At high temperatures $\alpha > g(E)$ is expected. If one defines

$$\Delta E_F / \Delta n_I = \Gamma \tag{51}$$

as the gross doping effect of a given impurity in a-Si:H, it follows from Eqs. (44), (46), and (49) that

$$\Gamma = \beta\eta/\alpha. \tag{52}$$

The position of the Fermi level close to the film surface has been directly measured by photoelectron spectroscopy (Williams *et al.*, 1979; von Roedern *et al.*, 1979; Gruntz, 1981). Indirectly, a Fermi-level shift in the bulk can be determined by conductivity or thermopower measurements. In the literature, doping effects in a-Si:H are often quoted by changes of the room temperature conductivity or of the conductivity activation energy E_σ^*. We note, however, that according to Section 4, E_σ^* is a poor measure of the Fermi energy since it can be strongly affected by a temperature shift of the Fermi level relative to the band edge. Changes of the conductivity at a given temperature T, on the other hand, are a measure of ΔE_F only if transport takes place energetically distant from E_F and if the transport path itself is not altered by doping. In this case

$$\Delta E_F = kT\, \Delta \ln \sigma, \tag{53}$$

i.e., from changes of σ upon doping, the quantity Γ can be determined. As discussed in Part II, the combined measurement of conductivity σ and thermopower S and the determination of the quantity $Q = \ln(\sigma\, \Omega\, \mathrm{cm}) + (q/k)S$ can provide information if the above conditions are fulfilled.

11. Dopants in a-Si:H

In this and the following section, experimental results on doping effects are reviewed for glow-discharge a-Si:H films prepared under optimized conditions, i.e., at a low rf power, at low pressure, and at a substrate temperature $T_s > 200°\mathrm{C}$. Also, these results refer mainly to films with a thickness

$\geq 0.5\ \mu m$. The influence of a lower substrate temperature and of a reduced film thickness on the doping effect will be discussed in Section 13.

a. Substitutional Doping from the Gas Phase

Results for phosphorus and boron doping are shown in Fig. 6. Plotted are the room temperature values of the dark conductivity σ as well as those of the quantity Q (see Section 5) for glow-discharge a-Si:H (M–I) films as a function of phosphine or diborane concentration (Beyer *et al.*, 1977a; Beyer and Mell, 1977). For comparison, results of Spear and LeComber (1976) are shown as well as those of Jan *et al.* (1980) and Tsai *et al.* (1977) for the case of boron doping. Also included in the figure are data of Jan *et al.* (1980) for arsenic doping. For phosphorus doping, an increase in the conductivity by more than four orders of magnitude is obtained compared with undoped material. The maximum conductivity is reached for 10^3-10^4 ppm PH_3, followed by a decrease of σ_{RT} at higher doping levels. Similar results are obtained for arsenic doping for which the maximum conductivity, however, is about an order of magnitude below the phosphorus data. In the case of boron doping, the conductivity first decreases with rising dopant level and then increases. Since undoped a-Si:H films are usually n-type, this minimum indicates that the Fermi level is moved through midgap and that a change takes place from electron to hole conduction. By boron doping, conductivity values almost as high as for phosphorus doping can be ob-

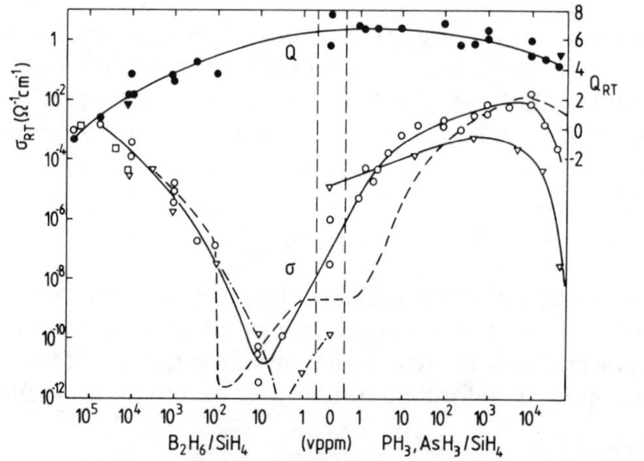

FIG. 6. Conductivity σ (open symbols) and quantity Q (closed symbols) at 300°K as a function of dopant concentration for gas-phase doping of a-Si:H: ○, P-, B-doped films. [Data from Beyer *et al.* (1977a) and Beyer and Mell (1977).] Dashed curve, P-, B-doped films. [From Spear and LeComber (1976).] ▽, As-, B-doped films. [From Jan *et al.* (1979, 1980).] □, B-doped films. [Data from Tsai *et al.* (1977).]

tained. However, no clear indication of saturation and subsequent decay of σ_{RT} at high doping levels is observed. In the range 10^2–10^4 ppm the results of different laboratories agree quite well for both B and P doping. Major differences occur at low doping levels, in particular the position of the conductivity minimum differs considerably. For all three dopants, the magnitude of Q_{RT} suggests the predominance of extended-states transport modified by the presence of potential fluctuations or by a second transport path (see Section 6). With rising doping level Q_{RT} decreases. The relatively low values of Q_{RT} for high doping levels of boron indicate some change in the transport process, presumably towards an increased hopping contribution.

The results for (M–I) compensated a-Si:H films containing both phosphorus and boron (Beyer *et al.*, 1981a) are shown in Fig. 7. Also included are data of Street *et al.* (1981). For this series of films the phosphorus doping level has been kept at 1000 ppm and the boron concentration varies. Both σ_{RT} and Q (measured at $T = 500°K$) show a sharp minimum near 1000 ppm B_2H_6. The agreement of the results of different laboratories is remarkable. Allan *et al.* (1977) and Street *et al.* (1981) also observed electrical compensation, i.e., the change from *n*- to *p*-type conduction within a factor of ~2 near the point of nominal compensation.

Doping from the gas phase has also been tried using $Sb(CH_3)_3$ and $Bi(CH_3)_3$ (Carlson and Wronski, 1979). Although in the latter case the conductivity of $\sigma_{RT} \sim 10^{-11}$–$10^{-9}\ \Omega^{-1}\ cm^{-1}$ for 0.1% doping suggests no doping action, in the former case a weak doping effect [$\sigma_{RT} \sim 10^{-9}$–$10^{-5}\ \Omega^{-1}\ cm^{-1}$ for 0.1% $Sb(CH_3)_3$] cannot be ruled out.

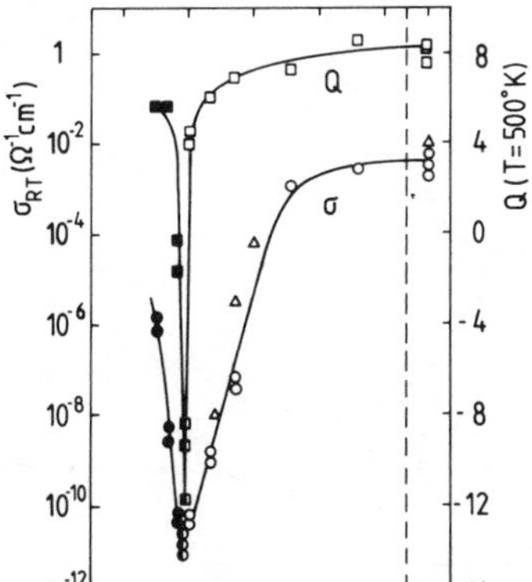

FIG. 7. Room temperature conductivity σ_{RT} (circles) and Q at 500°K (squares) of compensated a-Si:H as a function of diborane concentration in the gas phase. [Data from Beyer *et al.* (1981a).] △, σ_{RT}. [Data of Street *et al.* (1981).] Open symbols, *n*-type; closed symbols, *p*-type. $[PH_3] = 10^3$ vppm.

Plasma-chemical doping methods (Bauer and Bilger, 1982) have yielded room temperature conductivity values of up to $10^{-7}\ \Omega^{-1}\ cm^{-1}$ for nitrogen and boron, $10^{-4}\ \Omega^{-1}\ cm^{-1}$ for antimony, and $10^{-3}\ \Omega^{-1}\ cm^{-1}$ for indium, compared with $\sigma_{RT} \sim 10^{-10}\ \Omega^{-1}\ cm^{-1}$ for the undoped material.

b. Substitutional Doping by Ion Implantation

The results of implantation of all elements of group III and most elements of group V of the periodic system into glow-discharge a-Si:H (Kalbitzer *et al.*, 1980) are summarized in Fig. 8. The implantation temperature was 280°C. The increase of σ_{RT} suggests that a doping effect is present in all cases. Since there is no Q_{RT} data, some contribution of hopping conduction related to implantation defects (see Section 8) cannot be excluded. A comparison with Fig. 6 shows that the same boron concentration, introduced by ion implantation or by gas-phase doping with B_2H_6 (assuming that $p \sim 1$, i.e., that 10^3 ppm B_2H_6 in the gas phase leads to a boron concentration of $\sim 10^{20}\ cm^{-3}$ in the film) results in similar conductivity values. In the case of phosphorus and arsenic doping, on the other hand, a considerably higher impurity concentration is necessary for ion implantation than for gas-phase doping to reach comparable σ_{RT} data. Kalbitzer and co-workers (1980) have suggested that the incorporation of a dopant atom into fourfold coordination may be more difficult for implanted impurities than for those from the gas phase. Photoconductivity measurements indicate that nondoping configurations give rise to additional states in the gap acting as recombination

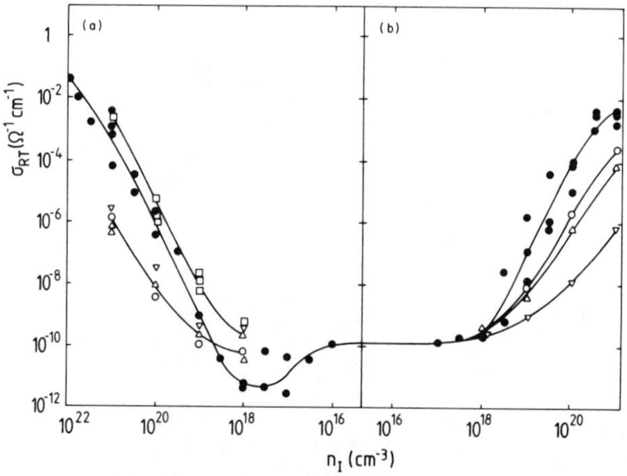

FIG. 8. Room temperature conductivity σ_{RT} as a function of implanted impurity concentration n_I for elements of groups III and V. (a) ●, B; ○, Al; △, Ga; ▽, In; □, Tl. (b) ●, P; ○, As; △, Sb; ▽, Bi. [After Kalbitzer *et al.* (1980).]

centers (Müller and LeComber, 1981). Furthermore, in spite of the relatively high implantation temperature used, the formation of implantation defects cannot be ruled out (Kalbitzer *et al.,* 1980).

c. Interstitial Doping

The results of interstitial doping by stepwise lithium implantation (Beyer *et al.,* 1979b,c) are displayed at the right hand side of Fig. 9. In this case, the implantation was carried out at room temperature and the film was annealed subsequently to remove the implantation defects. The quantity Q shows that this procedure was successful: The data lie well in the range of gas-phase doping. Yet, as with phosphorus implantation, a considerably higher dopant concentration is necessary, compared with gas-phase doping, to reach the same σ_{RT} values. The reason could be that part of the implanted lithium precipitates on annealing at 300°C. Also shown in Fig. 9 are results of Kalbitzer *et al.* (1980) for the implanation of the heavier alkali elements, suggesting that they all act as dopants in a-Si:H. Although the maximum conductivity obtained ($\sigma_{RT} \sim 5 \times 10^{-2}\ \Omega^{-1}\ cm^{-1}$) is higher than for lithium implantation and even for gas-phase phosphorus doping, the conductivity data for medium and low doping levels lie slightly below those for lithium. Kalbitzer *et al.* (1980) also investigated the influence of halogen implantation on σ_{RT} (left hand side of Fig. 9): No doping effect was observed. The

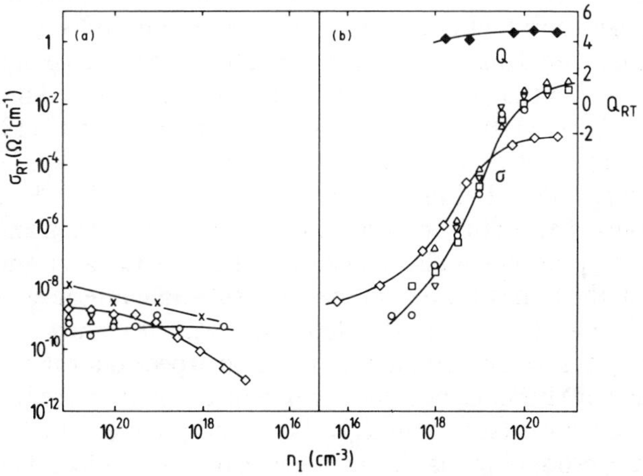

FIG. 9. Room temperature conductivity σ_{RT} and Q_{RT} of a-Si:H as a function of implanted impurity concentration n_I for elements of groups I, IV, and VII. [Li data from Beyer *et al.* (1979b); Si data from Spear *et al.* (1979).] (a) ◇, F; ○, Cl; △, Br; ▽, I; ×, Si. (b) ◇ and ◆, Li; ○, Na; △, K; ▽, Rb; □, Cs. [After Kalbitzer *et al.* (1980).]

slight increase of the room temperature conductivity with rising ion dose is probably related to an increasing concentration of implantation defects. Implantation with silicon (Spear *et al.*, 1979) leads to rather similar conductivity changes. The absence of a doping effect for implanted fluorine atoms has also been verified by Beyer *et al.* (1980) monitoring both conductivity and thermopower.

d. Unintentional Doping

The wide variation of σ_{RT} for nominally undoped a-Si : H films (see Fig. 6) suggests that doping by unintentionally incorporated dopants may be involved. Naturally, these dopants can originate from a contamination of the reaction chamber due to a previous production of doped material (Kuwano and Ohnishi, 1981). On the other hand, it has been observed that the addition of nitrogen, air, oxygen, and hydrogen can lead to a considerable enhancement of both dark- and photoconductivities. Room temperature conductivity values as high as $2 \times 10^{-2}\ \Omega^{-1}\ cm^{-1}$ have been obtained for sputtered films upon incorporation of nitrogen (Baixeras *et al.*, 1978) and of hydrogen (Nguyen Van Dong *et al.*, 1981). In the case of glow-discharge a-Si : H, σ_{RT} values up to $2 \times 10^{-3}\ \Omega^{-1}\ cm^{-1}$ were reached upon addition of oxygen (Xu *et al.*, 1981) and up to $10^{-4}\ \Omega^{-1}\ cm^{-1}$ upon admission of air (Griffith *et al.*, 1980). Still, the question remains as to whether we are dealing here with a true doping effect of amorphous silicon. Since all of these impurities presumably can saturate dangling bonds, it is conceivable that a reduction in the density of gap states leads either to an enhancement of the doping effect of other (unintentionally) incorporated impurities or to a shift of E_F due to a changing density-of-states distribution. This interpretation is supported by the fact that doping effects of the above mentioned impurities are not always observed. In the case of nitrogen, Pietruzco *et al.* (1981) did not find a doping effect upon addition of $10 - 5 \times 10^4$ ppm N_2. High-quality glow-discharge a-Si : H films are known to contain 5 – 15 at. % of hydrogen yet no doping effect of this impurity is observed. For nitrogen and nitrogen – oxygen incorporation it is also conceivable that a carrier accumulation layer is formed at the film surface or at the film – substrate interface. All reported conductivity data (Griffith *et al.*, 1980; Xu *et al.*, 1981) show a tailing of the conductivity at low temperatures as would be expected for the contribution of variable range hopping near the Fermi level or for the presence of a carrier accumulation layer (see Section 7). Since the photoconductivity is high, variable range hopping via defect states can be excluded. Thin layers of silicon nitride or silicon oxynitride could form at the film surface. These substances are known to contain a high concentration of positive charges (Hezel and Schörner, 1981) and such charges could cause a carrier accumulation layer. Still, these problems are far from being settled, and considerably more work appears necessary.

12. Incorporation Ratio and Doping Efficiency

The reported data on the dopant incorporation ratio as defined in Section 10 for singly doped films lie between $p \sim 0.5$ (Spear and LeComber, 1976) and $p = 5.2$ (Street *et al.*, 1981) for phosphorus doping. For arsenic doping, $p \sim 1.7$ has been reported by Knights *et al.* (1977) for arsine concentrations of 0.5 – 1%. In the case of boron doping, the reported data range from $p \leq 0.5$ (Spear and LeComber, 1976) to $p = 4.8$ (Zesch *et al.*, 1980), several investigators, however, found $p \sim 1$ (see Chevallier and Beyer, 1981). For compensated a-Si:H:P:B films, rather controversial results have been obtained for the incorporation ratio of boron. Chevallier and Beyer (1981) report $p \sim 1$ from nuclear analysis experiments, whereas Street *et al.* (1981) found values between 4 and 23, employing secondary ion mass spectrometry (SIMS). The origin of this discrepancy has not yet been clarified. Different deposition conditions are unlikely to be the reason since the conductivity data (see Fig. 7) agree well. Matrix effects of SIMS (Williams, 1979) could also be responsible.

The doping efficiency for arsenic in a-Si:H has been determined by Knights *et al.* (1977) using EXAFS. They obtained $\eta \sim 0.2$ for a doping level of 1% AsH_3. For phosphorus-doped films, Spear and LeComber (1976) have estimated the doping efficiency according to Eqs. (51)–(53). Using experimental data for the incorporation ratio p from ion-probe analysis and determining α according to Eq. (50) using the field-effect density of states they obtained $\eta \sim 0.3$–0.4. Street (1982) estimated η for boron and phosphorus doping by equating the concentration of impurity atoms in fourfold coordination to the density of compensating charge defects measured, e.g., by luminescence, a procedure that appears to us to be not without problems. He found a doping efficiency that drops with rising dopant gas concentration according to $\eta \propto (n_I^g)^{-1/2}$ with $\eta \sim 6 \times 10^{-2}$ for 10 ppm and $\eta \sim 10^{-3}$ for 10^4 ppm.

Following the lines of Spear and LeComber (1976) we plot in Fig. 10 the room temperature doping effect Γ_{RT} as a function of σ_{RT} for (M–I) phosphorus- and boron-doped as well as compensated a-Si:H films. The quantity Γ_{RT} has been obtained as the derivative of the smooth curves in Figs. 6 and 7, assuming $p = 1$. The room-temperature Fermi-level position is determined by inverting σ_{RT} according to

$$|E_t - E_F| = kT \ln(\sigma_0/\sigma_{RT}) \tag{54}$$

with a common $\sigma_0 = 2000\ \Omega^{-1}\ cm^{-1}$ (Overhof and Beyer, 1983). In Fig. 10,

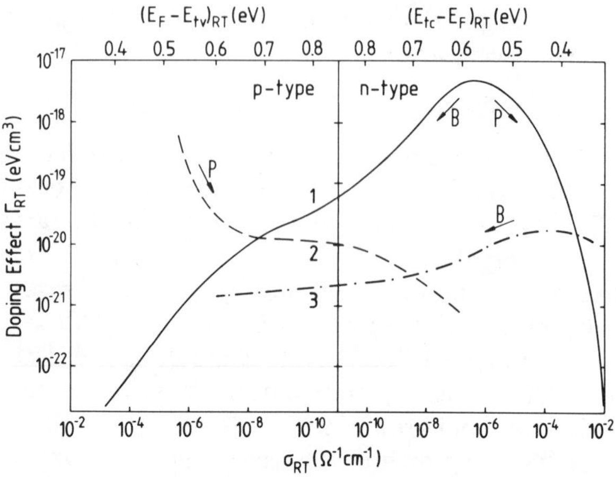

FIG. 10. Doping effect at room temperature $\Gamma_{RT} = \Delta E_F/\Delta n_I$. (1) Singly phosphorus- and boron-doped samples; (2), compensated samples, $[B_2H_6] = 1000$ ppm; (3), compensated samples, $[PH_3] = 1000$ ppm. The directions of increasing phosphorus and boron concentration, respectively, are indicated by arrows.

E_{tc} and E_{tv} denote the energies of the transport paths in the conduction and valence bands, respectively. For singly doped films (curve 1) Γ shows a maximum near $E_{tc} - E_F = 0.6$ eV and an asymmetric decrease toward both band edges. This behavior reminds us strongly of a mirror image of the energy distribution of the density of states. Indeed, this is what is expected since $\Gamma = \beta\eta/\alpha$ [Eq. (52)] and $\alpha = g(E)$ [Eq. (50)] are valid in the low temperature limit. Yet, the data of Fig. 10 refer to 300°K. Therefore, equating α with $g(E)$ would certainly lead to a considerable error, in particular for E_F close to the band edges. It must be mentioned, furthermore, that Γ will be strongly affected if compensating defects are created concomitantly with a dopant incorporation. For singly doped films, this effect will invariably lead to an increase in α, i.e., a drop in Γ, as is observed for high doping levels. For compensated films, the doping effect Γ can also be increased by this mechanism. Still, both series of compensated films (curves 2 and 3) yield for E_F near the gap center an almost constant Γ of 10^{-21}–10^{-20} eV cm^3. This result suggests that for compensated films also, Γ reflects the density of localized states. The density of states $g(E)$ comes out to be considerably higher than that for singly doped films at a low doping level. This is a further evidence for defect creation by both phosphorus and boron impurities in a-Si : H. In fact, the reproducible production of electrically compensated films from doping gases mixed with an accuracy of ~100 ppm would be impossible without this reduced doping effect.

13. Structural Aspect of Doping and Influence of the Preparation Conditions

a. Doping-Induced Changes in Hydrogen Evolution and in Hydrogen Content

Hydrogen evolution experiments have been proven a sensitive tool for the investigation of the bonding of hydrogen as well as of the processes governing its release from a-Si:H (Biegelsen *et al.*, 1979). The results displayed in Fig. 11 (Beyer *et al.*, 1981b; Beyer and Wagner, 1981) show that boron doping affects dramatically the hydrogen evolution spectra, whereas phosphorus doping causes only minor changes. Even low boron-doping levels result in a sizable shift of the hydrogen evolution peak to lower temperatures. The substrate potential has also been found to play an important role (Chen and Fritzsche, 1982). An analysis of the evolution peak position as a function of heating rate and of film thickness shows that boron doping induces a change in the process limiting the evolution rate. Volume diffusion of hydrogen is active in undoped a-Si:H and surface desorption of hydrogen is dominant in boron-doped material (Beyer and Wagner, 1982). This result demonstrates that gas-phase doping by boron can by no means be considered an effect involving only a Fermi-level shift. Amorphous silicon is changed by boron doping from an essentially compact material to a substance presumably dominated by grain boundaries. Results of Müller *et al.* (1980) indicating a doping-dependent hydrogen content for both boron and phosphorus gas-phase doping let us suspect that phosphorus doping may cause structural changes as well.

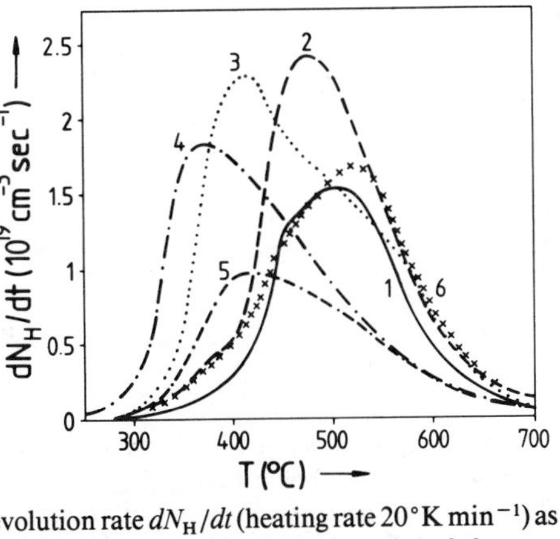

FIG. 11. Hydrogen evolution rate dN_H/dt (heating rate 20°K min^{-1}) as a function of temperature for doped and undoped a-Si:H. Film thickness 0.4–0.6 μm; substrate temperature 300°C. (1) Undoped, (2)–(5) [B_2H_6] = 10, 10^2, 10^3, 10^4 vppm, respectively, and (6) [PH_3] = 10^4 vppm. [After Beyer and Wagner (1981).]

b. Doping Effect on the Photoconductivity

The influence of doping on the photoconductivity also suggests that structural effects may be involved (Beyer and Hoheisel, 1983). For undoped and lightly phosphorus- and boron-doped films a unique correlation between the photoconductivity $\Delta\sigma_{ph}$ and the dark conductivity σ has been observed (curve 1 in Fig. 12). High doping levels of phosphorus and boron lead to a slight and heavy drop, respectively, in $\Delta\sigma_{ph}$, resulting in the well-known difference in $\Delta\sigma_{ph}$ between phosphorus- and boron-doped a-Si:H (Rehm *et al.*, 1977; Anderson and Spear, 1977). The same low $\Delta\sigma_{ph}$ as observed for singly boron-doped (*p*-type) a-Si:H is found for compensated *n*-type material suggesting that it is not the conduction by holes that is the origin of the reduced photoconductivity but the presence of boron.

c. Effect of the Substrate Temperature

The experimental results presented so far refer to a-Si:H films prepared at substrate temperatures $T_s > 200\,°C$. The influence of T_s on the room temperature conductivity of doped and undoped a-Si:H films is displayed in Fig. 13. The results show a reduction of the doping effect with decreasing T_s, in particular for phosphorus-doped material. This decay is accompanied by an increase in the dangling-bond spin density (Stuke, 1977) as well as in the hydrogen concentration. Hydrogen evolution experiments suggest that a void-rich structure is formed that differs for boron-doped and undoped or phosphorus-doped material (Beyer and Wagner, 1981).

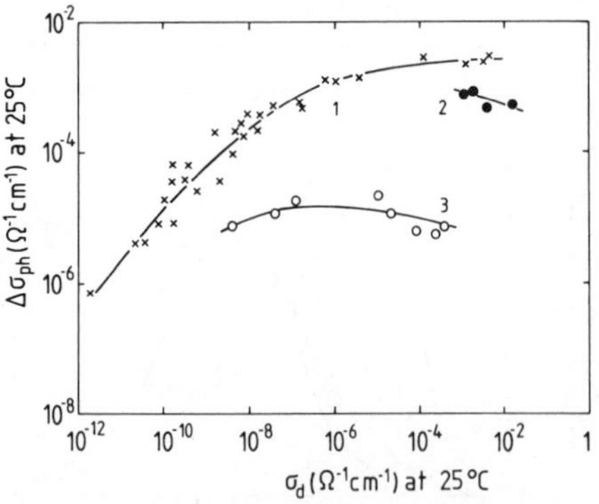

FIG. 12. Photoconductivity $\Delta\sigma_{ph}$ as a function of dark conductivity σ_d at room temperature for doped and undoped a-Si:H. (1) 0–10 ppm PH_3, B_2H_6; (2) 10^2–10^4 ppm PH_3; and (3) 10^2–10^4 ppm B_2H_6. [Reprinted with permission from *Solid State Communications,* Vol. 47, W. Beyer and B. Hoheisel, Photoconductivity and dark conductivity of hydrogenated amorphous silicon, Copyright 1983, Pergamon Press, Ltd.]

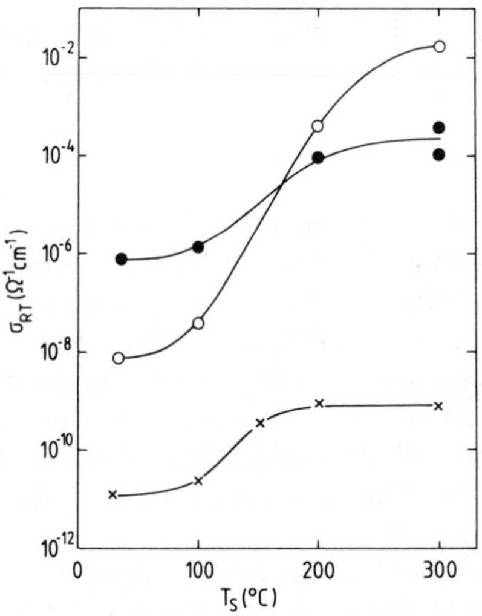

FIG. 13. Room temperature conductivity σ_{RT} as a function of substrate temperature T_s for doped and undoped a-Si:H. ×, undoped; ●, 1% B_2H_6; ○, 1% PH_3.

d. Effect of the Film Thickness

An influence of the film thickness on the dark conductivity of phosphorus-doped a-Si:H was first demonstrated by Ast and Brodsky (1980). For films doped with 500 ppm PH_3, σ_{RT} was found to drop by a factor of more than 20 when the film thickness was decreased from 0.5 to 0.1 μm. An analysis of measurements on both planar and sandwich structures suggests that a highly resistive layer (of ~0.2 μm thickness) is present also in thick films, presumably near the film–substrate interface. There is considerable experimental evidence that the hydrogen concentration in plasma-deposited a-Si:H changes considerably as a function of film thickness as well as within a film near the film surface and the film–substrate interface (Müller *et al.*, 1980; Ast and Brodsky, 1980; Beyer and Wagner, 1981). It is tempting to associate the decreased doping effect with the changed hydrogen concentration. Presumably, structural effects influence the film properties close to the film–substrate interface.

14. TRANSPORT MEASUREMENTS AS A FUNCTION OF TEMPERATURE

As discussed in Part II, measurements of transport properties as a function of temperature are necessary for the identification of the transport mechanism as well as for the detection of changes in the transport process by doping. In this section, the results of such measurements on doped a-Si:H films are presented. The samples were all M–I films deposited at a substrate temperature of 300°C.

a. Singly Doped Films

The results of conductivity and thermopower measurements as a function of temperature are shown in Fig. 14 for phosphorus-doped a-Si:H films (Beyer *et al.*, 1977a) and in Fig. 15 for boron-doped material (Beyer and Mell, 1977). The plots present log σ and S versus reciprocal temperature. We note that the data for phosphorus doping agree qualitatively well with results obtained for a variety of *n*-type dopants, including As, Li, K, and Na obtained by the present authors and by others (Beyer *et al.*, 1979b,c, 1980; Jan *et al.*, 1979). Our results for boron doping are similar to those of Jan *et al.* (1980) and Tsai *et al.* (1977).

A most striking result is the wide range of apparent conductivity prefactors σ_0^* obtained by extrapolation of the experimental data to $1/T = 0$. They range from < 10 to $10^3\ \Omega^{-1}\ \text{cm}^{-1}$ and vary both as a function of the doping level and of the temperature. Furthermore, the temperature dependence of both log σ and S reveals for all phosphorus-doped films a conspicuous kink near 400°K, whereas for boron-doped films at high doping levels a gradually decreasing slope of log σ towards low temperatures is observed. The question arises as to whether the spread of the σ_0^* values and the observation of structures in the temperature dependence of σ indicate the presence of different transport processes. According to Section 5, the plot of $Q = \ln(\sigma\ \Omega\ \text{cm}) + |(e/k)S|$ versus $1/T$ can give an answer since for unipolar con-

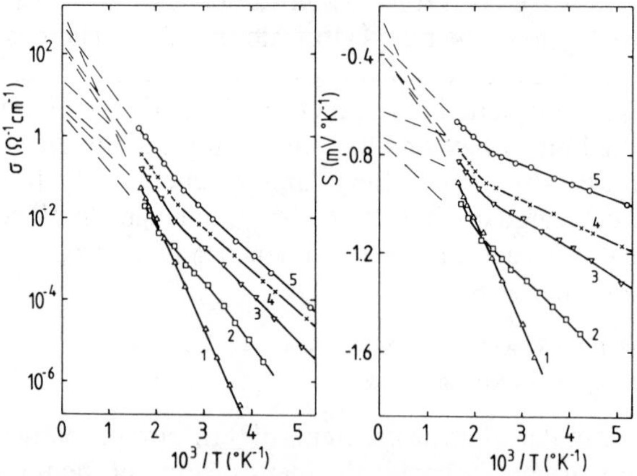

FIG. 14. Conductivity and thermopower as a function of inverse temperature for phosphorus-doped a-Si:H. $[PH_3]/[SiH_4]$ (vppm) = 1 (1), 3 (2), 250 (3), 10^3 (4), and 10^4 (5). [After Beyer *et al.* (1977a).]

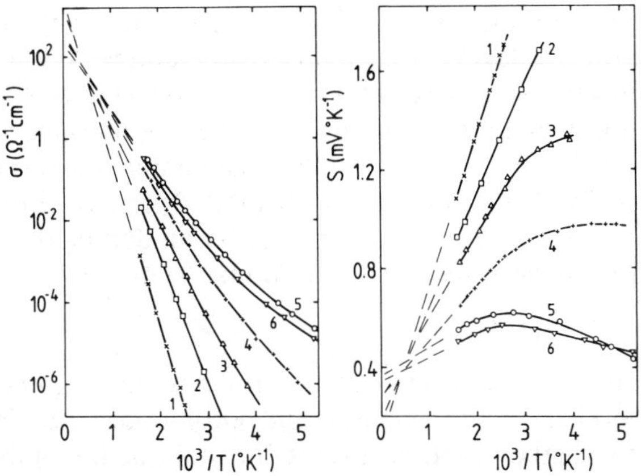

FIG. 15. Conductivity and thermopower as a function of inverse temperature for boron-doped a-Si:H from data of Beyer and Mell (1977). $[B_2H_6]/[SiH_4]$ (vppm) = 10 (1), 10^2 (2), 10^3 (3), 10^4 (4), 5×10^4 (5), and 2×10^5 (6).

duction and nondegenerate statistics the function $Q(T)$ is independent of the Fermi energy. In Fig. 16, Q is plotted for the films of Figs. 14 and 15. The results show straight lines in the case of phosphorus-doping (Fig. 16a), i.e., a disappearance of the kinks. When fitting this data with

$$Q = Q_0^* - E_Q^*/kT, \tag{55}$$

one obtains $Q_0^* = 9-11$ and $E_Q^* = E_\sigma^* - E_S^* = 0.05-0.15$ eV. We note that we have observed such behavior rather generally for high quality a-Si:H, undoped as well as doped by a multitude of different n-type dopants (Beyer *et*

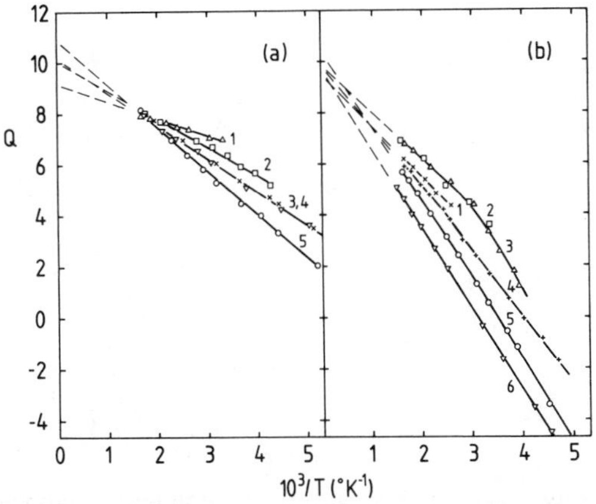

FIG. 16. Q as a function of inverse temperature. (a) Phosphorus-doped films of Fig. 14. (Key as for Fig. 14.) (b) Boron-doped films of Fig. 15. (Key as for Fig. 15.)

al., 1979b,c; Overhof and Beyer, 1980) (see also Fig. 26b). Boron-doped films (Fig. 16b) also follow Eq. (55) at high temperatures. At low T, the slope of Q tends to increase in some cases. For equal doping levels, E_Q^* is generally higher for boron-doped films than for those that are phosphorus doped.

The Hall mobility (Fig. 17a,b) exhibits behavior qualitatively similar to Q (Beyer *et al.*, 1977a, Beyer and Mell, 1977): Plotted versus reciprocal temperature, considerably steeper curves result for boron doping than for doping with phosphorus, in agreement with LeComber *et al.* (1977).

b. Compensated a-Si:H Films

Both the transport parameter Q and the Hall mobility μ_H suggest a difference in transport between phosphorus- and boron-doped a-Si:H. The question arises as to whether this difference is due to the action of the particular dopant or is merely a consequence of the fact that we are dealing with electron and hole conduction, respectively. Answers can be obtained from the investigation of compensated a-Si:H films containing both phosphorus and boron (Beyer and Mell, 1981; Beyer *et al.*, 1981a). Furthermore, such measurements can also yield valuable information on other important parameters of a-Si:H, namely, the width of the conductivity gap E_G^σ and the ratio of apparent conductivity prefactors for electrons and holes, $\sigma_{0n}^*/\sigma_{0p}^*$, when E_F is close to midgap. Two series of compensated films have been studied. In series A, the B_2H_6 concentration is kept constant at 1000 ppm and the PH_3 concentration is varied; in series B, the phosphine concentration is held at 1000 ppm and the boron concentration is changed. Results for series A in Fig. 18a show that Q is practically independent of the phosphorus concentration as well as of the sign of the charge carriers. In the other case

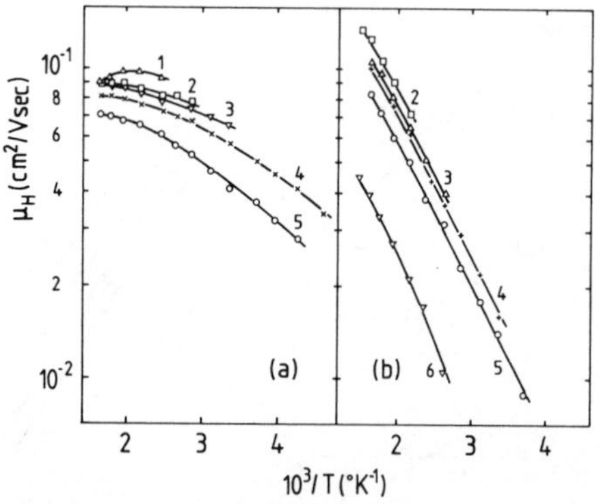

FIG. 17. Hall mobility μ_H as a function of inverse temperature. (a) Phosphorus-doped films of Fig. 14. (Key as for Fig. 14.) [After Beyer *et al.* (1977a).] (b) Boron-doped films of Fig. 15. (Key as for Fig. 15.) [Data from Beyer and Mell (1977).]

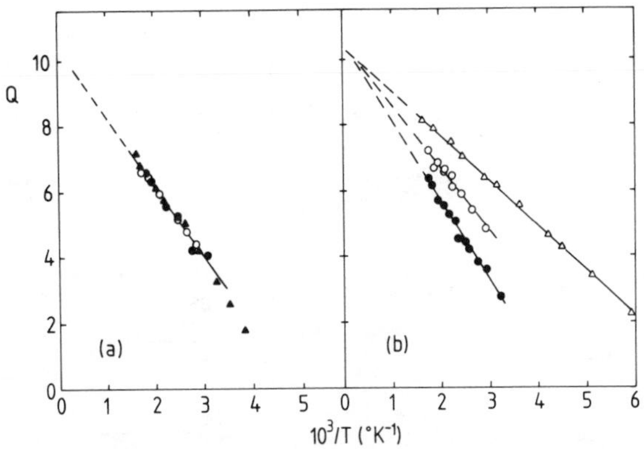

FIG. 18. Q as a function of inverse temperature for compensated a Si:H films (a) $[B_2H_6] = 1000$ vppm; $[PH_3] = 0$ (▲), 200 (●), 3000 (○) vppm. (b) $[PH_3] = 1000$ vppm. $[B_2H_6] = 0$ (△), 200 (○), 3000 (●) vppm. [Adapted with permission from *Solid State Communications,* Vol. 38, W. Beyer and H. Mell, Influence of boron doping on the transport properties of a-Si:H films, Copyright 1981, Pergamon Press, Ltd.]

(Fig. 18b) the slope of Q versus $1/T$ increases steadily with rising boron concentration, i.e., there is a gradual transition from the behavior typical of singly phosphorus-doped films to a behavior typical of singly boron-doped material, independent of the sign of the charge carriers. These results demonstrate again that boron doping causes changes in the electronic transport at both band edges, in agreement with the photoconductivity measurements discussed in Section 13.

In Fig. 19, the conductivity as a function of $1/T$ and the quantity eST as a function of temperature are shown for films of series B. Samples prepared

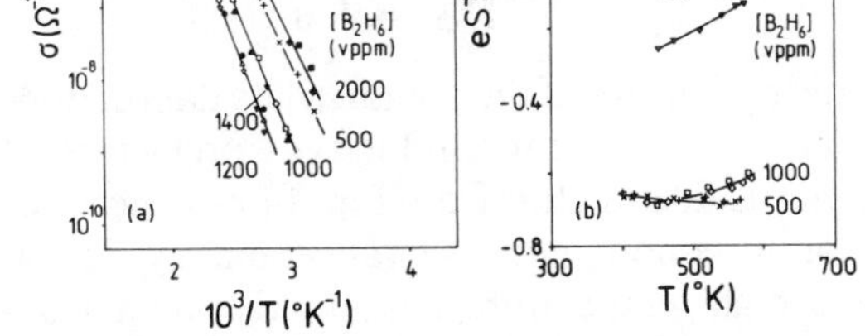

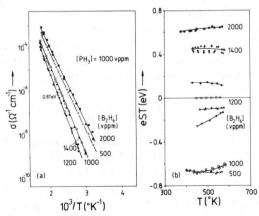

FIG. 19. Conductivity σ versus $1/T$ and quantity eST versus T for compensated a-Si:H films. [From Beyer *et al.* (1981a).]

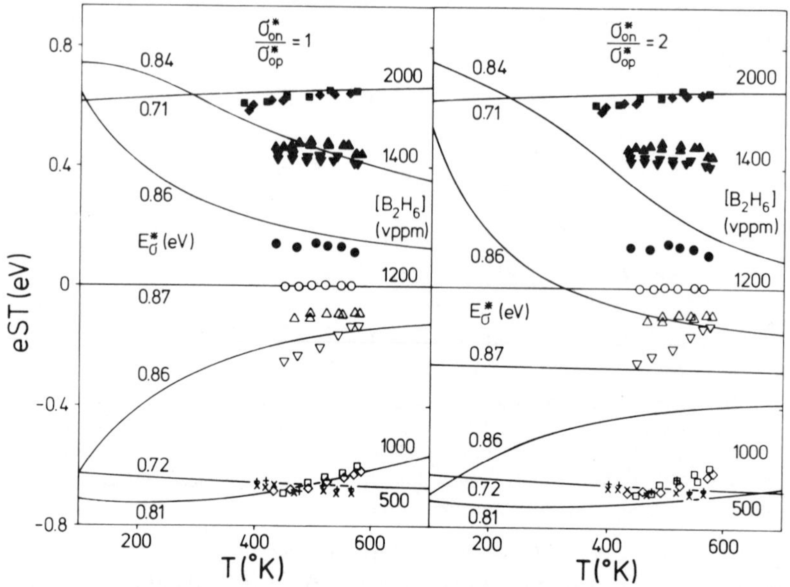

FIG. 20. Fit of thermopower data of Fig. 19, using different ratios of conductivity prefactors. [From Beyer *et al.* (1981a).]

with 1000 ppm PH_3 and 1200 ppm B_2H_6 turned out to be highly electrically compensated according to the almost vanishing thermopower. Since the Fermi level for these latter samples lies close to midgap, the width of the conductivity gap can be taken to be twice the conductivity activation energy. One obtains $E_G^\sigma \sim 1.74$ eV. Information on the ratio of the apparent conductivity prefactors of electrons and holes (for E_F close to midgap) is obtained by calculating the quantity eST from the conductivity data according to Eq. (41) using $\sigma_{0n}^*/\sigma_{0p}^*$ as a free parameter. The results show a good agreement between calculated and experimental eST data for equal conductivity prefactors (Fig. 20). The assumption of a ratio of two leads to a rather poor fit.

V. Discussion

One of the striking features of the conductivity data of doped a-Si : H films is the wide variation of apparent conductivity prefactors σ_0^*. In Fig. 21 σ_0^* values of phosphorus-doped films from Fig. 14 have been plotted as a function of the apparent conductivity activation energy E_σ^*, both for the low temperature and high temperature region. Also included are data for Li doping (Beyer and Overhof, 1979). The results agree well with those of Rehm *et al.* (1977), Carlson and Wronski (1979), and Anderson and Paul (1982). A

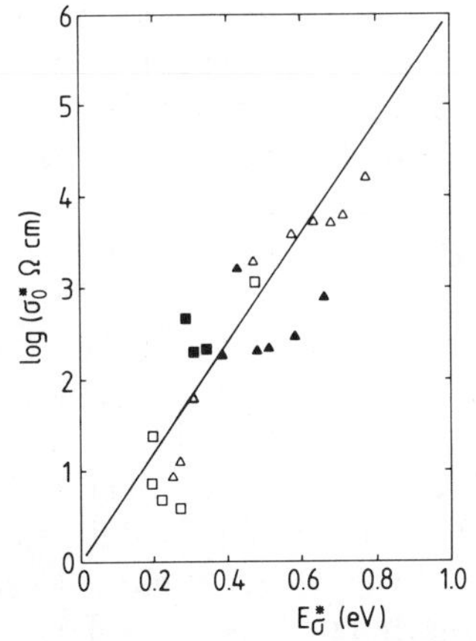

FIG. 21. Apparent conductivity prefactor σ_0^* as a function of conductivity activation energy E_σ^* for phosphorus-doped films of Fig. 14 (squares) and for Li doping (triangles). [Data from Beyer *et al.* (1979b).] Open symbols, low temperature range; closed symbols, high temperature range.

variation of σ_0^* with E_σ^* was reported by Meyer and Neldel in their study of baked semiconductor powders as early as 1937 (Meyer and Neldel, 1937). They find that upon heat treatment both parameters change according to

$$\sigma_0^* = C \exp(E_\sigma^*/E_0) \tag{56}$$

with $C \sim 1\ \Omega^{-1}\ \text{cm}^{-1}$ and $E_0 \sim 0.1$ eV. The striking similarity of the behavior of a-Si:H films with that of baked powders, (C and E_0 are found to be virtually the same) could be considered as an indication that there is a structural similarity as well. A transport model like that proposed by Anderson and Paul (1981) in fact reflects several aspects of a baked semiconductor powder system.

We shall show in the following that the opposite is likely to be true in hydrogenated amorphous silicon: The films are found to be rather insensitive to doping and to the preparation conditions if we consider $Q(T)$. Accordingly, the wide variation of σ_0^* must be ascribed to effects that do not alter the transport path.

15. THE TRANSPORT MECHANISM

We have seen in Section 14 that for most a-Si:H samples prepared under optimized conditions the general shape of $Q(T)$ is essentially the same. One observes $Q = Q_0^* - E_Q^*/kT$ [Eq. (55)] with $0.05 < E_Q^* < 0.25$ eV and Q_0^* close to 10. Note that similar $Q(T)$ curves are obtained for a-Ge:H (Hauschildt *et al.*, 1982) and for a variety of chalcogenide glasses (Overhof and Beyer, 1984). This rather general feature makes all transport models quite unlikely for which the nonzero slope of Q is caused by the transition from one dominant transport path to another as the temperature is altered. For the two-carrier model proposed by the Dundee group, we have seen in Fig. 3 that the shape of $Q(T)$ changes unsystematically upon doping. At least for moderately and highly doped samples in which the experimental $Q(T)$ data cover a wide temperature range and in which $E_Q^* > 0.1$ eV, the two-carrier model can be excluded.

As discussed in Section 6c, Döhler's model of a "soft" mobility edge is in principle capable of explaining a difference between E_σ^* and E_S^* and hence a nonzero slope E_Q^*. The extreme sensibility of the shape of $Q(T)$ on $\sigma(E)$, however, makes this transport model rather unlikely. The different E_Q^* values observed in the experiment require different $\sigma(E)$ functions; yet the observed $Q(T)$ data are always linear if plotted versus $1/T$, in contrast to what is obtained if we interpolate between different $\sigma(E)$ curves. The same shortcoming is found in the VRH model of Grünewald and Thomas (1979) (see Section 6d). Again, it is unlikely that $g(E)$ is always changed by doping in such way as to retain the linearity of the Q versus $1/T$ curves.

The observation that the general shape of $Q(T)$ does not change by doping and that only E_Q^* is altered suggests strongly that the explanation of a doping dependence of $Q(T)$ cannot involve the change of some function, or the simultaneous modification of several parameters preserving the dependence Eq. (55) by coincidence only. It must be related to a single parameter instead.

We have shown in Section 6b that transport in extended states above a mobility edge can account for the observed $Q(T)$ curves if long-range static potential fluctuations are included. Such a transport process leaves the general shape of $Q(T)$ unchanged if the form of the fluctuating potential is modified. The slope E_Q^* is proportional to the width Δ of the fluctuations. The presence of such fluctuations in a-Si : H has been postulated by several authors. Very recently direct experimental support has been found by core-level broadening in photoemission (Ley *et al.*, 1982). This latter result has been interpreted in terms of short-range charge fluctuations, but we believe that on the basis of the experimental data a decision between long- and short-range fluctuations is not entirely possible. According to our transport model, a reduced temperature $T' = kT/\Delta$ can be introduced leading to a scaling of the $\hat{Q}$ versus $1/T'$ curves shown in Fig. 2. Since, in the experiment, Δ is unknown, we cannot scale the experimental data of Q directly. Taking from the model calculations that at low temperatures $E_Q^* = 1.25\ \Delta$, we can determine Δ from the slope E_Q^*. The resulting graph Fig. 22 shows data of Q versus $1/T'$ for series B of compensated a-Si : H samples (see Section 14b) that are n-type (open symbols) and p-type (closed symbols). It is seen that not even the change of the sign of the carriers modifies $Q(1/T')$. Data for a-Ge : H (Hauschildt *et al.*, 1982) and for chalcogenide glasses (Overhof and Beyer, 1984) practically coincide with the data points of Fig. 22.

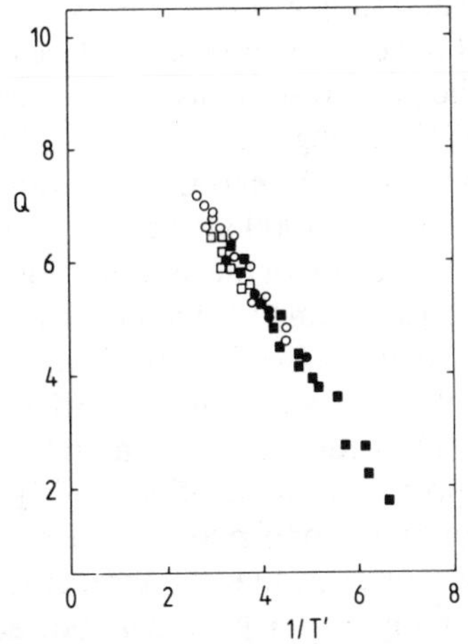

FIG. 22. Q as a function of inverse reduced temperature for compensated a-Si:H films. $[PH_3] = 1000$ vppm; $[B_2H_6] = 200$ (○), 500 (□), 2000 (●), 3000 (■) vppm. [From Overhof and Beyer (1983).]

Anderson and Paul (1982) have published $Q(1/T)$ data for doped sputtered a-Si:H films. They observed in one case a slight bending of $Q(1/T)$ at high temperatures. We note that this behavior is in perfect agreement with the calculated curves of Fig. 2. In another case they found a steplike structure in $Q(1/T)$. As shown in Fig. 4, this behavior could be explained by a carrier accumulation layer.

In some cases, in particular if the samples are highly boron-doped or if part of the hydrogen has been driven out from the samples, a hopping contribution to the conductivity is observed. This is revealed by a rapid decrease of Q at low temperatures and confirmed by the observation of an anomalous magnetoresistance (Weller *et al.*, 1981) that can be understood only in terms of a VRH model (Movaghar and Schweitzer, 1978). Although a hopping contribution can usually be subtracted for undoped a-Si:H, this proves difficult for doped samples: The distinction between activated transport and hopping is not clearcut if VRH takes place in a tail close to E_F, and neither a strict $\ln \sigma \sim T^{-1/4}$ law nor a vanishing hopping thermopower can be expected.

We return to the discussion of Section 5 concerning the modification of σ_0 by the electron–phonon coupling. From Fig. 22 we see that Q_0^* and hence σ_0 are practically equal for electrons and holes. The same conclusion is reached from the results of compensated a-Si:H films with E_F close to midgap (Section 14b) since for this Fermi-level position $\sigma_0^* = \sigma_0$ in our transport model. From the concept of minimum metallic conductivity we expect σ_{00} to be approximately equal for both carriers, leaving us with the conclusion that the shifts of both mobility edges with respect to the states at the Fermi level are equal in magnitude and rather independent of the Fermi-level position. This is surprising at first but can be understood if we recall that the density of states distribution in the gap presumably can be related to a single defect (Overhof and Beyer, 1981b) giving rise to the $g = 2.0055$ dangling-bond spin resonance signal. By comparing the extrapolated value of $Q_0^{*\prime} \sim 10.5$ in Fig. 22 with Eq. (24), we obtain $\sigma_0 \sim 2000\ \Omega^{-1}\ \mathrm{cm}^{-1}$, assuming that $A = 1$. This means that if the temperature shift of the mobility edges is taken into account we end up with $\sigma_{00} \sim 200\ \Omega^{-1}\ \mathrm{cm}^{-1}$, which is just what is expected for the minimum metallic conductivity (see Section 6a).

Surprisingly, we do not observe from our $Q(T)$ data, even for doping in the one-percent region, any indication of the formation of a donor band affecting the electronic transport. LeComber *et al.* (1977) have inferred the presence of a donor band in highly phosphorus-doped a-Si:H from their Hall mobility data. We note, however, that the temperature dependence of the Hall mobility as shown in Fig. 17 can be understood according to Eq. (27) by the presence of potential fluctuations. A possible explanation why a donor band is not observed in transport could be that the donor states are hidden by the distribution of tail states.

It is surprising that structural effects, as mentioned in Section 13, do not seem to influence strongly the shape of $Q(T)$. They will contribute to the fluctuations, hence increasing E_Q^*, but the influence on Q_0^* is apparently small. This could be due to the fact that regions in space that differ in conductivity considerably from the rest of the sample are not detected if these regions make up only a small volume fraction.

16. Density of States Distribution and Fermi-Level Shift

Having shown that σ_0 is not at all altered by various preparation and doping conditions in high-quality a-Si:H, we must explain the observation of the Meyer–Neldel rule otherwise. A possible explanation is a shift of either E_t or E_F with temperature. In the following, we confine ourselves to discussing the statistical shift of the Fermi level. This shift originates from the neutrality condition requiring that the total density of electrons n in a sample is independent of temperature. Because this density can be expressed as an integral over the density-of-states distribution times the Fermi distribution function, we obtain the integral equation

$$n = \int_{-\infty}^{+\infty} g(E) f_F(E_F, T)\, dE. \tag{57}$$

In, e.g., the exhaustion regime of crystalline semiconductors, E_F moves toward midgap with rising temperature because the density of electrons in the conduction band is kept equal to the donor density. In amorphous semiconductors the shift of the Fermi energy is large if the density of states at the Fermi energy is small and if there is a larger density of states above the Fermi level (for hole conduction, below). For low temperatures the structure in $g(E)$ near the Fermi level determines the statistical shift whereas at elevated temperatures the density of states of the nearest tail becomes important.

In principle, the statistical shift can be calculated from Eq. (57) if the density-of-states distribution is known. Calculations based on the $g(E)$ curves obtained by the Dundee group (Madan *et al.*, 1976) resulted in a statistical shift that was insignificant except for highly *n*-doped samples (Jones *et al.*, 1977). This was not sufficient to explain the transport data (Beyer *et al.*, 1977a). In the meantime, density-of-states measurements by Grünewald *et al.* (1981), Goodman (1982), Mackenzie *et al.* (1982), and Lang *et al.* (1982) estimated the density of states in the gap to be smaller than that in the earlier experiments. Accordingly, a better agreement with our transport data is expected.

As discussed in the preceding section, $\sigma_0 \sim 2000\ \Omega^{-1}\ \mathrm{cm}^{-1}$ may be assumed for transport at the band edges of a-Si:H. Hence $E_F(T)$ can be determined from the conductivity data by

$$|E_t - E_F(T)| = kT \ln[\sigma_0/\sigma(T)], \tag{58}$$

and we should stress that this method of obtaining $E_t - E_F$ is certainly more appropriate than is taking E_σ^*. We note, moreover, that according to Eq. (25) $\ln(\sigma_0/\sigma)$ is only slightly affected by potential fluctuations.

In Fig. 23 we show experimental data points of $E_{tc} - E_F$ as a function of temperature for a sample doped stepwise by lithium implantation (Overhof and Beyer, 1980, 1981b). We discuss this experiment because we are dealing here with a single specimen at different doping levels and, therefore, we have some hope that the density-of-states distribution is not drastically altered at each doping step. Taking the Fermi-level shift for a particular doping level from Fig. 23 we can, by trial and error, construct a $g(E)$ curve that gives rise to this shift. The bold line in Fig. 24 shows this $g(E)$ relation, whereas the curves in Fig. 23 demonstrate to what extent the experimental data can be matched. The fit is excellent except for highly doped samples at high temperatures. If we take the shift of E_F in the $300°\mathrm{K} < T < 400°\mathrm{K}$ temperature range and calculate σ_0^* as a function of E_σ^* we obtain the curve displayed in Fig. 25. For comparison we show also the data points used by Carlson and Wronski (1979) to illustrate the Meyer–Neldel rule in a-Si:H, and a good agreement is observed. We note that the rather large temperature shifts of the Fermi level required by our interpretation have recently been confirmed by photoemission experiments (Gruntz, 1981; Ley, 1984). The $E_F(T)$ dependence obtained is quite similar to ours.

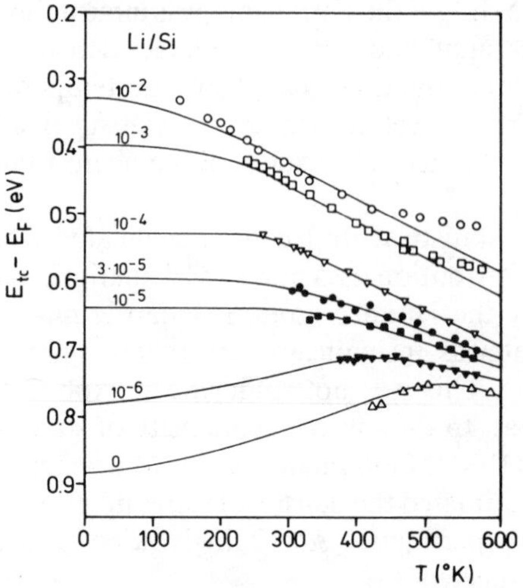

FIG. 23. Fermi-level position E_F (with respect to energy E_{tc} of the conduction-band transport path) for Li-doped a-Si : H. [After Overhof and Beyer (1980). Copyright North-Holland Publ. Co., Amsterdam, 1980.]

We return to the kinks in the Arrhenius plots of σ and S. These kinks show up in Fig. 23 as kinks of $E_F(T)$. We see that for undoped and lightly doped n-type samples a kink occurs when the statistical shift of E_F toward the conduction band reverses to a movement toward midgap. The excellent fit of calculated and experimental data in Fig. 23 is of course in part self-evident

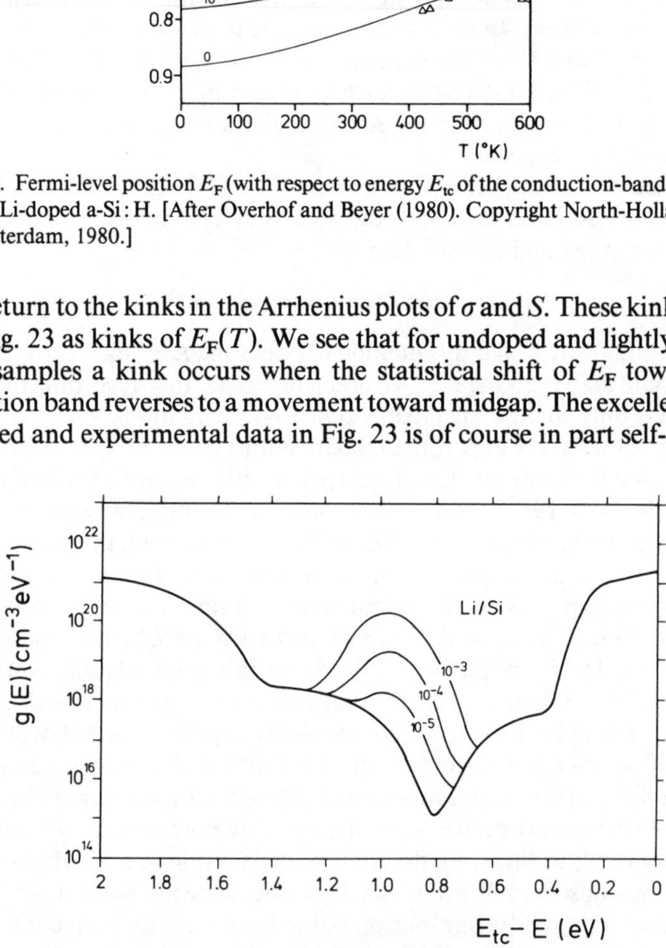

FIG. 24. Density-of-states distribution $g(E)$ determined from data of Fig. 23. [After Overhof and Beyer (1980). Copyright North-Holland Publ. Co., Amsterdam, 1980.]

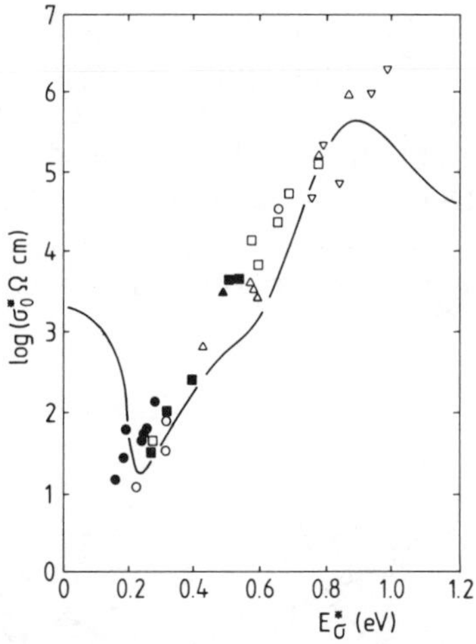

FIG. 25. Calculated curve of apparent conductivity prefactor σ_0^* versus conductivity activation energy E_σ^* for extrapolation from the 300°K < T < 400°K range. Experimental data for comparison—●, ○: PH_3; ■, □: AsH_3; ▲, △: $Sb(CH_3)_3$; ▽, $Bi(CH_3)_3$. Closed symbols, 1% doping; open symbols, 0.1% doping. [Experimental data from Carlson and Wronski (1979); figure from Overhof and Beyer (1983).]

since $g(E)$ was designed to give the experimental $E_F(T)$ curve. There are, however, very tight limits to the shifts that can be obtained by arbitrary density-of-states distributions. If we evaluate the Fermi-level shift on the basis of Döhler's soft mobility edge (Section 6c) we obtain curves for $E_F(T)$ that cannot be fitted by any reasonable choice of $g(E)$. The same holds for the VRH model of Grünewald and Thomas (Section 6d).

For highly doped samples, the experimental data of $E_F(T)$ are not satisfactorily reproduced by the calculated curves. Further assumptions have to be made to explain the leveling off of $E_F(T)$ at high temperatures (Fig. 23). Formally, this structure in $E_F(T)$ is in agreement with an increasing doping effect of the incorporated donors toward higher temperatures. We have proposed earlier that it can be understood by the assumption that a density-of-states peak near midgap is introduced by doping (Beyer *et al.,* 1977a). Indeed, much experimental evidence has accumulated recently for additional midgap states created by phosphorus doping [see, e.g., Jackson and Amer (1982), Lang *et al.* (1982), Wronski *et al.* (1982), and Tanaka and Yamasaki (1982)]. In particular, the doping-induced density-of-states distribution found by the latter from optical experiments is similar to what we have proposed for lithium doping (Overhof and Beyer, 1980). Figure 24 shows by thin lines how the density-of-states distribution has to be altered for highly Li-doped samples in order to fit the structure in $E_F(T)$.

We stress the point that a structure in $E_F(T)$ leading to a decreasing Fermi-level shift at higher temperatures is present in practically all highly doped a-Si : H samples. This is demonstrated by Fig. 26a showing $E_{tc} - E_F$ as a function of temperature for a-Si : H films doped with a variety of impurities. Corresponding $Q(1/T)$ curves displayed in Fig. 26b show the usual structureless pattern. So far we know only one exception in which the kink in $E_t - E_F$ is not observed: This is phosphorus-doped a-Si prepared by low

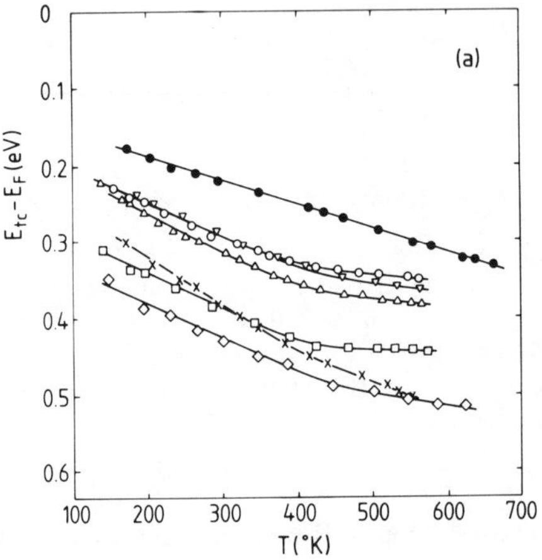

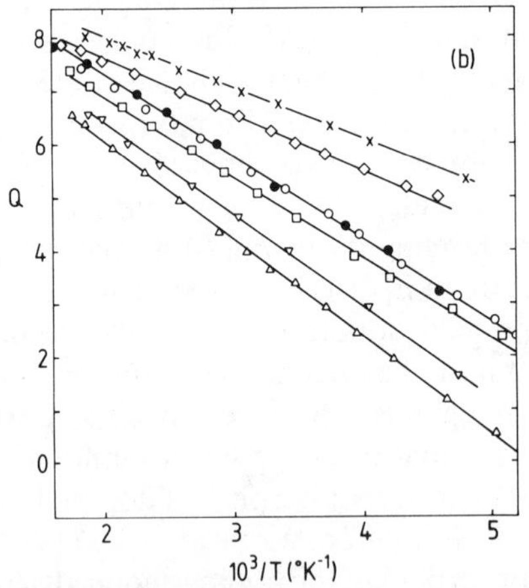

FIG. 26. (a) Fermi energy as a function of temperature and (b) Q as a function of inverse temperature for a-Si:H films doped with various n-type impurities. LPCVD a-Si shown for comparison. GD samples: ○, 1% PH_3; ×, 0.001% PH_3; ◇, 0.05% AsH_3; □, 1% Li implanted; △, 1% Na implanted; ▽, 1% K implanted; LPCVD sample: ●, 0.3% PH_3. [After Beyer and Overhof (1983).]

pressure chemical vapor deposition (LPCVD) (Beyer and Overhof, 1983). In this case, a straight line is obtained for $E_{tc} - E_F(T)$ as plotted in Fig. 26a. It is interesting to note that the doping effect in such films is considerably higher than in glow-discharge a-Si:H samples. Room temperature conductivity values as high as $\sim 1\ \Omega^{-1}\ cm^{-1}$ have been obtained (Sol *et al.*, 1980).

17. The Doping Effect

The doping effect in a-Si:H, i.e., the Fermi-level shift in dopant incorporation is at low doping levels no doubt strongly affected by the density of gap states $g(E)$ of the undoped material. Details of $g(E)$ apparently depend sensitively on the preparation conditions (Madan *et al.*, 1976; Lang *et al.*, 1982), and this effect can explain the wide variation of the room-temperature conductivity data of undoped and lightly doped films (Section 11). At high doping levels, on the other hand, there is a good agreement between different laboratories, suggesting that $g(E)$ is of minor importance only. Still we see that the doping effect in a-Si:H drops rapidly at high dopant concentrations (Section 12) and in the case of phosphorus and arsenic incorporation from the gas phase, even a decrease of σ_{RT} with rising doping level is observed. Clearly, the formation of a dopant-related density-of-states peak near midgap could explain this decay of the doping effect. As has been discussed in Section 16, the assumption of such a peak can also explain the temperature dependence of the conductivity and thermopower of a-Si:H at high doping levels.

The presence of a doping-related density-of-states peak in a-Si:H appears to be well established from optical measurements, at least for phosphorus doping. Yet, its nature is far from clear. Street *et al.* (1981) proposed that defect production is an inherent consequence of the shift of the Fermi energy. In an autocompensation process defects or defect–impurity complexes are created because they tend to move the Fermi energy toward midgap, thus reducing the total electronic energy. Yet, we note that a decrease of the doping effect is also observed for compensated films in which the Fermi energy lies close to midgap. The ESR results of Cohen *et al.* (1982) suggest that we are dealing with silicon dangling bonds. In this case, however, the question arises as to why these dangling bonds are not saturated by hydrogen, which is present in such films at a concentration of 10–15 at. %. A possible explanation could be that nondoping impurities cause strain in the silicon lattice so that silicon bonds are broken and voids are formed. In the vicinity of the silicon surface the Si–H bonding is likely to be rather unstable due to H_2 formation. Thus, the presence of a high concentration of silicon dangling bonds can be explained easily even for a material of high hydrogen content. In this model, the decay of the doping effect is directly related to the dopant concentration but not to the position of the Fermi level. Also the hydrogen content appears to play an important role in void formation. As mentioned previously, phosphorus-doped LPCVD a-Si films containing little hydrogen do not show any indication for a midgap defect peak according to our transport measurements, and the phosphorus doping effect is also considerably higher than for glow-discharge a-Si : H. It is conceivable that hydrogen, by breaking Si – Si bonds, eases the relaxation of the silicon lattice.

VI. Future Work

As we have seen, doping in a-Si : H is accompanied by defect production. Indeed, our transport data suggest that the creation of defects is a rather general feature of dopants in glow-discharge a-Si : H. Since the principal intention of doping is to obtain control of the electronic properties exclusively by shifting the Fermi energy, future work must be devoted to the elimination of this unwanted side effect. The data for low hydrogen LPCVD a-Si indicate that this may in fact be possible by more sophisticated film preparation and doping methods. It would be of considerable interest to study the energetic position of impurity levels in a-Si : H, in particular of impurities in a nondoping configuration. Also the interaction of dopants with hydrogen and other dangling-bond passivators should be investigated to a greater extent, as well as the interaction of impurities with dangling bonds. No doubt the understanding of the doping effect in a-Si : H would be extremely helpful for obtaining a high quality electronic material for exciting applications.

References

Allan, D., LeComber, P. G., and Spear, W. E. (1977). *In* "Amorphous and Liquid Semiconductors" (W. E. Spear, ed.), p. 323. CICL, Edinburgh.
Anderson, D. A., and Paul, W. (1980). *J. Phys. Soc. Jpn.* **49,** *Suppl. A,* 1197.
Anderson, D. A., and Paul, W. (1981). *Philos. Mag. B* **44,** 187.
Anderson, D. A., and Paul, W. (1982). *Philos. Mag. B* **45,** 1.
Anderson, D. A., and Spear, W. E. (1977). *Philos. Mag.* **36,** 695.
Ast, D. G., and Brodsky, M. H. (1980). *Philos. Mag. B* **41,** 273.
Baixeras, J., Mencaraglia, D., and Andro, P. (1978). *Philos. Mag. B* **37,** 403.
Bauer, G. H., and Bilger, G. (1982). *Proc. EC Photovoltaic Solar Energy Conf., 4th* (W. H. Bloss and G. Grassi, eds.), p. 773. Reidel, Dordrecht.
Beyer, W. (1979). *Solid State Commun.* **29,** 291.
Beyer, W., and Fischer, R. (1977). *Appl. Phys. Lett.* **31,** 850.
Beyer, W., and Hoheisel, B. (1983). *Solid State Commun.* **47,** 573.
Beyer, W., and Mell, H. (1977). *In* "Amorphous and Liquid Semiconductors" (W. E. Spear, ed.), p. 333. CICL, Edinburgh.
Beyer, W., and Mell, H. (1981). *Solid State Commun.* **38,** 891.
Beyer, W., and Overhof, H. (1979). *Solid State Commun.* **31,** 1.
Beyer, W., and Overhof, H. (1983). *J. Non-Cryst. Solids* **59-60,** 301.

Beyer, W., and Stuke, J. (1972). *J. Non-Cryst. Solids* **8-10,** 321.
Beyer, W., and Stuke, J. (1975). *Phys. Status Solidi A* **30,** K 155.
Beyer, W., and Wagner, H. (1981). *J. Phys. Colloq. Orsay, Fr.* **42,** C4-783.
Beyer, W., and Wagner, H. (1982). *J. Appl. Phys.* **53,** 8745.
Beyer, W., Stuke, J., and Wagner, H. (1975). *Phys. Status Solidi A* **30,** 231.
Beyer, W., Mell, H., and Overhof, H. (1977a). *In* "Amorphous and Liquid Semiconductors" (W. E. Spear, ed.), p. 328. CICL, Edinburgh.
Beyer, W., Medeisis, A., and Mell, H. (1977b). *Commun. Phys.* **2,** 121.
Beyer, W., Barna, A., and Wagner, H. (1979a). *Appl. Phys. Lett.* **35,** 539.
Beyer, W., Fischer, R., and Overhof, H. (1979b). *Philos. Mag. B* **39,** 205.
Beyer, W., Fischer, R., and Wagner, H. (1979c). *J. Electron. Mater.* **8,** 127.
Beyer, W., Stritzker, B., and Wagner, H. (1980). *J. Non-Cryst. Solids* **35-36,** 321.
Beyer, W., Mell, H., and Overhof, H. (1981a). *J. Phys. Colloq. Orsay, Fr.* **42,** C4-103.
Beyer, W., Wagner, H., and Mell, H. (1981b). *Solid State Commun.* **39,** 375.
Biegelsen, D. K., Street, R. A., Tsai, C. C., and Knights, J. C. (1979). *Phys. Rev. B* **20,** 4839.
Butcher, P. N., and Friedman, L. (1977). *J. Phys. C* **10,** 3803.
Carlson, D. E. (1980). *J. Non-Cryst. Solids* **35-36,** 707.
Carlson, D. E., and Wronski, C. R. (1979). *In* "Amorphous Semiconductors" (M. H. Brodsky, ed.), Chap. 10. Springer-Verlag, Berlin and New York.
Chen, Kun-ji, and Fritzsche, H. (1982). *Sol. Energy Mater.* **8,** 205.
Chevallier, J., and Beyer, W. (1981). *Solid State Commun.* **40,** 771.
Chittick, R. C., Alexander, J. H., and Sterling, H. F. (1969). *J. Electrochem. Soc.* **116,** 77.
Cohen, J. D., Harbison, J. P., and Wecht, K. W. (1982). *Phys. Rev. Lett.* **48,** 109.
Cohen, M. H. (1970). *J. Non-Cryst. Solids* **4,** 391.
Cohen, M. H., Fritzsche, H., and Ovshinsky, S. R. (1969). *Phys. Rev. Lett.* **22,** 1065.
Connell, G. A. N. (1972). *Phys. Status Solidi B* **53,** 213.
Davis, E. A., and Mott, N. F. (1970). *Philos. Mag.* **22,** 903.
Döhler, G. (1979). *Phys. Rev. B* **19,** 2083.
Emin, D. (1977a). *Philos. Mag.* **35,** 1189.
Emin, D. (1977b). *Solid State Commun.* **22,** 409.
Friedman, L. (1971). *J. Non-Cryst. Solids* **6,** 329.
Fritzsche, H. (1971a). *J. Non-Cryst. Solids* **6,** 49.
Fritzsche, H. (1971b). *Solid State Commun.* **9,** 1813.
Fritzsche, H. (1980). *Sol. Energy Mater.* **3,** 447.
Goodman, N. B. (1982). *Philos. Mag. B* **45,** 407.
Griffith, R. W., Kampas, F. J., Vanier, P. E., and Hirsch, M. D. (1980). *J. Non-Cryst. Solids* **35-36,** 391.
Grigorovici, R. (1968). *Mater. Res. Bull.* **3,** 13.
Grigorovici, R. Croitoru, N., and Dévényi, A. (1967). *Phys. Status Solidi* **23,** 621.
Grünewald, M., and Thomas, P. (1979). *Phys. Status Solidi B* **94,** 125.
Grünewald, M., Weber, K., Fuhs, W., and Thomas, P. (1981). *J. Phys. Colloq. Orsay, Fr.* **42,** C4-523.
Gruntz, K. J. (1981). Thesis, Univ. of Stuttgart, Stuttgart.
Hauschildt, D., Stutzmann, M., Stuke, J., and Dersch, H. (1982). *Solar Energy Mater.* **8,** 319.
Hezel, R., and Schörner, R. (1981). *J. Appl. Phys.* **52,** 3076.
Hindley, N. K. (1970). *J. Non-Cryst. Solids* **5,** 17.
Hirose, M. (1981). *J. Phys. Colloq. Orsay, Fr.* **42,** C4-705.
Holzenkämpfer, E., Stuke, J., and Fischer, R. (1982). *Proc. EC Photovoltaic Solar Energy Conf., 4th* (W. H. Bloss and G. Grassi, eds.), p. 778. Reidel, Dordrecht.
Jackson, W. B., and Amer, N. M. (1982). *Phys. Rev. B* **25,** 5559.

Jan, Z. S., Bube, R. H., and Knights, J. C. (1979). *J. Electron. Mater.* **8,** 47.
Jan, Z. S., Bube, R. H., and Knights, J. C. (1980). *J. Appl. Phys.* **51,** 3278.
Jones, D. I., LeComber, P. G., and Spear, W. E. (1977). *Philos. Mag.* **36,** 541.
Jones, D. I., Gibson, R. A., LeComber, P. G., and Spear, W. E. (1979). *Solar Energy Mater.* **2,** 93.
Kalbitzer, S., Müller, G., LeComber, P. G., and Spear, W. E. (1980). *Philos. Mag. B* **41,** 439.
Knights, J. C., Hayes, T. M., and Mikkelsen, J. C. (1977). *Phys. Rev. Lett.* **39,** 712.
Kolomiets, B. T. (1964). *Phys. Status Solidi* **7,** 713.
Kuwano, Y., and Ohnishi, M. (1981). *J. Phys. Colloq. Orsay, Fr.* **42,** C4-1155.
Lang, D. V., Cohen, J. D., and Harbison, J. P. (1982). *Phys. Rev. B* **25,** 5285.
LeComber, P. G., and Spear, W. E. (1970). *Phys. Rev. Lett.* **25,** 509.
LeComber, P. G., Jones, D. I., and Spear, W. E. (1977). *Philos. Mag.* **35,** 1173.
Ley, L. (1984). *In* "The Physics of Amorphous Silicon and its Applications" (J. D. Joannopoulos and G. Lucovsky, eds.). Springer-Verlag, Berlin and New York. To be published.
Ley, L., Reichardt, J., and Johnson, R. L. (1982). *Phys. Rev. Lett.* **49,** 1664.
Mackenzie, K. D., LeComber, P. G., and Spear, W. E. (1982). *Philos. Mag. B* **46,** 377.
Madan, A., LeComber, P. G., and Spear, W. E. (1976). *J. Non-Cryst. Solids* **20,** 239.
Madelung, O. (1957). *In* "Handbuch der Physik" (S. Flügge, ed.) Vol. 20, p. 1. Springer-Verlag, Berlin and New York.
Matsumura, H., Nakagome, Y., and Furukawa, S. (1980). *Appl. Phys. Lett.* **36,** 439.
Mayer, J. W., Eriksson, L., and Davies, J. A. (1970). "Ion Implantation in Semiconductors." Academic Press, New York.
Meyer, W., and Neldel, H. (1937). *Z. Tech. Phys.* **18,** 588.
Miller, A., and Abrahams, E. (1960). *Phys. Rev.* **120,** 745.
Moore, A. R. (1977). *Appl. Phys. Lett.* **31,** 762.
Mott, N. F. (1967). *Adv. Phys.* **16,** 49.
Mott, N. F. (1969). *Philos. Mag.* **19,** 835.
Mott, N. F. (1972). *Philos. Mag.* **26,** 1015.
Mott, N. F. (1981). *Philos. Mag. B* **44,** 265.
Mott, N. F., and Kaveh, M. (1983a). *Philos. Mag. B* **47,** L9.
Mott, N. F., and Kaveh, M. (1983b). *Philos. Mag. B* **47,** L17.
Mott, N. F., Davis, E. A., and Street, R. A. (1975). *Philos. Mag.* **32,** 961.
Movaghar, B. (1981). *J. Phys. Colloq. Orsay, Fr.* **42,** C4-73.
Movaghar, B., and Schweitzer, L. (1978). *J. Phys. C* **11,** 125.
Müller, G., LeComber, P. G. (1981). *Philos. Mag. B* **43,** 419.
Müller, G., Kalbitzer, S., Spear, W. E., and LeComber, P. G. (1977). *In* "Amorphous and Liquid Semiconductors" (W. E. Spear, ed.), p. 442. CICL, Edinburgh.
Müller, G., Demond, F., Kalbitzer, S., Damjantschitsch, H., Mannsperger, H., Spear, W. E., LeComber, P. G., and Gibson, R. A. (1980). *Philos. Mag. B* **41,** 571.
Nagels, P. (1979). *In* "Amorphous Semiconductors" (M. H. Brodsky, ed.), Chap. 5. Springer-Verlag, Berlin and New York.
Nguyen Van Dong, Fournier, Y., and Le Ny, J. Y. (1981). *J. Phys. Colloq. Orsay, Fr.* **42,** C4-647.
Overhof, H. (1975). *Phys. Status Solidi B* **67,** 709.
Overhof, H. (1981). *Philos. Mag. B* **44,** 312.
Overhof, H., and Beyer, W. (1980). *J. Non-Cryst. Solids* **35-36,** 375.
Overhof, H., and Beyer, W. (1981a). *Philos. Mag. B* **43,** 433.
Overhof, H., and Beyer, W. (1981b). *Phys. Status Solidi B* **107,** 207.
Overhof, H., and Beyer, W. (1983). *Philos. Mag. B* **47,** 377.
Overhof, H., and Beyer, W. (1984). *Philos. Mag. B,* **49,** L9.

Pietruszko, S. M., Narasimhan, K. L., and Guha, S. (1981). *Philos. Mag. B* **43,** 357.
Rehm, W., Fischer, R., Stuke, J., and Wagner, H. (1977). *Phys. Status Solidi B* **79,** 539.
Roilos, M. (1978). *Philos. Mag. B* **38,** 477.
Sol, N., Kaplan, D., Dieumegard, D., and Dubreuil, D. (1980). *J. Non-Cryst. Solids* **35-36,** 291.
Spear, W. E., and LeComber, P. G. (1975). *Solid State Commun.* **17,** 1193.
Spear, W. E., and LeComber, P. G. (1976). *Philos. Mag.* **33,** 935.
Spear, W. E., LeComber, P. G., Kalbitzer, S., and Müller, G. (1979). *Philos. Mag. B* **39,** 159.
Staebler, D. L., and Wronski, C. R. (1977). *Appl. Phys. Lett.* **31,** 292.
Staebler, D. L., and Wronski, C. R. (1980). *J. Appl. Phys.* **51,** 3262.
Sterling, H. F., and Swann, R. C. G. (1965). *Solid-State Electron.* **8,** 653.
Street, R. A. (1982). *Phys. Rev. Lett.* **49,** 1187.
Street, R. A., Biegelsen, D. K., and Knights, J. C. (1981). *Phys. Rev. B* **24,** 969.
Stuke, J. (1977). *In* "Amorphous and Liquid Semiconductors" (W. E. Spear, ed.), p. 406. CICL, Edinburgh.
Tanaka, K., and Yamasaki, S. (1982). *Solar. Energy Mater.* **8,** 277.
Tanielian, M., Fritzsche, H., Tsai, C. C., and Symbalisty, E. (1978). *Appl. Phys. Lett.* **33,** 353.
Tawada, Y., Kondo, M., Okamoto, H., and Hamakawa, Y. (1981). *J. Phys. Colloq. Orsay, Fr.* **42,** C4-471.
Tsai, C. C., Fritzsche, H. Tanielian, M. H., Gaczi, P. J., Persans, P. D., and Vasaghi, M. A. (1977). *In* "Amorphous and Liquid Semiconductors" (W. E. Spear, ed.), p. 339. CICL, Edinburgh.
von Roedern, B., Ley, L., and Cardona, M. (1979). *Solid State Commun.* **29,** 415.
Weller, D., Mell, H., Schweitzer, L., and Stuke, J. (1981). *J. Phys. Colloq. Orsay, Fr.* **42,** C4-143.
Williams, P. (1979). *Surf. Sci.* **90,** 588.
Williams, R. H., Varma, R. R., Spear, W. E., and LeComber, P. G. (1979). *J. Phys. C* **12,** 2209.
Wronski, C. R., Carlson, D. E., and Daniel, R. E. (1976). *Appl. Phys. Lett.* **29,** 602.
Wronski, C. R., Abeles, B., Tiedje, T., and Cody, G. D. (1982). *Solid State Commun.* **44,** 1423.
Xu, W. Y., Sun, Z. L., Wang, Z. P., and Lee, D. L. (1981). *J. Phys. Colloq. Orsay, Fr.* **42,** C4-695.
Yang, S. H., and Lee, C. (1983). *Solid State Commun.* **45,** 591.
Zesch, J. C., Lujan, R. A., and Deline, V. R. (1980). *J. Non-Cryst. Solids* **35-36,** 273.
Zvyagin, I. P., (1973). *Phys. Status Solidi B* **58,** 443.

CHAPTER 9

Electronic Properties of Surfaces in a-Si : H

H. Fritzsche

DEPARTMENT OF PHYSICS AND JAMES FRANCK INSTITUTE
UNIVERSITY OF CHICAGO
CHICAGO, ILLINOIS

I. Introduction

Semiconductor research can traditionally be divided into two categories: one involves studies of the bulk properties; the other deals with surface and interface phenomena. The effects of the latter extend to a depth of the diffusion length or the width of the space-charge region. Such a convenient division of attention and labor is not possible when one studies thin films of a high-quality (low-defect-density) semiconductor such as a-Si : H. The high semiconductor quality of a-Si : H was actually first demonstrated (Spear and LeComber, 1972) by showing that the conductance of the films can be strongly modulated by raising or lowering its surface or interface potential by

ISBN 0-12-752149-6

a transverse gate voltage in a field-effect transistor geometry. These and later measurements of the density of gap states showed that the space-charge region in accumulation or depletion can be 0.5 – 1 μm wide. The electronic bulk properties of a thin film semiconductor can therefore be obtained only when there is no band bending, or in other words, when the surface (or interface) potential, which produces space-charge layers, is zero. Such a unique condition is difficult to meet because the semiconductor film has to come to electronic equilibrium not only with its surface and interface states, but under normal circumstances also with its substrate, its surface oxide layers, and various chemisorbed species. All these effects produce nonzero potentials, which moreover change with temperature and light exposure and which affect and often dominate the electronic properties of a-Si:H films.

In view of their importance, it is indeed surprising that surface effects are too often neglected in interpreting the electronic transport properties of a-Si:H films. The reasons for this omission are perhaps the present difficulty of characterizing the surface and of measuring the surface and interface potentials and the fact that our understanding of the bulk properties is sufficiently vague that the omission is not immediately obvious.

In this chapter we shall try to summarize the effects of various surface conditions on the transport properties and the experiments that are needed to separate the surface and bulk properties. The experimental results and quantitative values of surface parameters quoted here will change as our experimental tools for characterizing amorphous surfaces improve. Nevertheless, a good number of results are presented here to illustrate the influence of surface effects on a large variety of transport phenomena.

The physics of the surface effects as well as the measuring techniques needed to explore them are essentially the same for crystalline and amorphous silicon. In explaining the basic principles we shall therefore borrow heavily from the extensive work on crystalline semiconductors. One major difference between a-Si:H and crystalline semiconductors needs special emphasis, however: a-Si:H films are often not homogeneous (Fritzsche, 1982). In the context of understanding surface and interface effects, it is most disturbing that the first as well as the last few hundred atomic layers near the substrate and the surface may have a composition and structure that are quite different from that of the bulk. During the first minutes of film growth various impurities desorbed or sputtered from the walls of the deposition chamber may become incorporated in the growing film. The top layers, on the other hand, may be enriched or partially depleted of hydrogen, depending on out-diffusion and the hydrogen equilibration process. Such vertical heterogeneities were revealed by various depth-probing analytical techniques and have been pointed out in the literature (Carlson *et al.*, 1979; Magee and Carlson, 1980; Magee and Botnick, 1981; Brodsky *et al.*, 1980;

Mailhiot *et al.*, 1980; Müller *et al.*, 1980; Demond *et al.*, 1982). We shall largely neglect these inhomogeneities and concentrate on the effect of surfaces and interfaces on *homogeneous* films of a-Si:H, because an understanding of the homogeneous case is needed before heterogeneities can be included in the analysis. We should keep in mind nevertheless that the experimental results discussed may have been performed on films having vertical composition gradients.

II. The Surface and the Oxide Interface

The surface chemistries of a-Si and c-Si are very similar. A major difference arises from the presence of hydrogen in a-Si:H, which saturates most dangling bonds, reduces the surface reactivity, and affects the oxidation process. The experimental determination of the surface-state density has been much more successful with c-Si than with a-Si:H, however, because of the much lower density of gap states in c-Si. We shall compare results on crystalline and amorphous Si surfaces in the following.

1. Surface States

A cleaved c-Si (111) surface has about $g_s = 8 \times 10^{14}$ eV^{-1} cm^{-2} surface states within a 1-eV-wide filled band, which pins E_F about 0.35 eV above E_v (Wagner and Spicer, 1974). These dangling bonds react readily with oxygen, and a surface covered with one or a few monolayers of oxide has only $N_s \simeq 10^{12}$ dangling bonds cm^{-2}. This incomplete saturation may be related to interfacial strain (Laughlin *et al.*, 1980). The number can be further reduced by a factor of 10 or more by annealing in a hydrogen atmosphere. An example of the interface-state density distribution of a thermally oxidized c-Si (111) surface before annealing, taken from the work of Johnson *et al.* (1983), is shown in Fig. 1. They found that the peaks at $E_v + 0.3$ eV and $E_c - 0.25$ eV are correlated: both peaks as well as the density of spins follow the same annealing kinetics and disappear with annealing at 700°K. The observed *g*-tensor components identify these spins as Si dangling bonds. Their density was found to be $N_s \sim 1.5 \times 10^{12}$ cm^{-2} before annealing. These observations suggest that the lower and upper peaks correspond to the one and two electron states of the silicon dangling-bond defect with a correlation energy of 0.6 eV.

We now turn to a-Si:H. Street and Knights (1981) studied a-Si:H with columnar growth morphology. The large internal surface area of these films facilitates the reaction studies of the surface, and we can assume that the same processes occur on the top surface of films that do not have these internal surfaces. The freshly deposited films had a very small spin density because of the efficient bond saturation with hydrogen. During exposure to

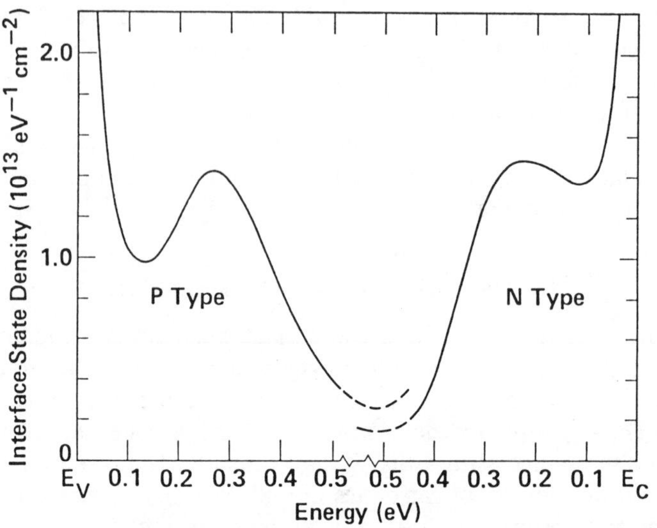

FIG. 1. Interface-state density of c-Si measured by deep-level transient current spectroscopy on thermally oxidized, unannealed, (111)-oriented silicon. [From Johnson *et al.* (1983).]

air for 70 days the internal surfaces oxidized and the spin density increased gradually by a factor of five. Subsequent annealing to 425°K for 1 hr in air lowered the spin density to its original low value, but the excess spins surprisingly reappeared within several hours at 300°K after annealing. By using an estimate of the internal surface area, Street and Knights obtained a maximum surface spin density of $N_s \sim 3 \times 10^{11}$ cm^{-2}, a value that is somewhat smaller than that of oxidized c-Si surfaces presumably because of the presence of hydrogen, which reduces the number of unsaturated bonds.

The annealing and the time dependence of the surface states will be of great importance for understanding long-term variations in the sheet conductance of a-Si : H films after annealing and after applying various surface treatments (Tanielian *et al.,* 1978, 1980; Tanielian, 1982). It seems that during the oxidation process a fraction of Si atoms is bonded neither to oxygen nor to hydrogen. Annealing heals these dangling bonds by local reconstruction (Street and Knights, 1981), but they reappear at lower temperatures, perhaps due to increased strain.

The interface-state density of a-Si : H measured by Suzuki *et al.* (1982) is shown in Fig. 2. The peaks observed in Fig. 1 have not yet been found in a-Si : H, and the correlation of the states in the lower and upper half of the gap has still to be investigated, but the similarity of the shapes of $g_s(E)$ and the *g*-value of the surface spins suggest that the origin of these states is the same for crystalline and amorphous Si.

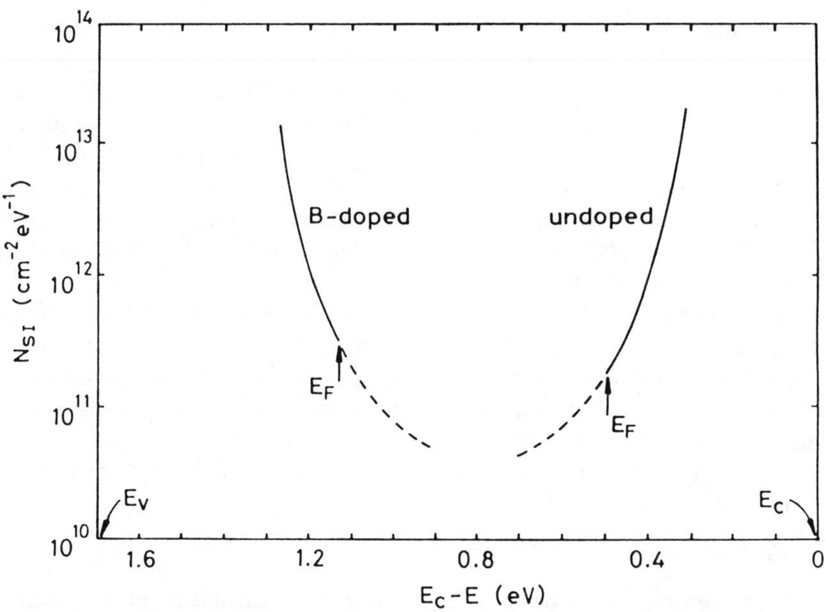

FIG. 2. Interface-state density of a-Si:H deposited on thermally oxidized SiO_2 at 300°C. Densities were measured by Suzuki *et al.* (1982) by combining the field-effect and capacitance–voltage methods. [From Suzuki *et al.* (1982).]

The magnitude of the surface state density depends on the conditions of the oxide growth and on the thermal history. Doping and the position of E_F may also have an effect by changing the equilibrium energy and therefore the concentration of surface defects, as indicated by the arguments presented by Street (1982) for bulk defects. In addition, surface impurities such as carbon before and during oxidation are likely to increase the density of surface states. Values as large as $g_s = 0.5-2 \times 10^{13}$ eV^{-1} cm^{-2} near the gap center were reported by Viktorovitch (1981) for sputtered a-Si:H.

The saturation of Si bonds by hydrogen and oxygen will, of course, be different at the substrate interface and at the interface between a-Si:H and its native oxide. Weisfield *et al.* (1981) found, for instance, that an excessive number of substrate interface states, which prevented measurements of the field effect in sputtered a-Si:H, could be eliminated by depositing first a 200-Å-thick SiO_x layer and then the a-Si:H film without extinguishing the sputtering plasma.

If the surface state density $g_s(E)$ were appreciably larger than the product $g(E)\lambda$ of the bulk density and the screening width, then E_F at the surface would lag behind when E_F in the bulk is shifted through the gap by doping. The bulk Fermi energy can be deduced with an acceptable error margin from the conductivity activation energy E_a in the case of transport in extended states. The surface value can be measured with respect to the energy of the Si 2p level by photoemission, which only probes a 10–20-Å-thick layer near the surface, which is determined by the small escape depth of the photoemitted electrons. Williams *et al.* (1979) found in this manner that E_F at the oxidized surface and in the bulk follow each other over a 1.2-eV shift of E_F with doping. Von Roedern *et al.* (1979) observed the same effect on oxide-free a-Si : H surfaces. These results suggest surface-state densities not much larger than those shown in Fig. 2.

Somewhat larger surface defect densities of $N_s \simeq 10^{12}\ cm^{-2}$ were found by Knights *et al.* (1977) and by Jackson *et al.* (1983a), who measured the thickness dependence of the spin density and of the sub-band-gap absorption ($h\nu < 1.5$ eV). It is not clear whether the defects measured in these experiments are near the substrate or at the free surface. The measured value is probably the sum of both contributions, so that we are back at defect concentrations near $5 \times 10^{11}\ cm^{-2}$.

2. Porosity

A pronounced columnar or pebblelike microstructure of dimensions varying between 100 and 2000 Å is observed in films prepared at high plasma power (Knights 1979, 1980; Knights and Lujan, 1979) when inert gases of high atomic number are present in the plasma (Knights *et al.*, 1981) and in boron-doped samples at certain plasma conditions (Schiff *et al.*, 1981). These films are 20–40% less dense than c-Si and are naturally quite porous; they show progressive oxidation of their internal surfaces (Street and Knights, 1981). The preparation conditions of high-quality films are usually chosen to avoid such growth morphologies because of their gross heterogeneities. Normal a-Si : H films are not porous. Water and other gases are adsorbed on the outer surface only as shown by measuring the weight increase due to adsorption by means of a quartz-crystal microbalance (Fritzsche and Tsai, 1979). The observation that water adsorption may decrease the conductance of an oxidized film and increase it after etching is further proof that these changes are not associated with water penetrating the film (Aker and Fritzsche, 1983). Nevertheless, porosity is often observed in films prepared at high rates and at temperatures below 260°C. Porosity tests should be carried out on films prepared by new techniques and on those containing more than 10 at. % hydrogen or high doping concentrations.

3. Oxidation Rate

The a-Si : H surface does not react as readily with oxygen as a c-Si surface because of the bond saturation with hydrogen. Freshly grown a-Si : H films contain 40–50 at. % hydrogen in the first two or three atomic layers, which is three or four times the amount in the bulk (Ley, 1984). This is due to the growth process from the silicon subhydrides of the plasma. Molecular oxygen does not adsorb on a-Si:H for exposures of up to 10^6 langmuir ($1\ L = 10^{-6}$ Torr sec) in contrast to c-Si, whose surface is saturated with a monolayer of O_2 after 5×10^4 L (Ley *et al.*, 1981; Kasupke and Henzler, 1980). The lowered reactivity of the a-Si:H surface may allow transfer of samples between chambers without oxidation (Gimzewski and Vepřek, 1983). In the presence of atomic oxygen on the other hand, oxidation of a-Si:H proceeds immediately because sufficient energy is available to replace the bonded hydrogen with oxygen.

Figure 3 compares the oxidation rate of glow-discharge a-Si:H at 300°K in air as measured by Ponpon and Bourdon (1982) with that of c-Si (Lukes, 1972). The oxide thicknesses were measured by ellipsometry and assuming constant values of the optical properties of the oxide. Whereas on crystalline Si the oxide growth follows a logarithmic dependence on time, on a-Si:H it follows quite closely a parabolic time dependence.

$$d_{\text{oxide}} = (Dt)^{1/2}, \tag{1}$$

which suggests that diffusion is the rate-limiting process and not the reaction kinetics at the interface as for crystalline Si at room temperature. The diffusion coefficient has the value $D = 2 \times 10^{-20}\ \text{cm}^2\ \text{sec}^{-1}$. These measurements were carried out on undoped a-Si:H. It will be interesting to see whether

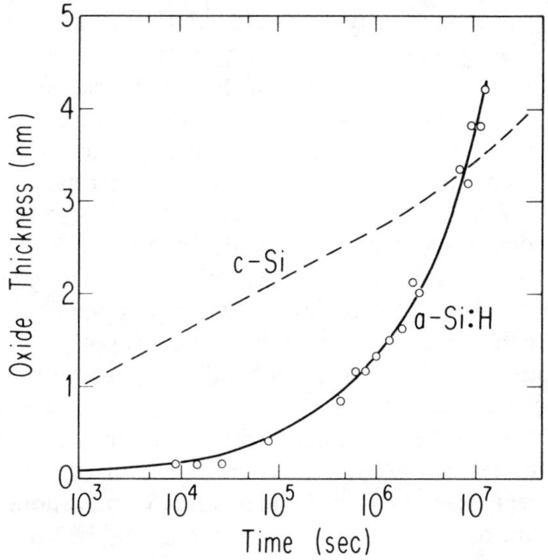

FIG. 3. Room-temperature oxidation of a-Si:H [after Ponpon and Bourdon (1982)] and of c-Si [after Lukes (1972)].

doping affects the rate. If it is indeed diffusion limited, the rate should be quite independent of doping.

The measurements of Ponpon and Bourdon (1982) give an estimate for the oxide thicknesses on films exposed to room ambients. It should be kept in mind, however, that the stoichiometry and structure of such oxides are quite uncertain and less reproducible than for oxide layers grown under controlled conditions at higher temperatures (Goldstein *et al.*, 1982).

III. Theory of Space-Charge Layers

We review briefly the expressions and concepts needed to describe the electronic phenomena of surfaces and space charges. We refer the reader to the extensive literature on crystalline semiconductors (Many *et al.*, 1965; Sze, 1969; Many, 1974; Goetzberger *et al.*, 1976). It will be shown that the continuous distribution of gap states in amorphous semiconductors requires some changes in the expressions for the conduction and potential distribution in the space-charge region and that the quasi-two-dimensional conduction in a narrow potential well of a space-charge layer introduces some new and as yet unresolved problems.

4. Definition of Terms

Figure 4 shows in two diagrams the basic ideas with which we are concerned. Figure 4a represents a semiconductor whose insulating overlayer is negatively charged, giving rise to an upward band bending, i.e., to an electron depletion or hole accumulation space-charge region. The negative charge is in localized interface states or further away from the surface in the insulating overlayer (the surface oxide or another insulator). The distance from the surface determines the rate at which these states equilibrate with the semiconductor at a given temperature. It is convenient to distinguish between fast and slow states. Moreover, the energy separation of these states from the Fermi level E_F determines whether they are fixed charge states or whether their occupancy depends on and changes with E_F, the surface potential V_s, and illumination during the experiment. One can imagine a position of E_F that would make the surface neutral. If this neutrality potential of the surface layer differs from E_F in the bulk, band bending occurs and a space-charge layer forms in the semiconductor.

The space charge in the amorphous semiconductor, which is positive in this example, resides nearly exclusively in the localized gap states whose occupancy has been changed by the band bending $V(x)$. Although the change in the concentrations of mobile carriers above E_c and below E_v govern the observed change in the conductance, their contribution to the space charge is usually negligible (in contrast to crystalline semiconductors). Let χ be the

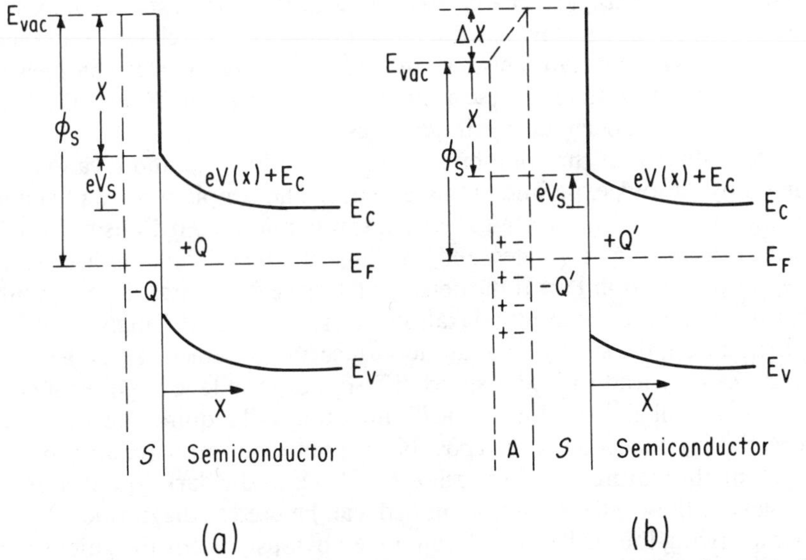

FIG. 4. (a) Band bending at semiconductor surface caused by charge $-Q$ in surface states. (b) Effect of adsorbates (A) having a dipole moment and changing the charge $+Q'$ of the space-charge layer. S = surface layer, χ = surface electron affinity, ϕ_s = work function, and E_{vac} = vacuum level. The potential $V(x)$ is negative for positive space charge; $eV(x)$ is positive.

surface electron affinity and ϕ_s the work function. If one could neutralize the surface layer, χ would remain the same and V_s would be zero.

Figure 4b shows a surface after chemisorption. Chemisorbed atoms or molecules will in general (1) introduce new surface states, (2) change the total charge of the surface $-Q'$ and of the space charge $+Q'$, and (3) add a dipole layer to the surface. The electron affinity χ at the surface and the work function ϕ_s will be reduced by

$$\Delta\chi = 4\pi M, \quad (2)$$

where M is the dipole moment per unit area pointing into the surface. The change of the work function by eV_s can alternatively be expressed in terms of the dipole layer formed by the charge in the space-charge region and the net charge of the surface layer, but the present treatment is more convenient. As we shall see later, the rate of the chemisorption (or desorption) process and of the resulting change in V_s can be slow if tunneling through an oxide layer limits the equilibration between the semiconductor and the adsorbate.

The substrate interface looks similar to Fig. 4a. The average energy to remove an electron from the semiconductor to the conduction band of the insulating substrate in this case will be represented by an internal work function ϕ_i. The trapped charge $-Q'$ resides either in fast states adjacent to the interface or in the slow states at larger distances. The relevant states may be intrinsic gap states in the substrate, impurity or defect states in the substrate, or interface states. These again may be fixed when far removed from E_F or variable in occupancy with changes in E_F and V_s.

One realizes that the equilibrium depicted in Figs. 4a and b can be disturbed by various experimental parameters. The simple picture of chemisorption may be disturbed by gases such as water or oxygen diffusing into the oxide, introducing new states, and reacting with the silicon interface. Moreover, as discussed in Part II the density of surface states may depend on the thermal history and exhibit relaxation effects (Street and Knights, 1981).

In addition, the strong fields at the surface and interface at elevated temperatures may lead to field-assisted diffusion of mobile ion species such as protons and alkali ions, and strong illumination will require a new quasistatic equilibration, accompanied possibly by photodesorption (or adsorption), which might require long-time relaxation back to the dark equilibrium.

Many of these effects, once identified, can be used as diagnostic tools for characterizing the surface condition. One obviously needs the guidance of experiments to sort out the various chemical and electronic processes that yield a specific surface condition.

5. Space-Charge Potential

The calculation of the potential $V(x)$ and the conductance G in the space-charge region is the same as that needed for understanding the field effect because it obviously does not matter at which external charges the field lines from the semiconductor space charges terminate. We assume that the film thickness is large compared to the combined widths of the space-charge layers at the surface and the substrate interface so that they are separated by an undisturbed flat-band region. Matching of the boundary conditions at both sides of the film are required to obtain solutions for $V(x)$ and the conductance $G_\square$ in thin films (Redfield 1983; Powell and Pritchard, 1983).

Since the band-bending potential $V(x)$ changes the occupancy of the gap states, whose density of states is $g(E)$, the local space-charge density ρ is directly related to $V(x)$ as

$$\rho(V) = -q \int_{-\infty}^{\infty} dE\, g(E)[f(E - qV) - f(E)]. \tag{3}$$

Here, $f(E)$ is the Fermi–Dirac distribution function. With the identity

$$\frac{d^2V}{dx^2} = \frac{1}{2}\frac{d}{dV}\left(\frac{dV}{dx}\right)^2 \tag{4}$$

and Poisson's equation

$$\frac{d^2V}{dx^2} = -\frac{\rho(V)}{\varepsilon\varepsilon_0}, \tag{5}$$

one obtains

$$\left(\frac{dV}{dx}\right)^2 = -\frac{2}{\varepsilon\varepsilon_0}\int_0^V \rho(V')\,dV', \tag{6}$$

$$\frac{dV}{dx} \equiv -F(V) = \mathrm{sgn}(V)\left|\frac{2}{\varepsilon\varepsilon_0}\int_0^V \rho(V')\,dV'\right|^{1/2}, \tag{7}$$

where we made use of the fact that $V(x) = 0$ and $dV/dx = 0$ in the interior of the film, which we assumed to have no space charge. The sign of the electrostatic potential V is negative for a positive space charge. Integration of Eq. (7) yields

$$\int_0^x dx = -\int_{V_s}^V dV'\,\frac{1}{F(V')} = D(V) - D(V_s). \tag{8}$$

Here, V_s is the potential at the surface $x = 0$ and $F(V)$ the field in the space-charge region (F is positive when the field is pointing in the x direction). For two arbitrary points in the space-charge region we obtain

$$D(V_2) - x_2 = D(V_1) - x_1 = D(V_s). \tag{9}$$

This important result implies that the potential profile for a different V_s is obtained by rigidly shifting $V(x)$ along x until the new V_s occurs at $x = 0$:

$$\Delta V_s = -F(V_s)\,\Delta x \tag{10}$$

and

$$V(x, V_s + \Delta V_s) = V(x - \Delta x, V_s). \tag{11}$$

6. Conductance

We assume that conduction takes place in the transport states near E_c and E_v, sufficiently removed from E_F that Boltzmann statistics applies (see Chapter 8 by Beyer and Overhof). The conductivity is the sum of the electron and hole contributions

$$\sigma(x) = \sigma_n \exp[qV(x)/kT] + \sigma_p \exp[-qV(x)/kT], \tag{12}$$

where

$$\sigma_n = \sigma_{0n} \exp(-E_{an}/kT) \tag{13}$$

and

$$\sigma_p = \sigma_{0p} \exp(-E_{ap}/kT) \tag{13a}$$

are the electron (n) and hole (p) conductivities in the flat-band region. Instead of the real conductance G, we use in the following the sheet conductance $G_\square = Gl/w$, where l and w are the separation and lengths of the coplanar electrodes. This conductance is then

$$G_\square(V_s) = \int_0^d \sigma(x)\,dx = G_\square(0) + \sigma_n \int_0^{V_s} dV\,[\exp(qV/kT) - 1]/F(V)$$
$$+ \sigma_p \int_0^{V_s} dV\,[\exp(-qV/kT) - 1]/F(V), \tag{14}$$

where $G_\square(0)$ is the sheet conductance for zero band bending. An increment of the surface potential ΔV_s yields

$$G_\square(V_s + \Delta V_s) = G(V_s) + [\sigma(V_s) - \sigma(0)]\,\Delta x. \tag{15}$$

Since ΔV_s and Δx are related by Eq. (10) one obtains

$$\partial G_\square/\partial V_s = -[\sigma(V_s) - \sigma(0)]/F(V_s)$$
$$= -\{\sigma_n[\exp(qV_s/kT) - 1] + \sigma_p[\exp(-qV/kT) - 1]\}/F(V_s). \tag{16}$$

Hence $G_\square$ has a minimum value $V_s = V_s^{\min}$ when

$$qV_s^{\min}/kT = \ln \sigma_p/\sigma_n. \tag{17}$$

In deriving these results (Abelson and de Rosny, 1983), we have made several assumptions which will be discussed in Section 8.

7. Approximate Solutions

We have shown that for a given density-of-states distribution $g(E)$ the space-charge potential $V(x)$ is a unique curve that terminates at the surface such that V_s, $F(V_s)$, or the total charge Q in the space-charge region satisfies the experimental situation. A general $g(E)$, determined for instance by capacitance, field-effect, or deep-level transient spectroscopy measurements, requires numerical integrations. In many cases it is instructive, however, to explore the results obtained with simpler forms of $g(E)$ that permit integration to be carried out. We shall first use an exponentially increasing density of states

$$g(E) = g(E_F) \exp[(E - E_F)/E_0] \tag{18}$$

and then a constant $g(E_F)$ that results by letting E_0 become very large (Abelson and de Rosny, 1983). We also approximate $f(E - qV)$ and $f(E)$ in Eq. (3) by step functions. The errors introduced by such $T = 0$ statistics have been analyzed by Powell (1981). Equation (3) then yields

$$\rho(V) = -qE_0 g(E_F)[\exp(qV/E_0) - 1]. \tag{19}$$

By integrating Eq. (7), we obtain

$$F(V) = \mathrm{sgn}(V)\sqrt{2}(E_0/\lambda q)[\exp(qV/E_0) - 1 - qV/E_0]^{1/2}, \tag{20}$$

where λ is the screening length

$$\lambda = [\varepsilon\varepsilon_0/q^2 g(E_F)]^{1/2}. \tag{21}$$

Equation (14) yields

$$G_\square(V_s) = G_\square(0) + \lambda(\sigma_n I_n + \sigma_p I_p) \tag{22}$$

and

$$I_{n,p} = \mathrm{sgn}(V_s)\frac{1}{\sqrt{2}}\int_0^{\beta/\gamma} du\,[\exp(\pm\gamma u) - 1][\exp(u) - u - 1]^{-1/2}, \tag{23}$$

where the plus sign belongs to the integral I_n and the minus sign to I_p, and

$$\gamma = E_0/kT \qquad \text{and} \qquad \beta = qV_s/kT. \tag{24}$$

In the limit of large E_0 one obtains the equations for a constant density of states:

$$\rho(V) = -q^2 V g(E_F), \tag{25}$$

$$F(V) = V/\lambda, \tag{26}$$

$$G_\square(V_s) = G_\square(0) + \lambda(\sigma_n I_n^c + \sigma_p I_p^c), \tag{27}$$

and

$$I_{n,p}^c = \mathrm{sgn}(V_s)\int_0^\beta du\,\frac{[\exp(\pm u) - 1]}{u} = \mathrm{sgn}(V_s)\sum_{n=1}^{\infty}\frac{1}{n \cdot n!}\left(\pm\frac{qV_s}{kT}\right)^n. \tag{28}$$

We should now examine more closely the assumptions made in our treatment of conduction in the space-charge layer.

8. Charge Transport in Space-Charge Layers

A wealth of new phenomena have been discovered in studies of charge transport in quasi-two-dimensional inversion layers in crystalline semiconductors. Even in simple accumulation layers, it is found in crystals that the mobility is affected by scattering in the potential well and thus is different from the bulk value. In deriving the conductance in the space-charge layer of amorphous semiconductors, as for instance in Eq. (14), we played down the fact that 90% of the current flows in a layer 50–100 Å thick, depending on V_s. Figure 5 shows an example of the current between the surface and the depth x as a function of x (Goodman and Fritzsche, 1980; Fritzsche 1980a). The reason is, of course, the fact that $V(x)$ is exponentiated in Eq. (14). One thus deals with charge motion in a potential well that is as narrow as those of inversion layers in crystalline silicon.

In a-Si:H one has not yet been able to observe the nonmetal–metal transition that is expected at large V_s when E_c is moved below or E_v above the Fermi level at the surface. There should not be any difficulty in principle except for the question of whether the mobility edges E_c and E_v indeed follow the internal potential $V(x)$. The mobility edges separate in three dimensions the localized and extended states. In a narrow 20–50-Å-wide slice of the potential well $V(x)$ near the surface, it is likely that at least the low-energy extended states become localized because of the decrease in dimensionality. As a consequence, E_c (or E_v) may not follow $V(x)$ up to the surface but instead meet the surface with a horizontal slope. This will decrease the conductance compared to the value calculated from Eq. (14).

The second major assumption made in calculating the space-charge conductance from Eq. (14) is that the conductivity prefactors σ_{0n} and σ_{0p} of Eq. (13) are independent of $V(x)$ or of the local separation between E_F and the respective mobility edge. These prefactors decrease by several orders of magnitude as E_F is moved by doping (or annealing the light-exposed state)

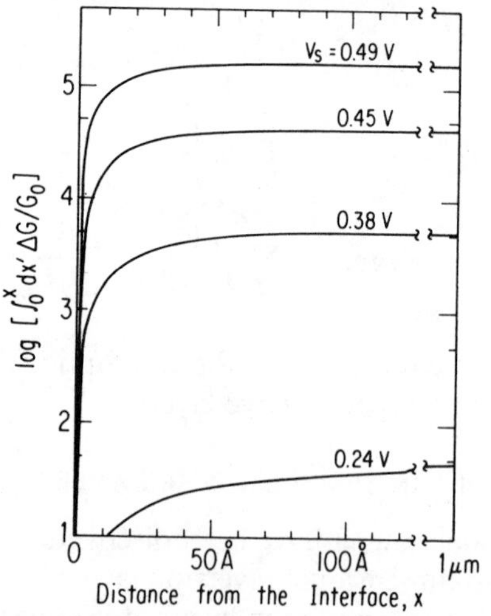

FIG. 5. Integrated current in the space-charge layer as a function of distance x from the surface for various surface potentials V_s. [From Goodman and Fritzsche (1980).]

from the gap center toward E_c or E_v (Fritzsche 1980a,b). The reason for this large decrease is not completely clear yet but the present explanation given for this would also predict a dependence of the local magnitude of σ_{0n} and σ_{0p} on $V(x)$ (see Chapter 8 by Beyer and Overhof). There is some indirect evidence that this indeed is the case. The evidence is the following.

In the absence of space-charge layers the conductivity σ is given by the sheet conductance $\sigma = G_\square/d$, where d is the sample thickness. When the sample conductance is dominated by the space charge, however, and the current flows according to Fig. 5 in an effective thickness $d_{eff} \ll d$, then the activation energy of $G_\square$ is decreased by eV_s relative to that in the bulk and an effective space-charge conductivity can be obtained from $\sigma_{eff} = G_\square/d_{eff}$. It turns out that the exponential prefactor of σ_{eff} is not equal to the value expected for E_a in the bulk but smaller and actually closer to the value expected for $E_a - |V_s|$, the activation energy in the space-charge region. An analysis of this kind is suggestive but not conclusive because of the growing evidence that the structure and composition of the layers close to the surface and to the substrate are different from those inside the film.

IV. Experimental Methods and Results: Surface Potential and Conductance

For calculating the space-charge conductance $G(V_s)$ we need to know the density of states $g(E)$ and the surface potential V_s. Under favorable conditions, V_s of a free surface can be obtained by the Kelvin probe method or from the saturation value of the surface photovoltage. These methods will be illustrated with some preliminary results obtained on a-Si:H. We then shall discuss metastable light-induced changes of the surface potential and conductance.

9. The Kelvin Probe Method

With the Kelvin probe or vibrating capacitor method one measures the work function difference $eV_{12} = \phi_2 - \phi_1$ between two electrodes that are externally connected and are at the same temperature (Gundry and Tompkins, 1968; Surplice and D'Arcy, 1970). If the capacitance C between the electrodes is altered periodically by vibrating one against the other, the net charge $Q = CV_{12}$ must change, thus producing an ac current dQ/dt in the external circuit. The work function difference is measured by applying an external bias equal and opposite to V_{12} so that the ac current is zero.

Any change in V_{12} caused by illuminating the sample surface or by adsorbates is equal to a change of V_s if there are no changes in any other parameters. Moreover, if strong illumination produces flat bands at the semiconductor surface, the saturation value of the surface photovoltage should yield

V_s. These conditions are very stringent and difficult to establish experimentally.

Vibrating Kelvin probe measurements are relatively easy to carry out except for two troublesome problems. First, it is difficult to find a stable and inert reference electrode. Gold is usually the first choice, but ϕ_{Au} can vary by ± 0.3 V, and is susceptible, as are the work functions of other metals, to adsorbates and thermal cycling. Stainless steel and conducting oxide layers have also been used as reference electrodes with varying success. There is no sure recipe available except the experience that aged reference surfaces are more stable (± 0.02 V) than fresh and clean ones.

Dipole layers of adsorbates on the sample surface present the second problem. As illustrated in Fig. 4b, adsorbates might alter ϕ_s not only by changing the charge in the semiconductor space-charge layer and hence V_s, but also by adding a dipole layer onto the surface, which changes χ according to Eq. (2).

In an elegant set of experiments Abelson and de Rosny (1983) measured simultaneously the change in the contact potential difference V_{12} and in the sheet conductance $G_\square$ of glow-discharge-deposited a-Si : H during H_2O adsorption (and desorption).

Figure 6 shows an example of their results. The undoped 0.5-μm-thick a-Si : H film was first annealed in vacuum and darkness to remove the effects of moisture and prior illumination. When moisture was admitted to the chamber at $t = 0$, the initial values were $G_\square = 3.5 \times 10^{-13}\ \Omega^{-1}$ and $V_{12} = 0.07$ V. The conductance dropped to $G^{min} = 2.8 \times 10^{-14}\ \Omega^{-1}$ and then increased, reaching the value $5 \times 10^{-13}\ \Omega^{-1}$ at $t = 17$ min. During this time

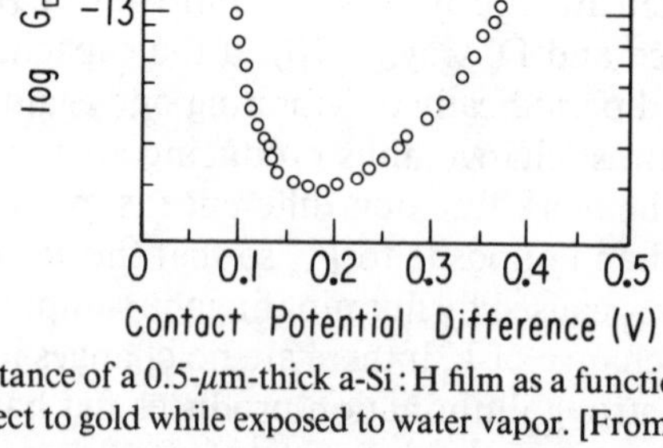

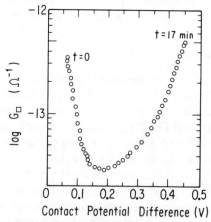

FIG. 6. Sheet conductance of a 0.5-μm-thick a-Si : H film as a function of the contact potential difference with respect to gold while exposed to water vapor. [From Abelson and de Rosny (1983).]

V_{12} increased monotonically from 0.07 to 0.46 V. The authors obtained an excellent fit of Eq. (22) to their experimental curve by choosing $V_{12}^{FB} = 0.28$ V as the flat-band contact potential difference and $G(0) = 4.5 \times 10^{-14}\ \Omega^{-1}$. Since the minimum conductance occurs at $V_{12}^{min} = 0.175$ V, one finds $V_s^{min} = V_{12}^{min} - V_{12}^{FB} = -0.105$ V. Substituting this value into Eq. (17) yields $\sigma_p/\sigma_n = 0.02$. The bulk of the sample is thus slightly n type. Before H_2O was admitted, the bands were bent upward, producing a hole accumulation layer that dominates the total conductance. The adsorbed H_2O layer acts as a donor, first diminishing the hole conduction (and $G_\square$) and then producing an electron accumulation layer with a surface potential $V_s = +0.18$ V. By fitting the approximate solution Eq. (22) to the experimental data of Fig. 6, Abelson and de Rosny also obtained $E_0 \simeq 0.026$ eV for the exponential slope of the conduction-band tail and $\lambda \geq 10^{-5}$ cm, from which it follows that $g(E_F) \leq 10^{17}\ \text{cm}^{-3}\ \text{eV}^{-1}$. These are very reasonable values.

We wish to stress here less these numerical values, which depend on a number of simplifying assumptions and approximations, and more the physical processes and measuring techniques needed to analyze them.

Figure 7 shows the annealing cycle carried out in vacuum after the film had been exposed to H_2O. Unfortunately, no measurements are available that follow V_{12} and $G_\square$ during evacuation, that is, between the $t = 17$ min point of Fig. 6 and the first data on Fig. 7. During that time V_{12} dropped from 0.46 to 0.01 V and the conduction returned to be dominated by a hole accumulation layer. If we disregard changes in ϕ_{Au} due to water desorption from the reference electrode as well as changes in the work function of the

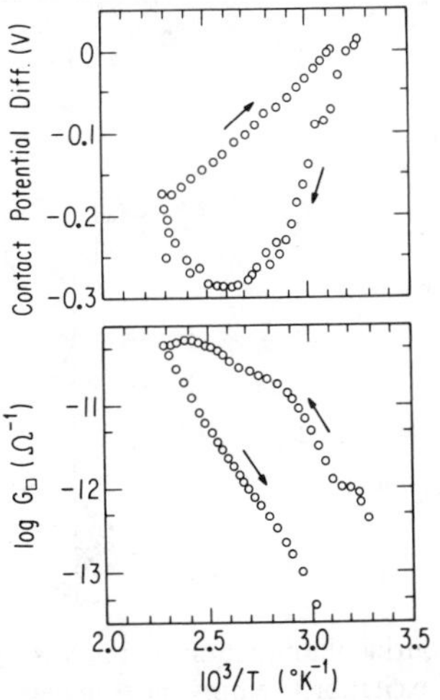

FIG. 7. Contact potential difference and sheet conductance of the sample for Fig. 6 during heating and cooling in vacuum. [From Abelson and de Rosny (1983).]

sample due to desorption of a dipole layer, then the data of Fig. 7 suggest (1) that the surface is strongly p type during the thermal cycle and (2) that the magnitude and the temperature dependence of $G_\square$ during cooling are strongly influenced by the temperature variation of V_s. We also note that after cooling the sheet conductance at 300°K is $G_\square \sim 5 \times 10^{-15}\ \Omega^{-1}$, which is less than the minimum conductance $G^{min} = 2.8 \times 10^{-14}\ \Omega^{-1}$. This low conductance $G_\square \sim 5 \times 10^{-15}\ \Omega^{-1}$ indicates, on the other hand, that space-charge conduction at the free surface and near the substrate interface must be negligible at 300°K and that the larger value $G_\square \sim 2.8 \times 10^{-14}\ \Omega^{-1}$ must be attributed to conduction near the interface that is diminished by annealing. The changes in the contact potential V_{12} during annealing might then be due to changes in the gold work function ϕ_{Au} and to changes in the electron affinity χ.

10. Surface Photovoltage

If strong illumination produces flat bands at the surface without changing χ, that is, without affecting the surface dipole layer, then the saturation value of the surface photovoltage measured by the Kelvin method ΔV_{12} should be equal to V_s. This cannot be taken for granted, however, because of the following considerations.

Figure 8a illustrates two processes that lead to a decrease of the negative surface charge: (1) direct optical excitation of trapped electrons into the conduction band and (2) capture of photoexcited holes from the valence band (Goldstein and Szostak, 1980). At high light intensities this produces a neutral surface and $\Delta V_{12} = V_s$. However, depending on the relative capture

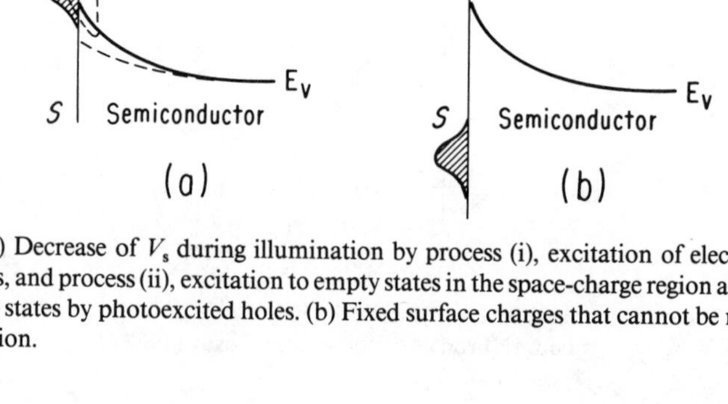

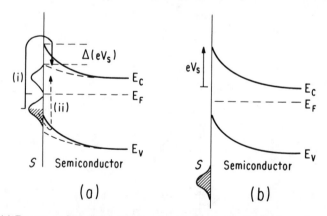

FIG. 8. (a) Decrease of V_s during illumination by process (i), excitation of electrons from surface states, and process (ii), excitation to empty states in the space-charge region and neutralizing surface states by photoexcited holes. (b) Fixed surface charges that cannot be neutralized by illumination.

cross sections of the surface states, one can also envision that illumination increases the surface charge. This is called *photovoltage inversion* (Lagowski *et al.*, 1971) and, when present in the background of other electronic transitions, *photovoltage quenching* (Gatos and Lagowski, 1973). Moreover, Fig. 8b shows a case in which the semiconductor space-charge layer is produced by fixed surface states whose occupancy may not change under illumination because they lie outside the gap region of the semiconductor. Hence the saturation value of the surface photovoltage does not necessarily yield V_s.

It is instructive to ask whether strong illumination is likely to flatten the bands in an a-Si:H field-effect transistor. In this case, the charges that produce V_s and the space charge in the semiconductor are on the gate electrode and not in surface states nearby. Light can produce a flat-band condition in the semiconductor for an arbitrary value of the gate voltage only when the total charge of the space-charge region moves into interface states as a result of interface recombination and optical transitions from and to interface states. This is a highly unlikely situation. Instead one finds that the space charge persists at high illumination levels, although its spatial distribution changes. This will be discussed in Section 13 on photoconduction in space-charge layers. Conversely, we cannot infer from the persistence of space-charge layers in the field-effect geometry that strong light cannot create neutrality and reduce V_s at the free surface. The steady-state equilibria under illumination of the surface and of an interface with an applied transverse field are quite different problems.

Goldstein and Szostak (1980) studied the surface photovoltage SPV of a-Si:H by means of the vibrating Kelvin probe method, where SPV $= \Delta V_{12}$ produced by illumination. The measurements were carried out in an ion-pumped vacuum of about 10^{-9} Torr after sputter cleaning the sample surface with 500-eV Ar ions and subsequently annealing at 200–250°C to remove the surface damage caused by sputtering. An example of their results is shown in Fig. 9. The saturation values of the SPV are plotted as a function of photon energy. The negative sign of the SPV means that the bands were bent upward (positive space charge) and that light decreased V_s. A maximum value SPV $= -0.2$ V was observed on clean and annealed surfaces. Goldstein and Szostak (1980) reported that sputter damage or O_2 physisorption reduced the SPV either by reducing the band bending or the ability of light to reduce the band bending. After oxidation of the surface in an ambient pressure of 5×10^{-4} Torr O_2 for several hours, the SPV response at 300°K nearly vanished. The reader is reminded that fresh a-Si:H surfaces are very inert to O_2 (see Part II). It might be that sputter etching depletes the surface of hydrogen and restores the reactivity to oxygen.

Much information about the distribution of surface states and the charge transfer process that (partially) neutralizes the surface can be obtained from

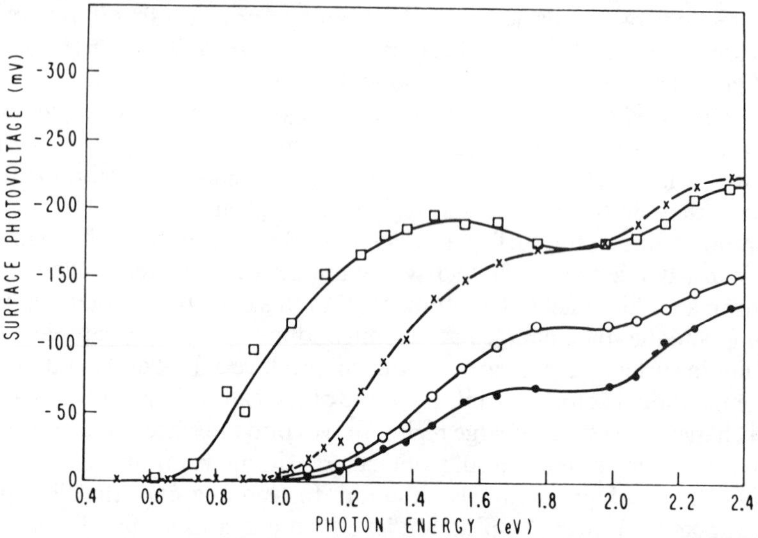

FIG. 9. Surface photovoltage spectra of undoped a-Si : H at 25°C and −168°C after various surface treatments: ●, $T = 25$°C, sputtered only; ×, $T = 25$°C, sputtered and annealed; ○, $T = 25$°C, sputtered and annealed with one monolayer physisorbed O_2; □, $T = -168$°C, sputtered and annealed. [From Goldstein and Szostak (1980). Copyright North-Holland Publ. Co., Amsterdam, 1980.]

the spectral dependence of the SPV and its time response as a function of temperature, as demonstrated by Goldstein and Szostak (1980) and discussed in general by Gatos and Lagowski (1973). It appears that direct excitation of electrons from a distribution of surface states to the conduction band, i.e., process (i) of Fig. 8a, dominates in their sample because the threshold photon energy (Fig. 9) decreases as the Fermi energy E_F is raised with doping. When the light is turned off, the empty surface states must refill. This process can be observed by measuring the decay of the SPV. The temperature dependence of the initial rate of decay suggests an activated process with an energy of 0.22 eV. This energy is high enough so that the SPV essentially does not decay at temperatures below 140°K. It should be realized that empty surface states cannot readily be refilled from occupied localized gap states. These gap states are located a considerable distance from the surface when the gap state density is small. Hence the tunneling probability between surface and bulk states can be very low. Consequently, the authors suggest that activation from the valence-band edge E_v to the bottom of the empty surface states or to the surface trap quasi-Fermi level during excitation is necessary. Figure 10b shows the energy diagram for a phosphorus-doped sample with slightly different energy values. It is suggested by Goldstein and Szostak (1980) that there are additional surface states below E_v that

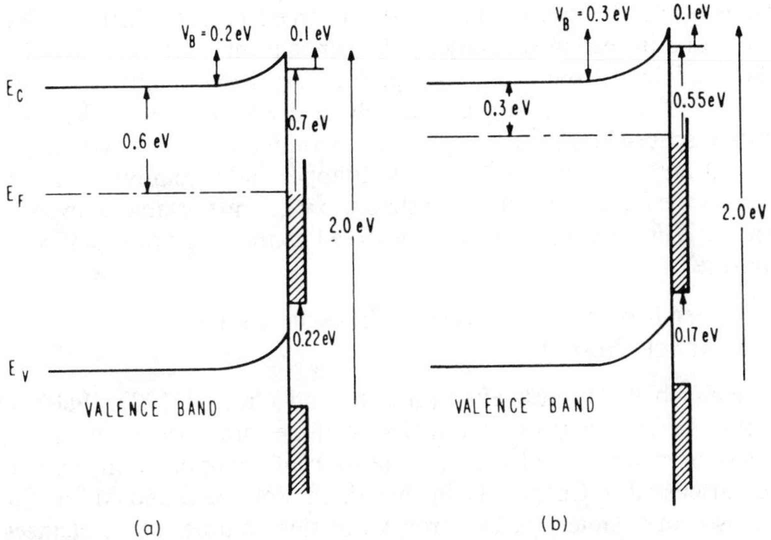

FIG. 10. Energy level diagram of the surface of (a) undoped and (b) phosphorus-doped a-Si:H deduced from surface photovoltage measurements. [From Goldstein and Szostak (1980). Copyright North-Holland Publ. Co., Amsterdam, 1980.]

require no activation energy for refilling. These states are used to explain a fast-decay component of the SPV that is observed for excitations with $h\nu \geq$ 2 eV. By equating the number of filled surface states with the charge in the space-charge layer, the authors estimate 1.5×10^{11} electrons cm^{-2} to within a factor of two.

The purpose of discussing this experiment in some detail is to explain the information that may be obtained from it, as well as the assumptions needed to extract this information. Our means for characterizing surfaces are insufficient to allow comparisons of results from different laboratories. We have measured the SPV of several a-Si:H films (Peng and Cai, 1983) and found that SPV = −0.17 V when the surface had been exposed to ambient atmosphere for several weeks and that SPV = −0.06 V after etching. We also found that care must be taken that the light does not reach the back contact, where an additional photovoltage might be created (Wey and Fritzsche, 1972; Wey, 1976). Experiments with sub-band-gap light yield much information about the distribution and dynamics of surface states (Gatos and Lagowski, 1973), but the back contacts have to be positioned in darkness because the light penetrates the thin film. This is not a trivial problem. We tried to measure the quite large change in the SPV that is expected when H_2O is adsorbed on the surface. With changes in the contact potential V_{12} of more than 0.4 V, as observed by Abelson and de Rosny (1983), one should find a change of the SPV from -0.2 to $+0.2$ V, an effect that is hard to miss. We have found instead only a very small change from -0.17 to -0.12 V.

We conclude from these very preliminary experiments that oxidized a-Si:H surfaces are often negatively charged with $V_s \simeq -0.2$ V, leaving a positive space-charge region in the semiconductor. The charge is greatly reduced but often does not entirely disappear after removal of the oxide layer. In the following section we shall show that metastable changes in the negative oxide charge have been observed as the result of prolonged light exposure.

11. Light-Induced Metastable Changes of the Surface Charge

The Staebler–Wronski effect (Staebler and Wronski, 1977, 1980), which manifests itself in a light-induced change of the dark Fermi level, of the bulk density of states, and of the concentration of recombination centers, has been discussed in Chapter 11 by Schade in Volume B and will be further discussed in Chapter 10, by Wronski, of this volume. These changes are stable at room temperature and disappear upon annealing above $\sim 430°$K. One does not know the detailed process underlying this effect, but it appears that recombination or trapping of carriers and not light absorption triggers this effect and that perhaps dangling-bond sites are created. It was suggested early on (Solomon *et al.*, 1978) that this effect might be associated with a metastable change of the charge equilibrium between the a-Si:H film and its substrate or surface and hence is a surface effect. There now is overwhelming evidence that one deals with a bulk process (see Chapter 10 by Wronski).

Nevertheless, in this section we present experimental evidence (Aker and Fritzsche, 1983) that the above-mentioned surface effect can occur and under certain circumstances, completely mask the bulk effect. It is obviously essential to take this surface effect into account in order to obtain reliable information about the bulk effect.

Figure 11 shows planar conductance curves of one of several boron-doped samples studied by Aker and Fritzsche (1983). The solid curve 59 is measured during cooling after annealing for 45 min at 450°K in vacuum and darkness. The slope of this curve decreases with decreasing temperature. Exposure to 50 mW cm^{-2} heat-filtered white light at 300°K increased the conductance by a factor of 50. This state is stable at 300°K but anneals along the dashed curve 59 back to the original conductance state. This sample had been exposed to the ambient for about nine months and hence was covered with an oxide layer of about 30-Å thickness as estimated from the oxidation rate. The oxide is negatively charged, and the planar conductance of the annealed and exposed sample 59 is dominated by a hole accumulation layer. Evidence for this is the very large decrease in G at 300°K when water is

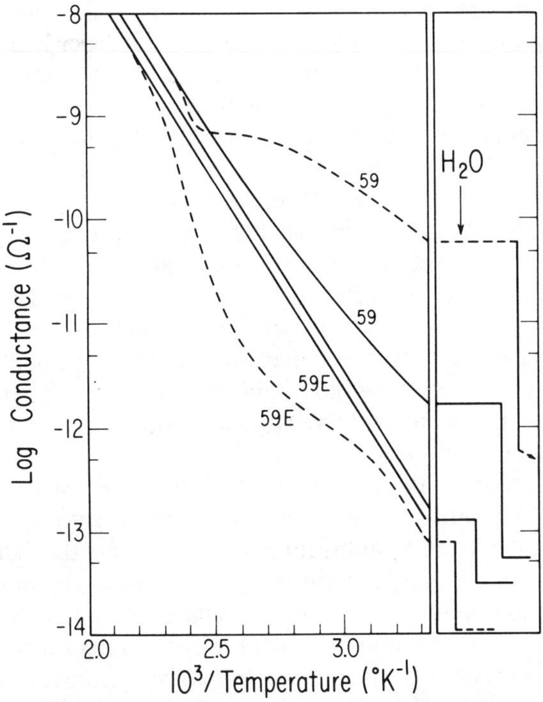

FIG. 11. Dark conductance as a function of inverse temperature of a lightly boron-doped ($B_2H_6/SiH_4 = 1.5 \times 10^{-5}$, $d = 1.62$ μm) a-Si:H film; 59 before etching and 59E after etching. Solid lines are cooling curves. Dashed lines are heating curves after 40 min light exposure at 300°K. The unlabeled solid line is the cooling curve of 59E after exposure to room air for one week. The right-hand panel shows the decrease of G at 300°K resulting from exposure to 22% relative humidity. [From Aker and Fritzsche (1983).]

adsorbed on the surface, as shown on the right-hand side of the figure. It should also be mentioned that the true conductance of the exposed sample 59 lies considerably higher than the measured dashed curve because of contact limitation caused by the tunneling resistance through the oxide layer between the semiconductor and the metal contacts (Aker and Fritzsche, 1983). This large increase in conductance is opposite in sign to the normal bulk Staebler–Wronski effect. It is caused entirely by a metastable increase in the negative charges of the oxide because, after the etching of the surface, the solid curve 59E was obtained after annealing and the dashed curve 59E after light exposure. The etched sample is not entirely free of a hole accumulation layer because water adsorption still causes a decrease in the 300°K conductance. However, the solid curve 59E is straight with an activation energy $E_a = 0.8$ eV and a conductivity prefactor $\sigma_0 = 1.8 \times 10^3$ Ω^{-1} cm^{-1}.

The unmarked solid curve between 59 and 59E was measured some days after etching, during which a 8–10-Å-thick oxide layer had grown.

A few more points are worth noticing in Fig. 11. The curvature of the solid curve 59 suggests that V_s decreases with increasing temperature, but the space-charge conduction and a finite V_s must persist to the highest temperatures because 59E (solid) remains below curve 59. Furthermore, the dashed annealing curve of the exposed 59 crosses and meets the solid curve 59 from below. This suggests that the metastable excess negative charge of the oxide anneals about 50°C earlier than the bulk effect that decreases the bulk conductance. Finally, the dashed curve 59E of the exposed etched sample is interesting. It seems that a positive surface effect that anneals early is superposed on a negative bulk effect that anneals later. A small surface effect remaining on the etched sample can be attributed to a thin oxide film that forms very soon after etching. Even a clean surface is not expected to be free of surface states.

What is the cause of the metastable increase in the negative oxide charge? It is possible that during photoexcitation electrons get trapped in oxide or surface states above the equilibrium Fermi level. After the light is turned off, these excess carriers must remain well isolated from the bulk states to account for the long-term stability of this metastable state at 300°K. As mentioned earlier, Goldstein and Szostak (1980) observed such a frozen-in disequilibrium between bulk and surface: the surface photovoltage disappeared below 140°K because holes trapped in surface states could no longer communicate with the bulk. In our present case, we would have to invoke extremely well isolated electron traps above E_F in or at the surface of a 10–30-Å-thick oxide. Another example of such charge disequilibria at 300°K is the photoenhanced injection of electrons or holes into the gate insulator of a field-effect transistor (Vaid and Fritzsche 1983).

Alternatively, one might suggest that the enhanced negative oxide charge after light exposure results from metastable defect states in the oxide that act as electron traps below E_F and that are created by the recombination processes similar to the defects created in bulk a-Si : H by the Staebler–Wronski effect. Ultraviolet bleaching and regeneration of dangling-bond defects at the surface of oxidized c-Si have been observed by Caplan *et al.* (1982). However, their photon energies were much larger than the $h\nu \simeq 2$ eV used by Aker and Fritzsche (1983).

It is not clear whether these same processes occur also in the oxide of *n*-type films or are limited to the oxide of *p*-type films that contain not only hydrogen but also boron. A preliminary study of the surface photovoltage in *n*-type films before and after prolonged illumination (Aker *et al.*, 1983) showed only a slight increase in band bending, which is much smaller than the large increase found in *p*-type films (Fig. 11).

V. Surface Effects on Photoconduction

The coplanar photoconductance is strongly influenced by space-charge layers and, even in their absence, by surface recombination. These effects can be studied by varying V_s and the space-charge conductance by a transverse field in a field-effect geometry (Goodman, 1982; Jackson and Thompson, 1983; Jackson *et al.*, 1983b; Vaid and Fritzsche, 1983) or by comparing the photoconductance of strongly absorbing light (absorption coefficient $\alpha \gg 1/d$) with that of nearly uniformly absorbed light ($\alpha \lesssim 1/d$). Since α reaches values larger than $5 \times 10^5\ \mathrm{cm}^{-1}$ at photon energies $h\nu > 2.6$ eV, most of the high-energy photons are absorbed in a depth of 600 Å or less. By illuminating the sample through the substrate (back-lit) or on the free surface (front-lit) and changing the surface conditions with ambients, etching, and heat treatments, information on the effects of space charges and of surface recombination can be obtained (Persans 1980; Fritzsche, 1980b). Only a few studies of this sort have been made, but they reveal the usefulness of these techniques.

12. Surface Recombination

Figures 12a and b compare in effect the spectral dependence of the normalized photocurrent of an undoped a-Si:H film illuminated from the front and through the Corning 7059 substrate (back-lit). The absorption coefficient α instead of the photon energy $h\nu$ is used as the abscissa in order to facilitate comparison with the theory of Schetzina (1979) on surface recombination. The sample was $d \simeq 2\ \mu$m thick; hence for $\alpha > 5 \times 10^3\ \mathrm{cm}^{-1}$ all incident photons are absorbed, after reflection losses are accounted for, and contribute to the photocurrent. A fraction e^{-1} of the available photons are absorbed in the absorption depth α^{-1}. The photocurrent is normalized by the incident photon flux, and this flux F was adjusted such that αF was constant to provide a nearly constant generation rate in the absorption depth. Since the photocurrent I_p was nearly proportional to the photon flux in this sample, actually $I_p \propto F^{0.9}$, we believe that this small nonlinearity is not the cause of the observed effects. One finds that I_p decreases by as much as a factor of 50 between $h\nu = 2$ and 3 eV or $\alpha = 10^4$ and $5 \times 10^5\ \mathrm{cm}^{-1}$. Curve A was measured in vacuum after the fresh sample had been transported through air. Curve B was taken after heat drying at 425°K for 1 hr and curve D after heat drying for 12 hr. Curve C was taken in vacuum after exposing the sample for 6 hr to a nitrogen flow containing 20% relative humidity. One finds that heat drying (B and D) affects the substrate interface slightly and the free surface appreciably. Exposure to H_2O essentially restores the initial condition A at the free surface but naturally does not affect the curve B measured prior to C in Fig. 12b. What do we know about the presence of

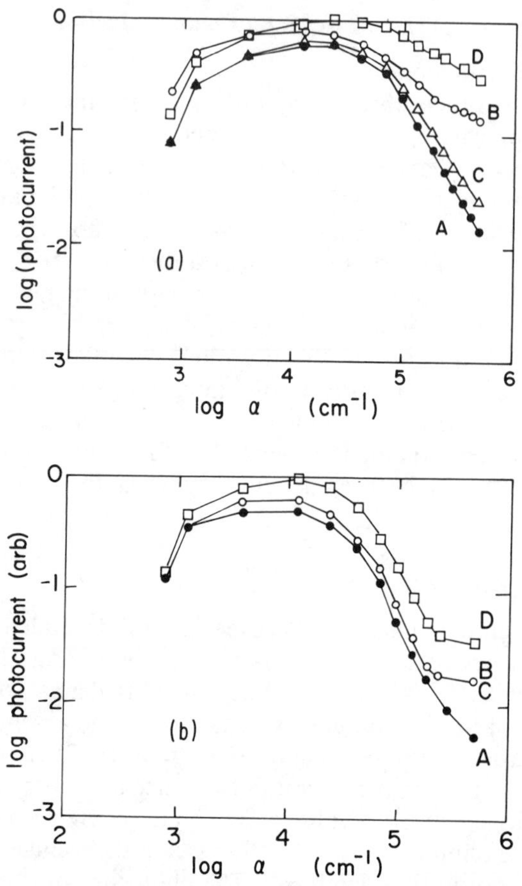

FIG. 12. Normalized photocurrent of 2-μm-thick undoped a-Si:H as a function of absorption coefficient $\alpha(h\nu)$. (a) Illumination of free surface (front-lit) and (b) illumination through glass substrate (back-lit). For explanations of surface and sample treatments A–D, see text. [After Persans (1980).]

space-charge layers? Since the sample had its lowest conductance in states A and C with an activation energy $E_a = 0.9$ eV, we conclude that the film has essentially flat bands under these conditions. The dark conductance of the annealed states B and D was a factor of 30 higher than that of A and C, which points to a small electron accumulation layer near the top surface.

The decrease in the normalized photocurrent at large α is probably caused by surface and interface recombination. The effective surface recombination states are, in turn, changed by band bending and surface treatments. It is not possible to exclude an influence of inhomogeneities along the film thickness, but it would be surprising to have similar compositional or structural differences in the film both at the front and back surfaces that yield identical curves A in Figs. 12a and b.

The analysis of the spectral dependence of I_p in the presence of surface recombination should yield the minority-carrier diffusion length L_p and S, the ratio of the surface recombination velocity to the bulk diffusion velocity. DeVore (1956) presented an analysis using a single carrier model in which local charge neutrality is maintained by fast dielectric relaxation. Schetzina (1979) suggested that this is not the case for a-Si:H and assumed instead that local neutrality is controlled by fast recombination. The two analyses differ only in the absolute magnitude of I_p but not in the spectral shape. We plotted the α dependence of the relative magnitude of I_p for various combinations of L_p and S as defined by Schetzina in Figs. 13a and b. The drop of I_p at large $h\nu$ or α occurs only for large L_p and S. However, when L_p approaches the sample thickness, I_p is decreased at all photon energies, because all photogenerated carriers can reach the surface and recombine if S is large enough. When S is increased at fixed L_p, on the other hand, I_p approaches a limit when all carriers within L_p recombine at the surface. The experimental curves A of Figs. 12a and b show a stronger drop at large α than the present theories predict. One of the reasons for this discrepancy might be that under steady-state illumination the surface is not neutral and the resultant space-charge field enhances drift to the surface and thus the overall recombination rate.

These experiments emphasize that an analysis of the photoconductivity requires consideration of surface recombination particularly when the film is thin and of high quality (large minority-carrier diffusion length). An example of the thickness dependence of the normalized photocurrent is shown in Fig. 14 (Persans, 1980).

13. Photoconduction in Space-Charge Layers

By illuminating an a-Si:H field-effect transistor, one can study the steady state and transient decay of photoconduction under controlled space-charge conditions. Figure 15 shows the conductance of a 1-μm-thick undoped a-Si:H film as a function of gate voltage applied across a 160-μm-thick suprasil 2 substrate (Vaid and Fritzsche, 1983). Positive V_g corresponds to a negative induced space charge in the semiconductor with +1 kV yielding 1.2×10^{11} electrons cm^{-2}. The sample had been previously illuminated to prevent light-induced changes (Staebler–Wronski effect) from occurring during the measurements.

Depending on the sign of V_g, one can produce an electron or a hole accumulation layer in the same sample. The small shift of the minimum conductance towards $-V_g$ shows that the ratio of the electron to the hole contribution to the conductivity under flat-band conditions increases

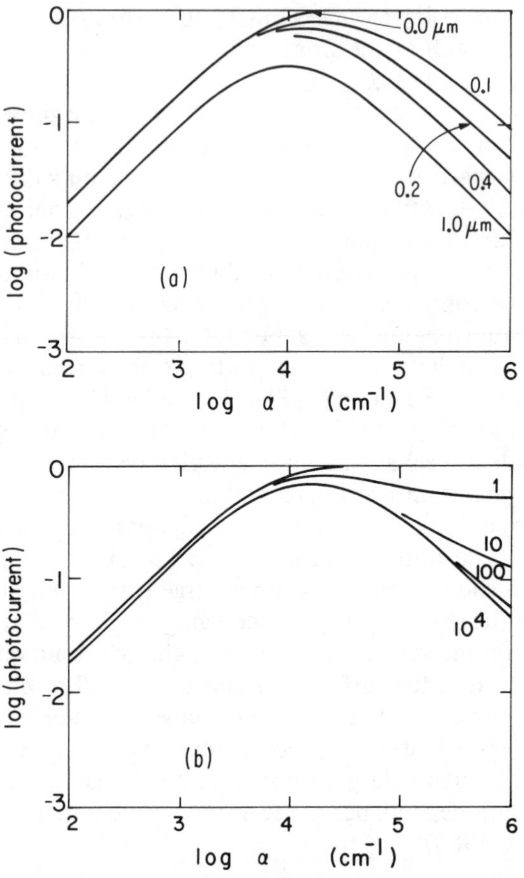

FIG. 13. Spectral dependence of the normalized photocurrent as calculated from the surface recombination model of Schetzina (1979) (a) for several L_p (the hole diffusion length) and (b) for several S (the surface recombination number) with $L_p = 0.2$ μm and film thickness 20 μm. [After Persans (1980).]

slightly with light intensity. The surprising result of these measurements is that the photoconductivity is greatly enhanced by the presence of either hole or electron accumulation layers, which are induced by the field effect. The early work of Solomon and Brodsky (1980) may be explained by this effect. Jackson and Thompson (1983) and Jackson *et al.* (1983b) have carried out careful measurements of the spectral and intensity dependence of the photoconductance as well as of the transient photocurrent as a function of the gate voltage. They find that all characteristic photoconductive parameters and decay times are different when the photoconductance is dominated by the space-charge layer, in part because the spatial charge distribution changes

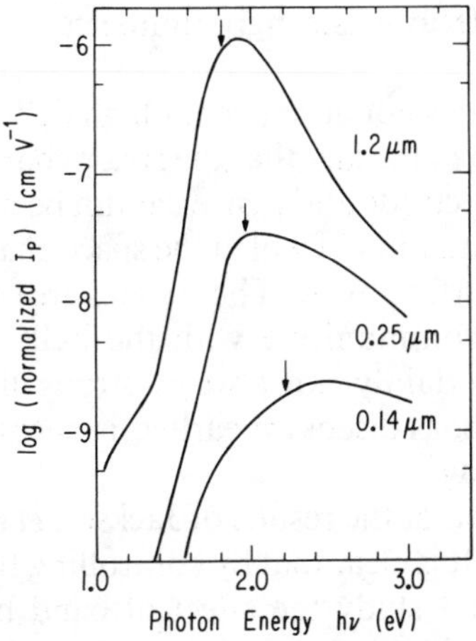

FIG. 14. Thickness dependence of the spectrum of the normalized photocurrent of undoped a-Si:H. The arrow points at $h\nu$ for which $\alpha d = 1$. [After Persans (1980).]

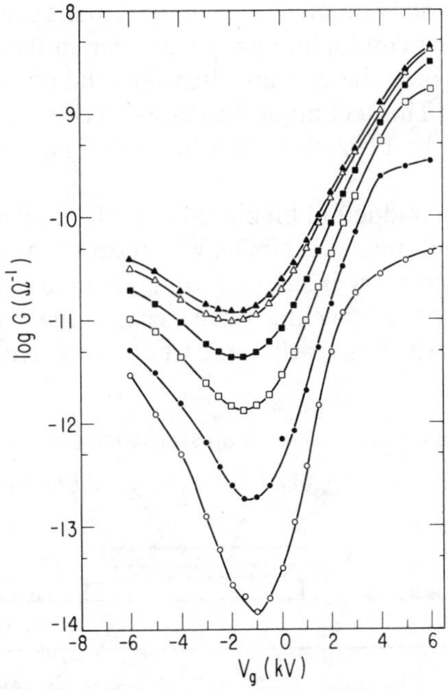

FIG. 15. Photo-field effect of undoped a-Si:H at 25°C for various heat-filtered white light intensities F: ▲, 1.0; △, 0.8; ■, 0.3; □, 0.1; ●, 0.01; ○, 0.001. $V_g = +1$ kV corresponds to a negative induced space charge of 1.2×10^{11} electrons cm^{-2}. [From Vaid and Fritzsche (1983).]

with light, and the decay is strongly influenced by the relaxation of this change.

The space charge does not disappear at high light levels. As mentioned in Section 10, this is due to the fact that the space charge is induced by charges that are on the gate electrode and hence cannot be neutralized by light. This is different from the situation in which the space charge is caused by states at the surface or in the surface oxide. These states are subject to the recombination traffic and can communicate with the bulk states. Under favorable circumstances, but certainly not always, strong light can then lead to a neutral surface, which, as discussed earlier, is necessary for getting V_s from the surface photovoltage.

A detailed discussion of the results of Jackson *et al.* would go beyond the scope of this chapter. It is clear that by controlling the space-charge layer by the field effect one can study the effect of band bending on a variety of transport processes and probably find that unsuspected space-charge layers can explain some of their anomalous behavior.

VI. Surface Studies by the Traveling Wave Method

In this part we briefly mention a new experimental tool that may become useful for analyzing conduction processes near surfaces and interfaces because it measures only these, even when the total dc conductance is dominated by the bulk. This technique was pioneered by Adler *et al.* (1981) and further developed by Fritzsche and Chen (1983) and Chen and Fritzsche (1983).

Consider a semiconductor film placed in the fringe field that exists above a piezoelectric crystal such as $LiNbO_3$, which carries a surface acoustic Rayleigh wave. The geometry is shown in Fig. 16. One can use the quasistatic approximation for the traveling wave because the wave travels with the sound velocity v_s, which is nearly static when compared with the light veloc-

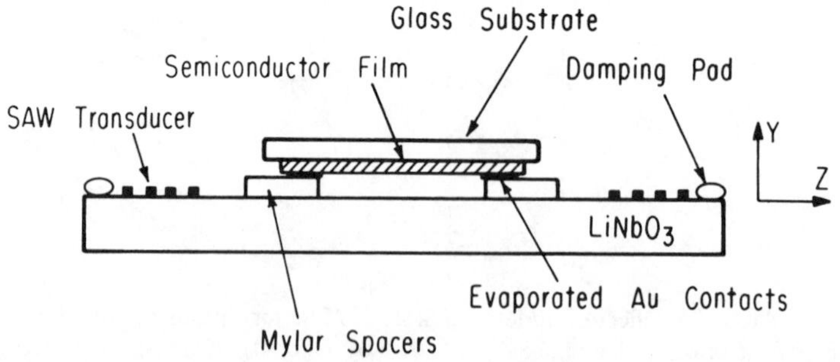

FIG. 16. Schematic diagram of traveling wave experiment. [From Adler *et al.* (1981).]

ity (Bløtekjaer, 1970). The field is thus obtained from a scalar potential

$$\mathbf{E} = -\nabla\phi. \tag{29}$$

Moreover, when the diffusion term is small compared to $\sigma\mathbf{E}$, no space-charge wave is produced and

$$\nabla^2\phi = 0 \tag{30}$$

in the semiconductor as well as in the air space and in the insulating substrate. We write the solution of Eq. (30) (the Laplace equation) as

$$\phi = \phi_0[A \cosh k(y-h) - B \sinh k(y-h)] \exp i(\omega t - kz). \tag{31}$$

The complex coefficients A and B are determined by the boundary conditions at the interfaces (Chen and Fritzsche, 1983; Fritzsche and Chen, 1983), and h is the separation between the free surface of the film (at $y = h$) and the piezoelectric crystal. The potential ϕ_0 at the crystal surface ($y = 0$) is determined by the acoustic wave power (Auld, 1973).

The electric field $E_y = -\partial\phi/\partial y$ produces a current σE_y in the film. This leads to two surface-charge waves of opposite phase at $y = h$ (surface 1) and $y = h + d$ (surface 2):

$$\delta_1 = -i\frac{\sigma}{\omega}\frac{\partial\phi}{\partial y}\bigg|_{y=h} \tag{32}$$

and

$$\delta_2 = i\frac{\sigma}{\omega}\frac{\partial\phi}{\partial y}\bigg|_{y=h+d} \tag{33}$$

The film thickness is d. By multiplying these surface charges by the local carrier mobility μ and $E_z = -\partial\phi/\partial z$, the local component of the field in the direction of the wave velocity, one obtains two surface currents. Since δ and E_z are proportional to ϕ, these surface currents are second order in ϕ and have finite time averages

$$I_1 = \sigma(\mu_1/v_s)wk(A \cdot B)\phi_0^2/2 \tag{34}$$

and

$$I_2 = -I_1\mu_2/\mu_1 + \sigma(\mu/v_s)dwk^2(|A|^2 + |B|^2)\phi_0^2/2, \tag{35}$$

where w is the width of the film, and $A \cdot B$ is the dot product of the complex coefficients A and B. The subscripts 1 and 2 refer to the film surfaces at $y = h$ and $y = h + d$, respectively. The open-circuit voltage V_{ae} between two electrodes on the film placed a distance L apart along the z direction is obtained by dividing the net current $I_1 + I_2$ by the film conductance G.

Under flat-band conditions, $\mu_1 = \mu_2$ in a homogeneous film, and the I_1 terms cancel in $I_1 + I_2$, so that

$$V_{ae} = \pm(\mu/v_s)Lk^2(|A^2| + |B|^2)\phi_0^2/2. \tag{36}$$

Since each surface current reverses direction with a change of sign of the majority carrier at that surface, the sign of V_{ae} in Eq. (36) yields the sign of the majority carrier.

How can we obtain information about the surface or the interface from this experiment? There are two answers. One concerns the mobility μ. In an amorphous semiconductor, in which charge transport occurs with multiple trapping, this quantity is the time-dependent drift mobility at $t = 1/v$, where v is the frequency of the wave. Only a fraction δ_f/δ of the carriers in the charge wave moves in transport states with a microscopic mobility μ_0. The rest δ_t/δ may be trapped in surface states and in localized bulk states near the surface. Here δ stands for either δ_1 or δ_2. A comparison of the surface drift mobility obtained here with the bulk drift mobility measured by the time-of-flight technique essentially compares μ_0 and the spectrum of traps near the surface and in the bulk. Experiments on n-type a-Si:H showed surprisingly good agreement of these quantities (Fritzsche and Chen, 1983; Chen and Fritzsche, 1983).

The second kind of information one can obtain from this experiment concerns space-charge layers, in particular inversion layers. Since each surface current, I_1 as well as I_2, reverses sign with the dominant charge carrier at that surface, the currents at both surfaces will flow in the same direction when one surface is n type and the other p type because of band bending. The n-type surface current will probably always dominate because the drift mobility of electrons is much larger than that of holes. By neglecting the hole current altogether under these circumstances, one finds that the larger term I_1 dominates either in Eq. (34) or (35), depending on whether the n-type surface is surface 1 or 2. The voltage measured is then

$$V_{ae} = \pm(\mu/v_s)(1/kd)Lk^2(A \cdot B)\phi_0^2/2. \tag{37}$$

The sign is negative when surface 1 is n type and surface 2 is p type. The sign is positive in the reverse situation. The attenuation factors $(A \cdot B)$ and $(|A|^2 + |B|^2)$ in Eqs. (36) and (37) take into account the screening of ϕ by the film having a conductivity σ and dielectric constant ε.

VII. Closing Comments

In this chapter we did not discuss the detailed studies of Tanielian *et al.* (1978, 1980) and Tanielian (1982) of the effect of adsorbates and insulating overlayers on the sheet conductance of a-Si:H. These results cannot be presented in condensed form because they deal with the central issues of the kinetics of the physicochemical surface reactions that still remain to be explored.

It is clear from the foregoing that the observed phenomena cannot be explained by a static distribution of surface states but that this distribution changes with time following chemical surface reactions and strain relaxations. The time scale of these reactions and relaxations can span many hours and even days as shown by the time dependence of surface spins (Street and Knights, 1981) and of the sheet conductance after annealing (Ast and Brodsky, 1979) and after adsorption (Tanielian *et al.*, 1978, 1980; Tanielian, 1982).

As an example we show in Fig. 17 the change of the dark current and of the photocurrent for strongly adsorbing light of undoped a-Si:H as a function of time after admitting to the chamber a constant flow of N_2 with 20% relative humidity. Figure 17b shows that the semiconductor charges relatively fast (2–10 min), but this is followed over a period of many hours by a gradual decrease in the excess charge, which had arisen in adsorption. The photocur-

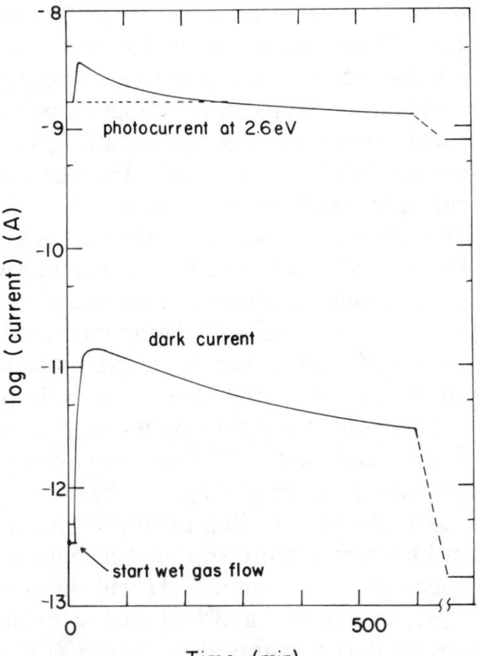

FIG. 17. Time dependence of the photocurrent near the surface due to strongly absorbing light, $h\nu = 2.6$ eV, and the dark current of undoped a-Si:H during exposure to a flow of N_2 with 20% relative humidity.

rent, which is influenced by surface recombination in this spectral range, as discussed in Section 12, shows in Fig. 17a a similar rise and slow decay (Persans, 1980). It crosses its original value, however, at a time when the dark current is still enhanced by a factor of 50. After a long time in flowing nitrogen containing 20% relative humidity, a new equilibrium state is reached, which is chemically quite different but electronically rather similar to the original one. Tanielian (1982) showed that the slow decay is not affected by an interruption of the H_2O exposure for 1 hr.

Again, such fast and long-term relaxation processes associated with adsorption are not new phenomena but are quite common on surfaces of crystalline semiconductors (Many *et al.*, 1965; Morrison, 1957; Jäntsch, 1965; Kiselev *et al.*, 1968; Kozlov *et al.*, 1971). Moreover, such time-dependent processes are not limited to H_2O adsorption but are observed with many other adsorbates. But let us take water as an example. The high ionization energy of H_2O molecules makes it unlikely that H_2O^+ ions form at room temperature and donate an electron to the semiconductor. Tanielian (1982) never observed a charge exchange larger than about 1.5×10^{11} electrons cm^{-2} on a surface saturated with at least 10^{15} H_2O molecules cm^{-2}. This low efficiency suggests that the excess carriers in the space charge come from a small part of coordination-bound molecules closest to the silicon interface. Alternatively, it is possible that unoccupied localized gap states or hydrogenated Si atoms provide free orbitals for coordination bonding. This may be investigated by studying the water reactions as a function of doping or by fluorinating the surface by etching with HF. The question is whether the surface can be made inert to adsorbates by proper bond saturation.

But instead of speculating, it would be most valuable to study the surface and bonding character of adsorbates as a function of time with the variety of surface analytical tools available. Moreover, the rate and amount of adsorption (and desorption) may be studied by volumetric methods or microbalance techniques. Bonding information may also be obtained from the heats of adsorption and desorption. Infrared multiple reflection spectroscopy should reveal the characteristic bending modes of adsorbed molecules and allow a quantitative determination of the surface coverage as a function of prior surface treatments and of annealing.

Unfortunately, a-Si:H surfaces will probably always lack the reproducibility of cleaved or high-temperature annealed crystalline surfaces. The results of surface chemical studies on a-Si:H will depend on whether the surface is freshly prepared in situ and hydrated or oxidized. Dehydrated surfaces obtained by sputtering or annealing above 300°C are expected to be more reactive than hydrogenated or fluorinated surfaces.

Simultaneous measurements of several of the desired parameters while the surface is subjected to known amounts of adsorbates appear to be particularly useful. These include the time dependence of the surface potential V_s, the space-charge conductance $G_\square - G_\square(0)$, the charge density Q_s in slow and in fast surface states, and the surface recombination velocity. Moreover, it would be of great value to compare the charge relaxation processes observed after adsorption with the time response of charging and discharging the surface in a field-effect experiment with the gate electrode above the free surface. Such comparison will tell us whether the time dependence is limited by chemical processes or by electron tunneling into slow surface states.

The effect of light on adsorption and desorption and on the light-induced creation of surface states are other quite unexplored research problems. One example that suggests such creation of surface states was discussed in Section 11. Moreover, it has been claimed (Jackson *et al.,* 1983b) that field-effect measurements are governed by interface states and are hardly sensitive to the bulk density of states. We do not believe that this is so, but if it is true, then the changes in field-effect conductance observed after annealing (Goodman and Fritzsche, 1981) and after prolonged light exposure (Tanielian *et al.,* 1981; Goodman, 1982) must be interpreted in terms of changes in the interface state density.

ACKNOWLEDGMENTS

I wish to thank my students and colleagues for many helpful discussions. I am grateful to J. Abelson, G. de Rosny, S. Veprek, W. B. Jackson, R. A. Street, M. J. Thompson, D. K. Biegelsen, and L. Ley for sending me preprints of their work and to P. Persans for providing me with his unpublished results. The work was supported by the National Science Foundation under grant No. DMR-8009225 and by the Materials Research Laboratory of the University of Chicago under National Science Foundation Grant No. DMR-7924007.

REFERENCES

Abelson, J., and de Rosny, G. (1983). *J. Phys. Orsay, Fr.* **44,** 993.
Adler, R., Janes D., Hunsinger B. J., and Datta S. (1981). *Appl. Phys. Lett.* **38,** 102.
Aker, B., and Fritzsche H. (1983). *J. Appl. Phys.,* **54,** 6628.
Aker, B., Peng, S.-Q., Cai, S.-Y., and Fritzsche, H. (1983). *J. Non-Cryst. Solids* **59/60,** 509.
Ast, D. G., and Brodsky, M. H. (1979). *Conf. Ser. Inst. Phys.* **43,** 1159.
Auld, B. A. (1973). "Acoustic Fields and Waves in Solids," Vol. II. Wiley (Interscience), New York.
Bløtekjaer, K. (1970). *IEEE Trans. Electron Devices* **ED-17,** 30.
Brodsky, M. H., Evangelisti, F., Fischer, R., Johnson, R. W., Renter, W., and Solomon, I. (1980). *Solar Cells* **2,** 401.
Caplan, P. J., Poindexter, E. H., and Morrison, C. R. (1982). *J. Appl. Phys.* **53,** 541.
Carlson, D. E., Magee, C. W., and Triano, A. R. (1979). *J. Electrochem. Soc.* **126,** 688.
Chen, K.-J., and Fritzsche, H. (1983). *J. Non-Cryst. Solids* **59/60,** 441.
Demond, F. J., Müller, G., Damjantschitsch H., Mannsperger, H., Kalbitzer, S., LeComber, P. G., and Spear, W. E. (1982). *J. Phys. Colloq. Orsay, Fr.* **42,** C4-779.
DeVore, H. B. (1956). *Phys. Rev.* **102,** 86.

Fritzsche, H. (1980a). *Solar Energy Mat.* **3,** 447.

Fritzsche, H. (1980b). *Solar Cells* **2,** 289.

Fritzsche, H. (1982). *Thin Solid Films* **90,** 119.

Fritzsche, H., and Chen K.- J. (1983). *Phys. Rev. B* **28,** 4900.

Fritzsche, H., and Tsai, C. C. (1979). *Solar Energy Mat.* **1,** 471.

Gatos, H. C., and Lagowski, J. (1973). *J. Vac. Sci. Technol.* **10,** 130.

Gimzewski, J. K., and Vepřek, S. (1983). *Solid State Commun.* **47,** 747.

Goetzberger, A., Klausman, E., and Schulz, M. J. (1976). *CRC Crit. Rev. Solid State Sci.* **6,** 1.

Goldstein, B., and Szostak, D. J. (1980). *Surf. Sci.* **99,** 235.

Goldstein, Y., Abeles, B., Wronski, C. R., Keleman, S. R., and Witzke, H. (1982). *J. Electron. Mater.* **11,** 191.

Goodman, N. (1982). *Philos. Mag. B* **45,** 407.

Goodman, N. B., and Fritzsche, H. (1980). *Philos. Mag. B* **42,** 149.

Goodman, N. B., and Fritzsche, H. (1981). *AIP Conf. Proc.* **73,** 176.

Gundry, P. M., and Tompkins, F. C. (1968). *In* "Experimental Methods in Catalytic Research" (R. B. Anderson, ed.), p. 100. Academic Press, New York.

Jackson, W. B., and Thompson, M. J. (1983). *Physica Amsterdam* **117B/118B,** 883.

Jackson, W. B., Biegelsen, D. K., Nemanich, R. J., and Knights, J. C. (1983a). *Appl. Phys. Lett.* **42,** 105.

Jackson, W. B., Street, R. A., and Thompson, M. J. (1983b). *Solid State Commun.* **47,** 435.

Jäntsch, O. (1965). *J. Phys. Chem. Solids* **26,** 1233.

Johnson, N. M., Biegelsen, D. K., Moyer, M. D., Chang, S. T., Poindexter, E. H., and Caplan, P. J. (1983). *Appl. Phys. Lett.* **43,** 563.

Kasupke, N., and Henzler, M. (1980). *Surf. Sci.* **92,** 402.

Kiselev, V. F., Kozlov, S. N., Novototskii-Vlasov, Y. F., and Prudnikov, R. N. (1968). *Surf. Sci.* **11,** 111.

Knights, J. (1979). *Jpn. J. Appl. Phys.* **18,** 101.

Knights, J. (1980). *J. Non-Cryst. Solids* **35/36,** 159.

Knights, J., and Lujan, R. A. (1979). *Appl. Phys. Lett.* **35,** 244.

Knights, J. C., Biegelsen, D. K., and Solomon, I. (1977). *Solid State Commun.* **22,** 133.

Knights, J. C., Lujan, R. A., Rosenblum, M. P., Street, R. A., Biegelsen, D. K., and Reimer, J. A. (1981). *Appl. Phys. Lett.* **38,** 331.

Kozlov, S. N., Kiselev, V. F., and Novototskii-Vlasov, Y. F. (1971). *Surf. Sci.* **28,** 395.

Lagowski, J., Balestra, C. L., and Gatos, H. C. (1971). *Surf. Sci.* **27,** 547.

Laughlin, R. B., Joannopoulos, J. D., and Chadi, D. J. (1980). *Phys. Rev. B* **21,** 5733.

Ley, L. (1984). *In* "Hydrogenated Amorphous Silicon" (J. Joannopoulos and G. Lucovsky, eds.), Vol. 1. Springer-Verlag, Berlin and New York.

Ley, L., Richter, H., Kärcher, R., Johnson, R. L., and Reichardt, J. (1981). *J. Phys. Colloq. Orsay, Fr.* **42,** C4-753.

Lukes, F. (1972). *Surf. Sci.* **30,** 91.

Magee, C. W., and Botnick, E. M. (1981). *J. Vac. Sci. Technol.* **19,** 47.

Magee, C. W., and Carlson, D. E. (1980). *Solar Cells* **2,** 365.

Mailhiot, C., Currie, J. F., Sapicha, S., Wertheimer, M. R., and Yelon, A. (1980). *J. Non-Cryst. Solids* **35–36,** 207.

Many, A. (1974). *CRC Crit. Rev. Solid State Sci.* **4,** 515.

Many, A., Goldstein, Y., and Grover, N. B. (1965). "Semiconductor Surfaces." North-Holland Publ., Amsterdam.

Morrison, S. R. (1957). *In* "Semiconductor Surface Physics" (R. H. Kingston, ed.), p. 159. Univ. of Pennsylvania Press, Philadelphia, Pa.

Müller, G., Demond, F., Kalbitzer, S., Damjantschitsch, H., Mannsperger, H., Spear, W. E., LeComber, P. G., and Gibson, R. A. (1980). *Philos. Mag. B* **41,** 571.
Peng, S.-Q., and Cai, S.-Y. (1983). Private communication.
Persans, P. D. (1980). Private communication.
Ponpon, J. P., and Bourdon, B. (1982). *Solid State Electron.* **25,** 875.
Powell, M. J. (1981). *Philos. Mag. B* **43,** 93.
Powell, M. J., and Pritchard, J., (1983). *J. Appl. Phys.* **54,** 3244.
Redfield, D. (1983). *J. Appl. Phys.* **54,** 2860.
Schetzina, J. F. (1979). *Phys. Rev. B* **19,** 3313.
Schiff, E. A., Persans, P. D., Fritzsche, H., and Akopyan, V. (1981). *Appl. Phys. Lett.* **38,** 92.
Solomon, I., and Brodsky, M. H. (1980). *J. Appl. Phys.* **51,** 4548.
Solomon, I., Dietl, T., and Kaplan, D. (1978). *J. Phys. Orsay, Fr.* **39,** 1241.
Spear, W. E., and LeComber, P. G. (1972). *J. Non-Cryst. Solids* **8–10,** 727.
Staebler, D. L., and Wronski, C. R. (1977). *Appl. Phys. Lett.* **31,** 292.
Staebler, D. L., and Wronski, C. R. (1980). *J. Appl. Phys.* **51,** 3262.
Street, R. A. (1982). *Phys. Rev. Lett.* **49,** 1187.
Street, R. A., and Knights, J. C. (1981). *Philos. Mag. B* **43,** 1091.
Surplice, N. A., and D'Arcy, R. J. (1970). *J. Phys. E* **3,** 477.
Suzuki, T., Hirose, M., Ueda, M., and Osaka, Y. (1982). *Solar Energy Mat.* **8,** 285.
Sze, S. M. (1969). "Physics of Semiconductor Devices." Wiley (Interscience), New York.
Tanielian, M. (1982). *Philos. Mag. B* **45,** 435.
Tanielian, M., Fritzsche, H., Tsai, C. C., and Symbalisty, E. (1978). *Appl. Phys. Lett.* **33,** 353.
Tanielian, M., Chatani, M., Fritzsche, H., Šmid, V., and Persans, P. D. (1980). *J. Non-Cryst. Solids* **35–36,** 575.
Tanielian, M. H., Goodman, N. B., and Fritzsche, H. (1981). *J. Phys. Colloq. Orsay, Fr.* **42,** C4-375.
Vaid, J., and Fritzsche, H. (1983). *J. Appl. Phys.* **55,** 440.
Viktorovitch, P. (1981). *J. Appl. Phys.* **52,** 1392.
Von Roedern, B., Ley, L., and Cardona, M. (1979). *Solid State Commun.* **29,** 415.
Wagner, L. P., and Spicer, W. E. (1974). *Phys. Rev. B* **9,** 1512.
Weisfield, R., Viktorovitch, P., Anderson, D. A., and Paul, W. (1981). *Appl. Phys. Lett.* **39,** 263.
Wey, H.-Y. (1976). *Phys. Rev. B* **13,** 3495.
Wey, H.-Y., and Fritzsche, H. (1972). *Proc. Int. Conf. Phys. Semicond., 11th, Warsaw, 1972,* p. 555.
Williams, R. H., Varma, R. R., Spear, W. E., and LeComber, P. G. (1979). *J. Phys. C* **12,** L209.

CHAPTER 10

The Staebler–Wronski Effect

C. R. Wronski

CORPORATE RESEARCH SCIENCE LABORATORIES
EXXON RESEARCH AND ENGINEERING COMPANY
ANNANDALE, NEW JERSEY

I. Introduction

In 1977, Staebler and Wronski discovered reversible, light-induced changes in the photoelectronic properties of a-Si : H. They found that prolonged exposure to light of undoped and doped a-Si : H films decreases both the dark conductivity and the photoconductivity and that these conductivities can be restored to their original state by subsequently annealing the films in the dark. Such photoinduced changes, commonly called the Staebler–Wronski (S–W) effect, are unique in that (1) they are perfectly reversible, (2) they can involve enormous conductivity changes, and (3) they are perfectly stable at room temperature. Because these changes have strong implications for the physical nature of a-Si : H and its technological applications, considerable interest has arisen in the S–W effect.

In the original studies of the S–W effect on undoped and doped a-Si : H films, the principal characteristics of the S–W effect and the presence of metastable defects in a-Si : H were established (Staebler and Wronski, 1977; Wronski, 1978; Staebler and Wronski, 1980). It was found that the large light-induced conductivity changes are a bulk phenomenon that occurs between what may be considered a thermally stable state A and a new metastable conductivity state B. State A is perfectly reproducible and is independent of previous exposures to light. It is obtained after the a-Si : H film is annealed (in the dark) at temperatures above $\sim 150°C$ and then cooled to room temperature. The annealing time required for state A depends on the

ISBN 0-12-752149-6

temperature. These relaxation times exhibit high activation energies. State B is also reproducible and is stable at room temperature; however, it depends on the light exposure used, so that, in contrast to state A, which has a clearly defined conductivity, the metastable state B can have a range of conductivities.

Reversible photoinduced conductivity changes have been observed on a-Si:H films fabricated under a wide range of deposition conditions. This includes films produced by plasma decomposition of SiH_4 (Staebler and Wronski, 1977, 1980), SiH_4 diluted with H_2 (Hirabayashi *et al.*, 1980), SiH_4 diluted with Ar (Tanielian *et al.*, 1981), SiH_4 diluted with SiF_4 (Johnson *et al.*, 1981), reactive sputtering (Maruska *et al.*, 1983), and homogenous chemical vapor deposition (Scott *et al.*, 1982). Such changes have also been observed in hydrogenated amorphous silicon–germanium alloys (Nakamura *et al.*, 1983). Although detailed studies of the conductivity changes have not been reported for all of these films, it appears that they all exhibit the characteristics associated with the S–W effect. Reversible light-induced changes have also been observed in the photoluminescence of a-Si:H films (Morigaki *et al.*, 1980; Pankove and Berkeyheiser, 1980) and in the photovoltaic properties of a-Si:H (Wronski, 1978; Staebler and Wronski, 1980; Jousse *et al.*, 1980). These changes are also associated with the S–W effect since they exhibit both light exposure and annealing characteristics similar to those of the conductivity changes.

There have been a large number of studies on reversible light-induced changes in various a-Si:H properties including the changes in gap states. The wide range of results that have been reported by investigators using different experimental techniques clearly indicate that, just as in the case of the electronic properties, the photoinduced reversible changes depend on the fabrication conditions, the doping, and the impurities in the different films. However, the absence of detailed characterization of the different films has left many unanswered questions about the mechanisms and the defects responsible for the S–W effects. This includes the possibility that there may be several different defects that are responsible for the wide range of photoinduced changes observed in different a-Si:H films. Rather than attempt to review all the work that has been reported on photoinduced metastable changes in a-Si:H films, this chapter will focus on the characteristics of the photoinduced changes that are most directly associated with the S–W effect.

II. Dark Conductivity Changes

Light-induced changes in the dark conductivities of a-Si:H have been observed in a large variety of films and are generally used to compare the S–W effect in different films. Under illumination the transitions in dark conductivities and photoconductivities from the thermally stable state A to a metastable state B take the form of a continuous rather than abrupt change in which the rate of these transitions depends on the a-Si:H film and on the level of illumination. This rate has been found to depend linearly or sublinearly on light intensity, and no evidence has been found for a threshold, so that changes in conductivity occur even in room light. It is therefore important to take such changes into account in order to avoid significant scatter in the experimental results. In the original and many of the subsequent studies, the intensities used to characterize photoinduced changes were between 100 and 200 mW cm^{-2} of red light. Such intensities are used because, not only do they correspond to illuminations of 1 to 2 suns, but also in many cases several hours of such illumination can result in a "light-soaked" state B. One knows that the material is in such a state B when the dark conductivity changes very little, if at all, even after much longer exposures to light.

Examples of conductivity changes that accompany the S–W effect are illustrated in Fig. 1 with results obtained with two a-Si:H films made by the rf-discharge decomposition of silane onto substrates at ~ 320°C. Figure 1 shows the decrease in photoconductivity (solid curves) and the dark conductivity (dashed curves) during illumination with 200 mW cm^{-2} of filtered tungsten light (0.6 to 0.7 μm). Both films exhibit similar characteristics even though the film in Fig. 1a was made from pure SiH_4 and the film in Fig. 1b was made from SiH_4 containing 0.1% PH_3. The dark conductivities, measured by periodically stopping the illumination, exhibit new values, which are stable for days after the illumination is interrupted. This type of conductivity decrease clearly indicates that the films undergo a gradual and continuous change; it is not a sudden change of state. The results in Fig. 1 also show that the dark conductivities decrease rapidly at first and then appear to saturate after about 4 hr of illumination. The rate of decay depends on the illumination intensity. Any given state B can be obtained by using different intensities of bulk-absorbed light by suitably adjusting the exposure time.

In spite of the similarity of the results for the two films shown in Fig. 1, large differences have been observed in the sensitivity of the dark conductivities to prior light exposure. A wide range in both rate and magnitude of dark conductivity changes has been observed for a-Si:H films deposited under different conditions. When the reported changes are expressed as the ratio of conductivity in state A to that in state B, the values of σ_A/σ_B range from virtually zero to 10^5. Despite these large differences, there have been very few studies of the dynamics of these changes; hence there is only limited information available about the possible mechanisms that determine the rates at which the conductivities change. The absence of studies of the dynamics of the changes in conductivity also introduces ambiguities into the comparison of results reported by different investigators using different ex-

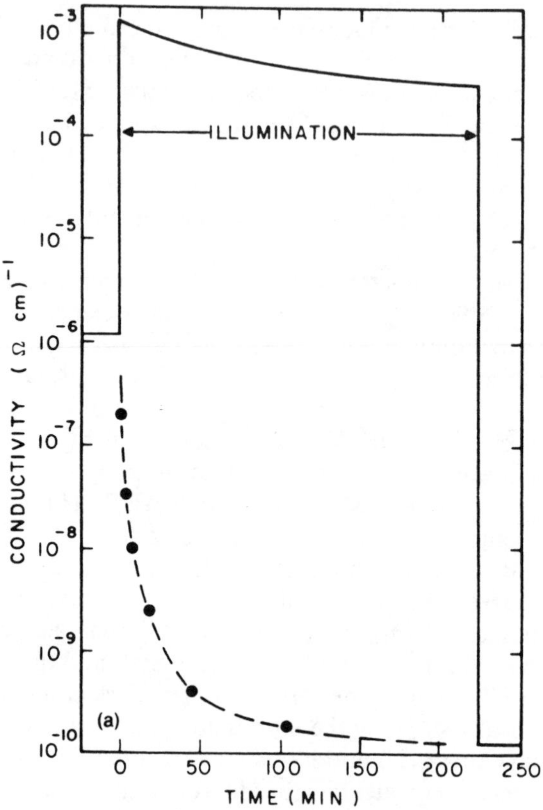

FIG. 1. Decrease in the photoconductivity (solid line) and dark conductivity (dashed line) of a-Si : H films during illumination with 200-mW-cm^{-2} filtered tungsten light. (a) Undoped and (b) doped with 0.1% of PH_3. [From Staebler and Wronski (1980).]

perimental conditions on different films. For a quantitative comparison between different films it is important to establish that the reported state B is truly light soaked and represents the maximum change in the dark conductivity of the film. Because of the wide range of illumination used and the different rates of conductivity change in different films, state B may not be achieved after only a few hours of illumination. There are cases in which much longer exposures to the 1–2-sun illumination is required, particularly when the rates of the conductivity changes are significantly smaller than those illustrated in Fig. 1.

The results shown in Fig. 1 represent the changes of the conductivities that occur in the bulk of the a-Si : H films. The fact that the S–W effect is a bulk effect was established by the results obtained on films of different thicknesses and by using different electrode configurations (Wronski, 1978; Staebler and

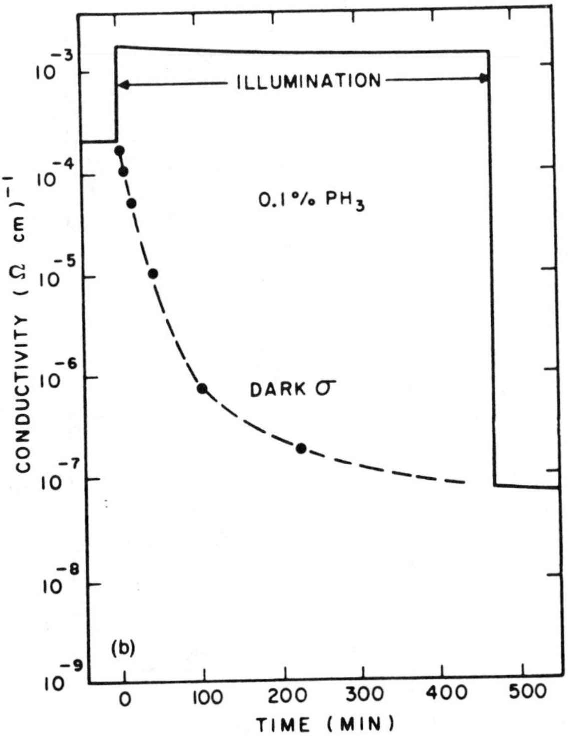

FIG. 1. *(continued)*

Wronski, 1980). By using transverse rather than parallel electrode geometry, the changes in bulk conductivity can be clearly distinguished. This is illustrated by the results in Fig. 2, which were obtained on an *n*-type a-Si:H (200 Å)/undoped a-Si:H (1 μm)/Pd (100 Å) Schottky-barrier structure. The undoped a-Si:H in this structure was deposited by dc discharge decomposition of SiH_4 onto a substrate at 300°C. Figure 2 shows the current densities obtained under far-forward bias of the diode in the region in which these currents are limited by the bulk resistance of the undoped a-Si:H. The ohmic dark currents and photocurrents (generated by 100 mW cm^{-2} of white light) are for states A and B obtained with an exposure similar to that used in the results of Fig. 1. The results obtained with this structure for the conductivity changes were in agreement with those obtained with coplanar electrodes on an undoped film codeposited onto quartz.

In characterizing the conductivity changes in a-Si:H films it is important to establish that the results reflect bulk changes. This is not always the case when measurements are carried out by using coplanar electrodes with thin a-Si:H films deposited onto insulating substrates. The results obtained by

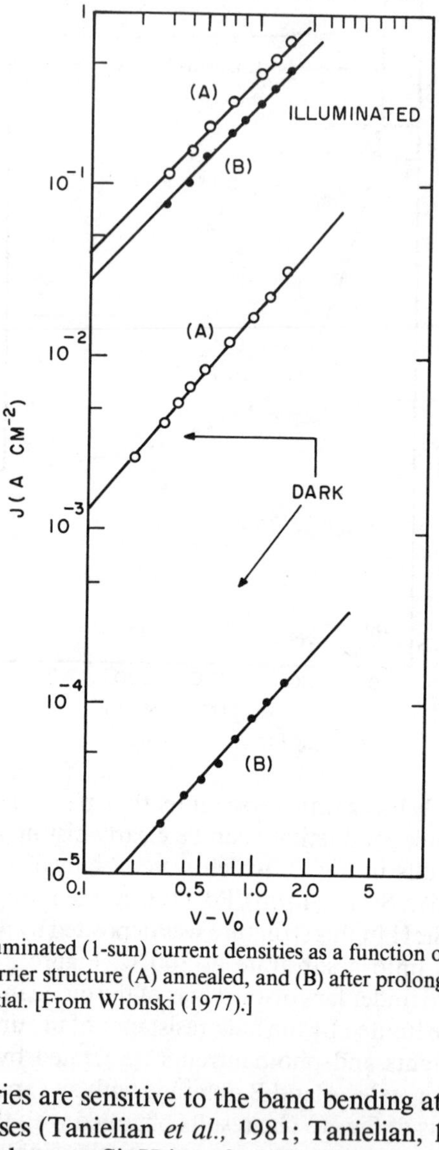

FIG. 2. Dark and illuminated (1-sun) current densities as a function of forward bias for an a-Si : H/Pd Schottky-barrier structure (A) annealed, and (B) after prolonged exposure to light. V_0 is the built-in potential. [From Wronski (1977).]

using such geometries are sensitive to the band bending at the free surface due to adsorbed gases (Tanielian *et al.,* 1981; Tanielian, 1982) and to the charging of the insulator–a-Si : H interface (Solomon *et al.,* 1978). This is particularly true for a-Si : H films, which have low densities of gap states, so that even small band bending results in a space-charge layer that extends over several tenths of a micrometer. In such cases, any light-induced change in band bending has a large effect on the conductivity measured parallel to the film and must be taken into account. Although these effects can be minimized, it is advantageous to carry out measurements using transverse electrode geometries with truly ohmic contacts.

The continuous decrease in dark conductivity that occurs under illumination is primarily due to a decrease in the density of free carriers. The movement of the Fermi level required for such a change is suggested by the observation that as the conductivity decreases there is a corresponding increase in activation energy. This is illustrated in Fig. 3, in which the temperature dependence of dark conductivity for the undoped and phosphorus-doped films of Fig. 1 is shown for different exposures to light. The A lines correspond to state A and the other lines to the different states B obtained after exposure to 200 mW cm^{-2} of red light for the number of minutes indicated in Fig. 3. The duration and temperature during the measurements

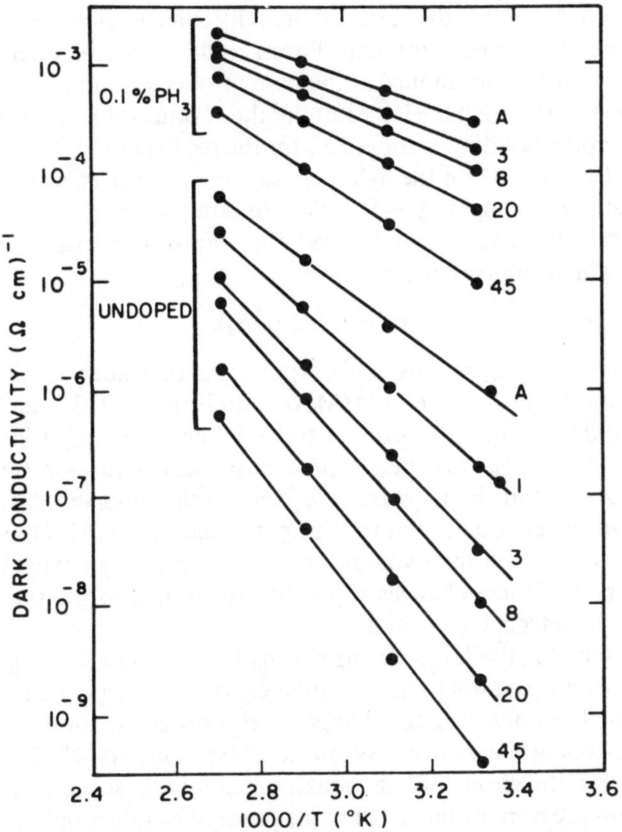

FIG. 3. Temperature dependence of the dark conductivity for the two a-Si:H films of Fig. 1. The A lines are for after annealing and the others for after optical exposure. The numbers indicate the exposure time in minutes. [From Staebler and Wronski (1980).]

were too low to cause any change in state B. Over the temperature range shown, all the conductivities σ follow the form

$$\sigma(T) = \sigma_0 \exp(-E_a/kt), \quad (1)$$

where σ_0 (the pre-exponential) and E_a (the activation energy) are independent of temperature, k is the Boltzmann factor, and T is the absolute temperature. These results show that as the conductivities decrease there is a continuous increase in the activation energy, which is given by the slope of the lines. The values of E_a in Fig. 3 undergo very large changes, from ~ 0.55 to 0.9 eV for the undoped film and from 0.3 to 0.55 eV for the phosphorus-doped film. Since both films are n-type, these changes are consistent with the displacement of the Fermi level away from the electron mobility edge by as much as several tenths of an electron volt.

A direct correlation of the dark conductivity changes induced by light with the corresponding movement of the Fermi level is, however, complicated by the presence of the simultaneous increase in both E_a and σ_0. This is illustrated in Fig. 4, in which the logarithm of the various values of σ_0 are plotted versus the corresponding values of E_a for the results of Fig. 4. These results show that for a given film the light exposure can change σ_0 by five orders of magnitude as E_a changes by ~ 0.4 eV. The straight line drawn through the results of both the undoped and doped film corresponds to a dependence of σ_0 on E_a, which can be written

$$\sigma_0 = C \exp(DE_a), \quad (2)$$

where C and D are constants. This exponential dependence of σ_0 and E_a, known as the Meyer–Neldel rule (Meyer and Neldel, 1937), can be seen in Fig. 4 to hold for values of σ_0 from 10 to 10^7 $(\Omega\ \text{cm})^{-1}$ and E_a from 0.3 to 1.0 eV. The simultaneous decrease in free-carrier concentration (reflected in increases of E_a) and the increase in σ_0 lead to the observed "asymptotic" approach of the conductivities to the light-soaked state. Also the opposing effect of σ_0 and E_a on the dynamics of the conductivity changes does not allow the rate of these changes to be directly related to the rate at which light-induced defects are created.

The reasons for the large changes in σ_0 brought about by light and its dependence on E_a are not clear. A similar exponential dependence of σ_0 and E_a have also been observed for changes in E_a brought about by both n-type and p-type doping (Carlson and Wronski, 1979; Tanielian, 1982). This similarity between the effects of light soaking and doping suggests that similar mechanisms are responsible for the wide range of σ_0 values obtained in both cases. However, the results that have been reported have such a large scatter that the fit of the Meyer–Neldel rule can result in a wide range of values for the constants C and D in Eq. (2). Also, there is no general agreement about

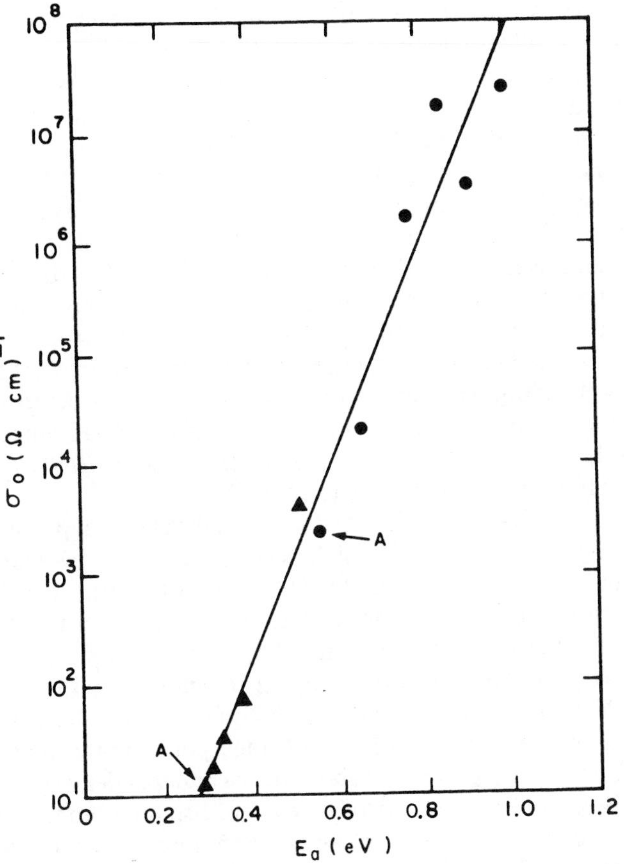

FIG. 4. The pre-exponential factor σ_0 versus the activation energy E_a for the data shown in Fig. 3. The points marked A are for after annealing and the others for after optical exposure. ●, undoped; ▲, 0.17. PH_3. [From Staebler and Wronski (1980).]

the mechanisms that can cause the experimentally observed large range of σ_0 values, and it has been suggested that inhomogeneities of defects in the a-Si:H have an important role (Fritzche, 1980). Hauschildt *et al.* (1982) have explained their results on the photoinduced changes in free-electron transport in phosphorus-doped a-Si:H films by the introduction of light-induced defects that are inhomogenously distributed. However, they also report that the photoinduced changes were quite different from those observed as a result of phosphorus doping.

There is general agreement that the changes in E_a that accompany the photoinduced changes in the dark conductivity reflect displacements of the Fermi level brought about by the introduction of metastable defects into the gap of a-Si : H. Although the results shown here indicate similar displacements of the Fermi level for both films, large differences in the displacements can be obtained, as indicated by the wide range of σ_A/σ_B values that have been reported. These displacements are determined not only by the energies and densities of the light-induced defects but also by the densities of gap states and the location of the Fermi level in state A. There is still a large degree of uncertainty about the interpretation of the various results in terms of the nature of the photoinduced defect and its location in the gap of a-Si : H. There is evidence, however, that in a variety of a-Si : H films the ligh-induced defects responsible for the conductivity changes have densities of $10^{16}-10^{17}\ cm^{-3}$ (Wronski and Daniel, 1980). Such densities located near midgap can cause large displacements of the Fermi level provided that the Fermi level is not pinned by larger densities of donors or acceptors. Since for undoped and lightly doped a-Si : H films the densities of gap states below the Fermi level of state A are also $10^{16}-10^{17}\ cm^{-3}$, comparable densities of defects can result in the changes shown in Fig. 4.

The results that have been reported so far on the effects of phosphorus and boron doping on the photoinduced dark conductivity changes are also consistent with this general model. Heavy doping with both phosphorus and boron drastically reduces the conductivity changes to values that cannot be detected even after very long exposure to high levels of illumination (Staebler and Wronski, 1980). On the other hand, Tanielian *et al.* (1981) found that over a wide range of *n*-type and *p*-type conductivities there was a large dependence on the low levels of dopants incorporated into the films. They not only found σ_A/σ_B values that showed a clear dependence on the activation energies of the conductivity of state A but also reported that on slightly *n*-type films dark conductivities can *increase* rather than decrease upon exposure to light. These increases are difficult to measure because they are small in samples having very low conductivities. Nevertheless they correspond to a minimum in the reported values of σ_A/σ_B, which increase as the a-Si : H films become more *n* and *p* type. This minimum corresponds to films that have activation energies of conductivities of ~ 1 eV and Fermi levels near midgap. The above results indicate that there is a "disappearance" of light-induced conductivity changes in both heavily doped and highly resistive a-Si : H films. In the case of the heavily doped films the densities of photoinduced defects are too small to overcome the pinning of the Fermi level by the significantly larger densities of donors and acceptors. The absence of large changes in the highly resistive films and the symmetric behavior of *n*- and *p*-type samples (Tanielian *et al.*, 1981) suggest that both donor and acceptor states are created in the photoinduced effect. The small photoinduced changes in dark conductivity observed in the highly resistive films can occur, even if the densities of gap states are low, because the Fermi level is near midgap and there may be a self-compensating effect within the light-induced states.

The reversible photoinduced changes in dark conductivities are consistent with metastable changes in the gap states of a-Si:H, but it is not yet clear whether a unique type of light-induced defect can account for the wide range of conductivity changes that have been reported for a-Si:H films. This question can be answered only after a more detailed characterization has been carried out on both the different photoinduced conductivity changes and the gap states.

III. Reversal of Light-Induced Changes

The photoinduced changes that are associated with the S–W effect have the striking characteristic that they are stable at room temperature and yet can be reversed by annealing in the dark at moderate temperatures. The early studies of the light-induced changes in the dark conductivities and photoconductivities of a-Si:H films showed that these changes can be reproducibly reversed by annealing the films in the dark for about an hour or less at temperatures above ~ 150°C (Staebler and Wronski, 1977). Subsequently, this type of annealing was found to reverse the photoinduced changes in a variety of a-Si:H properties, including changes in photovoltaic properties (Wronski, 1978; Staebler and Wronski, 1980; Jousse *et al.*, 1980), photoluminescence (Morigaki *et al.*, 1980; Pankove and Berkeyheiser, 1980), electron spin density (Dersch *et al.*, 1981), and density of gap states (Tanielian *et al.*, 1981; Lang *et al.*, 1982). The ability to reverse all of these photoinduced changes by the same kind of thermal treatment suggests that they result from the same kinds of metastable defects. However, only a few studies have been reported on the kinetics of the annealing process, which would establish similarities or differences that would add further insights into the differences in the light-induced changes that are observed in different a-Si:H films.

A detailed investigation of the kinetics of the conductivity changes from any given metastable state B to the thermally stable state A has been reported for undoped a-Si:H films (Staebler and Wronski, 1977, 1980). It was found in these studies that just as the rate of the transition from state A to state B depends on the intensity of the light exposure, the rate at which the conductivities revert back to state A depends on the annealing temperature. It was also found that the time required to obtain state A at a given temperature depends on prior illumination and that the recovery rate becomes very slow as the annealing temperature falls below ~ 150°C. These results clearly indicate that the transition from state B to state A is a thermally activated process. This is illustrated in Fig. 5, in which the temperature dependence of the annealing process from state B to state A is shown for the undoped

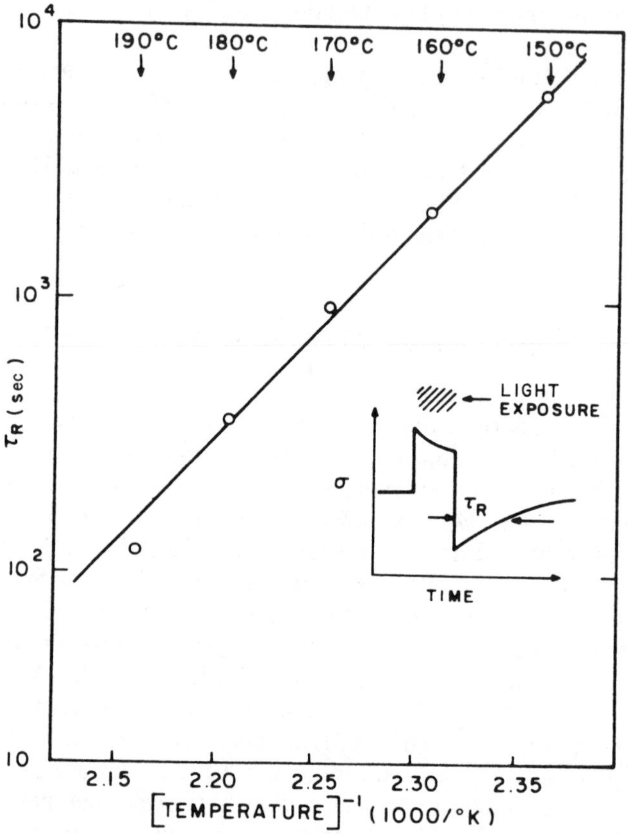

FIG. 5. Thermal relaxation time as a function of temperature. The measurements were made at the indicated temperatures as shown in the inset. [From Staebler and Wronski (1980).]

sample of Fig. 1. The annealing kinetics in these experiments were measured by exposing the films to penetrating light at temperatures below 150°C and then allowing them to revert to their original state A. (This is illustrated in the inset of Fig. 5, in which τ_r is defined as the time required to remove half of the light-induced effect.) The relaxation times measured in this way agreed with the values obtained by exposing the films at room temperature and then rapidly heating them to the higher annealing temperatures. In Fig. 5 the relaxation time τ_r obtained at the different temperatures is plotted versus the inverse temperature of the experiment. The slope of the line gives an activation energy of 1.5 eV. It can be noted here that the activation energy found for these films has a striking similarity to the 1.5-eV activation energy obtained for the diffusion constant of hydrogen and deuterium in a similar a-Si:H film (Carlson and Magee, 1979).

The lack of systematic studies on the kinetics of the annealing process for different a-Si : H films is partly due to the difficulties that can be encountered with measurements such as those already described (Staebler and Wronski, 1980). The differences observed by Staebler and Wronski (1980) and the absence of results on a-Si : H films having different light-induced conductivity changes therefore allow for variation in activation energies that could result from different mechanisms and defects. However, results have been reported on the annealing characteristics of light-induced changes in photoluminescence (Pankove and Berkeyheiser, 1980) and electron spin resonance (Dersch *et al.*, 1981) that indicate that similar kinds of light-induced defects contribute to the changes in the different properties of undoped a-Si : H films.

Pankove and Berkeyheiser (1980) reported on a study of the temperature dependence of reversible changes in photoluminescence intensity for undoped, glow-discharge-produced a-Si : H films. There is a change in photoluminescence analogous to the reversible light-induced decrease in bulk photoconductivity. It is found that prolonged illumination with the high intensity laser light, which is used to generate the photoluminescence, decreases the efficiency at the 1.2-eV peak, that this fatigue persists at room temperature for many hours, and that it can be reproducibly erased by a short anneal at 200°C. Pankove and Berkeyheiser also found that over a wide temperature range there was a direct inverse relation between the photoluminescence efficiency of the 1.2-eV peak and that of the 0.8-eV peak that corresponds to a radiative recombination center. The results that they obtained on the light-induced 0.8-eV peak with isochronal, 10-min anneals in the dark at temperatures above 100°C are shown in Fig. 6, in which the relative decrease in the efficiency of the 0.8-eV peak $\Delta L/L$ is plotted versus annealing temperature T_A. The light-induced 0.8 eV peak was obtained by exposure to prolonged laser irradiation at the temperature T_F of 200°C. Even though no result was reported on corresponding conductivity changes, the similarity between these results and the kinetics discussed earlier is consistent with their suggestion that the deep centers responsible for the 0.8-eV peak also contribute to the decrease in carrier lifetimes observed in light-induced photoconductivity changes.

Dersch *et al.* (1981) reported results on the annealing characteristics of light-induced dangling bonds in undoped, glow-discharge-produced a-Si : H films that have a striking similarity to the kinetics observed in the light-induced changes of both the conductivity and photoluminescence. They found that the electron spin densities in their film increased under prolonged illumination with light and that the light-induced spins disappear after annealing at 220°C. The annealing characteristics they report are shown in Fig. 7. In Fig. 7a is shown the decrease in the spin density N_s as a function of time

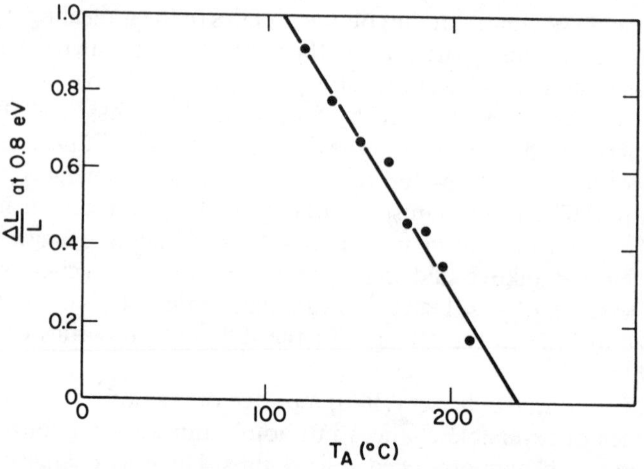

FIG. 6. Isochronal anneal of the 0.8-eV photoluminescence peak as a function of T_A. measured at 78°K after an annealing time of 10 min. The photoluminescence peak was generated at $T_f = 200$°C. [From Pankove and Berkeyheiser (1980).]

at a temperature of 220°C. The similarity to the relaxation of the dark conductivities shown in the inset of Fig. 5 can be clearly seen. In Fig. 7b is shown the decrease in N_s obtained with isochronal anneals of 15 min at the different temperatures. The similarlity of these results to those shown in Fig. 6 are also evident. The densities of spins, which are associated with the densities of dangling bonds, in Fig. 7b are larger than in Fig. 7a as a result of a longer exposure to light, so that it is not clear how close to the light-soaked

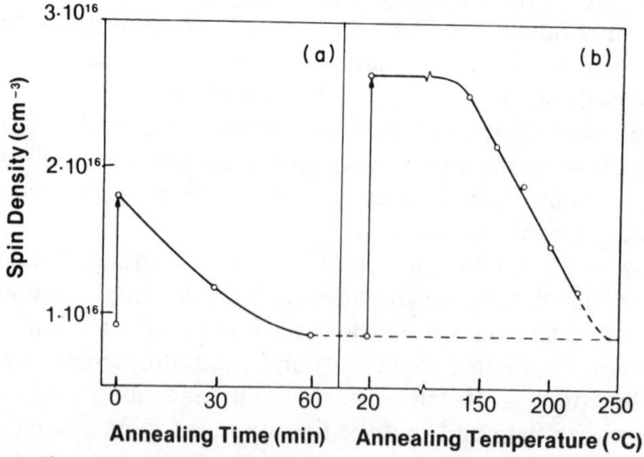

FIG. 7. Annealing characteristics of the light-induced spin densities: (a) isothermal anneal at 220°C and (b) isochronal anneal with an annealing time of 15 min. [From Dersch *et al.* (1981).]

state the a-Si:H film was in these experiments. However, it is noteworthy that the light-induced increases in the spins are only $\sim 2 \times 10^{16}$ cm^{-3}, which corresponds to the small changes in gap states that were discussed earlier. The observed similarity of the annealing characteristics of the conductivity, photoluminescence, and changes in the electron spin resonance clearly illustrates that dangling bonds and deep-lying defects are involved in the S–W effect. However, further studies are required to fully characterize the effect of these defects as well as of other states in the gap.

IV. Photoconductivity Changes

The reversible light-induced changes in photoconductivity that have been observed in a variety of a-Si:H films are also often used to compare the S–W effect in different films. The wide range of photoconductivity changes that have been reported exhibit dynamics during light exposure that are similar to those shown earlier in Fig. 1. However, results such as those shown in Fig. 1 are not adequate in evaluating the light-induced photoconductivity changes of different films because of the large differences in the photoconductivities of these a-Si:H films and the dependence of the photoconductivities on the intensity of illumination. A detailed evaluation that is necessary to characterize the photoconductivity changes however also allows valuable information to be obtained about the light-induced changes in the gap states of a-Si:H.

Photoconductivity σ_p is determined by the generation rate of free carriers by light f and their lifetimes τ. In the usual case in which photoconductivity is due to electrons, the photoconductivity is given by

$$\sigma_p = q\eta\mu_n n = q\eta\mu_n\tau_n f, \tag{3}$$

where n is the density of free electrons and μ_n their mobility in extended states. The generation rate of the carriers is determined by the number of photons absorbed per unit volume f and the photogeneration efficiency η, which in most cases is very close to unity. The free-carrier lifetimes are determined by the recombination kinetics, which depend strongly on the densities and types of states in the gap of the a-Si:H films. The effect of the photoinduced defects on the photoconductivity is not evident from results such as those shown in Fig. 1, because the same changes in photoconductivities can result from significantly different changes in the recombination kinetics (Wronski and Daniel, 1980). The recombination kinetics are reflected in the dependence of the photoconductivity σ_p on the intensity of illumination F. It is commonly found that

$$\sigma_p \propto F^{\gamma}, \tag{4}$$

where the values of γ are typically between 0.5 and 1. However, not only are different values of γ obtained in different films, but also the photoconductivities of the same film often exhibit two values of γ for different regions of illumination intensity. This is due to the differences in recombination kinetics that are present at the different densities of photogenerated carriers. The presence of such photoconductivity characteristics must be taken into account in evaluating the photoconductivity changes between state A and state B. The common practice of reporting photoconductivity changes at only one level of illumination represents an arbitrary criterion and also does not indicate the extent of the changes in the recombination that result from the exposure to light.

Detailed investigations of the light-induced photoconductivity changes have shown that large changes can occur in the recombination kinetics between state A and state B (Staebler and Wronski, 1977, 1980; Wronski and Daniel, 1980). An example of the changes that occur due to prolonged illumination is illustrated in Fig. 8, which uses the undoped film of Fig. 1. The photoconductivities, obtained by using bulk-absorbed red light, are for the film in state A and successive red light exposures of 200 mW cm^{-2} up to the maximum of 4 hr. The results after the different exposures are plotted versus the incident light intensity F from 10^{12} to 10^{17} photons cm^{-2} sec^{-1}. In Fig. 8 a continuous change in γ can be seen to occur as the photoconductivity changes with exposure to light. The values of $\gamma = 0.8$ and 0.5 in state A convert to a single $\gamma = 0.89$ after prolonged exposure to light. These changes clearly indicate that not only is there a decrease in the recombination lifetime, but also that over a wide range of illumination levels there are significant changes in the recombination kinetics. Such transitions from $\gamma \simeq 0.5$ to higher values are often, but not always, found to accompany the light-induced changes in photoconductivity. Their presence makes the photoconductivity changes much smaller for the higher levels of illumination. In the case shown here, the ratio of photoconductivities is 10^3 at $F = 10^{12}$ photons cm^{-2} sec^{-1}, but it is less than 10 at $F = 10^{17}$ photons cm^{-2} sec^{-1} (corresponding to ~100 mW cm^{-2} of red light). It is important to note that this intensity is comparable to the levels of illumination used to obtain state B. This makes the photoconductivity changes that are observed during exposure appear to be much less sensitive to light than the dark conductivity.

The light-induced changes in photoconductivity can in principle result from changes in the generation efficiency η, the mobility μ, or the free-carrier lifetime τ. However, the contributions of any changes in η and μ appear negligible. A photoinduced decrease in η might be expected to change the spectral dependence of the photoconductivity present in state A and state B. No change in the spectral response of photoconductivity has been observed for photon energy between 1.5 and 2.0 eV even though the absolute values of the photoconductivity changed by orders of magnitude (Staebler and Wronski, 1980). A change in mobility as a major contribution to the decrease in photoconductivity can also be ruled out by the results shown in Fig. 8. The initial decreases in photoconductivity due to the light exposure are in the $\gamma = 0.8$ portion of state A, leaving the $\gamma = 0.5$ portion virtually unchanged. Photoconductivity changes in which a decrease in mobility is important would, right from the start, lead to parallel downward displacements of the curves in Fig. 8, which clearly is not the case. This absence of any significant effect of light exposure on the mobility is also indicated by drift mobility measurements in which the same results are obtained for states A and B (Fuhs *et al.*, 1979). The light-induced changes in photoconductivity are therefore primarily due to the decrease in the carrier lifetime resulting from the introduction of metastable states into the gap.

The introduction of these light-induced states affects not only the carrier lifetimes but also the photoconductive response time τ_0. Because of the continuous distribution of gap states in a-Si:H, there are many more states that can trap photogenerated carriers close to the mobility edges than those that act as recombination centers. The response time depends on the density

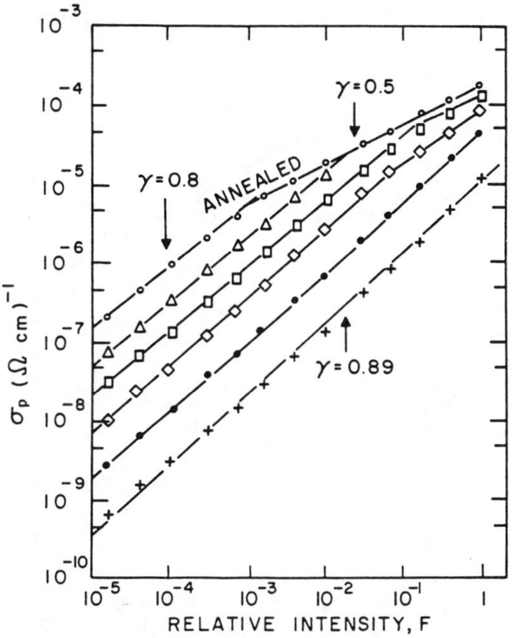

FIG. 8. Photoconductivity σ_p generated by penetrating light as a function of relative light intensity for the undoped film of Fig. 1 after annealing and subsequent prolonged exposures to light to a maximum of 4 hr. [From Staebler and Wronski (1980).]

of carriers trapped in these states, n_t for an n-type photoconductor, where τ_0 is given by

$$\tau_0 = (n_t/n)\tau_n. \tag{5}$$

Photoinduced changes in both τ_n and τ_0 have been observed in undoped a-Si:H_x films (Fuhs *et al.*, 1979; Wronski and Daniel, 1980), and an example is shown for an undoped, n-type a-Si:H film similar to that of Fig. 8. The results in Fig. 9 are for state A, and those in Fig. 10 are for state B obtained after a 4-hr exposure to red light at 200 mW cm^{-2}. In both figures the electron lifetimes τ_n [obtained by using Eq. (3) with $\eta\mu_n = 1$ cm^2 V^{-1} sec^{-1}] and the response times τ_0 are plotted versus the volume generation rate of carriers. A striking difference can be seen between Figs. 9 and 10 even though the carrier generation rates are the same in both figures. The prolonged exposure to light has greatly reduced the lifetimes and has changed the recombination from the bimolecular type ($\tau_n \propto f^{-0.5}$) to monomolecular, in which the lifetime is virtually independent of f. It has also significantly changed the photoconductive response times and the trapping kinetics reflected by the slopes of the corresponding lines in Figs. 9 and 10. The results shown here illustrate the large differences that can occur in both the recombination and trapping kinetics, which in many cases are much less evident. The

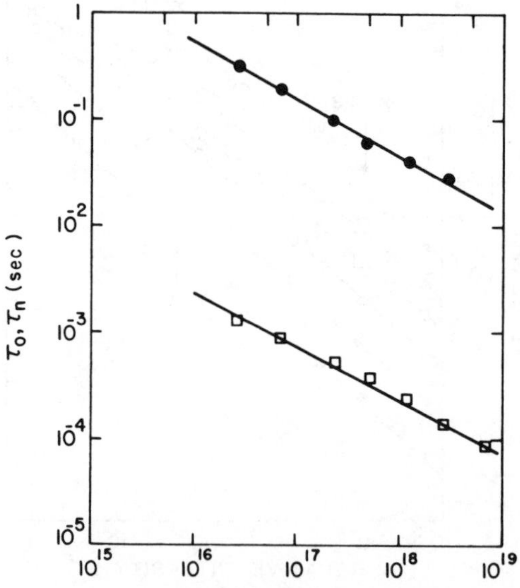

FIG. 9. Response time τ_0 and electron lifetime τ_n versus photogeneration rate for an undoped a-Si:H film in state A. ●, $\tau_0 \propto f^{-0.54}$; □, $\tau_n \propto f^{-0.5}$. [From Wronski and Daniel (1980).]

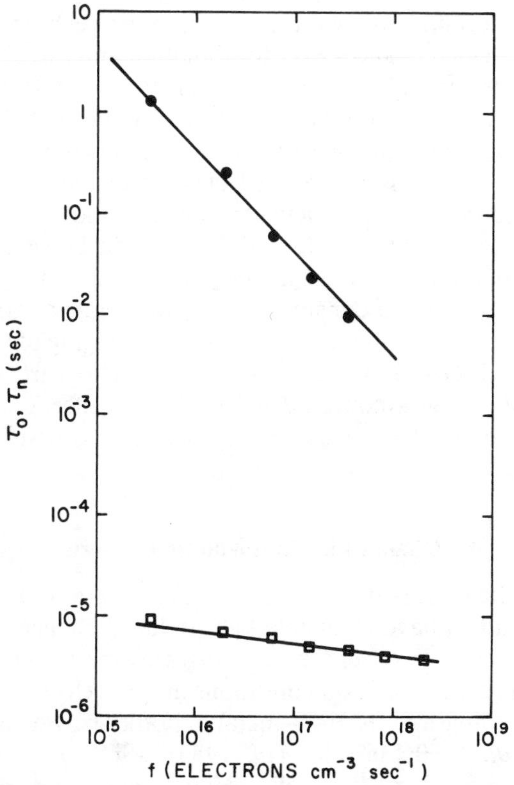

FIG. 10. Response time τ_0 and electron lifetime τ_n versus photogeneration rate for the same film as in Fig. 9 but after prolonged exposure to light. ●, $\tau_0 \propto f^{-1}$; □, $\tau_n \propto f^{-0.1}$. [From Wronski and Daniel (1980).]

variations that occur between different films stem from the sensitivity of these kinetics to the densities, carrier-capture cross sections, and energies of all the states in the gap of a-Si:H.

Light-induced changes in photoconductivity during light exposure have been reported for many different a-Si:H films, but there are few results on the effects of phosphorus and boron doping and of structural differences in the a-Si:H films. The few results available indicate that the largest light-induced photoconductivity changes are present in undoped, *n*-type a-Si:H films and that in all cases there is a decrease in photoconductivity (Tanielian, 1982). The more detailed studies of undoped a-Si:H films show that light introduces deep-lying centers that have electron capture cross sections significantly different from those that determine the carrier recombination in state A (Wronski and Daniel, 1980; Vanier, 1982). The photoinduced defects and their effect on photoconductivity have not yet been well characterized, but the observed changes in recombination and trapping kinetics are consistent with the introduction of $\sim 10^{16}$–10^{17}-cm^{-3} defects near midgap that is indicated by the results on the light-induced dark conductivity changes. Such densities can cause the large changes in the recombination lifetimes illustrated in Figs. 9 and 10 because of the large difference in electron capture cross section between the light-induced recombination centers and those present in state A (Wronski and Daniel, 1980; Vanier, 1982). Even though these centers are deep in the gap, the decreases in the carrier lifetimes affect the densities of photogenerated carriers and thus change not only the recombination but also the trapping of carriers in shallow states (Rose, 1960). However, in order to obtain a clearer understanding of the nature of the light-induced defects and their effect on photoconductivity, it will be necessary to have more detailed results on well-characterized a-Si : H films.

V. Changes in Photovoltaic Properties

The photovoltaic properties and the efficiency of a-Si : H solar cells are commonly found to have reversible light-induced changes that are often associated with the S – W effect. These changes are present in cells fabricated over a wide range of deposition conditions and in cells that have different structures, including a-Si : H *n–i–*metal Schottky-barrier cells (Wronski, 1978; Jousse *et al.,* 1980; Maruska *et al.,* 1983), a-Si : H *n–i–p* cells (Staebler *et al.,* 1981), a-Si : H *p–i–n* cells (Hanak and Korsun, 1982), and *p*(a-SiC : H)–*i*(a-Si : H)–*n*(a-Si : H) cells (Tsuda *et al.,* 1983). A variety of results has been reported for the light-induced changes in the cell characteristics of the different solar cells. Very often the changes are quite small, but in all cases they correspond to a deterioration in one or more of the cell parameters short-circuit current J_{sc}, open-circuit voltage V_{oc}, and the fill factor *FF*. It has been difficult to identify and evaluate all the mechanisms responsible for these changes because of the complexities resulting from multilayered structures and the commonly found significant differences between cells fabricated under similar conditions. Consequently, all the causes of the observed changes in cell characteristics are not well understood, but there is general agreement that the light-induced changes in the thick (undoped) *i* regions are important. The possible contributions of these regions to the reversible light-induced changes in cell characteristics can be most conveniently investigated by studying a-Si : H *n–i–*metal Schottky-barrier cell structures.

The light-induced changes in conductivity and gap states can in principle have very large effects on photovoltaic properties. The changes in bulk conductivity directly affect the series resistance of the cells. The changes in the gap states and the carrier recombination can have a serious effect on the junction properties and on the collection of free carriers. However, early studies (Wronski, 1978; Staebler and Wronski, 1980) showed that the large reversible light-induced changes in conductivity need not greatly affect the photovoltaic properties and performance of solar cells. It was found that even after prolonged light exposure, there were no discernible changes in the diode characteristics and barrier heights; that the changes in conductivity have a large effect only on the junction series resistance in the absence of bulk-absorbed light; and that the large changes in the recombination kinetics found in photoconductivity are not reflected in the collection of carriers across the junction of a photovoltaic device.

The junction properties of a photovoltaic device are important for the collection of photogenerated carriers and also determine the open-circuit voltages. These properties are often characterized by the diode characteristics under forward bias and by the short-circuit current J_{sc} and the open-circuit voltage V_{oc} obtained with different light intensities. The diode characteristics for voltages greater than ~ 50 mV are given by

$$J = J_0 \exp(qV/nkT) \tag{6}$$

in the dark, and under illumination by

$$J_{sc} = J_0 \exp(qV_{oc}/n'kT), \tag{7}$$

where J is the current density under forward bias voltage V, J_0 is the saturation current density, and n and n' are the diode quality factors obtained in the dark and under illumination, respectively. The changes in the $I(V)$ characteristics that are commonly found in a-Si:H n–i–metal Schottky barriers, even in the presence of large conductivity changes, are illustrated in Fig. 11.

The $J(V)$ and the white light $J_{sc}(V_{oc})$ diode characteristics observed in the a-Si:H n–i–Pd Schottky-barrier cell structure of Fig. 2 are shown in Fig. 11 for the annealed and light-soaked states. All the results in Fig. 11 exhibit the exponential regimes corresponding to Eqs. (6) and (7) with n and n' values close to unity, indicating in both cases that J_0 is not determined by carrier recombination in the junction. The exponential region for the $J(V)$ characteristics is affected only by the approximately 100-fold light-induced decrease in the dark conductivity and the series resistance of the bulk (indicated by the results in Fig. 2), which only reduces the region over which Eq. (6) is valid. The light exposure did not affect either the value of the diode quality factor or J_0, which indicates that there is no effect on the barrier height of the junction (Wronski and Carlson, 1977). It also had no effect at all on the $J_{sc}(V_{oc})$ characteristics, as indicated in Fig. 11 by the results represented by a

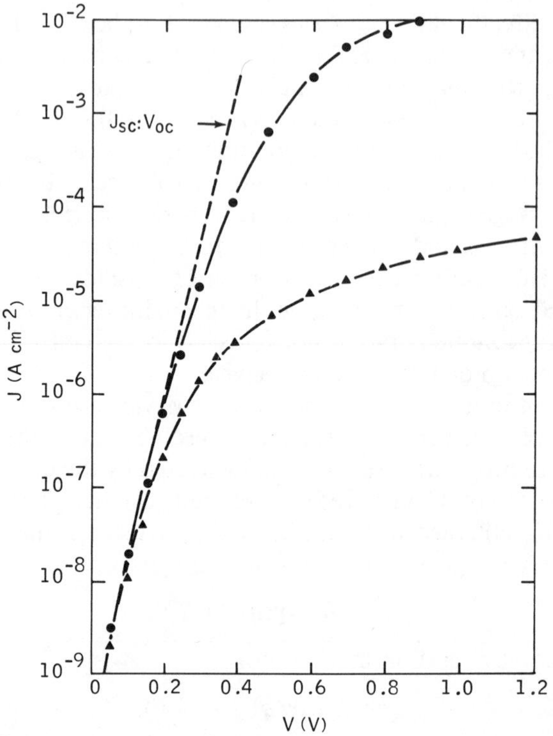

FIG. 11. The dark current density as a function of forward bias for (●) annealed and (▲) light-soaked states, and J_{sc} versus V_{oc} (dashed line) for an a-Si:H/Pd Schottky-barrier cell structure. [From Wronski (1977).]

single dashed line. The $J_{sc}(V_{oc})$ results also show that the J_0 values are the same under illumination as in the dark, but there is no longer a series resistance effect limiting the exponential behavior even at the highest illumination level of 1 sun. The absence of such a limitation is due to the fact that under white-light illumination the series resistance is determined by the photoconductivity that is generated by the red photons absorbed in the bulk and is still high enough in the light-soaked state to eliminate any significant effect on series resistance. The effect of light exposure on the junction characteristics, such as those illustrated here, is very important to solar-cell operation because, as given by Eq. (6), these characteristics determine the open-circuit voltage. The stability of V_{oc} to prolonged illumination that is indicated by the results presented here is observed in the wide variety of solar cells that have been studied. Also, it is often found that the small changes in V_{oc} are not due to the junction properties of the undoped i layer. The absence of series resistance effects in the diode characteristics under illumination even in the light-soaked state is also highly significant; together with the voltage dependence of the photogenerated carrier collection efficiencies, the series resistance determines the fill factor.

There are no limitations imposed on solar-cell operation by the changes in series resistance of the bulk, provided that the photoconductivity generated by the $\sim 10^{17}$ photons cm^{-2} sec^{-1} of red light is still sufficiently high (Wronski, 1977). Fortunately, these high intensities correspond to levels of illumination for which the differences between the state A and the light-soaked state B are generaly small, as illustrated by the results shown in Fig. 8. Consequently, light-induced conductivity changes have a large effect on the photovoltaic properties only when the light is not bulk absorbed. This is illustrated in Fig. 12 in which the light $I(V)$ characteristics of a ~ 3-μm Pt Schottky-barrier solar cell are shown for states A and B, which corresponds to a 4-hr exposure of 200-mW-cm^{-2} red light. The results in Fig. 12 are for simulated sunlight and weak blue light (produced by passing the sunlight through a blue filter). These results clearly show that the light-induced changes affect the blue-light characteristics. This happens because blue light does not penetrate far into the film and the dark conductivity of the a-Si:H film dominates the series resistance of the cell. The long exposure to bulk-absorbed red light increases this resistance and thus flattens out the $I(V)$ curve for the blue light. The $I(V)$ curve in white light does not change because the red light penetrating into the bulk is sufficient to lower the series resistance without affecting the $I(V)$ curves. These results show that light-in-

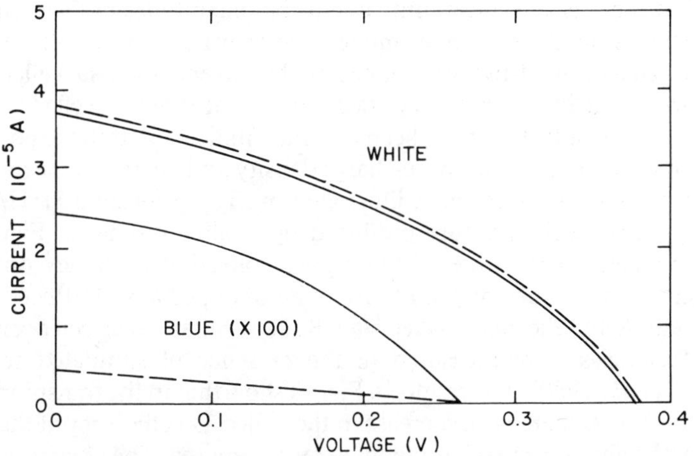

FIG. 12. Illuminated $I(V)$ curves for ~ 3-μm-thick a-Si:H/Pt Schottky-barrier solar cell. The curves taken with blue light are shown on $\times 100$-expanded current scale. The light exposure was 4 hr of 200-mW-cm^{-2} red light. ———, after anneal (state A); — — —, after exposure (state B). [From Staebler and Wronski (1980).]

duced conductivity changes need not be a serious limitation on the photovoltaic properties of a-Si : H.

The short-circuit currents and white-light $I(V)$ characteristic shown in Fig. 12 are unchanged by the prolonged exposure to light. (The difference in the figure is the experimental error.) But this is not always the case, and a wide range of changes has been observed. The reasons for these large differences are not clear, but in many cases they can be attributed not to light-induced changes in series resistance, but rather to the changes in the collection of photogenerated carriers. The results obtained on a large variety of solar cells have shown that the changes in J_{sc} and FF are highly sensitive to the fabrication conditions (which affect the density of gap states in the *i* layer) and to the cell structure (which affects the collection of photogenerated carriers) (Staebler *et al.*, 1981; Hanak and Korsun, 1982; Moller *et al.*, 1982; Nakano *et al.*, 1982; Tsuda *et al.*, 1983). However, only a few attempts have been made at correlating these changes with the S–W effect and the light-induced defects. Such studies are complicated by the dependence of the cell photocurrent not only on the efficient collection of both carriers, but also on the internal electric field generated by the junction (Wronski, 1977; Gutkowicz-Krusin *et al.*, 1981; Okomoto *et al.*, 1981). This means that, in evaluating the effects of prolonged exposure to light on solar-cell photocurrent, it is necessary to take into account the dependence of photocurrent carrier collection efficiencies on the recombination lifetimes of both *electrons* and *holes*, the space-charge density in the junction, and the thickness of the cells. Evidence for light-induced changes in ambipolar lifetimes and junction electric fields has been found by several investigators (Staebler *et al.*, 1981; Jousse *et al.*, 1980), but there is still uncertainty about the magnitudes of these changes because the values depend on the model used to evaluate the results. Also, in characterizing the light-induced changes in the *i* layer of the solar cell, it is not clear how to take into account the effects due to the different cell structures.

Most of the results that have been obtained indicate that the exposure to light increases the junction space-charge density and decreases the ambipolar lifetime. The effect of this is illustrated in Fig. 13 for an a-Si : H/metal Schottky-barrier cell structure produced by rf glow discharge. Figure 13 shows the spectral response of the carrier collection efficiency obtained under short-circuit current conditions for an undoped a-Si : H/Pt cell structure in state A and the light-soaked state B. To simulate solar-cell operation, these efficiencies were measured in the presence of simulated sunlight (Wronski *et al.*, 1980). The results in Fig. 13 show that in the transition from state A to state B there is an increase in the collection efficiency at the short wavelengths and a decrease at the longer wavelengths. The change at short wavelengths results from an increase in the surface field and space-charge density; the change at the long wavelengths results from a decrease in hole

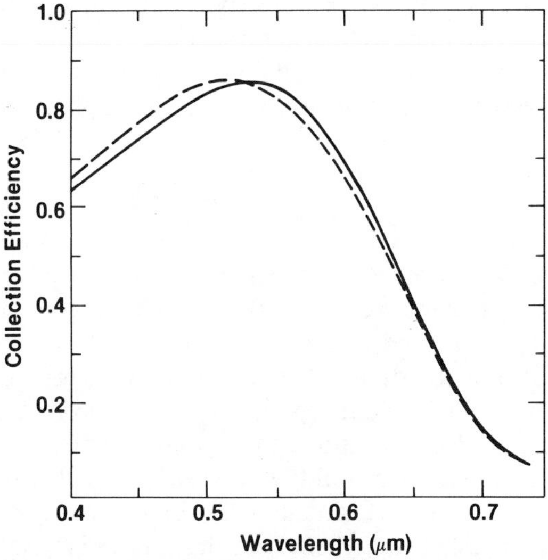

FIG. 13. The short-circuit current carrier-collection efficiencies of an a-Si:H/Pt Schottky-barrier cell in the annealed state A (solid line) and in the light-soaked state B (broken line). The measurements were carried out with 1-sun bias illumination.

diffusion length and ambipolar lifetime (Gutkowicz-Krusin, 1981). The small changes illustrated here have virtually no effect on the short-circuit current and cause only a very small change in the fill factor. On the other hand, the transition from state A to state B decreases the photoconductivity under 1 sun illumination by a factor of 5 and the dark conductivity by about a factor of 100. These changes in both conductivity and the carrier collection efficiency are consistent with the introduction of a low density of deep-lying, light-induced defects that displace the Fermi level and increase the junction space charge. The commonly found asymmetry between the changes in photoconductivity and J_{sc} is due to the different effects that these defects have on the photoconductive electron lifetime and the ambipolar hole lifetime (Wronski and Daniel, 1980).

Similar effects of prolonged illumination on hole diffusion length, hole lifetime, and electric field have been observed in undoped a-Si:H films by using surface photovoltage measurements (Dresner *et al.*, 1980; Moore, 1982; McMahon *et al.*, 1982). Also, surface-photovoltage profiling experiments carried out by Goldstein *et al.* (1981) showed that the prolonged illumination of a-Si:H p–i–n cells increases the space-charge density of the i layer. It has become evident that the light-induced metastable defects responsible for the conductivity changes in undoped a-Si:H films can also contribute to the changes in photovoltaic properties of a different solar-cell structure. However, because of the sensitivity of the light-induced changes to fabrication conditions and cell structure, more extensive studies must be carried out in order to understand all the mechanisms and the nature of the light-induced defects responsible for the changes in photovoltaic properties of a-Si : H.

VI. Discussion

The many results on the S–W effect in the wide variety of a-Si : H films clearly indicate that light creates metastable changes in the gap states of a-Si : H. Such light-induced changes in gap states have been reported in different a-Si : H films by different investigators who used different experimental techniques. This includes changes that have been measured by using photoconductivity (Wronski and Daniel, 1980; Vanier, 1982), field effect (Tanielian *et al.,* 1981; Grumwald *et al.,* 1981; Powell *et al.,* 1982), deep-level transient spectroscopy (Crandall, 1981; Lang *et al.,* 1982), electron spin resonance (Hirabayashi *et al.,* 1980; Dersch *et al.,* 1981), isothermal capacitance transient spectroscopy (Okushi *et al.,* 1983), and photoluminescence (Morigaki *et al.,* 1980; Pankove and Berkeyheiser, 1980). However, there are large differences in the conclusions that are reached by the different investigators about the nature, density, and energy of the observed light-induced defects. Also, Wronski (1983) and Han and Fritzsche (1983) have reported results which indicate the presence of at least two types of light-induced defects in undoped a-Si : H. Consequently, there are still many uncertainities about the low density of photo-induced defects that cause the large changes in different a-Si : H properties, and it is not clear whether the same metastable defects are responsible for all the reported results.

These uncertainties have not helped to establish a generally accepted model for the S–W effect and the metastable defects created by light. Staebler and Wronski (1977, 1980) proposed two general models for the light-induced changes in the conductivities of a-Si : H films. In the first model they suggested that localized defects undergo a bond reorientation or atomic displacement that could involve small fractions of the hydrogen present in the films. In the second model they suggested that there is a transfer of electrons to centers that are in poor communication with the extended states that could involve large changes in carrier capture cross sections and electron traps in microvoids surround by a repulsive barrier. In the last few years, a variety of more specific models for the metastable defects have been proposed for a-Si : H films that are related to not only Si–Si and Si–H, but also to different complexes such as those discussed by Carlson (1982). Even in the case of Si–Si and Si–H-related defects, quite different models have been proposed. Elliott (1979) suggested that photogenerated excitons can become self-trapped and result in intimate, oppositely charged dangling bonds that are stabilized by their Coulomb interaction and a "soft" lattice that occurs near weak bonds. Adler (1981) has proposed that such intimate pairs of defects occur naturally at neutral dandling-bond sites where the Coulomb attraction results in a negative correlation energy. Wautelet *et al.* (1981), on the other hand, suggested that the light-induced metastable centers result from two neutral configurations of a single dangling bond. Morigaki *et al.* (1980) and Pankove and Berkeyheiser (1980) suggested that the metastable defects are simply the breaking of weak Si–Si bonds. Dersch *et al.* (1981), on the other hand, proposed that the metastable defects are created not only by the breaking of the weak Si–Si bonds but also by the subsequent transfer of hydrogen atoms to these broken bonds from extended clusters of Si–H bonds, which can occur at surfaces of internal voids.

In order to answer the many questions that still exist about the origin and properties of light-induced defects in a-Si : H, it is necessary to carry out more extensive investigations of both the light-induced changes themselves and the electronic properties of the different a-Si:H films studied. It is very important to obtain an understanding of the observed sensitivity in light-induced changes to small differences in fabrication conditions and intentional or unintentional low levels of dopants. This will require detailed studies of light-induced changes in different properties of the same a-Si : H film, as well as the detailed characterization of different films. Such studies will help to answer these questions as well as questions about the real limitations imposed by the S–W effect on the present and future technological applications of a-Si : H.

REFERENCES

Adler, D. (1981). *J. Phys. Colloq. Orsay, Fr.* **42,** *Suppl. 10,* C4-3.

Carlson, D. E. (1982). *Solar Energy Mat.* **8,** 129.

Carlson, D. E., and Magee, C. W. (1979). *Appl. Phys. Lett.* **39,** 81.

Carlson, D. E., and Wronski, C. R. (1979). "Amorphous Semiconductors, Topics in Applied Physics" (M. H. Brodsky, ed.), Vol. 36. pp. 287–328. Springer-Verlag, Berlin and Heidelberg.

Crandall, R. S. (1981). *Phys. Rev. B* **42,** 7457.

Dersch, H., Stuke, J., and Beichler, J. (1981). *Appl. Phys. Lett.* **38,** 456.

Dresner, J., Goldstein, B., and Szostak, D. (1980). *Appl. Phys. Lett.* **38,** 998.

Elliott, S. R. (1979). *Philos. Mag.* **29,** 349.

Fritzche, H. (1980). *Solar Energy Mat.* **3,** 447.

Fuhs, W., Meleville, M., and Stuke, J. (1979). *Phys. Status Solidi B* **89,** 4839.

Goldstein, B., Redfield, D., and Szostak, D. J., and Carr, L. A. (1981). *Appl. Phys. Lett.* **39,** 258.

Grumwald, M., Weber, K., Fuhs, W., and Thomas, P. (1981). *J. Phys. Colloq. Orsay, Fr.* **42,** *Suppl. 10,* C4-523.

Gutkowicz-Krusin, D. (1981). *J. Appl. Phys.* **52,** 5370.

Gutkowicz-Krusin, D., Wronski, C. R., and Tiedje, T. (1981). *Appl. Phys. Lett.* **38,** 87.
Han, D., and Fritzsche, H. (1983). *J. Non-Cryst. Solids* **59/60,** 397.
Hanak, J. J., and Korsun, V. (1982). *Conf. Rec. IEEE Photovoltaic Spec. II Conf., 16th, San Diego,* p. 1381. IEEE, New York.
Hauschildt, D., Fuhs, W., and Mell, H. (1982). *Phys. Status Solidi B* **111,** 171.
Hirabayashi, I., Morigaki, K., and Nitta, S. (1980). *Jpn. J. Appl. Phys.* **19,** L357.
Johnson, G. O., McMillan, J. A., and Peterson, E. M. (1981). *AIP Conf. Proc.* **79,** 329.
Jousse, D., Basset, R., Delionibus, S., and Bourdon, B. (1980). *Appl. Phys. Lett.* **15,** 208.
Lang, D. V., Cohen, J. D., Harbison, J. P., and Sergent, A. M. (1982). *Appl. Phys. Lett.* **40,** 474.
McMahon, H. (1982). *Conf. Rec. IEEE Photovoltaic Spec. Conf., 16th, San Diego,* p. 1389. IEEE, New York.
Maruska, H. P., Moustakas, T. D., and Hicks, M. C. (1983). *Solar Cells* **9,** 37.
Meyer, W., Neldel, H. (1937). *Z. Tech. Phys.* **12,** 588.
Moller, M., Rauscher, B., Kruhler, W., Plattner, R., and Pfleiderer, H. (1982). *Conf. Rec. IEEE Photovoltaic Spec. Conf., 16th, San Diego,* p. 1376. IEEE, New York.
Moore, A. R. (1982). *Appl. Phys. Lett.* **40,** 403.
Morigaki, R., Hirabayashi, I., and Nakayama, M. (1980). *Solid State Commun.* **33,** 851.
Nakamura, G., Sato, K., and Yukimoto, Y. (1983). *Solar Cells,* (In Press).
Nakano, S., Fukatsu, T., Wishiwaki, H., Shibuya, H., Tsukamoto, K., and Kuwano, V. (1982). *Conf. Rec. IEEE Photovoltaic Spec. Conf., 16th,* San Diego, p. 1124. IEEE, New York.
Okomoto, H., Yamaguchi, T., Nonomura, S., and Hamakawa, Y. (1981). *J. Phys. Colloq. Orsay, Fr.* **42,** *Suppl. 10,* C4-507.
Okushi, H., Miyagawa, M., Tokumaru, Y., Yamasaki, S., Oheda, H., and Tanaka, K. (1983). *Appl. Phys. Lett.* **42,** 895.
Pankove, J. I., and Berkeyheiser, J. E. (1980). *Appl. Phys. Lett.* **37,** 705.
Powell, M. J., Easton, B. C., and Nicholls, D. H. (1982). *J. Appl. Phys.* **53,** 5068.
Rose, A. (1960). "Concepts in Photoconductivity and Allied Problems." Wiley (Interscience), New York.
Scott, B. A., Reiner, J. A., Plecenik, R. M., Simonyi, E. E., and Reuter, W. (1982). *Appl. Phys. Lett.* **40,** 973.
Solomon, I., Dietl, T., and Kaplan, D. (1978). *J. Phys. Colloq. Orsay, Fr.* **39,** 1241.
Staebler, D. L., and Wronski, C. R. (1977). *Appl. Phys. Lett.* **31,** 292.
Staebler, D. L., and Wronski, C. R. (1980). *J. Appl. Phys.* **51,** 3262.
Staebler, D. L., Crandall, R. S., and Williams, R. (1981). *Appl. Phys. Lett.* **39,** 733.
Tanielian, M. (1982). *Philos. Mag. B* **45,** 435.
Tanielian, M., Goodman, N. B., and Frizche, H. (1981). *J. Phys. Colloq. Orsay, Fr.* **42,** *Suppl. 10,* C4-375.
Tsuda, S., Nakamura, N., Takahama, T., Nishinaki, H., Ohnishi, M., and Kuwano, Y. (1983). *Solar Cells.* **9,** 25.
Vanier, P. E. (1982). *Appl. Phys. Lett.* **41,** 986.
Wautelet, M., Laude, L. D., and Failly-Lovoto, M. (1981). *Solid State Commun.* **39,** 979.
Wronski, C. R. (1977). *IEEE Trans. Electron Devices* **24,** 351.
Wronski, C. R. (1978). *Conf. Rec. IEEE Photovoltaic Spec. Conf., 13th, Washington, D.C.,* p. 744. IEEE, New York.
Wronski, C. R., (1983). *J. Non-Cryst. Solids* **59/60,** 401.
Wronski, C. R., and Carlson, D. E. (1977). *Solid State Commun.* **23,** 421.
Wronski, C. R., and Daniel, R. (1980). *Phys. Rev. B* **23,** 794.
Wronski, C. R., Abeles, and Cody, G. D. (1980). *Solar Cells,* **23,** 421.

CHAPTER 11

Schottky Barriers on a-Si : H

R. J. Nemanich

XEROX CORPORATION
PALO ALTO RESEARCH CENTER
PALO ALTO, CALIFORNIA

I. Introduction

The Schottky barrier is a fundamental property of metal/semiconductor interfaces. Although initial models used to describe the effect were posed as long ago as the 1930s and the problem has been addressed by the most up-to-date surface sensitive probes, a complete understanding has eluded us. Even though the details of barrier formation are not understood, much is known about the Schottky barrier on crystalline semiconductors. The transport properties are well described by various theoretical models, which are somewhat material dependent. For reviews of the properties of Schottky barriers on crystalline semiconductors, see Rhoderick (1978) or Sze (1969).

Because the Schottky barrier on crystalline Si is not understood in detail, the understanding of the barrier on a-Si : H is at a very fundamental level. Most progress has been made by empirical studies, which used techniques well tested on studies of crystalline Si. What has emerged from this is a rudimentary understanding that allows crude applications of rectifying and ohmic contacts to experimental and device configurations.

Since a-Si : H is a unique semiconductor in its own right, the properties of metal contacts on a-Si : H do present a series of questions at the basic science

ISBN 0-12-752149-6

level. The aspects that are clearly identifiable are the role of band-tail states, the defects in the band gap, and the fact that the most useful material may be the undoped a-Si : H, which exhibits a near-mid gap Fermi energy. Of course, the most remarkable aspect of a-Si : H is the ability to shift the Fermi energy through "substitutional" doping. Thus the properties of metal contacts on doped a-Si : H are of significant importance.

Schottky diodes on a-Si : H were first produced in 1974 (Carlson, 1980) with the first report of well-characterized barriers by Wronski *et al.* in 1976. Although there have been significant efforts since the first study, it has only been since then that some studies have focused on the properties of the barrier specifically rather than on the properties of a-Si : H.

The second part of this chapter will describe some of the general aspects of the Schottky barrier and properties specific to a-Si : H. The third part is devoted to the transport mechanisms and measurements, and the fourth describes the effects due to atomic structural properties of the interface. Next, the effects due to doping are addressed, and following that the origin of the Schottky barrier is discussed. By presenting the question of the origin of the barrier near the end of this chapter, we imply that it is an unsolved problem. The chapter is concluded with comments about applications and important problems to be addressed for further understanding.

II. Aspects of the Schottky Barrier

Once it is recognized that electronic states of amorphous Si can be modeled by energy bands, it is a small step to realize that Schottky-barrier formation is likely. There are, however, significant differences in the properties of the Schottky barrier on a-Si : H from those for crystalline Si. A basic difference is that the most common Schottky diode of a-Si : H is on undoped material because the carrier lifetime is significantly longer than that of doped a-Si : H. This leads to a metal/semiconductor junction with the Fermi energy at the middle of the semiconductor gap as opposed to the band edges for crystalline Si. The steps of barrier formation as a metal and a-Si : H films are brought together are illustrated in Fig. 1. The solid lines that would represent the band edges for a crystalline semiconductor represent the energy at which extended-state conductivity can occur. The schematic assumes a density of surface states for a-Si : H. This possibility is supported by several experiments and is discussed in more detail in Chapter 9 by Fritzsche. Thus, even when the semiconductor and metal are electrically separated as in Fig. 1a, there will be band bending at the surface. An unusual aspect of this configuration is that the electrical neutrality is maintained by ionized centers near midgap. This is to be contrasted with crystalline Si, in which the ionized donors or acceptor states lie near the band edges.

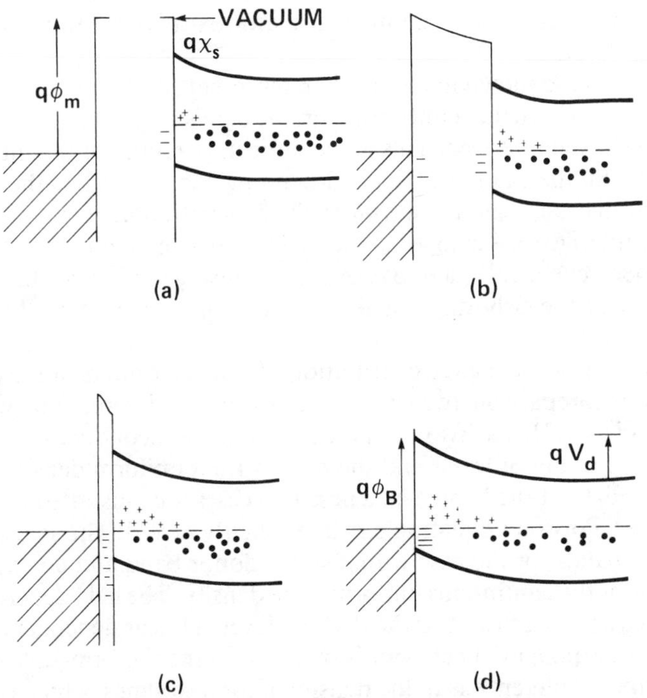

FIG. 1. A schematic of Schottky-barrier formation for a metal and a-Si:H. In (a) the metal is isolated from the a-Si:H whereas in (b) and (c) they are in electrical equilibrium. In (d), the two materials are in intimate contact.

The formation of the Schottky barrier occurs when the metal and semiconductor are brought into electrical equilibrium and then into intimate contact. This is considered in stages in Fig. 1b–c. In Fig. 1b, the metal and semiconductor are well separated but are in electrical equilibrium. This can be accomplished conceptually by assuming that a wire connects the two. Electrons will then be transported to the metal surface, which will establish a potential (which varies linearly with distance) such that the Fermi energy of the two materials is equivalent. Note that for large distances few charges must be transferred to the metal surface. The slanted line between the metal and semiconductor represents the potential due to the charges. These charges are balanced by additional ionized centers that will occur near the semiconductor surface. As the distance between the metal and semiconductor is decreased, more charge is transferred to the metal surface. When the magnitude of the transferred charge becomes comparable to the surface defect density on the semicondutor, then additional band bending will occur. The configuration shown in Fig. 1c may equally well represent the case in which an oxide or another insulator exists between the metal and semiconductor.

The schematic shown in Fig. 1d then represents the Schottky barrier after intimate contact. In the semiconductor, the bands are bent due to the ionized centers near the interface. This is the built-in potential of the interface. In crystalline Si, the extent of the band-bending region is called the depletion (or accumulation) region, referring to the concentration of majority carriers near the interface. Of course, as the metal and semiconductor are brought into contact, chemical reactions can occur, causing significant changes in the properties of the Schottky barrier. These properties will be discussed at length in Part IV.

Consider now the charge distribution of ionized centers in the depletion region. For comparison, two cases can illustrate the difference between crystalline and a-Si : H. The two cases are for *n*-type semiconductors, one with a single band of donor levels and the other with a uniform density of donor levels throughout the band gap. These two cases are illustrated in Fig. 2. In these cases, Poisson's equation can be explicitly solved. The solutions yield parabolic bands for the case with a single donor band and an exponential behavior for the continuous uniform-state density. The field dependences of the two cases differ also. For the discrete level, a linear dependence results, whereas an exponential behavior is obtained in the uniform-state case. The most striking difference is in the density of ionized states, which is uniform

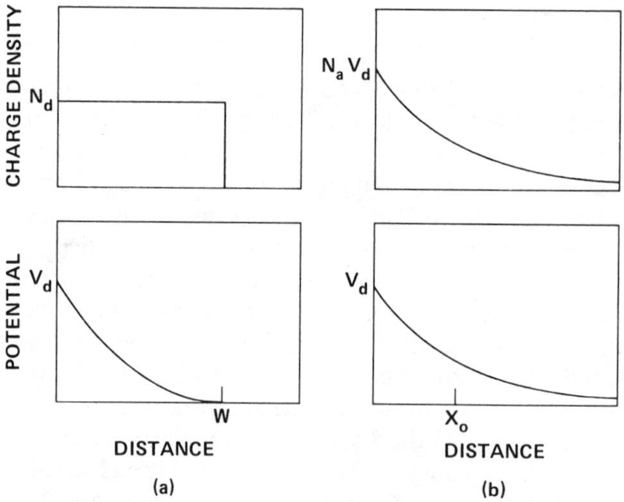

FIG. 2. Schematic of the charge-density and potential profiles of the depletion region for a semiconductor with (a) a single donor band and (b) a uniform distribution of donors throughout the gap.

over the depletion region for the discrete level. Again an exponential dependence is observed in the continuous case. The depletion region is no longer well defined, but often the $1/e$ point is used to demark the depletion region.

It should be noted that a-Si:H exhibits a density of states that is neither constant nor discrete. Deep-level transient spectroscopy (DLTS) (Lang *et al.*, 1982), optical absorption (Jackson and Amer, 1982), and photoconductivity (Jackson *et al.*, 1983b) measurements have indicated that a rather broad defect band lies in the gap and is centered below the Fermi energy. This means that for cases in which the built-in potential V_d is less than the width of the defect band, the exponential behavior will be expected. Experimental verification of this is discussed in Section 4.

One aspect of the Schottky barrier that has been of concern for crystalline semiconductors is image force lowering. This occurs when an electron transits the depletion region. The image potential causes the effective barrier to be reduced, and the potential maximum will occur at several nanometers into the semiconductor instead of the interface. Thus measurements like $J-V$ or internal photoemission will reflect the image force lowering of the barrier. In contrast, measurements like $C-V$ are not made in the presence of a transiting charge and will not be affected. Furthermore, because the barrier occurs away from the interface, weak effects due to bias variations are possible. This property is basically a small effect that may result in a change in the barrier of <0.02 eV. The measurements on a-Si:H Schottky barriers to date have not been concerned with such weak effects, which would provide only minor corrections to the observed properties.

III. Transport Mechanisms and Measurements

Several possible current transport mechanisms are illustrated in Fig. 3. The schematic represents a Schottky barrier on an undoped sample under forward bias. The three arrows for electron transport are drawn for comparison with crystalline semiconductors in which thermionic emission, tunneling via thermionic field emission, or field emission represent the usual mechanisms.

At room temperature it is usual to consider that electron transport over the barrier dominates the forward-bias current. For high-mobility materials such as crystalline Si, it is presumed that the electron population at any energy above the conduction-band minimum is independent of distance from the surface. The transport is then limited by the emission of electrons into the metal. For low-mobility materials it is possible that the current transport is limited by electron diffusion through the depletion region. In this case, the electron population at any energy will vary as a function of distance from the interface. Thus if the thermionic emission limits the transport, then

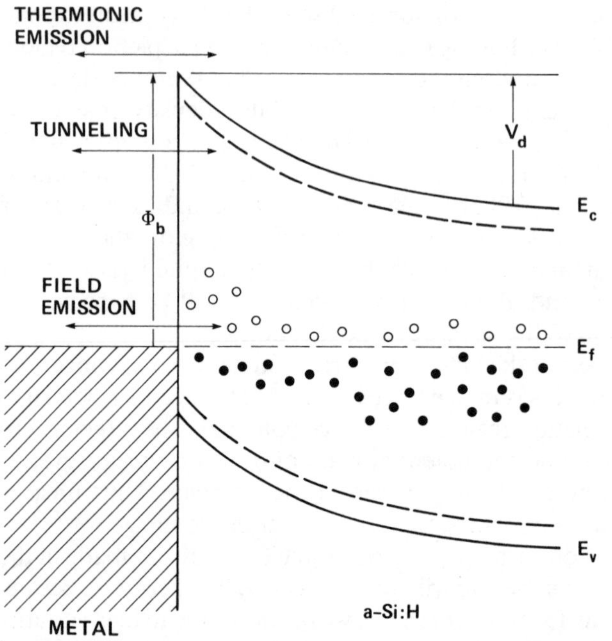

FIG. 3. Schematic of the Schottky barrier on a-Si : H showing the different transport mechanisms.

for an applied bias the Fermi energy will be constant throughout the depletion region and drop (or rise) to the metal Fermi energy inside the metal film. In contrast, for the diffusion-limited transport, the Fermi energy will vary in the depletion region. Because amorphous silicon has a low mobility of ~2 $(\Omega\ \mathrm{cm})^{-1}$, it is not clear which mechanism will dominate. In fact, both mechanisms are observed and are discussed in the following section.

At lower temperatures, tunneling currents may become more important. The tunneling of electrons from the conduction band into the metal is certainly observed in crystalline Si and has been reported for amorphous Si (Alkaisi and Thompson, 1979–1980). The process is more complicated in a-Si : H because the localized states near the conduction-band tail may significantly enhance the tunneling current.

The last possibility, which corresponds to field emission in crystalline semiconductors, is a much less likely current path in a-Si : H Schottky barriers under forward bias. Since the Fermi energy is near midgap, there is a relatively low density of states and the states are localized. Thus there are few states that can field emit into the metal. In reverse bias, field-emission breakdown can be observed because the emission is from the metal Fermi energy into the a-Si : H conduction band. Measurements and an analysis have been carried out by Snell *et al.* (1979). The analysis was based on the theoretical work of Padovani and Stratton (1966). For lightly doped a-Si:H, field emission is observed at reverse bias of 12.5 V, whereas more highly doped samples exhibit field emission at 4.5 V, and these results are consistent with the analysis.

Although Schottky diodes are usually considered majority-carrier devices, the minority carriers can play a significant role. Because the Fermi energy is near midgap, the hole current can become significant. This is especially true for low work-function metals. Although the minority-carrier contribution may be higher than in crystalline Si diodes, it can still be neglected for most analyses.

1. CURRENT–VOLTAGE CHARACTERISTICS

The forward-bias current–voltage characteristic is the most used method of characterizing the barrier. Following the formalism developed for crystalline semiconductors (Rhoderick, 1978), the current density J can be written as

$$J = J_0 \exp(eV/nkT)\{1 - \exp(-qV/kT)\}, \tag{1}$$

where J_0 is the reverse-bias saturation current, n the ideality parameter, V the applied voltage, T the temperature, q the electron charge, and k the Boltzmann constant. The equation is usually written as

$$J = J_0 \exp(qV/nkT), \tag{2}$$

which is a good approximation for V greater than $3kT/q$. The equations as written to this point are independent of whether the current transport is thermionic emission or diffusion limited.

If a uniform metal/semiconductor junction exists and the transport is by thermionic emission, then the ideality parameter n will be nearly 1. There are two obvious possibilities that can cause deviations from 1. These are (1) recombination in the depletion region and (2) a nonuniform junction. In the first case, if the current transport is diffusion limited, then significant recombination may occur, and this will lead to a value of n that can be as high as 2. It should be noted, however, that even if the current is diffusion limited but no recombination occurs, then n will still approach 1.

For the case of a nonuniform junction, the barrier height may vary from place to place. The J–V measurement will be dominated by the lower barrier-height regions, but the characteristics will exhibit a high ideality value (Freeouf *et al.*, 1982). The aspect of interface formation that can lead to or minimize these effects are described in Part IV.

Thus with proper preparation it is possible to produce diodes that exhibit nearly ideal behavior (Thompson *et al.*, 1981). An example of J–V measurements from such a sample is shown in Fig. 4. Once ideal characteristics are obtained, it is feasible to extract a barrier height. If the transport is limited by thermionic emission, then following the analysis of barriers on crystalline Si,

$$J_0 = A^{**} T^2 \exp(-q\,\Phi_B/kT), \tag{3}$$

where Φ_B is the effective barrier height and A^{**} Richardson's constant. Whereas A^{**} is a well-defined constant in crystalline materials, it is determined by the effective mass of the electron. Because the electron effective mass is not determined in a-Si : H and because of the disorder that leads to localized band-tail states, it is not possible to calculate A^{**} directly for a-Si:H. Thus to obtain the barrier height, it is necessary to measure the temperature dependence of J_0 and to use Eq.(3). Then plotting ln J_0/T^2 versus $1/T$ should result in a straight line. These characteristics verify thermionic-emission transport and the slope of the line is $-q\Phi_B/k$. An example of such a plot for Pd on a-Si : H (Thompson *et al.*, 1981) is shown in Fig. 5.

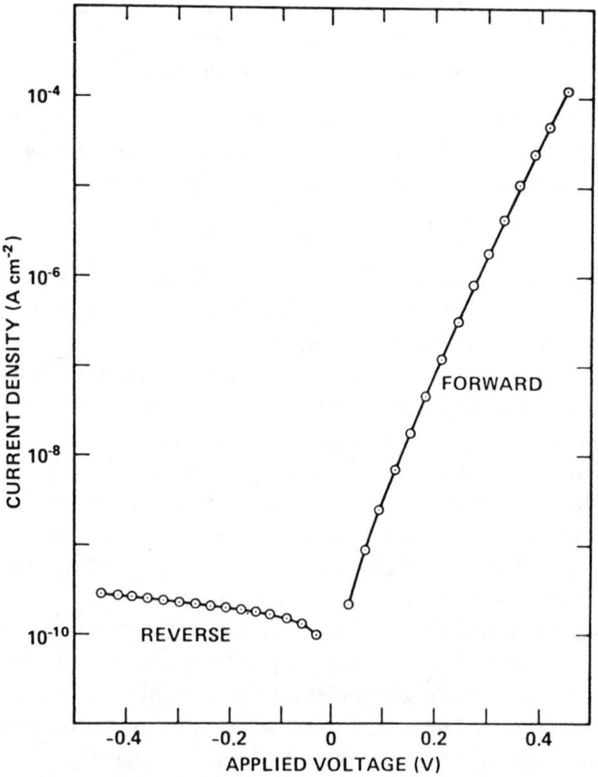

FIG. 4. The forward- and reverse-bias $J-V$ characteristics of a Pd/a-Si : H Schottky barrier. [From Thompson *et al.* (1981).]

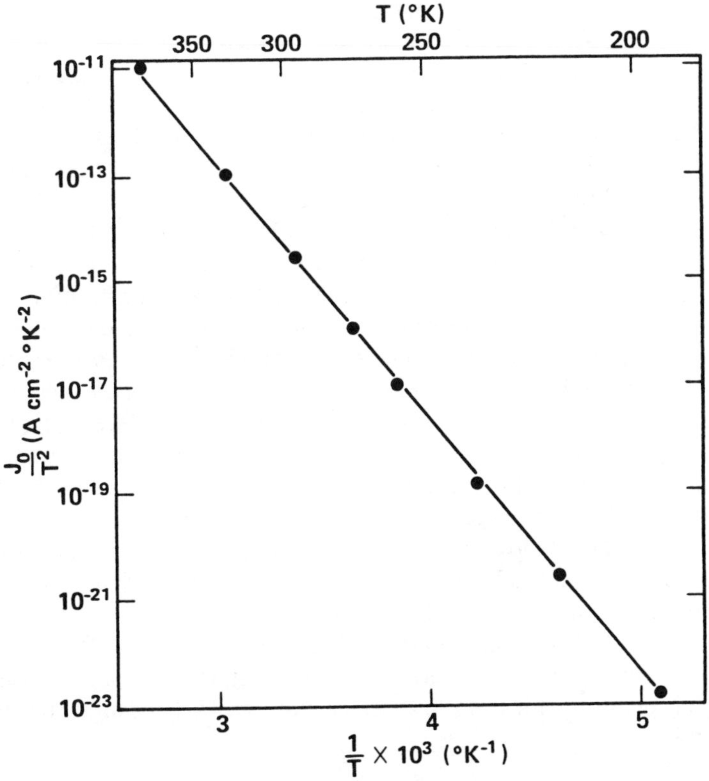

FIG. 5. Thermionic emission plot for an annealed Pd/a-Si:H Schottky barrier. The slope gives an effective barrier height of 0.97 eV. [From Thompson *et al.* (1981).]

Because of the low mobility of a-Si:H, it is likely that many barriers will exhibit diffusion-limited transport. Here again, for cases in which nearly ideal behavior is obtained, the temperature dependence of J_0 can be used to obtain the barrier height. The expression for diffusion-limited transport from Rhoderick (1978) is

$$J_0 = eN_c \mu F_\mu \exp(-\Phi_B/kT), \tag{4}$$

where μ is the electron mobility F_μ the maximum field, and N_c the effective density of states in the conduction band. Thus a plot of ln J_0 versus $1/T$ can be used to verify diffusion-limited transport. Other details of diffusion-limited transport have been modeled by Chen and Lee (1982).

It has not yet been determined whether Schottky diodes on a-Si:H exhibit predominantly thermionic-emission or diffusion characteristics. Although thermionic-emission characteristics have been reported for Pd and Pt diodes

(Thompson *et al.*, 1981; Nemanich *et al.*, 1983), Deneuville and Brodsky (1979) have shown that it is often difficult to distinguish the two characteristics over the limited temperature range available. Using both analyses, they could fit their data for Pt/a-Si:H diodes and found that the barrier height differed by only 0.02 eV. The temperature range of the measurement is limited to ~−100°C because at that temperature the transport is trap-limited and tunneling to localized levels is significant.

Thus the ideality value obtained from the room-temperature measurements proves to be a very useful tool to characterize the uniformity of the interface and the presence of recombination centers. Furthermore, it has been reported that A^{**} for Pt/a-Si:H diodes is approximately equal to the value obtained for crystalline Si (Deneuville and Brodsky, 1979). Hence the value of J_0 will give a good estimate of the barrier height for nearly ideal diodes. Furthermore, changes in J_0 can be interpreted, with reasonable confidence, as changes in barrier height. The results presented so far seem to indicate that the role of band-tail states is somewhat minimal in the depletion region. At low temperature and for material with higher defect densities, this is not the case. At low temperature, hopping currents involving tunneling to localized band-tail states have been observed (Madan *et al.*, 1982). Field-emission tunneling from localized states through the barrier has been observed in high-defect-density a-Si:H (Thompson *et al.*, 1980), and tunneling to extended electronic states can be the dominant transport mechanism at low temperatures for even low-defect material (Alkaisi and Thompson, 1979/1980).

2. Internal Photoemission

Internal photoemission is a component of the photoresponse of the diode. Several optical transitions contribute to the photoresponse, and they are illustrated in Fig. 6. Here transition A represents the internal photoemission. In this transition light is absorbed in the metal, and the electron is photoemitted into the a-Si:H. With the diode under reverse bias the electron transits the a-Si:H and is collected at the back contact. The internal photoemission "turns on" at energies less than the band-to-band absorption. Since the electron mean free path in a metal is less than 30 Å, it is necessary to illuminate the metal near the interface. This necessarily requires that significant light intensity exists in the a-Si:H. Thus the optical band-to-band absorption will occur, and since the carriers are excited to extended states, they will be collected and contribute to the photoresponse. This process is indicated by B. Two subgap absorptions that may contribute and mask the internal photoemission are indicated by C and D in Fig. 6. Here C is a transition from a localized state to an extended state, and D is a transition in the heavily doped back contact region. Transition C is observed in optical absorption, and since it occurs in the same energy region as the internal

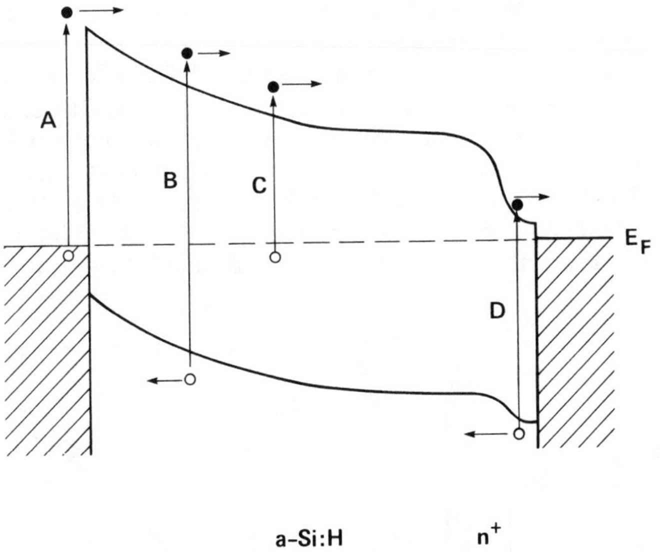

FIG. 6. Schematic of the optical transitions that contribute to the photoresponse of a-Si:H Schottky-barrier diodes. A, internal photoemission; B, optical band-to-band absorption; C, localized to extended state absorption; D, absorption in the doped layer.

photoemission, it might mask the photoemission. But because the excitation leaves a hole in a localized state, an electron is likely to recombine there before the hole can transit to the metal film. For all the excitations both the hole and electron must be collected to contribute to the dc response. Thus the transitions involving localized states are suppressed in this arrangement. The transitions in the heavily doped back contact involve extended states, and collection is more probable. Still, the hole must transit the sample, and since hole mobilities are lower, recombination with "photoemitted" electrons is likely, and transition D will be suppressed in the photoresponse. Another reason the back contact transitions are not observed is that the region is usually only 10–20 nm thick and has an absorption constant much less than the metal.

The theory for internal photoemission has been worked out originally by Fowler and is summarized in more recent texts [see Rhoderick (1978) or Sze (1969)]. Although the analysis was made for an epitaxial interface, many nonepitaxial metal/semiconductor interfaces show the same characteristics. The photoemission yield per incident photon Y exhibits a wavelength dependence described by

$$\begin{aligned} Y(h\nu) &= A(h\nu - \Phi_B)^2 \qquad (h\nu > \Phi_B), \\ Y(h\nu) &= 0 \qquad (h\nu < \Phi_B). \end{aligned} \tag{5}$$

Here, A is a constant determined by the absorption of the metal and the probability of photoemission into the semiconductor. Thus by plotting $Y^{1/2}$ versus photon energy, extrapolation to zero will display the barrier height. The first internal photoemission results for a-Si : H Schottky barriers were reported by Wronski *et al.* in 1980, and their results are shown in Fig. 7. The spectrum has two components: the strong absorption at >1.4 eV due to band-to-band transitions and the weaker feature at <1.4 eV due to the internal photoemission. By plotting the square root of the internal photoemission in Fig. 8, the barrier height can be obtained. As is evident from the

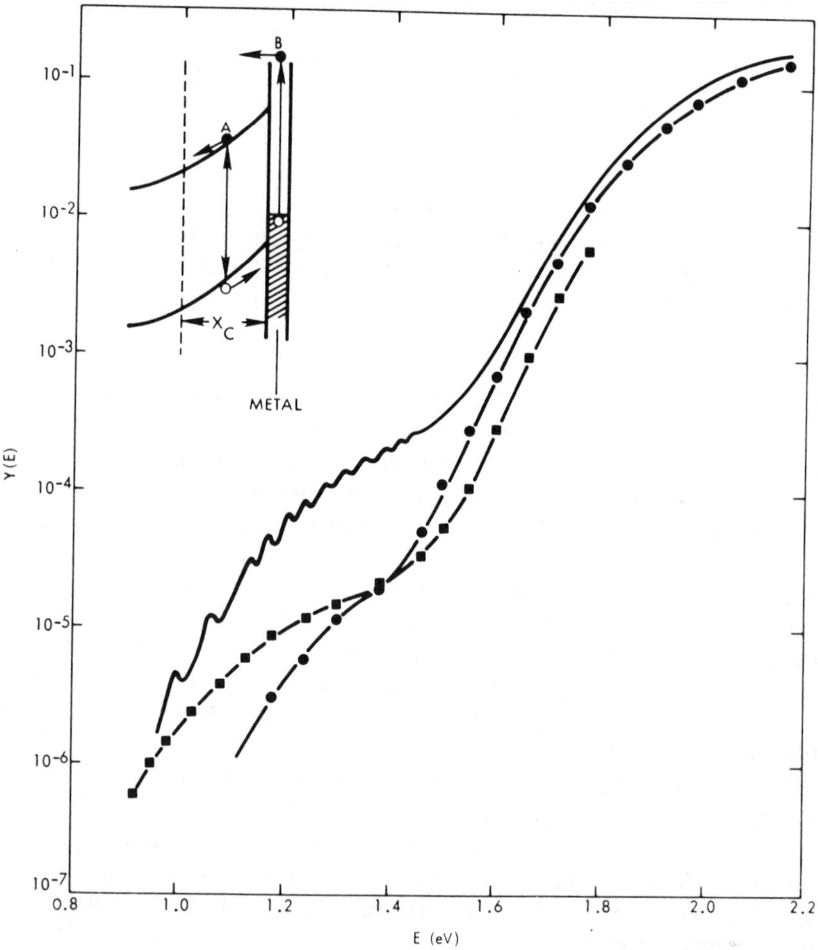

FIG. 7. The photoresponse of Schottky barrier diodes. Contact metal and measurement temperature: ■, Cr (200°K); ———, Pd (296°K); ●, Pt (296°K). [From Wronski *et al.* (1980).]

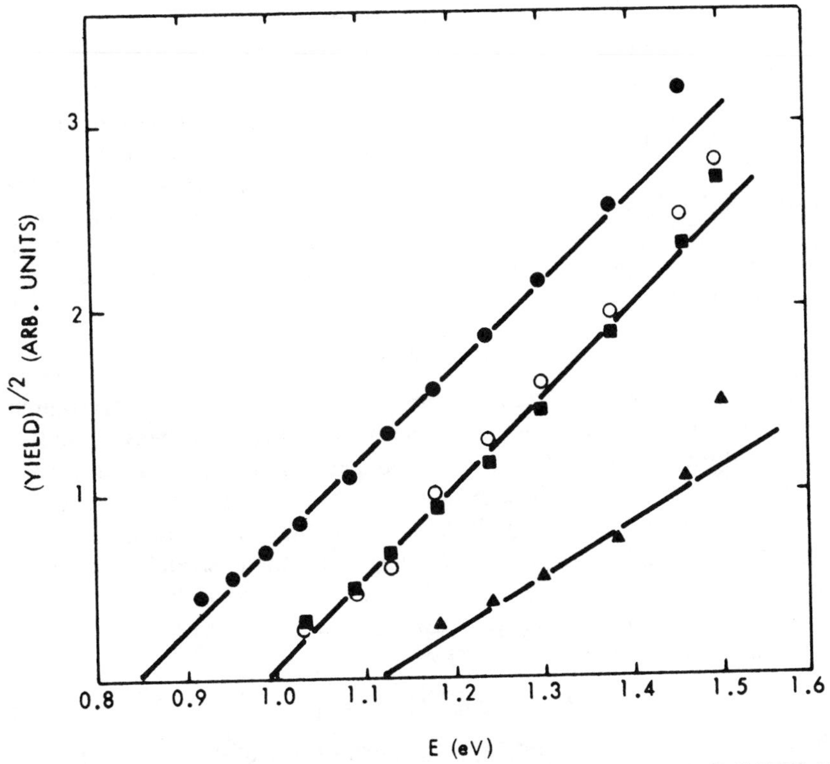

FIG. 8. The internal photoemission of the Schottky diodes shown in Fig. 7. ●, Cr (200°K); ■, Pd (296°K); ○, Pd (200°K); ▲, Pt (296°K). [From Wronski *et al.*, (1980).]

figure, for Cr and Pd barriers good straight line fits to the data were obtained, and the extrapolations agree with barrier-height determinations from $J-V$ characteristics. Nemanich *et al.* (1983) have re-examined the photoemission on well-characterized samples. They found for Pt/a-Si:H barriers that it was difficult to fit the $Y^{1/2}$ to a straight line. Because of the high barrier of Pt/a-Si:H junctions, the internal photoemission contribution does not turn on until near the band-to-band transitions. Thus these and other transitions tend to make the analysis more difficult.

The internal photoemission of Pd/a-Si:H diodes was also reported by Yamamoto *et al.* (1981). They also noted that the data did not fit the expected behavior. In fact, they found a better fit to a $Y^{1/3}$ dependence. This dependence is expected for a thin oxide or insulator film between the metal and semiconductor.

There may yet remain controversy over whether the measurements indeed represent the internal photoemission process, but the correlation with the $J-V$ measurement seems incontrovertible (Wronski *et al.*, 1980; Nemanich *et al.*, 1984). It should be noted that obtaining Φ_B from the temperature dependence of J_0 results in the zero-temperature limit, whereas the internal photoemission determines Φ_B at the measured temperature. It has been shown that the band gap of a-Si : H increases by ~ 0.1 eV when cooled to 4°K (Tsang and Street 1979; Cody *et al.*, 1981). Thus the similarity of the two barrier-height measurements indicates the barrier height is independent of temperature. Supporting this are the results of the Wronski *et al.* (1980) in which they found the internal photoemission independent of temperature down to 200°K.

3. Capacitance–Voltage Measurements

A commonly used measurement for characterizing barriers on crystalline Si is the differential capacitance as a function of voltage. The measurement is based on the fact that the depletion region changes for applied voltage; thus the amount of charge and the capacitance also change. The measurement is carried out by superposing a small ac voltage on the dc bias voltage. A bridge circuit that responds only to the ac component is then used to measure the differential capacitance versus applied voltage. For an ideal diode with parabolic bands (i.e., uniform charge density qN_d in the depletion region) it has been shown that (Rhoderick, 1978)

$$C = [q\varepsilon_s N_d/2(V_d + V)]^{1/2}, \tag{6}$$

where C is the capacitance and ε_s is the permittivity of the semiconductor. Thus plotting C^{-2} versus V can be used to obtain the built-in potential V_d from the intercept to the voltage axis.

The difficulties of using the standard analysis for a-Si : H were demonstrated by Snell *et al.* (1979) Their results are shown in Fig. 9 for barriers on a-Si : H with different doping levels. First, the extrapolation was difficult because the forward-bias capacitance could be measured to only ~0.2 V. Adding to this complication was the fact that the $1/C^2$ plots did not exhibit a straight-line dependence. For the depletion layer approximation the slope of curve yielded the donor density. Thus the observed deviation from linear behavior suggests that the charge density in the depletion region is not uniform.

Another aspect that makes routine barrier analysis difficult is that the states in the gap exhibit a wide range of response times. This has been demonstrated by several groups (Spear *et al.*, 1978; Viktorovitch *et al.*, 1979; Snell *et al.*, 1979; Snell *et al.*, 1980; Viktorovitch and Jousse, 1980; Beichler *et al.*, 1980; and Fernandez-Canque *et al.*, 1983). The frequency response reported by Beichler *et al.* (1980) for undoped a-Si : H is shown in Fig. 10. The capacitance was measured over a frequency range of 10^{-3}–10^{4} Hz. They

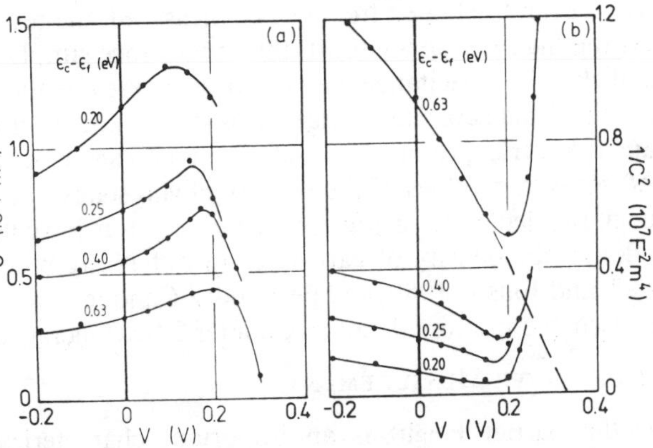

FIG. 9. (a) The differential capacitance versus voltage for Schottky barriers on a-Si:H with different doping levels. Shown in (b) is a plot of the same data as shown in (a) plotted as C^{-2} versus voltage. The dashed line extrapolation indicates the built-in potential. [From Snell *et al.* (1979).]

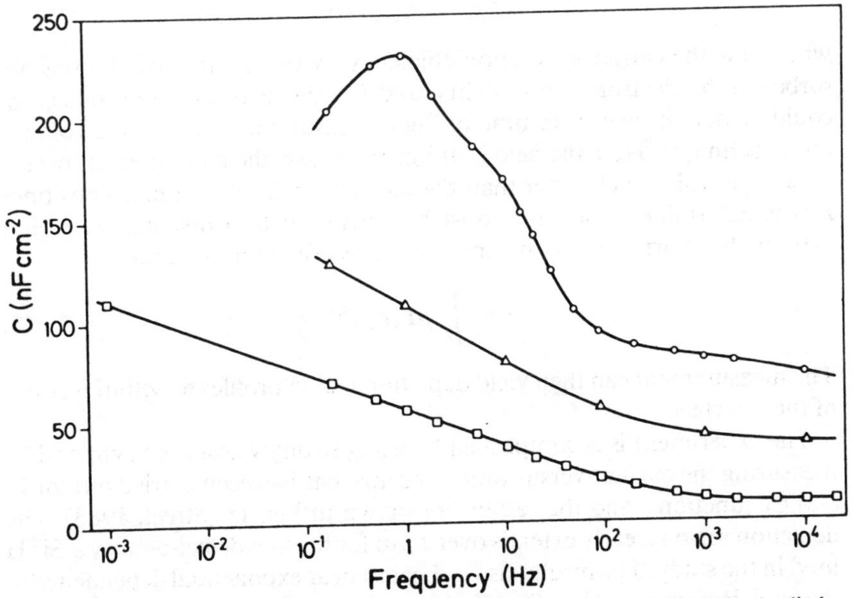

FIG. 10. The frequency response at 300°K of the capacitance of a-Si:H/Pt Schottky diodes. ○, 10-ppm $PH_3 \times 0.5$; △, 1-ppm PH_3; □, undoped. [From Beichler *et al.* (1980). Copyright North-Holland Publ. Co., Amsterdam, 1980.]

showed that at even the lowest frequencies measured, the response did not saturate. At the high frequency limit, the capacitance simply becomes the geometric, dielectric capacitance, i.e., $C = \varepsilon/d$, where d is the thickness.

The analysis of the capacitance data thus is dependent on the gap-state densities and the barrier profiles. An analysis method described by Snell *et al.* (1979) used measured values of the density of gap states to determine the field profile in the depletion region. Other investigators have used the $C-V$ results to derive the density of gap states in differently prepared a-Si:H (Viktorovitch and Jousse, 1980; and Fernandez-Canque *et al.*, 1983). These results are discussed in more detail in Chapter 2 by Cohen.

4. Depletion Layer Measurements

Whereas the barrier height is an important characterization of the Schottky barrier, the shape of the potential in the depletion region is also critical in understanding the transport properties. A technique using transient photoconductivity has been developed by Street (1983) to profile the depletion region. A light pulse is applied to the Schottky barrier, which is under dc reverse bias. The charge generated near the contact will drift because of the internal field, and the current through the sample will vary as a function of time by the relation

$$J(t) = \eta N q \mu F(t), \tag{7}$$

where η is the carrier generation efficiency, N the number of photons absorbed, μ the electron drift mobility, and $F(t)$ the field. The constants $\eta N q \mu$ could be determined from time-of-flight measurements that use a voltage-pulse technique. Here the field is uniform because the measurement occurs on a time scale much faster than the dielectric relaxation time. Thus once $F(t)$ is determined, the time must be converted to a distance scale. For carriers that start at $x = 0$ at time $t = 0$, x is related to the time by

$$x = \int_0^t \mu F(t')\, dt'. \tag{8}$$

The measurement can then yield depletion region profiles to within ~200 Å of the interface.

The experiment is accomplished by using strongly absorbed light and by measuring the current versus time. The analysis has been carried out for Pt and Cr junctions, and the results are shown in Fig. 11 (Street, 1983). The depletion region clearly extends over 1 μm for the low-defect-density a-Si:H used in the study. The results also indicate a near exponential dependence to the field. Referring back to Part II, this indicates that the Fermi energy is in a region of approximately constant density of states. The similar shape of the curves is consistent with nearly identical N_d for both junctions.

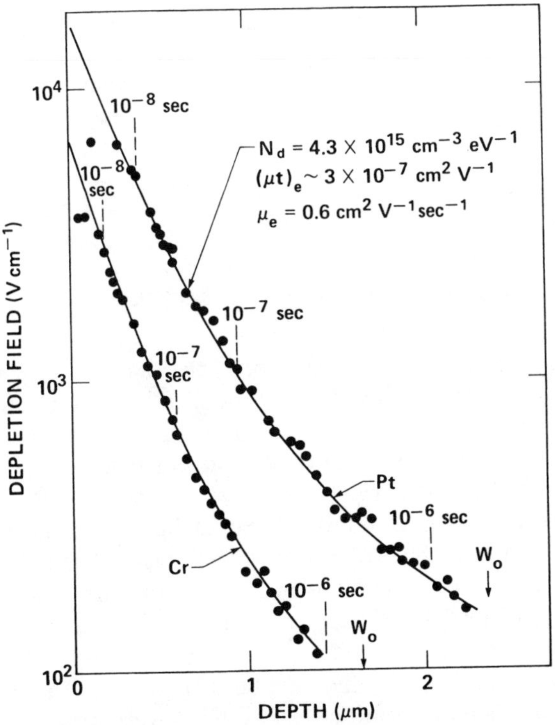

FIG. 11. The field profiles for the depletion-layer of a-Si:H deduced from transient photoconductivity. The $(\mu\tau_D)_e$ and μ_e are characteristic of the a-Si:H used for both samples. $N(F)$ was obtained from the fit to the data using a constant density of states throughout the gap.

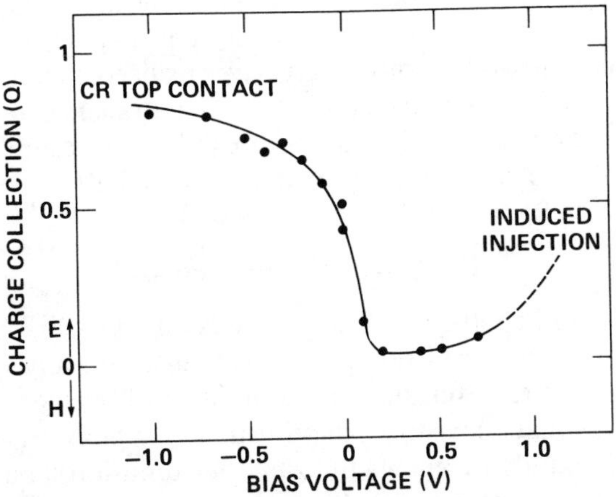

FIG. 12. The charge collection from a light pulse incident on a Cr contact to undoped a-Si:H 5.7 μm thick at 300°K as a function of dc bias. [From Street (1983).]

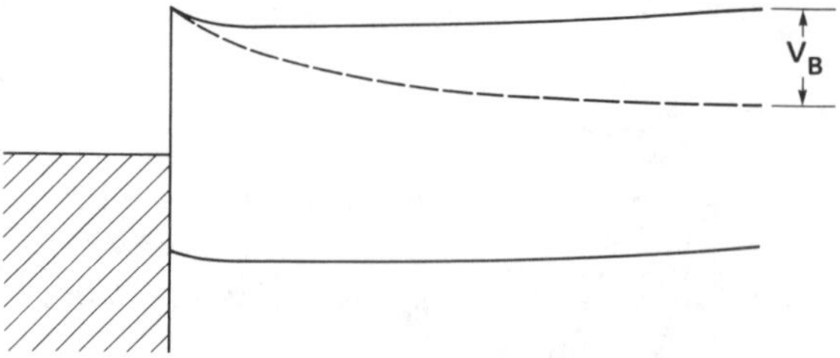

FIG. 13. The band potential when nearly flat-band dc bias V_B is applied.

Another variation of the measurement that can be used to obtain the built-in potential is to monitor the charge collection as a function of dc bias. When flat-band conditions are obtained, the charge collection will go to zero. Thus extrapolation to zero charge collection will yield the built-in potential. Results obtained by Street (1983) are shown in Fig. 12. For reverse bias the charge collection approaches unity efficiency but extrapolates to 0 at a forward bias of 0.2 V, which corresponds to the built-in potential. The increase of the charge collection at far forward bias is not fully understood but has been explained (Street, 1983) as injection induced by carriers collecting in a local potential minimum, as indicated in Fig. 13.

An alternative method by which the built-in potential can be determined is by monitoring dc photoconductivity as the forward bias is changed. In this case, when flat-band conditions are obtained, the current will change signs as the photoexcited carriers drift in the opposite direction. Results by Jackson *et al.* (1984) indicate that this method is successful for undoped a-Si : H. Both techniques are, however, limited by the voltage drop across the sample due to the resistance of the a-Si : H. This leads to the band diagram shown in Fig. 13. Thus since true flat-band conditions are never realized, the techniques give a slightly high value of the built-in potential. The problem can be minimized by using thin samples, and it may be possible to measure routinely the built-in potential to an accuracy of better than 0.03 eV.

IV. Structural Interactions

From studies of Schottky-barrier structures on crystalline Si, it has been shown that one of the least likely situations is an atomically sharp metal/Si interface with no interdiffusion. An examination of binary phase diagrams clearly indicates that coexisting adjacent metal/Si films are metastable at best. If no compounds are known to form, then diffusion should occur until the solubility limits are reached. If compounds exist in the binary phase diagram, then it is likely that they will form at the metal/Si interface.

Clearly, these interactions will affect the sensitive transport properties of the junction. For a review of thin-film reactions on crystalline Si, see Tu and Mayer (1978) and Ottaviani (1979).

Although structural and electrical properties have been correlated for barriers on crystalline semiconductors, most transport measurements on a-Si : H have ignored the structural properties. This in itself may account for the measurement variations and instabilities reported on a-Si : H. There have been some attempts to correlate the structural and electrical properties. The experimental probes that have proved most useful for examining the structural interactions are TEM, SEM, Auger, Rutherford ion backscattering (RBS), and interference-enhanced Raman scattering (IERS) (Connell *et al.*, 1980). Whereas most of the techniques are standard, the IERS technique is new and has proved useful in studying reactions on a-Si : H (Nemanich *et al.*, 1981).

The IERS technique employs a multilayer thin-film structure, which is shown schematically in Fig. 14a. The configuration utilizes optical interference to efficiently couple the light into the interfacial region. This is shown in Fig. 14b, in which the light intensity variation through the structure is dis-

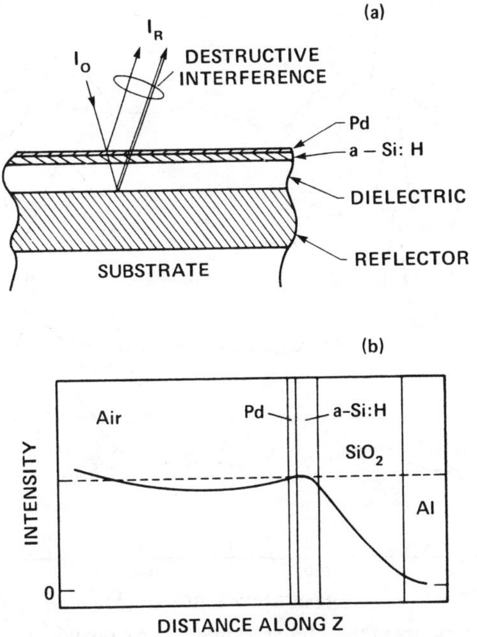

FIG. 14. (a) Schematic of the four-layer structure used to obtain interference-enhanced Raman scattering and (b) the light intensity in and near the structure for normally incident 5145-Å light. [From Nemanich *et al.* (1981).]

played. It is maximum at the interface and drops off into the sample. Because the Raman scattered light is shifted by only a small amount, the same interference conditions apply, and the light is coupled out of the film very efficiently.

The most dramatic interfacial interactions have been observed for Pd on a-Si : H (Nemanich *et al.*, 1981). A series of Raman spectra obtained by using IERS is displayed in Fig. 15. Because simple metals do not exhibit a first-order Raman spectrum, all the features detected are due to atomic vibrations of the a-Si : H and the reacted interface region. The broad features due to the a-Si network vibrations are shown in spectrum (a), but as shown in spectrum (b), immediately after the Pd deposition sharp features are observed. By

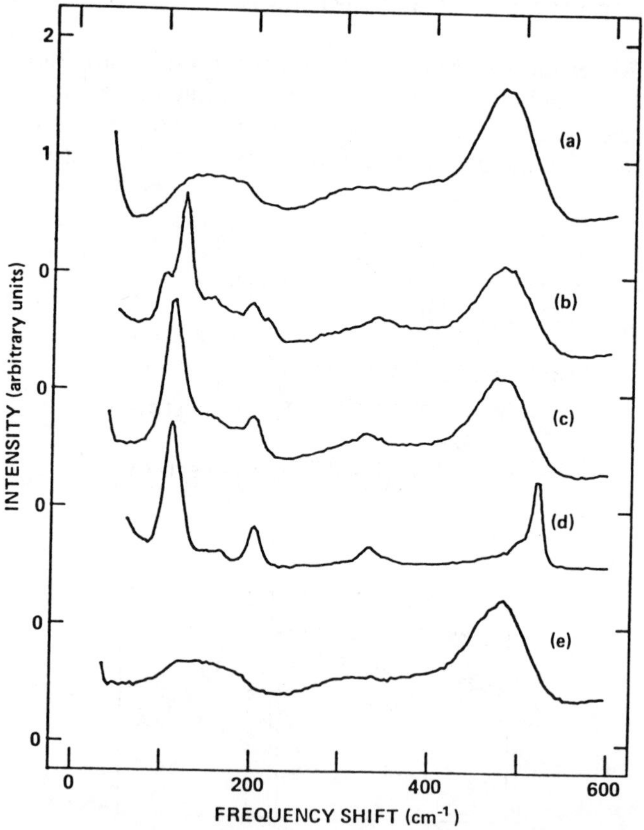

FIG. 15. The Raman spectra of the Pd/a-Si : H interface region. (a) a-Si : H as deposited, (b) Pd deposited on freshly prepared a-Si : H, (c) the sample annealed to 300°C, (d) the sample annealed to 550°C, and (e) Pd deposited on oxidized a-Si : H and annealed to 400°C. [From Nemanich *et al.* (1981).]

comparison with RBS and TEM results from thicker films, it was determined that ~20 Å of Pd reacted to form Pd_2Si. Annealing at <200°C causes further growth of that phase, whereas annealing at >200°C causes the formation of a second phase of Pd_2Si, which is indicated in spectrum (c). Further annealing to higher temperature induces the Si to crystallize as indicated by the sharp feature at 520 cm^{-1} in Fig. 15d. It was also found that when the Pd was deposited on a-Si:H that had been oxidized in air for several weeks, no sign of silicide formation was detected [spectrum (e)].

The electrical properties of the Pd/a-Si:H diodes also reflect the changes due to the structural interactions (Thompson *et al.*, 1981; Pietruszko *et al.*, 1982). The $J-V$ measurements taken before and after annealing are shown in Fig. 16 (Thompson *et al.*, 1981). Whereas in both cases good rectification is observed, the reverse-bias saturation current is reduced after annealing. More dramatic is the change in ideality parameter. As shown in the insert in Fig. 16, a progressive improvement with annealing is found. At 200°C all the

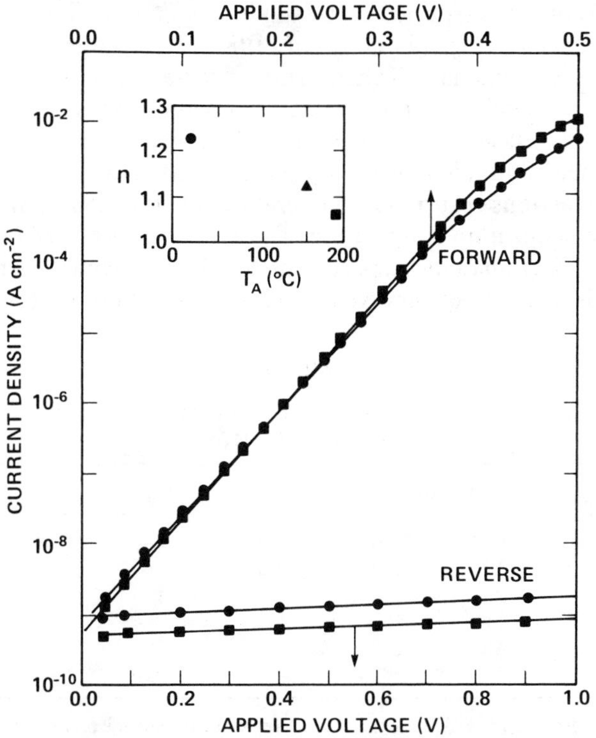

FIG. 16. The $J-V$ characteristics of a Pd/a-Si:H Schottky-barrier diode before (●) and after 180°C annealing (■) to form a uniform silicide. The insert shows the improvement of the ideality parameter for annealing at various temperatures. [From Thompson *et al.* (1981).]

metal is consumed and the ideality parameter reaches a value of 1.05. This is as low as is obtained on crystalline Si diodes. Thus the interface integrity is similar in both cases.

The same structural analysis techniques have been applied to study several other metals, including Cr, Ni, and Pt (Nemanich *et al.*, 1983) that are known to form silicides, and Al and Au (Tsai *et al.*, 1982, 1983), which do not. For the metals that form silicides, it was found that atomic interdiffusion, which resulted in a disordered phase, often preceded silicide formation. Then, silicide formation occurred at an elevated temperature. The results are summarized in Table I. For all these cases the internal photoemission showed no change in barrier height (<0.03 eV) after annealing, but the photoemission signal increased by a factor of two. For all the metals that form silicides it appears that the reactions occur at temperatures similar to those on crystalline Si. The disordered intermixed phase, however, has not been definitively observed for metals on crystalline Si.

The reactions at the Au or Al interface on a-Si:H exhibited different properties from those reported on crystalline Si, and the result is displayed most dramatically in Fig. 17 (Tsai *et al.*, 1982). Here, large (~ 1-μm) inclusions are observed that have been determined to be crystalline Si. The lateral extent indicates dendritic growth, which resulted from Si diffusion through the Au. Similar effects are observed for Al deposited on a-Si:H. In both cases, evidence of atomic interdiffusion was noted before the Si crystallization. These reactions occurred at temperatures of 200–250°C, which is well below the crystallization temperature of a-Si:H or the lowest eutectic in the binary phase diagram of the respective materials. Although the large crystalline Si inclusions are not observed for similar structures on crystalline Si,

TABLE I

THE TEMPERATURE OF INTERFACE INTERACTIONS FOR METALS DEPOSITED ON a-Si:H[a] (IN °C)

Metal	Intermixed	Silicide	Crystalline Si
Cr	350	400	—
Ni	150	200	—
Pd	—	RT	550
Pt	150	200	—
Au	100	—	200
Al	100	—	250

[a] The intermixed phases are disordered while the silicide phases refer to crystalline phases. The temperature of crystalline Si inclusions is also indicated.

FIG. 17. A TEM micrograph of a Au/a-Si:H structure after annealing to 200°C. The "snow-flakelike" islands are crystalline Si and are ~ 1 μm in diameter. [From Tsai *et al.* (1982).]

similar structures have been reported by Herd *et al.* (1972) for reactions on (unhydrogenated) a-Si.

The *J*–*V* characteristics before and after the crystallization are shown in Fig. 18 (Tsai *et al.,* 1983). Surprisingly, the Au/a-Si:H diodes showed an improved ideality value, whereas the Al exhibited degradation to nearly ohmic behavior. The nearly ideal characteristics of the annealed Au diodes indicate that the a-Si:H must not be in direct contact with the crystalline Si inclusions. For the annealed Al diodes, direct contact between the crystalline regions and the underlying a-Si:H may account for the observed "ohmic" behavior.

The results of the correlated structural and electrical measurements indicate generally more ideal behavior after atomic interdiffusion or silicide formation. This most likely results because of elimination of surface defects on the a-Si:H and the formation of a laterally uniform interface. The fact that the barrier height does not change during the reactions indicates that

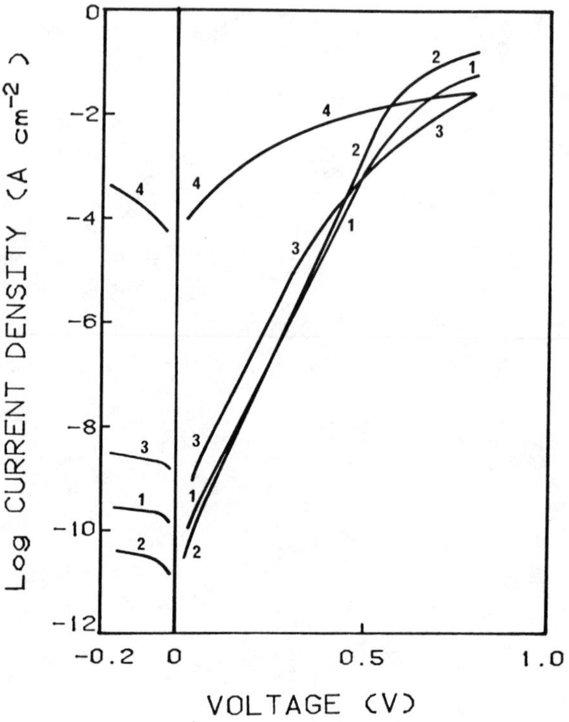

FIG. 18. The J–V characteristics of Au and Al Schottky diodes on a-Si:H before and after annealing to form crystalline Si islands. (1) Au/a-Si:H as deposited, (2) Au/a-Si:H after 200°C anneal, (3) Al/a-Si:H as deposited, and (4) Al/A-Si:H after 250°C anneal. [From Tsai *et al.* (1983). Copyright North-Holland Publ. Co., Amsterdam, 1983.]

some initial reactions must occur that resemble the structure of the final interface. In the case of Au/a-Si:H diodes it seems that the lateral nonuniformities must not extend to the interface. It should be noted that a host of interactions have been identified for metals deposited on crystalline Si. It appears that even more possibilities occur on a-Si:H. Thus it is clear that correlation of structural and electrical properties should be emphasized.

V. Barrier on Doped a-Si:H

Whereas the Schottky barrier on undoped a-Si:H is a useful configuration for experimental studies, the barrier on doped a-Si:H may prove more important technologically. Because thin-film transistor action has been demonstrated, these devices may well form the basis of a new technology (Snell *et al.*, 1981). An important aspect of this will be to form ohmic contacts to the doped regions. To accomplish this it is necessary to understand the doping dependence of the Schottky barrier.

Following the traditional analysis of Schottky barriers, it would be anticipated that the actual barrier height is an intrinsic property of the materials and will not change. However, upon moderate doping, the depletion region will become very narrow because the ionized donor levels will screen the interface. The sequence is shown schematically in Fig. 19. As the semiconductor doping is increased, the Fermi energy will move toward the conduction band, and the resultant Schottky barrier will exhibit an increased built-

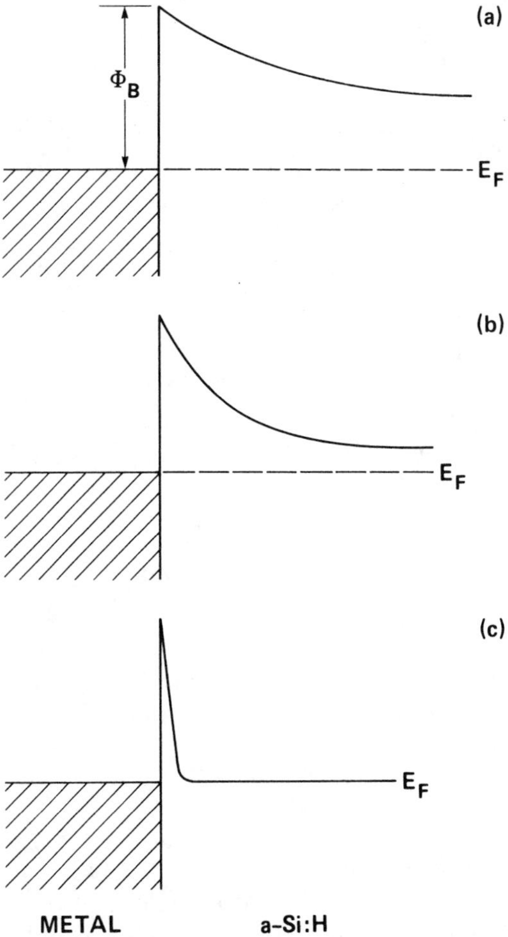

FIG. 19. Schematic of a possible sequence describing the effect of doping on the Schottky barrier. (a) Undoped a-Si:H, (b) $\sim 10^{-6}$ *n*-doped, and (c) $\sim 10^{-3}$ *n*-doped.

in potential. When the doping becomes very heavy ($\geq 10^{-3}$), the depletion region will collapse, and electrons will be able to tunnel through the barrier. This results in an effective ohmic contact.

There have been two recent studies addressing the changes in the Schottky barrier with doping. The initial study by Viktorovitch *et al.* (1981) noted that the barrier height for Pt/a-Si:H diodes dropped by ~0.3 eV for gas-phase doping levels of 2×10^{-4}. They also found that the conductivity exhibited an activation energy of ~0.3 eV. It was proposed that the data could be accounted for by a donor level ~0.3 eV below the conduction band. Thus the transport through the barrier would involve this level. In another study Madan *et al.* (1982) showed that large increases in the ideality parameter occurred for doped diodes measured at low temperature. The data were analyzed in terms of a recombination current due to a second transport mode. This mode could be described by hopping conduction through a level 0.3 eV below the conduction band.

The barrier height versus doping was studied by Nemanich *et al.* (1983).

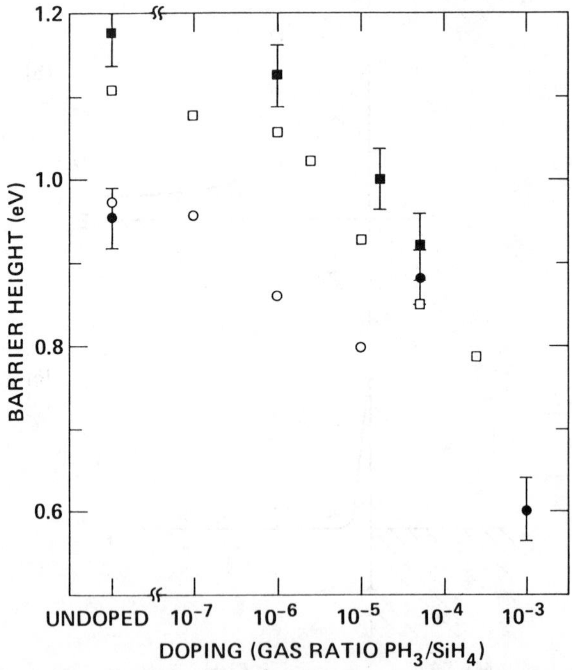

FIG. 20. Barrier height versus doping for Pd (circles) and Pt (squares) diodes determined from $J-V$ (open symbols) and internal photoemission measurements (filled symbols). [From Nemanich *et al.* (1983b). Copyright North-Holland Publ. Co., Amsterdam, 1983.]

They correlated structural and electronic properties and examined a wider range of dopants. The barrier height versus doping was measured by $J-V$ and internal photoemission, and the results are shown in Fig. 20. The first point to note is that good agreement was obtained between the two measurements. This was most likely due to the fact that steps were taken to ensure a uniform interface. Furthermore, the results showed very little dependence on doping up to gas-phase concentrations of 10^{-5}. After this, the effective barrier height dropped rapidly ~0.3 eV for both Pd and Pt diodes.

Although these results could be consistent with the model proposed by Viktorovitch *et al.* (1981), there have been no other measurements that have detected the level at 0.3 eV below the conduction band. The results are also consistent with the usual model of tunneling through a collapsed depletion region.

Another possibility that should also be considered is that the interface properties that are responsible for the barrier actually change with doping of the a-Si:H film. It has been shown that the surface states do increase with doping (Jackson *et al.,* 1983a). Furthermore, the structural studies have indicated that silicide formation occurs at slightly higher temperatures for Ni and Cr on doped a-Si:H (Nemanich *et al.,* 1983). Thus, if the interface properties actually change, then the barrier may change also.

Thus, more measurements are necessary to sort out the actual mechanism responsible for the transition from Schottky barrier to ohmic contact. Key among these would be depletion-layer profiling and built-in potential correlated with barrier-height variations. Furthermore, detailed field-dependent internal photoemission measurements may also help in solving the problem.

VI. Schottky-Barrier Origin

Although the origin of the Schottky barrier was alluded to in Part II, the actual mechanism was not specified. In crystalline Si several models have been considered, and these include interfacial defects and the metal work function. The classical model put forth by Schottky and Mott (see Rhoderick, 1978) suggests that the barrier height simply reflects the difference of the metal work function and the electron affinity of the semiconductor. Unfortunately, few systems satisfy this condition by using the pure metal work function. This is also true for a-Si:H. An alternative model is that the band bending is determined by surface states. If the surface-state density is very high, then the barrier will be independent of the metal work function because the charge transfer necessary to equilibrate the electron potential can be stored in the surface states.

It is more likely that surface states exist but that different metals can affect the band bending. An analysis based on this model has been carried out by

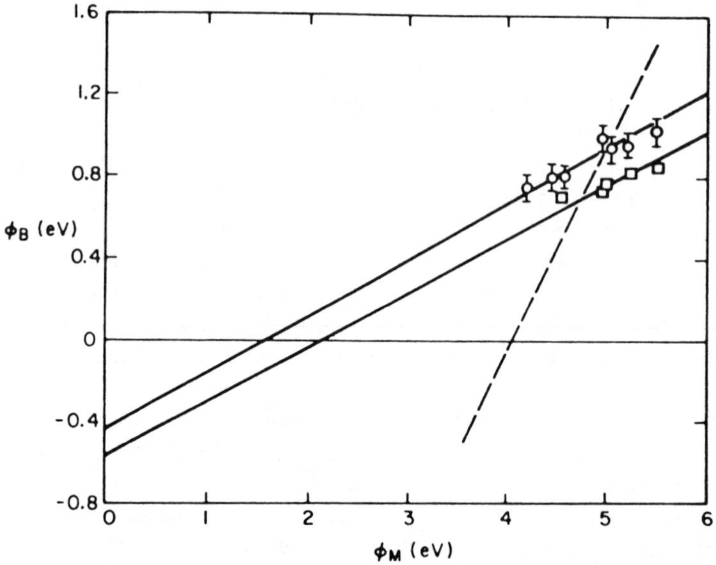

FIG. 21. The plot of Φ_B for Schottky barriers on a-Si : H (○) and *n*-doped crystalline Si (□) versus metal work function. The dashed line is the expected behavior for the metal work-function model ($\Phi_B = 1.00, \Phi_M - 4.05$) and the solid lines are fits to the data for models that include interface states. The fits give $\Phi_B = \Phi_M - 0.44$ for a-Si : H and $\Phi_B = \Phi_M - 0.55$ for *n*-doped c-Si. [Reprinted with permission from *Solid State Communications,* Vol. 23, C. R. Wronski and D. E. Carlson, Surface states and barrier heights of metal–amorphous silicon Schottky barriers. Copyright 1977, Pergamon Press, Ltd.]

Wronski and Carlson (1977). The results shown in Fig. 21 indicate that the barriers on several metals parallel those of crystalline Si. The analysis clearly shows that the unmodified work-function model cannot be applied directly, but a significant linear dependence on metal work function was observed. The analysis indicated an interface density of $\sim 2 \times 10^{13}\,\mathrm{cm}^{-2}$. This is similar to the surface-state density reported by Jackson *et al.* (1983a).

Because of the close parallel in barrier results of a-Si : H and crystalline Si, it is worthwhile to examine other models used for crystalline Si. One such model has been termed the effective work-function model (Freeouf, 1980). This model is based on the fact that atomic interactions occur for almost all metal–Si interfaces studied to date. Thus the work-function model should be modified to use the work function of the metallic states adjoining the semiconductor. This model thus must include the chemistry of the interface (Grunthaner *et al.,* 1980; Rubloff *et al.,* 1981). Since similar atomic interactions are observed for metals on a-Si : H as on crystalline Si, a similar model may apply. Unfortunately, the theoretical details of the model have not been sufficiently addressed to determine whether it is truly consistent with the data even for crystalline Si. But because of the similarity in atomic interactions and parallel barrier heights in a-Si : H and crystalline Si, it is likely that the Schottky barrier has a similar origin in both systems.

VII. Concluding Remarks

Given the current understanding of Schottky barriers on a-Si : H, it is appropriate to question whether there are any obvious technological applications. Most of the first a-Si : H solar-cell structures utilized Schottky-barrier configurations (see Wronski, 1977; Carlson, 1980). The best of these cells exhibited conversion efficiencies of ~ 7%, but they suffered from apparent intrinsic instabilities (Carlson and Magee, 1979). Although structural interactions are more fully understood now, the instability question was never addressed in detail because it was found that *p–i–n* solar cells exhibited a conversion efficiency of ~ 10% or more (Carlson, 1983). This was because the maximum open-circuit voltage of a Schottky-barrier cell is about 0.70 V, whereas *p–i–n* structures can achieve ~0.90 V by using a wider-band-gap *p*-layer like SiC. With improved understanding of the interface kinetics and instabilities, Schottky-barrier solar cells may still have some limited applications.

A technological area that seems to have moved to the forefront recently is the use of a-Si : H for large-area electronics. Although a-Si : H will never rival crystalline Si for many applications, there appears to be a need for large-area arrays of devices that operate at moderate to slow speeds. A typical example of such a device is a flat-panel display that replaces the CRT in some applications. Here, thin-film transistors (TFTs) are a most likely candidate, and TFT structures have been fabricated by several groups (Snell *et al.*, 1981; Powell *et al.*, 1981; Tuan *et al.*, 1982). A unique structure using a dual gate is shown in Fig. 22 (Tuan *et al.*, 1982). The major benefit of a-Si : H is that it is less costly to deposit arrays of a-Si : H on glass substrates than to bond together crystalline-Si wafers. Furthermore, a-Si : H TFTs seem to be more reliable than the polycrystalline-Si TFTs. If large-area a-Si : H-based electronics become important, Schottky-barrier structures will be utilized. But of even greater importance will be the formation of well-characterized ohmic contacts to other components like a TFT.

In the laboratory the Schottky-barrier structure will continue to be significant in many experimental structures. Of significant use is the Schottky barrier in DLTS and, obviously, capacitance and photoconductivity measurements, which have been described in this chapter and in Chapter 2 by Cohen.

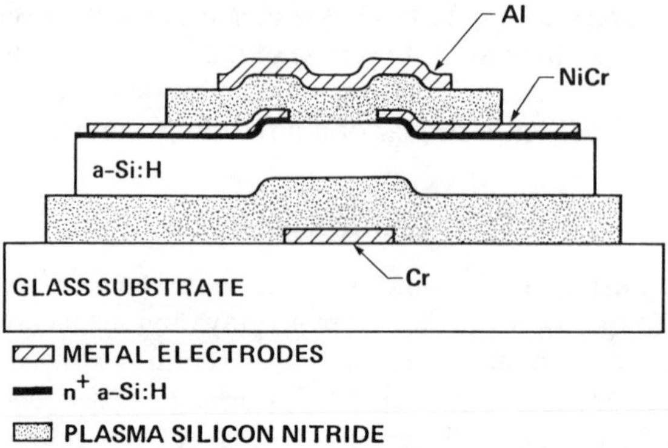

FIG. 22. A schematic of an a-Si : H thin-film transistor. [From H. C. Tuan, M. J. Thompson, N. M. Johnson, and R. A. Lujan, Dual-gate a-Si : H thin film transistors. *IEEE Electron Device Letters,* ©1982 IEEE.]

At this stage the critical issues regarding a-Si : H Schottky barriers are addressed. The first is whether the forward-bias transport of undoped diodes is limited by diffusion or thermionic emission. This point may prove crucial to interpreting other experiments such as DLTS or frequency-dependent $C-V$ to obtain gap-state densities. It seems from the experiments described to date, that well-prepared, well-characterized interfaces for at least some metals yield transport characteristics consistent with the thermionic emission theory.

The internal photoemission measurements appear to be on solid ground, but more detailed investigations, including the field dependence on well-characterized interfaces, would be important. Other experiments such as transient photoconductivity have been applied to depletion-layer profiling, and these appear to be quite useful.

The structural characterizations to date have shown that interactions similar to those of metals on crystalline Si are often observed. But different interactions such as disordered intermixed phases and dendritic Si island formation are also observed. Further work in this area will benefit from the UHV surface-sensitive tools used to investigate interfaces on crystalline semiconductors.

The properties of metals on doped a-Si : H are sure to be an important area of research. Conflicting models have already been discussed. Sorting this problem out will be important for both research applications and technology. Lastly, the actual details of Schottky-barrier formation remain a mystery. The previous studies seem to indicate a strong similarity in many aspects of Schottky barriers on a-Si : H and crystalline Si.

Thus it appears that research into the properties of Schottky barriers on a-Si:H will be an area of expanding interest. Many of the basic scientific questions remain that relate to barrier formation on a disordered solid. Also driving the field, though, is the rapid development of technologies involving a-Si:H.

REFERENCES

Alkaisi, M. M., and Thompson, M. J. (1979–1980). *Solar Cells* **1,** 91.

Beichler, J., Fuhs, W., Mell, H., and Welsch, H. M. (1980). *J. Non-Cryst. Solids* **35** and **36,** 587.

Carlson, D. E. (1980). *J. Non-Cryst. Solids* **35** and **36,** 707–717.

Carlson, D. E. (1983). Private communication.

Carlson, D. E., and Magee, C. W. (1979). *Conf. Proc. Eur. Conf. Photovoltaic Solar Energy, 2nd* (R. Van Overstraeten and W. Palz, eds.), p. 312. Reidel, Boston.

Chen, I., and Lee, S. J. (1982). *Appl. Phys.* **53,** 1045.

Cody, G. D., Tiedje, T., Abeles, B., Brooks, B., and Goldstein, Y. (1981). *Phys. Rev. Lett.* **47,** 1480.

Connell, G. A. N., Nemanich, R. J., and Tsai, C. C. (1980). *Appl. Phys. Lett.* **36,** 31.

Deneuville, A., and Brodsky, M. H. (1979). *J. Appl. Phys.* **50,** 1414.

Fernandez-Canque, H., Allison, J., and Thompson, M. J. (1983). *J. Appl. Phys.* **54,** 7025.

Freeouf, J. L. (1980). *Solid State Commun.* **33,** 1059.

Freeouf, J. L., Jackson, T. N., Laux, S. E., and Woodall, J. M. (1982). *J. Vac. Sci. Technol.* **21,** 570.

Grunthaner, P. J., Grunthaner, F. J., and Mayer, J. W. (1980). *J. Vac. Sci. Technol.* **17,** 924.

Herd, S. R., Chaudhari, P., and Brodsky, M. H. (1972). *J. Non-Cryst. Solids* **7,** 309.

Jackson, W. B., and Amer, N. M. (1982). *Phys. Rev. B* **25,** 5559.

Jackson, W. B., Biegelsen, D. K., Nemanich, R. J., and Knights, J. C. (1983a). *Appl. Phys. Lett.* **42,** 105.

Jackson, W. B., Nemanich, R. J., and Amer, N. M. (1983b). *Phys. Rev. B* **27,** 4861.

Jackson, W. B., Nemanich, R. J., and Thompson, M. J. (1984). To be published.

Lang, D. V., Cohen, J. D., and Harbison, J. P. (1982). *Phys. Rev. B* **25,** 5285.

Madan, A., Czubatyj, W., Yang, J., Shur, M. S., and Shaw, M. P. (1982). *Appl. Phys. Lett.* **40,** 234.

Nemanich, R. J., Tsai, C. C., Thompson, M. J., and Sigmon, T. W. (1981). *J. Vac. Sci. Technol.* **19,** 685.

Nemanich, R. J., Thompson, M. J., Jackson, W. B., Tsai, C. C., and Stafford, B. L. (1983a). *J. Vac. Sci. Technol.* **B1,** 519.

Nemanich, R. J., Thompson, M. J., Jackson, W. B., Tsai, C. C., and Stafford, B. L. (1983b). *J. Non-Cryst. Solids* **59/60,** 513.

Ottaviani, G. (1979). *J. Vac. Sci. Technol.* **16,** 1112.

Padovani, F. A., and Stratton, R. (1966). *Solid-State Electron.* **9,** 695.

Pietruszko, S. M., Narasimhnan, K. L., and Guha, S. (1982). *J. Vac. Sci. Technol.* **20,** 801.

Powell, M. J., Easton, B. C., and Hill, O. F. (1981). *Appl. Phys. Lett.* **38,** 794.

Rhoderick, E. H. (1978). "Metal–Semiconductor Contacts." Oxford Univ. Press (Clarendon), London and New York.

Rubloff, G. W., Ho, P. S., Freeouf, J. L., and Lewis, J. E. (1981). *Phys. Rev. B* **23,** 4183.

Snell, A. J., Mackenzie, K. D., LeComber, P. G., and Spear, W. E. (1979). *Philos. Mag. B* **40,** 1.

Snell, A. J., Mackenzie, K. D., LeComber, P. G., and Spear, W. E. (1980). *J. Non-Cryst. Solids* **35** and **36,** 593.
Snell, A. J., Mackenzie, K. D., Spear, W. E., and LeComber, P. G. (1981). *Appl. Phys.* **24,** 357.
Spear, W. E., LeComber, P. G., and Snell, A. J. (1978). *Philos. Mag. B* **38,** 303.
Street, R. A. (1983). *Phys. Rev. B* **27,** 4924.
Sze, S. M. (1969). *In* "Physics of Semiconductor Devices," Chap. 8. Wiley, New York.
Thompson, M. J., Alkaisi, M. M., and Allison, J. (1980). *Proc. Inst. Electr. Eng.* **127,** 213.
Thompson, M. J., Johnson, N. M., Nemanich, R. J., and Tsai, C. C. (1981). *Appl. Phys. Lett.* **39,** 274.
Tsai, C. C., Nemanich, R. J., and Thompson, M. J. (1982). *J. Vac. Sci. Technol.* **21,** 632.
Tsai, C. C., Nemanich, R. J., Thompson, M. J., and Stafford, B. L. (1983). *Physica* **117B** and **118B,** 953.
Tsang, C., and Street, R. A. (1979). *Phys. Rev. B* **19,** 3027.
Tu, K. N., and Mayer, J. W. (1978). *In* "Thin Films Interdiffusion and Reactions" (J. M. Poate, K. N. Tu, and J. W. Mayer, eds.), p. 359. Wiley, New York.
Tuan, H. C., Thompson, M. J., Johnson, N. M., and Lujan, R. A. (1982). *IEEE Electron Device Lett.* **EDL-3,** 357.
Viktorovitch, P., and Jousse, D. (1980). *J. Non-Cryst. Solids* **35** and **36,** 569.
Viktorovitch, P., Jousse, D., Chenevas-Paule, A., and Vieux-Rochas, L. (1979). *Rev. Phys. Appl.* **14,** 201.
Viktorovitch, P., Moddel, G., and Paul, W. (1981). *AIP Conf. Proc.* **73,** 186.
Wronski, C. R., Carlson, D. E., and Daniel, R. E. (1976). Appl. Phys. Lett. **29,** 602.
Wronski, C. R. (1978). *Jpn. J. Appl. Phys.* **17,** 299.
Wronski, C. R., and Carlson, D. E. (1977). *Solid State Commun.* **23,** 421.
Wronski, C. R., Abeles, B., Cody G. D., and Tiedje, T. (1980). *Appl. Phys. Lett.* **37,** 96.
Yamamoto, T., Mishima, Y., Hirose, M., and Osaka, Y. (1981). *Jpn. J. Appl. Phys.* **20,** *Suppl. 20-2,* 185.

CHAPTER 12

Amorphous Semiconductor Superlattices

B. Abeles and T. Tiedje

CORPORATE RESEARCH SCIENCE LABORATORIES
EXXON RESEARCH AND ENGINEERING COMPANY
ANNANDALE, NEW JERSEY

I. Introduction

Semiconductor superlattice structures fabricated from alternating layers of crystalline III–V materials exhibit many interesting transport and optical properties that are associated with quantum size effects (Esaki and Tsu, 1970; Ploog and Dohler, 1983). These superlattices can only be fabricated from materials that have a nearly perfect match in their lattice constants and that can be grown epitaxially on top of one another. Otherwise the density of defects is so great that the phenomena associated with quantum size effects are obscured. It has been shown (Abeles and Tiedje, 1983) that the range of materials from which superlattices can be fabricated can be extended to hydrogenated amorphous semiconductors. Superlattices were made from materials such as hydrogenated amorphous silicon (a-Si:H), a-Ge:H, a-SiN$_x$:H, and a-Si$_{1-x}$C$_x$:H that are neither lattice matched nor epitaxial and yet have interfaces that are essentially defect free and nearly atomically sharp. The stringent requirements for lattice matching in crystalline superlattices are relaxed in the amorphous case because of the nonperiodic structure and the ability of hydrogen to passivate coordination defects.

The electronic properties of the new superlattice materials help to shed light on some major fundamental questions in amorphous semiconductors. Quantum size effects in the case of crystalline superlattices raise the lowest allowed electron and hole energy levels and give rise to a density of states that increases in discrete steps. This structure in the density of states is reflected in

ISBN 0-12-752149-6

an increase in the optical gap and in discretely spaced absorption peaks. In amorphous superlattices the fine structure is absent, presumably because of disorder broadening; the evidence for quantum size effects comes from the observation of the large changes in the optical band gap and electrical resistivity. Quantum size effects in a-Si : H layers as thick as 40 Å suggest that the coherence of the wave functions is 40 Å or larger—heretofore it was generally thought to be of the order of an interatomic distance. The increase in the optical gap (Part III) and resistivity (Part V) with decreasing a-Si : H layer thickness in a-Si : H/a-SiN$_x$: H superlattices is explained by a free electron and hole quantum well model with effective masses of $0.2m$ for electrons and $1.0m$ for holes where m is the free electron mass. The change in the mobility edge with well thickness in this model is assumed to coincide with the change in the lowest free-electron energy level allowed by the quantum well. Theoretical work on localization in reduced dimensionality systems (Abrahams *et al.*, 1979) suggests that there might be additional large changes in the position of the mobility edge in amorphous semiconductors in going from three to two dimensions. No such changes are indicated in the behavior of the electrical resistivity in n-type a-Si : H layers in the thickness range <40 Å, where the conduction-band-edge electrons become two dimensional due to quantum size effects.

Another area in which the layered materials can be utilized to great advantage is in the study of interfaces. It is possible to extend the detection sensitivity of optical absorption measurements to very low density of interface states because of the large enhancement that can be achieved by using many layers. Roxlo *et al.* (1984a,b) utilized electroabsorption measurements in a-Si : H/a-SiN$_x$: H films to measure the charge distribution in the layers. From their results the authors infer that the electronic structure of the interfaces depends on the sequence of deposition (a-Si : H on a-SiN$_x$: H or a-SiN$_x$: H on a-Si : H).

We expect that the high degree of control of the electronic and mechanical properties achievable in the layered amorphous semiconductors will make it possible to engineer new materials with properties desirable in electronic devices and other applications. We briefly discuss some examples here. We have found that transfer of electrons from a-SiN$_x$: H layers into a-Si : H layers can result in material with resistivity (10^3 Ω cm) five orders of magnitude down from undoped bulk films and yet with a density of defects more than an order of magnitude lower than in the phosphorus-doped material of comparable resistivity (Tiedje and Abeles, 1984). This new "modulation" doping mechanism has potential applications in field-effect transistors (FETs) analogous to the transfer doped GaAs/Ga$_{1-x}$Al$_x$As heterostructure FET (Morkoc and Solomon, 1984). A large increase in room-temperature photoluminescence efficiency with decreasing a-Si : H film thickness suggests that the a-SiN_x:H barriers prevent the photogenerated electron–hole pairs from drifting apart and recombining nonradiatively. The enhancement in photoluminescence efficiency may find application in improving the performance of a-$Si_{1-x}C_x$:H electroluminescent devices (Munekata and Kukimoto, 1983) by forming superlattices of the alloy with insulators such as a-SiO_x:H or a-SiN_x:H. The ability to vary the band gap in the a-Si:H/a-$Si_{1-x}C_x$:H system in the range 1.7–2.2 eV and in a-Si:H/a-Ge:H from 1.1 to 1.7 eV has potential application in solar cells, for which there is need for high-efficiency solar-cell materials with band gaps of 1.2 and 2.1 eV to complement the optical gap of 1.7 eV of a-Si:H in three-stage tandem solar cells (Fan and Palm, 1983).

In this chapter we review the work done in our laboratory on the structure (Part II), optical absorption (Part III), photoluminescence (Part IV), and electrical transport (Part V) of a-Si:H/a-SiN_x:H superlattices. Results with single quantum well structures are discussed by Kukimoto in Chapter 12 of Volume 21D.

II. Film Deposition and Structure

The amorphous superlattice materials were fabricated in thin film form by a plasma-assisted chemical-vapor-deposition process in which the composition of the reactive gas in the plasma reactor is changed periodically without interrupting the plasma (Abeles and Tiedje, 1983). A schematic diagram of the capacitive plasma reactor used is shown in Fig. 1. The substrates are placed on the anode, and both cathode and anode are heated to 180–220°C. In order to achieve sharp interfaces between the layers, the residence time T_R of the gas molecules in the reactor must be short compared with the time T_M

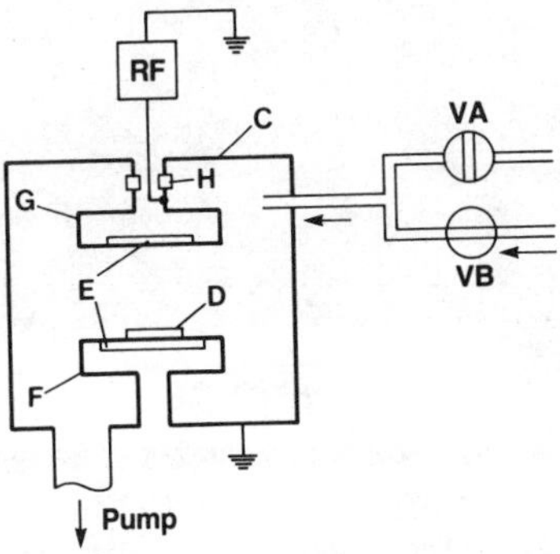

FIG. 1. Deposition system. Valves VA and VB open and close alternately to admit gases A or B; C, stainless steel bell jar; D, substrate; E, heaters; F, anode; G, cathode; H, insulator.

required to grow a monolayer. The residence time for the gas is

$$T_R = Vp/F_0p_0, \tag{1}$$

where V is the volume of the reactor, p the gas pressure, F_0 the flow rate at STP, and p_0 standard pressure. The condition $T_R < T_M$ is readily achievable. For instance, in our experiments we used $V = 30$ liter, $p = 30$ mTorr, $F_0 = 85\ \mathrm{cm^3/min}$, $T_R = 1$ sec, rf power-5 W at 13.6 MHz, and $T_M = 3-5$ sec. The thicknesses of the individual layers are determined by the flow duration of the gas mixtures and deposition rates of the layers.

Transmission electron microscopy (TEM) and x-ray diffraction provide complementary information on the structure of the superlattices. Transmission electron microscopy gives an image of the film cross section and provides information on localized defects; the interpretation of the data can be relatively simple, but the sectioning of the samples for microscopy is a difficult technical task. X-ray diffraction provides information on the struc-

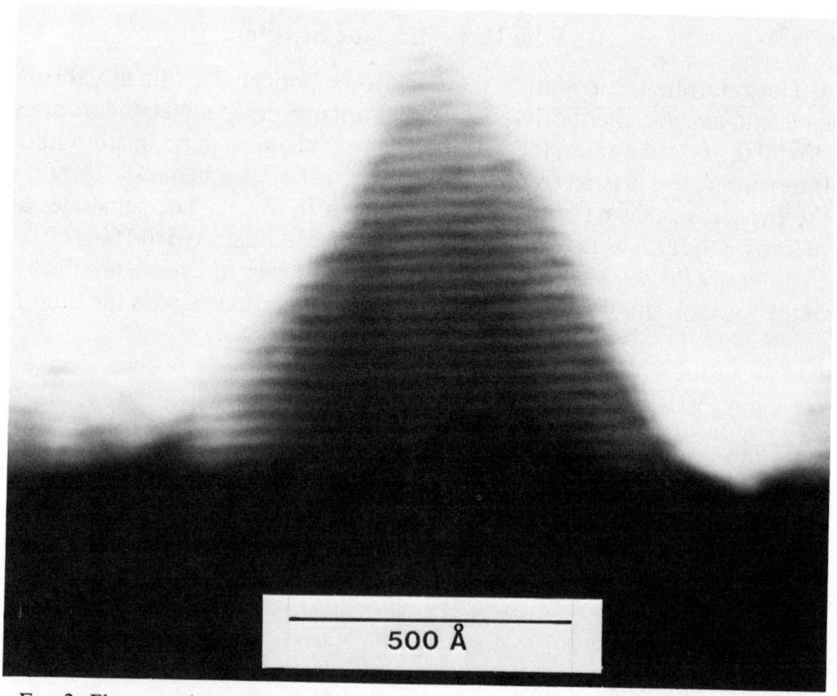

FIG. 2. Electron micrograph of a-Si:H/a-Ge:H superlattice. The dark stripes are a-Ge:H; the light stripes are a-Si:H layers. The period of the superlattice is 26 Å. [From Deckman *et al.* (1984).)]

ture averaged over the sample volume. Sample preparation and measurements are straightforward, but detailed interpretation of the data requires a mathematical analysis of structural models.

Deckman *et al.* (1984a,b) have developed a novel plasma etching technique for fabricating microscope samples in the shape of a small cone with its axis normal to the layers. Figure 2 (Abeles *et al.* 1984a) shows a TEM of such a cone consisting of alternating layers of a-Si : H and a-Ge : H deposited on a polished crystalline silicon substrate. The gas flow periods used for this sample were pure SiH_4 (20 sec) alternating with GeH_4 diluted 10 : 1 with hydrogen (80 sec). The period of the superlattice in Fig. 2 is 27 Å and the layers are smooth and parallel to better than 5 Å.

X-ray diffraction measurements were made on films deposited on $\frac{1}{2} \times \frac{1}{2}$-inch quartz suprasil substrates $\frac{1}{16}$-inch thick by using radiation at wavelength $\lambda = 1.54$ Å. In Fig. 3 are shown the first four diffraction peaks corresponding to a a-Si : H/a-SiN_x : H superlattice (Abeles and Tiedje, 1983). The layer spacing, 68 Å, determined from the position of the Bragg peaks in Fig. 3 is in agreement with the layer thickness determined from the deposition rate of the superlattice. The width of the first Bragg peak gives an indication of the uniformity of the layered structure. The measured full width at half maximum (FWHM) in the 2θ scan for this 41-period structure is 0.06° (diffrac-

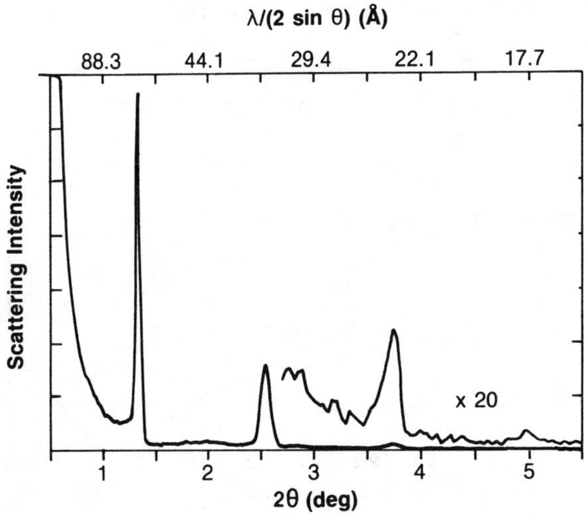

FIG. 3. X-ray diffraction pattern versus scattering angle (lower scale) and lattice spacing (upper scale) of an a-Si : H/a-SiN_x : H superlattice on a quartz substrate with 41 periods of 68 Å each. [From Abeles and Tiedje (1983).]

tometer resolution 0.02°) or within a factor of two of the value for an ideal, nonabsorbing, weakly scattering, layered sample. The relative intensities of the higher-order diffraction peaks are consistent with the form factor expected for a-SiN_x:H and a-Si:H layers with relative thicknesses in the ratio 0.59, with an rms fluctuation in layer thickness of 5.4 Å (Abeles *et al.*, 1984a).

III. Optical Absorption

Optical absorption has been studied in superlattices made of a-Si:H layers alternating with a-SiN_x:H, a-$Si_{1-x}C_x$:H, or a-Ge:H (Abeles and Tiedje, 1983; Tiedje *et al.*, 1984). Figure 4 shows the optical absorption coefficient α versus photon energy E of a series of a-Si:H/a-SiN_x:H superlattices about 1 μm thick, in which the thickness of the a-Si:H layer L_S is varied and the a-SiN_x:H layer thickness $L_N = 35$ Å is held fixed. The large blue shift in the optical absorption edge with decreasing L_S has been attributed to an increase in the optical band gap in the a-Si:H layers due to confinement of the charge carriers in quantum wells between the larger band gap a-SiN_x:H layers.

To derive the shift in the optical band gap, Tiedje *et al.* (1984) used the conduction and valence densities of states corresponding to the model of free electrons and holes in a one-dimensional periodic potential shown in Fig. 5a. The parameters of the model were chosen as follows: bulk amorphous silicon band gap 1.8 eV; conduction- and valence-band-edge discontinuities at the a-Si:H/a-SiN_x:H interfaces $U_c = 1.0$ eV and $U_v = 0.6$ eV, respectively

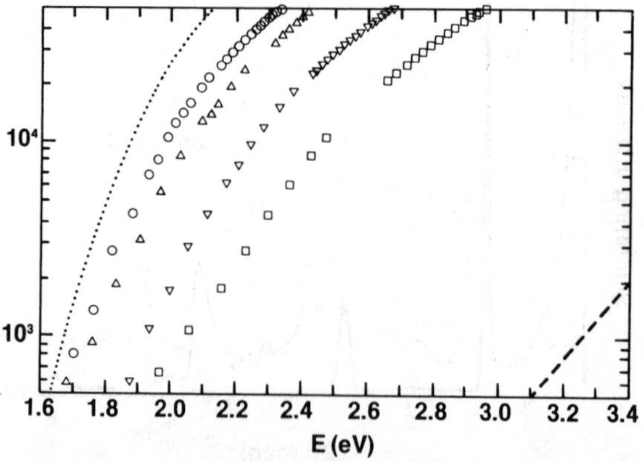

FIG. 4. Optical absorption coefficient α versus photon energy E for a series of a-Si:H/a-SiN_x:H superlattices for different a-Si:H layer thickness L_S with the a-SiN_x:H layer thickness $L_N = 35$ Å. L_c(Å) = 40 (○), 20 (△), 12 (▽), 8 (□). Included in the figure is α for bulk a-Si:H (· · ·) and a-SiN_x:H (———) films. [From Abeles and Tiedje (1983).]

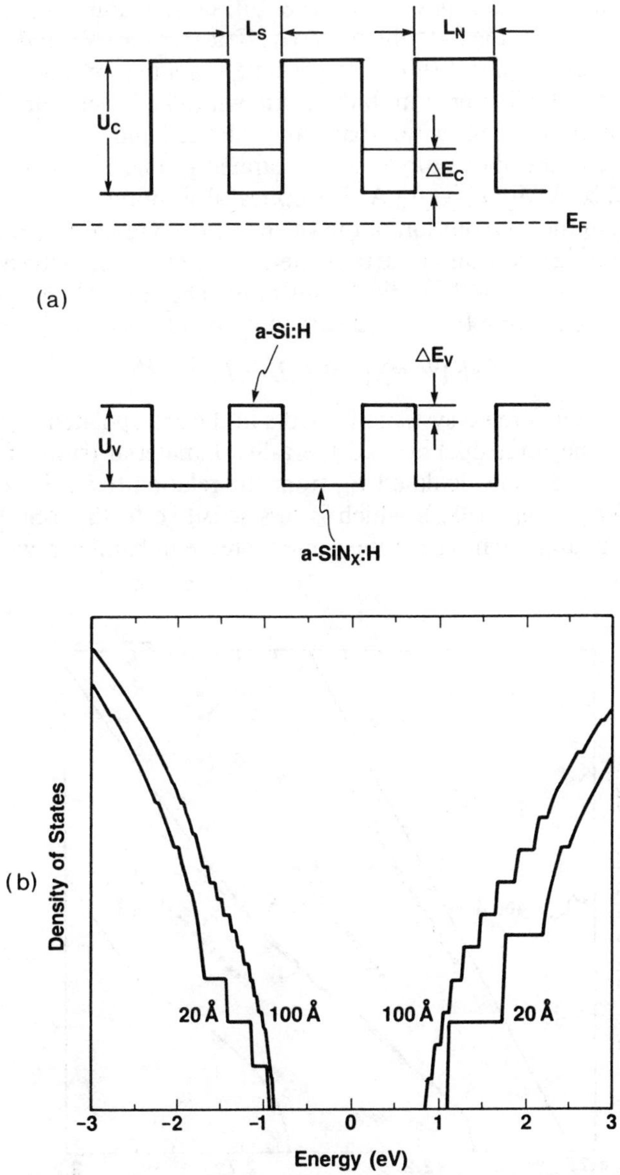

FIG. 5. (a) Potential well model for the a-Si : H/a-SiN_x : H superlattice. Indicated in the figure are the energy levels ΔE_c and ΔE_v for the lowest quantum states for electrons and holes; E_F is the Fermi level and U_c and U_v are the discontinuities in the conduction and valence bands, respectively. (b) Density of states for a free electron gas in the presence of the rectangular shape potential shown in Fig. 5a with effective mass 0.2m for electrons, 1.0m for holes, and well widths $L_S = 20$ and 100 Å. [From Tiedje *et al.* (1984).]

[from photoemission (Abeles *et al.*, 1984b)], conduction-band effective mass $m_c = 0.2m$ from the fit to the resistivity data (see Part V), and valence-band effective mass $m_v = 1.0m$. A larger hole mass than electron mass is reasonable since the free-hole mobility is known to be lower than the free-electron mobility. The conduction-band and valence-band densities of states corresponding to the above model are illustrated in Fig. 5b for the case of $L_S = 100$ and 20 Å with $L_N = 35$ Å. The optical absorption coefficient α was calculated from the convolution of the valence and conduction densities of states by assuming a constant matrix element and relaxation of the momentum conservation rule for the optical transitions. The optical band gap E_G is derived from α by fitting the optical data with the relation

$$(\alpha E)^{1/2} = \text{const} \cdot (E - E_G). \tag{2}$$

To check that there are no systematic errors in the extrapolated gap due to differences in the total thickness of the a-Si:H material (sum of all the a-Si:H layers), we also calculated E_G from the relation $(\alpha E)^{1/2} = \text{const} \times (E - E_G)$ (Cody *et al.*, 1982), which is less sensitive to the a-Si:H total thickness. Both approaches gave the same increase in band gap with layer thickness.

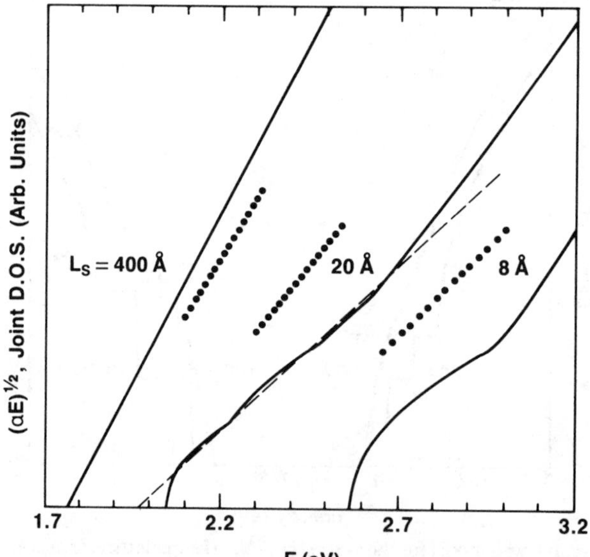

FIG. 6. Optical absorption coefficient α for the a-Si:H/a-SiN$_x$:H superlattice for values of well width $L_S = 400$, 20, and 8 Å plotted as $(\alpha E)^{1/2}$ versus photon energy E. The experimental data are given by the full circles. The solid curves were calculated for the density-of-states model shown in Fig. 5b. The extrapolation of the gap E_a is shown for the case of $L_S = 20$ Å by the dashed line. (From Tiedje *et al.* (1984).)

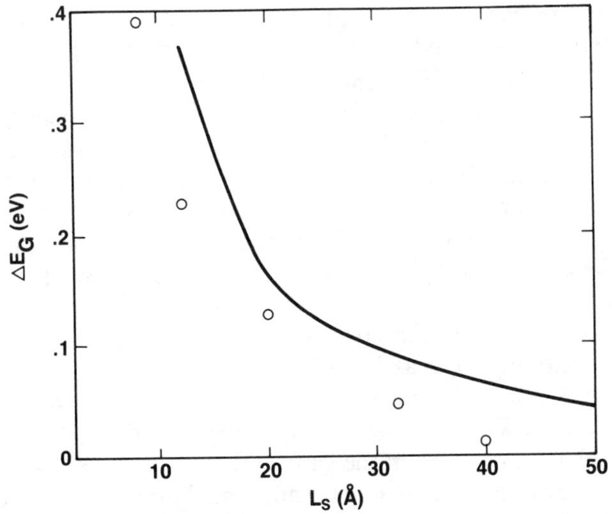

FIG. 7. Change in optical band gap ΔE_G as a function of a-Si : H layer thickness L_S. Solid line is the extrapolated gap from the model data in Fig. 6; open circles are extrapolated values of E_G derived from the experimental data in Fig. 4 and Eq. (2). [From Tiedje *et al.* (1984).]

The calculated optical absorption plotted as $(\alpha E)^{1/2}$ is shown in Fig. 6 for three values of L_S with $L_N = 35$ Å, $m_c = 0.2m$, and $m_v = 1.0m$. The calculated $(\alpha E)^{1/2}$ plot becomes progressively more nonlinear as L_S decreases because of the effects of the superlattice potential. Note that the extrapolated gap given by the dashed line is smaller than the energy at which $\alpha = 0$. The consequences of this are discussed in more detail in Part V. The dependence of the change in the extrapolated optical gap ΔE_G on a-Si : H layer thickness L_S is shown in Fig. 7. The theoretical curve stops at $L_S = 12$ Å because for $L_S < 12$ Å the model curves are not sufficiently linear [over the energy range in which the experimental data are fitted by Eq. (2)] to determine a unique extrapolated gap. The agreement of the model calculation with the experimental values for ΔE_G obtained from the data in Figs. 4 and 6 is good, except for an overall shift (see Fig. 7). Additional factors that may have to be considered in explaining the data in Fig. 7 are first that the quantum size effect may be absent for L_S greater than the coherence length of the electronic wave functions and second that the intrinsic gap of the material when it is in the form of thin layers may be reduced owing to disorder (Cody *et al.*, 1981) resulting from interface defects.

IV. Photoluminescence

The broadening of the absorption tail in Fig. 4 with decreasing layer thickness shows that the layered structure also affects the distribution of localized states in the mobility gap. In order to probe these states the photoluminescence was measured as a function of temperature in the same samples (Abeles and Tiedje, 1983). The photoluminescence spectrum was measured with a krypton ion laser (568-nm line), a cooled Ge detector, and an excitation intensity of 100 mW cm^{-2} chopped at 1 kHz. All of the samples showed the broad and featureless photoluminescence spectrum characteristic of amorphous semiconductors, with no anomalous structure, as might be expected from the strongly bound 2D exciton. In Fig. 8 we show the temperature dependence of the integrated photoluminescence emission intensity for three samples with different layer thicknesses. Note that the temperature dependence of the emission intensity decreases and the room-temperature emission intensity increases as $L_S \rightarrow 0$.

Although the photon energy of the peak emission in the photoluminescence increases with decreasing layer thickness, the magnitude of the change (0.1 eV) is smaller than the shift in the optical gap (0.4 eV) for the same series of samples, presumably because the photoluminescence is associated with localized states that are relatively insensitive to the layered structure. Since the width of the emission band also increases (from 0.3 to 0.5 eV FWHM) as the layer thickness is reduced, the magnitude of the shift of the

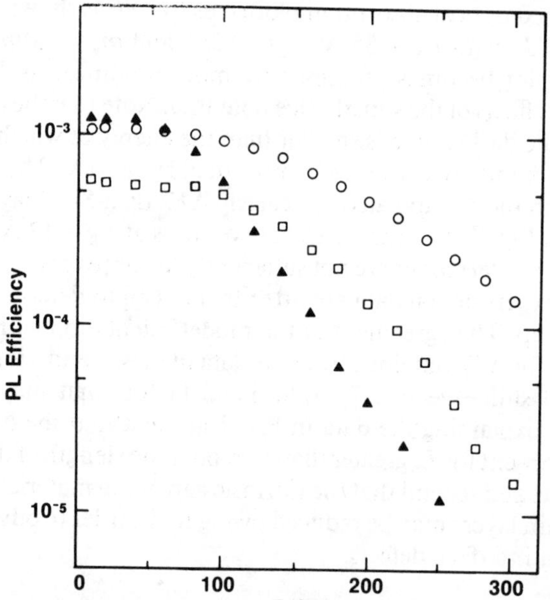

FIG. 8. Photoluminescence emission efficiency as a function of temperature for three representative layered films from the same series as in Fig. 4 with a-Si : H layer thickness L_S = 40 (▲), 12 (□), and 8 Å (O).

high-energy edge of the emission spectrum (0.2 eV) is in better agreement with the change in band gap. The exponential part of the absorption edge (Urbach edge) in amorphous semiconductors is believed to be due to optical transitions involving states in the band tails. Thus, if E_0 is the characteristic energy that describes the width of the Urbach edge, then E_0 is a measure of the width of the band-tail state distribution. Similarly, the exponential temperature dependence of the photoluminescence $[I_{PL} = I_0/(1 + \exp(T/T_0))]$ has been explained by a distribution of thermally activated nonradiative processes in which the thermal activation energy has an exponential distribution (Higashi and Kastner, 1979; Paul and Anderson, 1981). The characteristic width W of the distribution of thermal activation energies defines the thermal quenching parameter T_0 by $W = kT_0 \ln(\nu_0/\nu_r)$, where $\nu_0 = 10^{12}$–10^{13} sec^{-1} is the maximum nonradiative recombination rate and $\nu_r \sim 10^8$–10^9 sec^{-1} is a radiative rate. In Fig. 9 is shown the dependence of the photoluminescence parameter T_0 and the Urbach edge parameter E_0 on the a-Si:H thickness in the a-Si:H/a-SiN$_x$:H superlattice structures. Note that E_0 and T_0 both increase with decreasing layer thickness at about the same rate, which suggests that the enhanced room-temperature photoluminescence is related to a modification of the localized state distribution, and is not simply a double heterostructure confinement effect.

We conclude from these data that the distribution of localized states broadens as the layer thickness decreases. At this point, we do not have enough information to know whether this broadening of the localized-state distribution is due to new localized states associated with the interfaces or whether it is simply a consequence of the fact that the band-tail states become progressively more strongly localized farther from the band edges. In the latter picture, the superlattice potential has a relatively small effect on the energy of the strongly localized states that are deep in the band tails and a

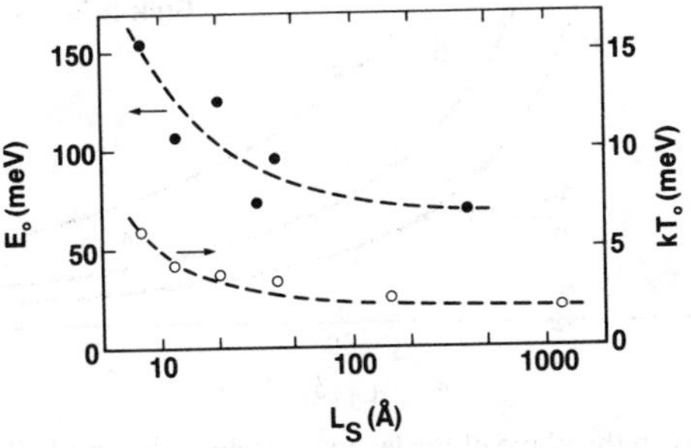

FIG. 9. Dependence of the Urbach slope parameter E_0 (left scale) and photoluminescence quenching parameter T_0 (right scale) on a-Si:H layer thickness L_S. [From Abeles and Tiedje (1983).]

large effect on the energy of the weakly localized shallow states. As a result, the distribution of localized states is stretched out by the superlattice.

V. Electrical Transport

Samples for resistivity measurements were deposited through a 5-mm-diameter circular opening in a metal contact mask on predeposited Cr fingers coated with 1% P-doped a-Si : H for ohmicity, spaced 3 mm apart on quartz substrates. This patterning scheme alleviates contacting problems associated with the extreme electrical anisotropy of the layered films, particularly those with relatively thick a-Si : H sublayers. As long as the insulating a-SiN_x : H material is a relatively small fraction of the total film thickness, the tapering of the layer thicknesses near the contact mask will connect all the silicon layers together electrically. Resistivity perpendicular to the plane of the film was measured on unpatterned samples deposited on conductive substrates with evaporated metal dots as top contacts.

The room-temperature resistivity of the superlattice material in the plane of the layers $\rho_{\parallel}$ is shown in Fig. 10 as a function of the a-Si : H layer thickness L_S for a fixed a-SiN_x : H layer thickness of 35 Å and a total film thickness of

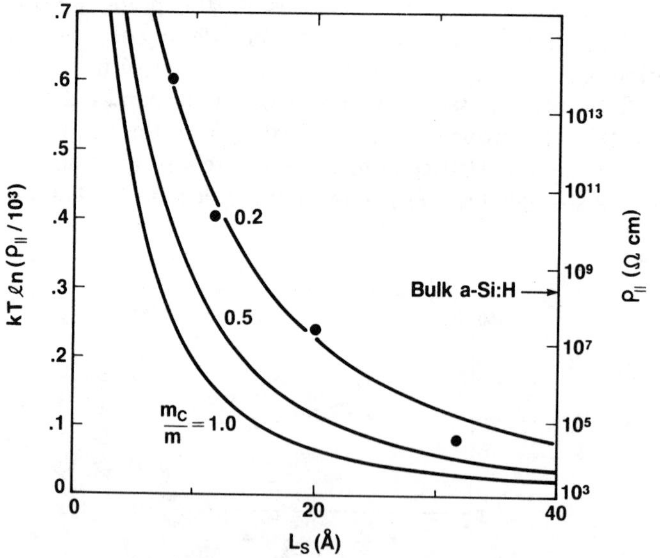

FIG. 10. Resistivity in the plane of the layers $\rho_{\parallel}$ (right scale) and $kT\ln(\rho_{\parallel}/10^3)$ (left scale) as a function of a-Si : H layer thickness L_S. The filled circles are measured values. The value of the conductivity for bulk a-Si : H film is indicated by the arrow; the lines are values of ΔE_c computed from the model in Fig. 5.

about 1 μm. The resistivity of a nonlayered a-Si:H film prepared under the same deposition conditions is indicated in Fig. 10 by an arrow. The large anisotropy in resistivity is shown in Fig. 11, in which we have plotted the ratio $\rho_{\parallel}/\rho_{\perp}$, where $\rho_{\perp}$ is the resistivity measured normal to the layers. Most of the samples had nonlinear $I-V$ characteristics, with $\rho_{\parallel}$ decreasing by a factor of 2–5 from 0 to 400 V; the data in Fig. 10 were determined from the 200–400-V resistivity. Similarly the $\rho_{\perp}$ data in Fig. 11 were obtained from the resistivity measured at 1–6-V applied bias.

Two striking effects are observed in Fig. 10. One is the large decrease in the resistivity of the layered material relative to the nonlayered material for large L_S. The low resistivity (10^3 Ω cm) persists even for relatively thick a-Si:H layers (1200 Å) and is of n-type conductivity from the sign of the thermoelectric power. The decrease in resistivity has been ascribed to transfer doping by the a-SiN_x:H layers (Tiedje and Abeles, 1984). The other noteworthy effect in Fig. 10 is the large increase in resistivity when L_S is reduced below 40 Å. It

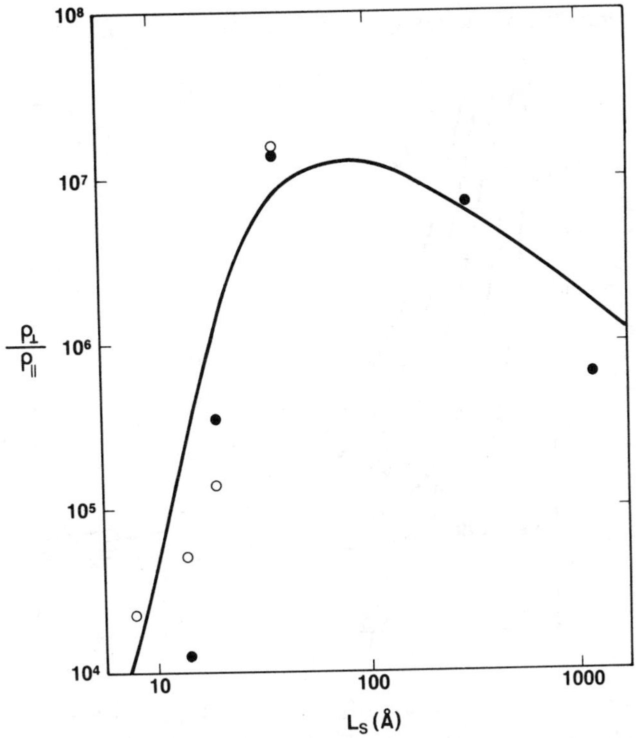

FIG. 11. Ratio of in-plane resistivity $\rho_{\parallel}$ to perpendicular resistivity $\rho_{\perp}$ as a function of a-Si:H layer thickness L_S. ●, dark; ○, 1-sun illumination.

has been shown that this increase can be explained by a shift in the conduction-band mobility edge resulting from quantum size effects (Tiedje *et al.*, 1984). We will discuss the transfer doping effect first.

To verify that the low resistivity is not caused by substitutional doping due to NH_3 contamination during the growth of the a-Si : H layers, we prepared an a-Si : H film from SiH_4 containing 5% NH_3 and found a resistivity of $10^9\ \Omega$ cm, similar to that of pure a-Si : H made under the same conditions. We concluded that the low resistivity is due to charge transfer doping of the a-Si : H layers by the a-SiN_x : H. The fact that the resistivity is independent of Si thickness for L_S between 80 and 1200 Å suggests that the Fermi level is pinned by a high density of gap states in the silicon nitride layers. To estimate this density, a series of films was prepared with fixed layer dimensions, 400 Å of a-Si : H and 35 Å of a-SiN_x : H, and a variable number of periods M. The resistivity as a function of M is shown in Fig. 12, along with the resistivities of a number of other films with various a-Si : H layer thicknesses and numbers of layers M. We note that $\rho_\parallel$ in Fig. 10 corresponds to films that have between 40 and 400 layers.

The dependence of the resistivity on the number of layers in Fig. 12 is

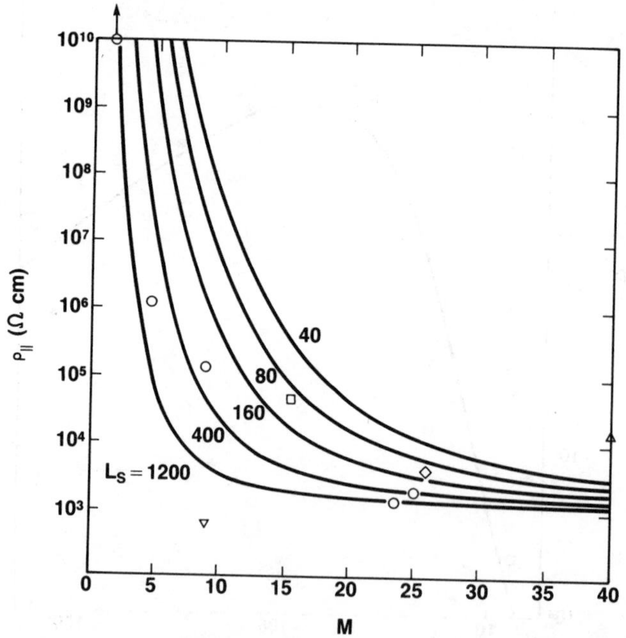

FIG. 12. Resistivity in the plane of layers, $\rho_\parallel$ as a function of number of periods M in a-Si : H/a-SiN_x : H superlattice for different a-Si : H layer spacing L_S. The full curves were computed from Eq. (7) for the energy-band model given in Fig. 13. Experimental L_s (Å) = 1200 (▽), 400 (○), 160 (□), 80 (◇), 40 (△). [From Tiedje and Abeles (1984).]

attributed to band bending caused by charge depletion from the part of the superlattice film next to the substrate. This is illustrated in Fig. 13, which shows the Fermi level pinned near midgap by a high density of states at the interface between the film and the quartz substrate; outside the depletion region, i.e. for $x > x_0$, the Fermi level is pinned by the silicon nitride layers. We assume that all the gap states in the superlattice are in the silicon nitride layers and the density of states in the nitride controls the width of the depletion layer.

The average space charge Q in the layered material is given by

$$Q = Q_N[L_N/(L_N + L_S)] \tag{3}$$

where Q_N is the charge density in the nitride. Similarly, the effective dielectric constant ε of the superlattice is

$$\varepsilon = (L_S + L_N)(L_N/\varepsilon_N + L_S/\varepsilon_S)^{-1}, \tag{4}$$

where ε_S and ε_N are the dc dielectric constants of a-Si:H and a-SiN_x:H, respectively. In the limit that the density of gap states n_t is distributed uniformly in the band gap of the a-SiN_x:H, the potential distribution in the depletion layer satisfies the Poisson equation

$$d^2\phi/dx^2 = (en_t/\varepsilon)[L_N/(L_N + L_S)]\phi, \tag{5}$$

which has the solution $\phi(x) = \phi_0 \exp(-x/x_0)$, where the net charge in the nitride $Q_N = n_t\phi$ is proportional to the band bending ϕ. The quantity ϕ_0 is the built-in potential, n_t the density of gap states per electron volt assumed to be entirely concentrated in the a-SiN_x:H, and x the distance from the interface (Fig. 13). The depletion width x_0 is given by

$$x_0 = \{[(L_N + L_S)/L_S]/(\varepsilon/n_t e)\}^{1/2}. \tag{6}$$

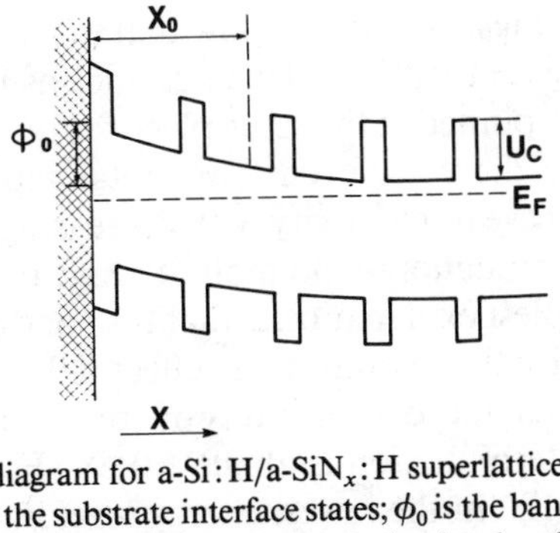

FIG. 13. Energy-band diagram for a-Si:H/a-SiN_x:H superlattice showing the effect of pinning the Fermi level E_F by the substrate interface states; ϕ_0 is the band-bending potential, x_0 the depletion width, and U_c the discontinuity of the conduction bands at the a-Si:H/a-SiN_x:H interface.

The in-plane conductivity $\sigma_\parallel$ is the sum of the conductivities of the individual layers, which depend exponentially on the band-bending potential ϕ. Thus,

$$\sigma = \frac{\sigma_0}{M} \sum_{m=1}^{M} \exp\left(-\frac{\phi_0}{kT} \exp\left\{ \frac{m[L_N + L_S(\varepsilon_N/\varepsilon_S)]L_N}{x_N} \right\} \right), \qquad (7)$$

where $x_N = (\varepsilon_N/en_t)^{1/2}$ is the depletion width in bulk a-SiN$_x$:H. We can now fit the resistivity as a function of M in Fig. 12 with Eq. (7) by using ϕ_0 and x_N as adjustable parameters with $\sigma_0 = 10^{-3}\ \Omega^{-1}\ \mathrm{cm}^{-1}$, $\varepsilon_N = 7.5$, and $\varepsilon_S = 12$. The values of the best fit parameters are $\phi_0 = 1.0$ eV and $x_N = 300$ Å. This value for x_N corresponds to a bulk density of states of $7 \times 10^{17}\ \mathrm{cm}^{-3}\ \mathrm{eV}^{-1}$ or, equivalently, to $10^{11}\ \mathrm{cm}^{-2}\ \mathrm{eV}^{-1}$ states per interface. This density of states is more than sufficient to pin the Fermi level in high-quality a-Si:H, where the density of gap states in the upper half of the gap is $< 10^{16}\ \mathrm{cm}^{-3}\ \mathrm{eV}^{-1}$. The fit, shown in Fig. 12, is excellent given the range of samples covered and the expected sensitivity of the band bending at the substrate interface to substrate preparation conditions.

The new transfer doping mechanism produces conductive material with a lower density of gap states than phosphorus-doped material of comparable resistivity, where the substitutional dopant always introduces extra defect states. Evidence for the lower density of defects comes from the magnitude of the low energy shoulder in the photoconductivity response spectrum shown in Fig. 14, where the absorption of the layered material at photon energies below 1.4 eV is more than an order of magnitude lower than the phosphorus-doped material of comparable dark resistivity. Furthermore, the photoconductivity of the transfer-doped material is large ($10^{-2}\ \Omega^{-1}\ \mathrm{cm}^{-1}$) compared with the photoconductivity achievable in heavily P-doped material under similar illumination.

Aside from band-bending effects at the film–substrate interface, the experimental results in Figs. 10 and 12 show that the a-SiN$_x$:H layer pins the Fermi level in the a-Si:H for a-Si:H layers as thick as 1200 Å. We expect the Fermi level to remain pinned at the same place for material with thin silicon layers (<40 Å), since there are even fewer total gap states in the thinner layers. Thus, the increase in resistivity with decreasing L_S in Fig. 10 must be due to a shift in the conduction-band mobility edge rather than a shift in the Fermi level. The simplest explanation is that the conduction-band states are shifted up in energy by the quantum size effect. Ultraviolet photoemission experiments show that the discontinuity in the conduction band at the a-Si:H/a-SiN$_x$:H heterojunction is about 1.0 eV. From this value for the well depth, the shift ΔE_c in the lowest energy level in the conduction-band well can be calculated, as is illustrated by the density of states given in Fig. 5b. If it is assumed for the moment that with decreasing L_S the mobility edge

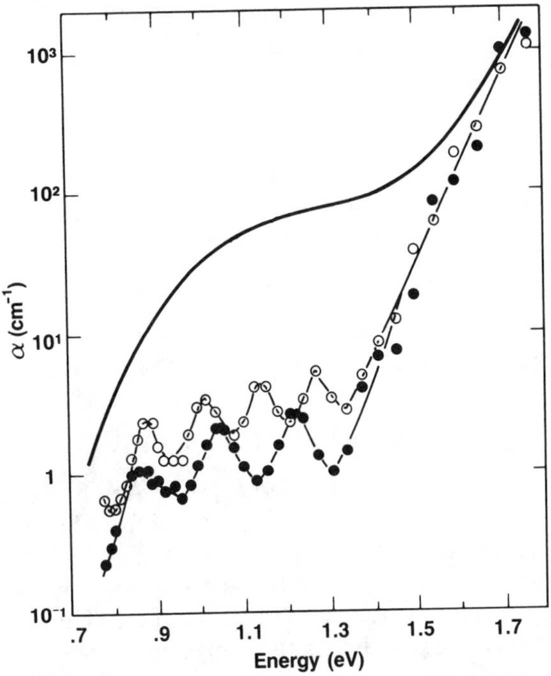

FIG. 14. Optical absorption coefficient α determined from photoconductivity for a-Si:H/a-SiN$_x$:H superlattice film 1.2 μm thick with alternating layers of 1200 Å a-Si:H and 35 Å a-SiN$_x$:H (○), compared with α for bulk undoped a-Si:H (●) and for P-doped a-Si:H (———) with same dark resistivity as the superlattice. [From Tiedje and Abeles (1984).]

moves at the same rate as the density of states, then the parallel resistivity $\rho_\parallel$ will be an exponential function of ΔE_c through the Boltzmann factor:

$$\rho_\parallel = \rho_0 \exp(\Delta E_c/kT). \tag{8}$$

The quantum shift ΔE_c depends on the well width L_S, the barrier height, and the electron effective mass m_c. In the case of an infinitely high barrier height

$$\Delta E_c = \hbar^2/2\, m_c L_S^2. \tag{9}$$

By treating the electron effective mass m_c as an adjustable parameter the fit indicated by the solid line in Fig. 10 is obtained with an effective mass of $0.2m$ and a conduction-band-edge discontinuity of $U_c = 1.0$ eV at the a-Si:H/a-SiN$_x$:H interface, both reasonable values.

The anisotropy in the electrical resistivity can be explained by similar arguments. In the simplest model the in-plane resistivity of the layered material is the reciprocal of the average conductivity or

$$\rho_\parallel = (L_S + L_N)(L_S/\rho_S + L_N/\rho_N)^{-1}, \tag{10}$$

where ρ_S and ρ_N are the resistivities of the a-Si:H and a-SiN$_x$:H layers, respectively. Similarly, the perpendicular resistivity is proportional to the sum of the resistances of the two layers, or

$$\rho_\perp = [1/(L_N + L_S)](L_N\rho_N + L_S\rho_S). \quad (11)$$

We assume that the charge transport between the a-SiN$_x$:H layers is by quantum tunneling so that to a first approximation $\rho_N = \rho_S/t$, where the transmission coefficient ($t \ll 1$) is given by

$$t = \exp[-\sqrt{2m_N(U_c - \Delta E_c)/\hbar^2}L_N]. \quad (12)$$

Here, m_N is the effective mass in the a-SiN$_x$:H layer, and U_c is the conduction-band edge-discontinuity. It follows that the anisotropy in the resistivity is given by

$$\rho_\parallel/\rho_\perp = L_S L_N/[t(L_S + L_N)^2]. \quad (13)$$

By using $m_N = 1.0m$, $U_c = 1.0$ eV, and $L_N = 35$ Å in Eqs. (12) and (13), we obtain the resistivity anisotropy as a function of the a-Si:H layer thickness L_S shown in Fig. 11. Although the model curve in Fig. 11 does not fit the data in detail, it does reproduce the overall shape and approximate magnitude of the anisotropy in the resistivity as a function of L_S. This semiquantitative agreement between the resistivity anisotropy and the one-dimensional tunneling model supports the quantum well interpretation of the optical data and confirms the structural integrity of the a-SiN$_x$:H layers.

It is interesting to compare the increase in the conduction-band mobility edge ΔE_c in Fig. 10 with the increase in the optical gap ΔE_G in Fig. 7 for the same series of samples. The change in optical gap is smaller than the change in the conduction-band edge alone, an apparently paradoxical result. In principle this result could be interpreted as evidence that the conduction-band mobility edge is a stronger function of L_S than the density of states. However, the data can also be explained by a more elementary interpretation based on the idea that since the optical gap is an extrapolation from high photon energies, it is a measure of the average energy-level position several tenths of an electron volt above the band edge, a quantity that is less sensitive to the superlattice potential than is the lowest energy level in the potential well. This is demonstrated in Fig. 6, which shows how the change in the extrapolated band gap E_G falls below $\Delta E_c + \Delta E_v$.

VI. Conclusion

In conclusion, we have reviewed some of the structural optical and electrical properties of amorphous silicon/amorphous silicon nitride superlattices. The regularity and smoothness of the layers, the abruptness of the interfaces, and the low density of defects gives these materials a number of novel properties, some similar to those found in crystalline superlattices—coherence of the wave functions extending over 40 Å, tunability of the optical band gap, enhanced photoluminescence, and transfer-doping effects. The ability to synthesize layered amorphous semiconductors with interesting and novel electronic properties is expected to have a major impact on the physics and technology of amorphous semiconductors.

ACKNOWLEDGMENTS

We wish to thank H. W. Deckman for the TEM, P. D. Persans for the quantum well model calculations and C. R. Wronski for the photoconductivity spectra.

REFERENCES

Abeles, B., and Tiedje, T. (1983). *Phys. Rev. Lett.* **51,** 2003.

Abeles, B., Tiedje, T., Liang, K. S., Deckman, H. W., Stasiewski, C. H., Scanlon, J. C., and Eisenberger, P. (1984a). *AIP Conf. Proc.* (in press).

Abeles, B., Wagner, I., Eberhardt, W., Stöhr, J., and Sette, F. (1984b). To be published.

Abrahams, E., Anderson, P. W., Licciardello, D., and Ramakrishnan, T. V. (1979). *Phys. Rev. Lett.* **42,** 673.

Cody, G. D., Tiedje, T., Abeles, B., Brooks, B., and Goldstein, Y. (1981). *Phys. Rev. Lett.* **47,** 1480.

Cody, G. D., Brooks, B. G., and Abeles, B. (1982). *Solar Energy Mat.* **8,** 231.

Deckman, H. W., Dunsmuir, S. H., and Abeles, B. (1984a). *Bull. Am. Phys. Soc.* **29,** 255.

Deckman, H. W., Dunsmuir, S. H., and Abeles, B. (1984b). *Appl. Phys. Lett.* To be published.

Esaki, L., and Tsu, R. (1970). *IBM J. Res. Dev.* **14,** 61.

Fan, J. C. C., and Palm, B. S. (1983). *Solar Cells* **10,** 81.

Higashi, G. S., and Kastner, M. A. (1979). *J. Phys. C.* **12,** L821.

Morkoc, H., and Solomon, P. M. (1984). *IEEE Spectrum,* **21,** 28.

Munekata, H., and Kukimoto, H. (1983). *Appl. Phys. Lett.* **42,** 432.

Paul, W., and Anderson, D. A. (1981). *Solar Energy Mat.* **5,** 229.

Ploog, K., and Dohler, G. H. (1983). *Adv. Phys.* **32,** 285.

Roxlo, C. B., Abeles, B., and Tiedje, T. (1984a). *Bull. Am. Phys. Soc.* **29,** 256.

Roxlo, C. B., Abeles, B., and Tiedje, T. (1984b). *Phys. Rev. Lett.* **52,** 1994.

Tiedje, T., and Abeles, B. (1984). *Appl. Phys. Lett.* To be published.

Tiedje, T., Abeles, B., Persans, P., Brooks, B. G., Cody, G. D. (1984). *AIP Conf. Proc.* (in press).

Index

E

F

H

I

K

L

M

N

O

P

Q

R

S

T

U

V

W

Z

Contents of Previous Volumes

Volume 4 **Physics of III–V Compounds**

Volume 5 **Infrared Detectors**

Volume 6 **Injection Phenomena**

Volume 7 **Application and Devices: Part A**

Volume 7 **Application and Devices: Part B**

Volume 8 **Transport and Optical Phenomena**

Volume 9 **Modulation Techniques**

Volume 10 **Transport Phenomena**

Volume 11 **Solar Cells**

Volume 12 **Infrared Detectors (II)**

Volume 13 **Cadmium Telluride**

Volume 14 **Lasers, Junctions, Transport**

Volume 15 **Contacts, Junctions, Emitters**

Volume 16 **Defects, (HgCd)Se, (HgCd)Te**

Volume 18 **Mercury Cadmium Telluride**

Volume 19 **Deep Levels, GaAs, Alloys, Photochemistry**

Volume 20 **Semi-Insulating GaAs**

Volume 21 **Hydrogenated Amorphous Silicon: Part A**

Volume 21 **Hydrogenated Amorphous Silicon: Part B**